Learning Materials in Biosciences

Learning Materials in Biosciences textbooks compactly and concisely discuss a specific biological, biomedical, biochemical, bioengineering or cell biologic topic. The textbooks in this series are based on lectures for upper-level undergraduates, master's and graduate students, presented and written by authoritative figures in the field at leading universities around the globe.

The titles are organized to guide the reader to a deeper understanding of the concepts covered.

Each textbook provides readers with fundamental insights into the subject and prepares them to independently pursue further thinking and research on the topic. Colored figures, step-by-step protocols and take-home messages offer an accessible approach to learning and understanding.

In addition to being designed to benefit students, Learning Materials textbooks represent a valuable tool for lecturers and teachers, helping them to prepare their own respective coursework.

More information about this series at http://www.springer.com/series/15430

Margarethe Geiger

*Editor*

# Fundamentals of Vascular Biology

 Springer

*Editor*
**Margarethe Geiger**
Department of Vascular Biology and Thrombosis Research
Center for Physiology and Pharmacology
Medical University of Vienna
Wien, Austria

ISSN 2509-6125          ISSN 2509-6133   (electronic)
Learning Materials in Biosciences
ISBN 978-3-030-12269-0        ISBN 978-3-030-12270-6   (eBook)
https://doi.org/10.1007/978-3-030-12270-6

This Springer imprint is published by the registered company Springer Nature Switzerland AG
The registered company address is: Gewerbestrasse 11, 6330 Cham, Switzerland

# Preface

Research in vascular biology has rapidly evolved during the last decades and is one of the major fields in biomedical research. It has been a common knowledge for quite some time that blood and lymphatic vessels cannot be considered as simple pipes for providing substances to organs and for discarding organ-derived waste material. Blood and lymphatic vessels can rather be considered as a very complex system, which is involved in a variety of physiological and pathophysiological processes in the organism.

Important study areas within the field of vascular biology are the research on atherosclerosis, a major disease in Western countries with millions of affected patients in Europe and the United States, and on thromboembolic diseases. Also, the analysis of the formation of new blood vessels either from existing vessels (angiogenesis) or from stem cells (vasculogenesis), respectively, is a major research area in vascular biology, since the formation of new blood vessels is essential for wound healing, for reproduction, for tumor growth, and for metastasis formation.

This book was written with the intention to provide basic knowledge in vascular biology for PhD students, diploma students, and other scientists working in different fields of vascular biology. In the first three chapters, the authors describe the morphology of different blood and lymphatic vessels and explain very basic principles of cardiovascular physiology. These chapters are addressed mainly to those readers without or with very little biomedical background. Following are chapters describing state-of-the-art knowledge of the biology of the vascular system, diseases of the vascular system and disease mechanisms, as well as experimental techniques and animal models used in vascular biology research.

I would like to thank all authors who contributed to this book. Their expertise, effort, and patience are very much appreciated! Many thanks also to the Springer staff, especially to Dr. Amrei Strehl, for their support and help.

Margarethe Geiger
Vienna, Austria

# Contents

# Contributors

Alice Assinger, PhD
Department of Vascular Biology and
Thrombosis Research
Center for Physiology and Pharmacology
Medical University of Vienna
Vienna, Austria
alice.assinger@meduniwien.ac.at

Helga Bergmeister, MD, PhD
Ludwig Boltzmann Cluster for Cardiovascular
Research at the Center for Biomedical Research
Medical University of Vienna
Vienna, Austria
helga.bergmeister@meduniwien.ac.at

David Bernhard, PhD
Medical Faculty, Center for Medical Research
Johannes Kepler University Linz
Linz, Austria
david.bernhard@jku.at

Christoph J. Binder, MD, PhD
Department of Laboratory Medicine
Medical University of Vienna
Vienna, Austria
christoph.binder@meduniwien.ac.at

Christine Brostjan, PhD
Department of Surgery
Medical University of Vienna
Vienna, Austria
christine.brostjan@meduniwien.ac.at

Rainer de Martin, PhD
Department of Vascular Biology and
Thrombosis Research
Center for Physiology and Pharmacology
Medical University of Vienna
Vienna, Austria
rainer.demartin@meduniwien.ac.at

Svitlana Demyanets, MD, PhD
Department of Laboratory Medicine
Medical University of Vienna
Vienna, Austria
svitlana.demyanets@meduniwien.ac.at

Thomas Gary, MD
Department of Angiology
Medical University Graz
Graz, Austria
thomas.gary@medunigraz.at

Margarethe Geiger, MD
Department of Vascular Biology and
Thrombosis Research
Center for Physiology and Pharmacology
Medical University of Vienna
Wien, Austria
margarethe.geiger@meduniwien.ac.at

Stefan H. Geyer, PhD
Department of Anatomy
Center for Anatomy and Cell Biology
Medical University of Vienna
Vienna, Austria
stefan.geyer@meduniwien.ac.at

Ouafa Hamza, MD
Ludwig Boltzmann Cluster for Cardiovascular
Research at the Center for Biomedical Research
Medical University of Vienna
Vienna, Austria
ouafa.hamza@meduniwien.ac.at

Brigitte Hantusch, PhD
Department of Pathology
Medical University of Vienna
Vienna, Austria
brigitte.hantusch@meduniwien.ac.at

**Thomas M. Hofbauer, MD**
Department of Cardiology, Internal Medicine II
Medical University of Vienna
Vienna, Austria
thomas.hofbauer@meduniwien.ac.at

**Attila Kiss, PhD**
Ludwig Boltzmann Cluster for Cardiovascular
Research at the Center for Biomedical Research
Medical University of Vienna
Vienna, Austria
attila.kiss@meduniwien.ac.at

**Julia B. Kral-Pointner, PhD**
Department of Molecular Medicine and Surgery
Karolinska Institute and University Hospital
Stockholm, Sweden

Department of Vascular Biology and
Thrombosis Research
Center for Physiology and Pharmacology
Medical University of Vienna
Vienna, Austria

**Norbert Leitinger, PhD**
Department of Pharmacology and Robert M.
Berne Cardiovascular Research Center
University of Virginia
Charlottesville, VA, USA
nl2q@virginia.edu

**Christine Mannhalter, PhD**
Department of Laboratory Medicine
Medical University of Vienna
Vienna, Austria
christine.mannhalter@meduniwien.ac.at

**Florian J. Mayer, MD**
Department of Laboratory Medicine
Medical University of Vienna
Vienna, Austria
florian.mayer@meduniwien.ac.at

**Barbara Messner, PhD**
Department of Surgery
Cardiac Surgery Research Laboratory
Medical University of Vienna
Vienna, Austria
barbara.messner@meduniwien.ac.at

**Marion Mussbacher, PhD**
Department of Vascular Biology and
Thrombosis Research
Center for Physiology and Pharmacology
Medical University of Vienna
Vienna, Austria
marion.mussbacher@meduniwien.ac.at

**Felix Nagel, MD**
Ludwig Boltzmann Cluster for Cardiovascular
Research at the Center for Biomedical Research
Medical University of Vienna
Vienna, Austria
felix.nagel@meduniwien.ac.at

**Patrick M. Pilz, MD**
Ludwig Boltzmann Cluster for Cardiovascular
Research at the Center for Biomedical Research
Medical University of Vienna
Vienna, Austria
patrick.pilz@meduniwien.ac.at

**Roberto Plasenzotti, PhD**
Ludwig Boltzmann Cluster for Cardiovascular
Research at the Center for Biomedical Research
Medical University of Vienna
Vienna, Austria
roberto.plasenzotti@meduniwien.ac.at

**Bruno K. Podesser, MD**
Ludwig Boltzmann Cluster for Cardiovascular
Research at the Center for Biomedical Research
Medical University of Vienna
Vienna, Austria
bruno.podesser@meduniwien.ac.at

Manuel Salzmann, MSc
Department of Vascular Biology and
Thrombosis Research
Center for Physiology and Pharmacology
Medical University of Vienna
Vienna, Austria
manuel.salzmann@meduniwien.ac.at

Diethart Schmid, MD
Institute of Physiology, Center of Physiology
and Pharmacology
Medical University of Vienna
Vienna, Austria
diethart.schmid@meduniwien.ac.at

Waltraud C. Schrottmaier, PhD
Department of Vascular Biology and
Thrombosis Research
Center for Physiology and Pharmacology
Medical University of Vienna
Vienna, Austria
waltraud.schrottmaier@meduniwien.ac.at

Adrian Türkcan, MD
Institute of Laboratory Medicine
University Hospital, Ludwig Maximilian
University of Munich
Munich, Germany
adrian.tuerkcan@med.uni-muenchen.de

Pavel Uhrin, PhD
Department of Vascular Biology and
Thrombosis Research
Center for Physiology and Pharmacology
Medical University of Vienna
Vienna, Austria
pavel.uhrin@meduniwien.ac.at

Ellen Umlauf, PhD
Department of Vascular Biology and
Thrombosis Research
Center for Physiology and Pharmacology
Medical University of Vienna
Vienna, Austria
ellen.umlauf@meduniwien.ac.at

Clint Upchurch
Department of Pharmacology and Robert
M. Berne Cardiovascular Research Center
University of Virginia
Charlottesville, VA, USA

Wolfgang J. Weninger, MD
Department of Anatomy
Center for Anatomy and Cell Biology
Medical University of Vienna
Vienna, Austria
wolfgang.weninger@meduniwien.ac.at

Johann Wojta, PhD
Department of Internal Medicine II
Medical University of Vienna
Vienna, Austria
johann.wojta@meduniwien.ac.at

Maria Zellner, PhD
Department of Vascular Biology and
Thrombosis Research
Center for Physiology and Pharmacology
Medical University of Vienna
Vienna, Austria
maria.zellner@meduniwien.ac.at

# Abbreviations

| | | | |
|---|---|---|---|
| 2D | Two-dimensional | cm H2O | Centimeters of water |
| 3D | Three-dimensional | CNP | C-type natriuretic peptide |
| | | CO | Cardiac output |
| ABTS | 2,2'-Azino-bis(3-ethylbenzothia-zoline-6-sulphonic acid) | CoA | Coenzyme A |
| | | CPI-17 | C-kinase-potentiated protein phosphatase-1 inhibitor |
| AC | Adenylyl cyclase | | |
| Ach | Acetylcholine | CPT | Carnitine palmitoyltransferase |
| ADP | Adenosine diphosphate | CYP enzyme family | Cytochrome P450 enzyme family |
| AMP | Adenosine monophosphate | | |
| AngII | Angiotensin II | | |
| ANP | Atrial natriuretic peptide | DAG | Diacylglycerol |
| AP | Action potential | DAPI | 4',6-Diamidino-2-phenylindole |
| ATP | Adenosine triphosphate | DDR2 | Discoidin domain receptor 2 |
| ATPase | Adenosine triphosphatase | DiI | 1,1'-dioctadecyl-3,3,3',3'-tetramethylindocarbocyanine perchlorate |
| AV | Atrioventricular | | |
| AVN | Atrioventricular node | | |
| | | DMEM | Dulbecco's modified Eagle's medium |
| BCA assay | Bicinchoninic acid assay | | |
| bFGF | Basic fibroblast growth factor | DMSO | Dimethyl sulfoxide |
| BNP | Brain natriuretic peptide | DNA | Deoxyribonucleic acid |
| Bpm | Beats per minute | | |
| BrdU assay | Bromodeoxyuridine/5-bromo-2'-deoxyuridine assay | E2 | Estrogen |
| | | ECG | Electrocardiogram |
| BSA | Bovine serum albumin | ECM | Extracellular matrix |
| | | ECs | Endothelial cells |
| $Ca^{2+}$ | Calcium | EDHF | Endothelial-derived hyperpolarization factor |
| CaCl2 | Calcium chloride | | |
| CaD | Caldesmon | EDV | End-diastolic volume |
| Calcein-AM | Calcein acetoxymethyl ester | EF | Ejection fraction |
| CaM | Calmodulin | ELISA | Enzyme-linked immunosorbent assay |
| cAMP | Cyclic adenosine monophosphate | | |
| | | ER | Estrogen receptor |
| CaP | Calponin | E-selectin | Endothelial leucocyte adhesion molecule |
| CCL2 | C-C motif ligand 2 | | |
| CFDA | 5-Chloromethylfluorescin diacetate | ESV | End-systolic volume |
| cGMP | Cyclic guanosine monophosphate | FAT | Fatty acid translocase |
| | | FATP | Fatty acid transport protein |
| CGRP | Calcitonin gene-related peptide | FBS | Foetal bovine serum |
| | | FCS | Foetal calf serum |
| CICR | Calcium-induced calcium release | FIBs | Fibroblasts |
| | | FSP-1 | Fibroblast-specific protein 1 |

| | | | |
|---|---|---|---|
| G-6-P | Glucose-6-phosphate | MTT | 3-(4,5-Dimethylthiazol-2-yl)-2,5-diphenyltetrazolium bromide |
| GC | Guanylyl cyclase | | |
| GDP | Guanosine diphosphate | mV | Millivolt |
| GLUT | Glucose transporter | MYPT | Myosin phosphatase target subunit |
| GMP | Guanosine monophosphate | | |
| GTP | Guanosine triphosphate | | |
| | | NA | Noradrenaline |
| HCl | Hydrochloric acid | $Na^+$ | Sodium |
| HR | Heart rate | NADH | Nicotinamide adenine dinucleotide |
| HRP | Horseradish peroxidase | | |
| HUVECs | Human umbilical vein endothelial cells | NADPH | Nicotinamide adenine dinucleotide phosphate |
| | | NCX | $Na^+/Ca^{2+}$ exchanger |
| IFN | Interferon | NCX1 | $Na^+/Ca^{2+}$ exchanger 1 |
| IICR | $IP_3$-induced calcium release | NF-κB | Nuclear factor kappa-light-chain-enhancer of activated B cells |
| IL | Interleukin | | |
| IL-8 | Interleukin-8 | | |
| $IP_3$ | Inositol-triphosphate | NO | Nitric oxide |
| $IP_3R$ | IP3 receptor | NSCC | Nonselective cation channel |
| iPSC | Induced pluripotent stem cells | OSM | Oncostatin M |
| $K^+$ | Potassium | | |
| kPa | Kilopascal | Pa | Pascal |
| | | PAI-1 | Plasminogen activator inhibitor-1 |
| L | Liter | | |
| LA | Left atrium | PBS | Phosphate-buffered saline |
| LDL | Low-density lipoprotein | PBS−/− | Phosphate-buffered saline without calcium and magnesium |
| LV | Left ventricle | | |
| | | PDE | Phosphodiesterase |
| MAP | Mean arterial pressure | PDGFα | Platelet-derived growth factor receptor-α |
| MAPK | Mitogen-activated protein kinase | | |
| | | PEDF | Pigment epithelium-derived factor |
| MCP1 | Monocyte chemoattractant protein 1 | | |
| | | $PGI_2$ | Prostacyclin |
| M-CSF | Macrophage colony-stimulating factor | $PIP_2$ | Phosphatidylinositol-biphosphate |
| MLC | Myosin light chain | PKA | Cyclic adenosine monophosphate-dependent protein kinase |
| $MLC_{20}$ | 20kD regulatory light chain of myosin | | |
| MLCK | Myosin light chain kinase | PKC | Protein kinase C |
| mmHg | Millimeter of mercury | PKG | Cyclic guanosine monophosphate-dependent protein kinase |
| MMP | Matrix metalloproteinases | | |
| MP | Myosin phosphatase | $PLA_2$ | Phospholipase $A_2$ |
| msec | Milliseconds | PLC | Phospholipase C |
| MTC | Mitochondrium | PM | Plasma membrane |

| | | | | |
|---|---|---|---|---|
| PMCA | Plasma membrane calcium ATPase | | TAC | Total arterial compliance |
| PP1Cδ | Catalytic subunit of type 1 phosphatase δ | | TAG | Triacylglycerol |
| | | | Tcf21 | Transcription factor 21 |
| | | | TGF | Transforming growth factor |
| RA | Right atrium | | TIMP | Tissue inhibitor of metalloproteinases |
| RhoA | Ras homolog gene family member A | | TNF | Tumor necrosis factor |
| ROCC | Receptor-operated calcium channel | | TNF-alpha | Tumour necrosis factor-alpha |
| | | | TPR | Total peripheral resistance |
| ROCK | Rho-associated protein kinase | | Trypsin-EDTA | Trypsin-ethylenediaminetet-raacetic acid |
| RV | Right ventricle | | | |
| RyR | Ryanodine receptor | | TUNEL assay | Terminal deoxynucleotidyl transferase dUTP nick end labelling assay |
| RyR2 | Ryanodine receptor type 2 | | | |
| SA | Sinoatrial | | VCAM-1 | Vascular cell adhesion molecule-1 |
| SAN | Sinoatrial node | | | |
| SERCA | Sarcoplasmic/endoplasmic calcium ATPase | | VE-cadherin | Vascular endothelial cadherin |
| | | | VEGF | Vascular endothelial growth factor |
| SGC | Soluble guanylyl cyclase | | | |
| SM22 alpha | Transgelin | | VOCC | Voltage-operated calcium channel |
| SM22α | Smooth muscle protein 22α | | | |
| SMCs | Smooth muscle cells | | VSMCs | Vascular smooth muscle cells |
| SMOCC | Second messenger-operated calcium channel | | | |
| SOCC | Store-operated calcium channel | | WT1 | Wilms tumor 1 |
| SR | Sarcoplasmic reticulum | | XTT | 2,3-Bis-(2-methoxy-4-nitro-5-sulfophenyl)-2H-tetrazolium-5-carboxanilide |
| SV | Stroke volume | | | |
| SVR | Systemic vascular resistance | | βOHB | Beta-hydroxybutyrate |

# Morphological and Functional Characteristics of Blood and Lymphatic Vessels

*Brigitte Hantusch*

© Springer Nature Switzerland AG 2019
M. Geiger (ed.), *Fundamentals of Vascular Biology*, Learning Materials in Biosciences,
https://doi.org/10.1007/978-3-030-12270-6_1

**What You Will Learn in This Chapter**

- There are two circulation systems in the body: the blood and the lymphatic vasculature.
- The closed nature of blood circulation was fully understood not until the seventeenth century, when the capillaries were traced by microscopes.
- The anatomic structure and role of the lymphatic vasculature remained similarly enigmatic but was also elucidated during the seventeenth century.
- The two vascular systems form anatomically distinct, separate networks.
- Blood circulation is a closed system of arteries and veins, which are connected by the capillaries.
- The lymphatic system is composed of a network of blind-ended "blood-less" vessels located in close proximity to the veins.

## 1.1 The Vasculature: Overview and History

*Blood vessels* constitute the blood-containing circulation, representing a closed system of arteries and veins, which are connected by the capillaries. The blood vascular system provides all tissues with oxygen, nutrients, and immune cells and undertakes the evacuation of metabolites. Arteries deliver oxygenated blood from the lung to the arterioles and further to the capillaries in the periphery of the body where a bidirectional exchange between blood and tissue occurs. The task of veins is to collect deoxygenated blood and to transport it back to the heart. The *lymphatic system* works auxiliary to the venous vasculature by collecting excess interstitial fluid, proteins, particles, and cells that have leaked into tissues, returning these into the blood. It takes over approximately 10% of the capillary filtrate residuum. Importantly, lymphatic vessels directly absorb dietary lipids from the gut and deliver them to the blood circulation. Before returning lymph fluid into the blood, it is passed through lymph nodes for presentation of foreign components to immune cells. Concomitantly, lymphatic vessels guide immune cells to the lymphoid organs.

## 1.1.1 Body Fluids Are Transported Within the Vasculature

■ **Body Fluid and Body Compartments**

Approximately 60% of a vertebrate organism's mass consist of aqueous solution in which life-sustaining substances are stored and transported [1]. This watery content is split between the *intracellular* and the *extracellular* compartment in a proportion of two- (40%) to one-third (20%) [2]. In order to maintain liquid homeostasis and to provide all body regions with necessary substances [3], vertebrates are pervaded by the vascular systems. These form a specialized and highly branched network of vessels that are lined by endothelial cells and extend through the whole body [4]. Hence, the fluid of the extracellular compartment itself is divided between the *interstitial* and the *intravasal* volume, again split in a ratio of two-third/one-third, respectively. Assuming as an example a body weight of 70 kg, the intracellular fluid volume comprises approx. 28 l (40%) and the extracellular approx.14 l (20%), which is subdivided into 9.3 l (13.3%) interstitial and 4.7 l (6.7%) intravascular fluid. There is constant fluid and substance exchange between these liquid entities that is driven by continuous circulation of its components.

■ **Vertebrates Have Two Vascular Systems**

Vertebrate bodies contain *two circulation systems*, the blood and the lymphatic vessels, serving as transport routes of liquids. Their detailed anatomy and function was uncovered and ascertained during the sixteenth to twentieth century (see ▶ Sects. 1.2 and 1.3). In both systems, large vessels sprout consecutively into smaller and smaller vessels which finally form smallest structures, the capillaries. They are two anatomically distinct, non-communicating types of vessel networks, which have different structures and fulfill different tasks [5, 6]. Essentially, the cardiovascular system consists of the heart and blood vessels (arteries, veins, and capillaries), while the lymphatic system comprises the lymphatic vessels (thoracic duct, collecting ducts, and capillaries) and lymphatic organs (lymph nodes, bone marrow, spleen, thymus, tonsils).

*Blood vessels* constitute the blood containing circulation, representing a closed system of arteries and veins, which are connected by the capillaries. The blood vascular system provides all tissues with oxygen, nutrients, and immune cells and undertakes the evacuation of metabolites [7]. Arteries deliver oxygenated blood from the lung to the arterioles and further to the capillaries in the periphery of the body where a bidirectional exchange between blood and tissue occurs [8]. The task of veins is to collect deoxygenated blood and to transport it back to the heart (◻ Fig. 1.1a).

*The lymphatic vascular system* represents a separate network of blind-ended lymphatic vessels, closely associated with the blood vessels. Lymphatic vessels collect and drain excess tissue fluid, extravasated plasma proteins, and cells from the interstitial compartment [11, 12]. Tissue fluid is filtered in the lymph nodes and transferred back into the blood circulation through the lymphatic ducts and finally transported into the venous system via the thoracic duct (◻ Fig. 1.1a). It is a thin-walled and fragile system, closely located to the venous vasculature, and its morphology is hardly preserved in morphological studies, which makes it difficult to visualize.

■ **Morphology and Molecular Markers of the Vasculature**

Until a few years ago, blood and lymphatic vessels residing in the tissues were distinguished according to morphological and histological criteria, thinner walls, no basement membranes (BM), and lack of pericytes of lymphatics. Highly detailed inspections of transmission electron microscopy (TEM) images largely contributed to the characterization of the vascular systems [13], without use of distinct molecular markers. It was evident, that the inner surface of both of these hollow tubes is covered by blood (BEC) or lymphatic (LEC) endothelial cells disposing of diverse morphology, indicating that they serve diverse functions and dispose of specific molecular equipments. However, uncertainty remained about the clear distinction of blood and lymphatic endothelial cells on a molecular level. During the last 10 to 20 years, progress in the field of molecular biology dramatically accelerated the identification of endothelial-specific markers, and a range of exclusive and non-exclusive endothelial cell-specific proteins was identified [14–18] that are used for specific labeling of vessels and the respective surrounding vascular basal lamina [13].

## 1.1.2 Discovery of the Blood Vasculature and Concept of Blood Circulation

■ **Understanding the Anatomy of the Blood Vasculature**

Our principal understanding of blood vessel structure and vascular function is based on the early works of ancient Greek anatomists. Outstanding among these are *Herophilus* and

**1**

**□ Fig. 1.1    a** Bodies contain two vascular systems. Scheme of the cardiovascular and lymphatic vascular circulation. (From: Jones et al., *Nat Rev Mol Cell Biol* 2001 [9]). **b** Emergence of the lymphatic vasculature during embryonic development. **b** Localization of the primary lymph sacs in a 9-week-old human embryo. The peripheral lymphatic system originates from the primary lymph sacs. Sabin, *Am. J. Anat.*, 1902. (Adapted from: Oliver, *Nat Rev Immunol* 2004 [10])

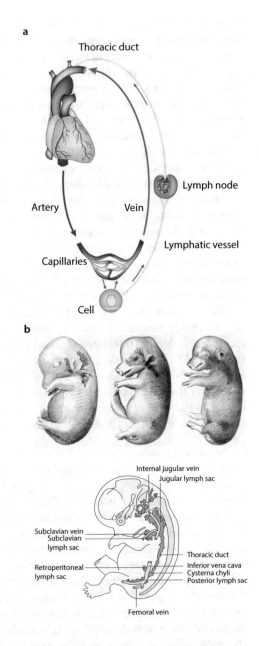

*Erasistratus*, who studied and published in Alexandria in the third century B.C. during the era of the Ptolemaic kingdoms. They performed autopsies and were the first to describe the blood vasculature, distinguishing arteries from veins and understanding that they formed a closed circulation. Unfortunately, their works were destroyed during the burnings of the Alexandrian library, but they were extensively cited approximately 450 years later in the book *Methodus medendi* by *Galenos of Bergamo* (~130–210), who worked during the second century A.C. in Rome as a famous physician. He undertook the first attempts to understand the circulation in its entirety, essentially assuming that blood is generated in the liver from the chyle (i.e., the intestinal lymph fluid) and consumed by the tissues, after it has absorbed vital spirits by the breath (pneuma) [19]. Blood was believed to pass from the right to the left side of the heart, and the venous system was seen as separated. This sight was sacrosanct until the mid of the seventeenth century. However, already during the period comprising the eighth to the sixteenth century, when scientific activity remained rather reclined in Europe, physicians in the Islamic world disagreed with Galen's dogma. The Persian universal scientist *Avicenna (Ibn Sina)* (980–1037) anatomically described arteries and deep/saphenous (Arab. *safin*) veins in his *Canon of Medicine*. *Ibn al-Nafis* (1213–1288), an Arab physician, postulated the lung circulation and emphasized that he could not trace any passage between the right and left cavity of the heart [19]. Not until the fifteenth and sixteenth century, highly detailed anatomic studies were undertaken in Europe, which at least provided us with the first accurate morphological studies of the blood vessels: Among these are drawings of the veins by *Leonardo da Vinci* (1452–1519) and descriptions of the venous valves by Italian anatomists (*G. Canano, L. Vassaeus, S. Ambianis*) in the mid of the sixteenth century. A. *Cesalpino* understood their function in 1559 (*De re anatomica*) by saying that they "prevent the blood from returning" [20].

- Understanding the Riddle of Systemic Circulation

The Flemish Andreas *Vesalius* (1514–1564), an ingenious renaissance anatomist and surgeon, founded our modern anatomic view by performing detailed research on the human blood vessel morphology. In his seven books *De humani corporis fabrica libri septem* (1538–1542), he gives a comprehensive and clear overview of the structures of the human body, illustrated by more than 200 xylographs. He describes the anatomy of the vasculature, including the venous valves. After all, it is the merits of William *Harvey* (1578–1657) to break with Galen's view of constant blood production and consumption and to give us a real idea of the principle of the blood circulation [21]. In his famous *De motu cordis* (1628), he showed by detailed experimental studies that blood constantly circulates and postulated a yet unobserved link between arteries and veins. This theory was finally proved by the discovery of the "missing link" between arteries and veins, namely, the finest blood capillaries, by Marcello *Malpighi* (1628–1694) in 1661, who made use of the first microscopes at that time [22].

## 1.1.3 Discovery of the Lymphatic System and Its Development

- Understanding Lymphatic Vessel Anatomy

Already the most famous ancient physician *Hippocrates of Kos* (460–370 B.C.) recognized mentions the lymphatic system and lymph nodes. After Herophilus and Erasistratus, the Roman physician *Rufus of Ephesus* (~80–150 AD) identified lymph nodes and the thymus. Again in the sixteenth century, which somehow was the golden age of anatomy, the lacteals and the thoracic duct were described by Gabriele *Falloppio* (1523–1562) and Bartolomeo *Eustachi* (1510–1574), respectively [23]. Just some decades later, the lymphatic vasculature was extensively described by Gaspare *Aselli* (1581–1626) in Milano, after he had discovered "milky veins" during vivisections in the intestine of dogs. His dissertation "De lactibus sive lacteis veni" [24] contains several colored xylographs, illustrating his findings in a very beautiful manner. The German anatomist Johann *Veslingius* (1598–1649), who was a professor in Padua, contributed to the issue with detailed sketches of the lacteals in humans in 1647 [25]. However, the key finding of lacteals being finally connected to the thoracic duct, where the chyle is flowing into the venous bloodstream and only then

**1**

arrives in the liver, was established by Jean *Pecquet* (1622–1674) in 1651, which again was a counterevidence to Galen's still prevalent dogma. Ultimately, the physicians Thomas *Bartholin* (1616–1680) and Olaus *Rudbeck* (1630–1702) in 1652 quasi simultaneously published their independently gained finding of lymphatic vessels ultimately closing the circulation by conferring lymph transport through the thoracic duct back to the venous system [26, 27]. It evoked a kind of early scientific battleship about the precedence of their findings, which ended up in a hateful controversy [28].

■  Understanding Lymphatic Function

After this period, knowledge about the lymphatic system and its function was further extended: The German physician and very fruitful medical writer Friedrich *Hoffman* (1660–1742) described it in his *Medicina rationalis systematica* (1718–1734) as an absorbent system, previous to the more famous physicians Alexander *Monro II* (1733–1817) and William *Hunter* (1718–1783). Felice *Fontana* (1730–1805), a natural scientist, anatomist, and physiologist working in Florence, founded a collection of anatomic wax models that were exhibited in the museum Reale Museo di Fisica e Storia Naturale, also called "La Specola." He was in charge of employing the artist Clemente *Susini* (1754–1814) to form these figures. Emperor *Josef II. of Austria* (1741–1790) visited the museum in 1780; he was fascinated by these objects and initiated the production and transport of a second collection of such wax models during 1784–1786 with pack mules to Vienna. These nearly 1200 wax figures are nowadays a precious part of the Museum of Medical History of Vienna, the so-called Josephinum. At the same time, the Italian scientist and physician Paolo *Mascagni* (1755–1815) used ink injections and microscopy to study the lymphatic vessels [29], and his findings were included in the formation of the Viennese wax models by Susini, which impressively show the equal occurrence of lymphatic vessels besides the established complexity of the blood vasculature. He published *Vasorum lymphaticorum corporis humani historia et iconographia* (1787), where the lymphatic system of the human body was described extensively. Mascagni, however, was overestimating the abundance of this vascular system by claiming that also hair, nails, and teeth were containing lymphatic vessels.

■  Understanding the Embryonic Development of the Lymphatic System

Due to the lack of visibility, exploration of the developmental steps of the lymphatic system has remained rather neglected, and the origin of this "blood-less" vessel system remained unexplained [30]. In the nineteenth century, a centripetal model was established, suggesting that the earliest lymphatic endothelial cells were deriving from the mesenchyme. Strikingly, in 1902, *Florence Sabin* (1871–1953) proposed a centrifugal model of lymphatic vasculature development [31, 32]: She injected ink into pig embryos of varying age and showed that lymphatic vessels have their origin by budding from the cardinal veins, forming the primary *lymph sacs* (◻ Fig. 1.1b and c). They then extend by endothelial sprouting into the surrounding tissues and organs, where local capillaries are formed. On the contrary, the anatomists George *Huntington* and Charles *McClure* [33] claimed that the primary lymph sacs are built in the mesenchyme, and afterward connections to the venous system are established. At least, this hypothesis has been proven true for avian lymphatic development [34]. Only at the end of the twentieth century, the Sabin hypothesis could be verified by novel genetic models and fluorescent staining techniques [35–37], and it is now generally accepted that in mammalians, lymphatic vessels arise only after the development of the cardiovascular system by transdifferentiation from veins.

## 1.2  Blood and Lymphatic Vessel Morphology

Blood and lymphatic vessels consist of vessel networks that are lined by blood and lymphatic endothelial cells (*BECs* and *LECs*), which descend from the same embryonic origin [6]. Complementary to extensive histological studies, vessel organization, size, and cellular features of the vasculature at different anatomical sites have been analyzed in high detail, revealing that lymphatic capillaries, blood capillaries, and bigger collecting lymphatic vessels show eminent structural differences. Along with the discovery of endothelial-specific markers, it became possible to clearly identify and distinguish blood from lymphatic vascular endothelium in morphology. The distribution and abundance of these molecular markers between BECs and LECs has been visualized and analyzed in detail. However, it was observed that their expression abundance and patterns might change in unphysiological conditions or disease.

- Development of the vascular system is called vasculogenesis.
- Arborization of blood and lymphatic vessels into smallest capillaries.
- In the fetus, first the cardiovascular system develops.
- Blood and lymphatic capillaries show eminent structural differences.
- Lymphatic vessels build a separate vascular system, and they transdifferentiate from veins.
- Ductus thoracicus and ductus lymphaticus dexter are still connected to veins.

### 1.2.1  Visualization and Marker Molecules

**Visualization of vessel morphology**    During the last two decades, by means of genetic modifications in mouse models, yet unknown blood and lymphatic vascular markers were identified. The according use of novel tissue staining and in vivo imaging techniques has led to a considerable progress in understanding vessel development (lymph−/angiogenesis), blood and lymphatic vessel role and functioning, and, not least, disease states. Many endothelial markers are expressed on both, blood and lymphatic vessels, which affirms the close structural and developmental relationship between the two vessel systems. These markers have fundamentally helped to improve our understanding of the formation, morphology, and function of blood and lymphatic vessels. Moreover, they have enabled the separation and cultivation of primary BECs and LECs in vitro to study their molecular features in high detail (◻ Fig. 1.2c).

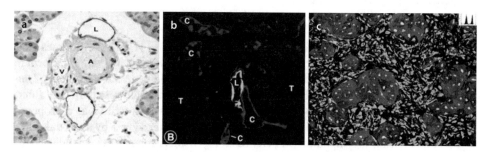

◻ **Fig. 1.2**   Visualization of blood and lymphatic capillaries and endothelial cells. **a** Immunohistochemical podoplanin staining of lymphatic capillaries (L) besides an arteriole (A) in human pancreas. (From: Kerjaschki et al. 2004). **b** Human kidney cortex labeled with anti-podoplanin (green) and blood vessel endothelium-specific anti-PAL-E (red) antibodies. L lymphatic vessel, C capillary, T tubule. (From: Breiteneder-Geleff et al., *Am J Pathol* 1999 [38]). **c** Immunofluorescence staining of LECs with podoplanin (red) surrounded by vWF (green) positive BECs. (From: Kriehuber et al., *J Ex Med* 2001 [39])

1

Marker molecules   Some endothelial cell markers are expressed in both vessel types. These are, for example, vascular endothelial cadherin (VE-cadherin, CD144) [40], platelet endothelial cell adhesion molecule (PECAM-1, CD31) [41], von Willebrand factor (vWF) [42], and thrombomodulin (CD141) [43] and are now important tools in diagnostic and scientific studies. Specific for blood vessels are melanoma cell adhesion molecule (MCAM, CD146, MUC18), Pathologische Anatomie Leiden-Endothelium (PAL-E) antigen (◘ Fig. 1.2b), endoglin (CD105), and multimerin 2 (MMRN2, endoglyx-1). Blood vessels show strong membrane expression of CD31, PAL-E, and CD34 molecule, while the basal lamina essentially contains collagen IV (COLIV) that is connected to the surrounding connective tissue by proteoglycans [13]. The most important lymphatic vessel markers are vascular endothelial growth factor receptor 3 (VEGFR-3/Flt-4) [44], prospero homeobox protein 1 (Prox-1) [35, 45], lymphatic vessel endothelial hyaluronan receptor 1 (Lyve-1) [46], and podoplanin (PDPN) [38] (◘ Fig. 1.2a–c). While Prox-1 is a transcription factor residing in LEC nuclei, VEGFR-3, LYVE-1, and podoplanin are located at the cell membranes.

## 1.2.2 Vasculogenesis

■ **Blood Vasculature Origin and Development**
In the fetus, first the cardiovascular system develops (◘ Fig. 1.3a). The de novo formation of blood vessels during embryogenesis from precursor cells is defined as *vasculogenesis* [47, 48]. Proliferation and differentiation of mesodermal cells lead to the formation of hemangioblasts and *angioblasts*, which are the precursors of blood endothelial cells (◘ Fig. 1.3b). They form primitive blood islands, which are composed of an inner structure of hematopoietic precursor cells, surrounded by an outer layer of angioblasts. Angioblasts differentiate to endothelial cells and form a primary vascular plexus of blood vessels, whereas the hematopoietic precursor cells differentiate to mature hematopoietic cells. The blood islands fuse to a primary capillary plexus which then leads to the formation of the primary vascular network, including the dorsal aorta and the cardinal vein. These are later covered by *vascular smooth muscle cells* (*vSMCs*). Expanding of the primary vascular plexus occurs by remodelling and proliferation [49, 50], ultimately generating the network of arteries, capillaries, and veins. Microvessels are closely interacting with *pericytes*, which are cells with presumable mesenchymal derived origin that are embedded in the basement membrane (◘ Fig. 1.3b).

■ **Lymphatic Vasculature Origin and Development**
Lymphatic vessels emerge by sprouting from embryonic veins [44, 51] or from mesodermal lymphangioblastic precursor cells that reside within tissues [36, 37, 52]. They form in a manner similar to blood vessels: LECs bud from veins and generate six *primary lymphatic sacs* near the junction of the subclavian and anterior *cardinal vein* [53]. The differentiation process of venous into lymphatic endothelial cells is described as a proposed four-step process, where upregulation of the lymphatic marker *Lyve-1* by a yet unknown signal or factor leads to lymphatic competence (◘ Fig. 1.4a). A consecutive cascade of lymphatic bias and specification leads to lymph sac formation. Starting from these primary lymphatic sacs, LECs form a primitive lymphatic vessel system alongside the main venous trunks, then growing into surrounding tissues and organs (◘ Fig. 1.1b). The final and main lymphatic vessel, the left *thoracic duct* (*ductus thoracicus*) develops from the *cisterna chyli*, a dilated sac at the lower end of this trunk. Interestingly, it derives from a pair of lymphatic

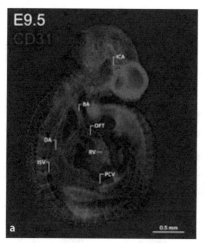

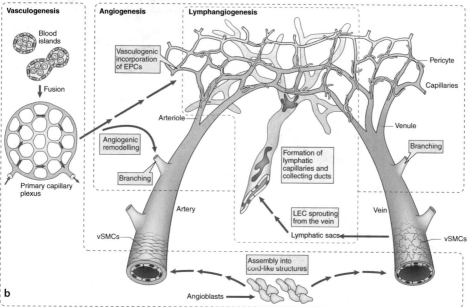

**Fig. 1.3** Vasculogenesis. **a** Murine embryonic vasculature at day 9.5, visualized with CD31 (PECAM-1) immunofluorescence staining. (From: Coultas et al., *Nature* 2005 [41]). **b** Vasculogenesis and blood vasculature assembly. (From: Adams & Alitalo, *Nat Rev Mol Cell Biol* 2007 [4])

trunks and is formed by anastomosis and a remnant of the right lymphatic trunk, which results in numerous variations of its structure in adults. The thoracic duct remains as a final connection to the venous blood circulation and is the main provider of lymph fluid [55]. Despite this site, blood and lymphatic vasculature are strictly separated.

One milestone in the recent upturn of lymphatic vessel research was the discovery of the small mucin-like protein *podoplanin* as a specific LEC surface marker, enabling the particular visualization of lymphatic vessels [38]. Importantly, the final separation step between blood and lymphatic vasculature is exerted by the lymphatic-specific surface protein podoplanin, which serves as an ultimate closing device toward incoming blood by

1

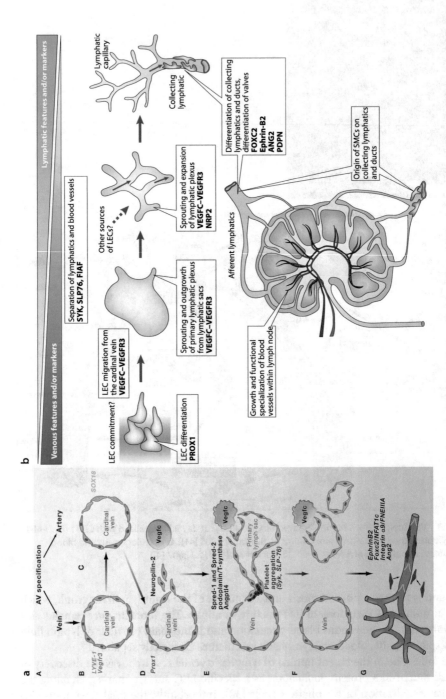

◻ **Fig. 1.4** Four-step model of lymphangiogenesis and subsequent development of the lymphatic vasculature. **a** The proposed four-step model of lymphatic differentiation: (1) Lymphatic competence in venous endothelial cells, responding to an unknown inductive signal. (2) LEC bias toward LEC determination by expression of transcription factor Prox-1. (3) LEC specification by expression of additional specific markers. (4) Lymph sac formation, followed by further LEC differentiation and lymphatic vessel maturation. (From: Tammela & Alitalo, *Cell* 2010 [54]). **b** Formation of lymphatic vessel network and their connection to lymph nodes. (From: Adams & Alitalo, *Nat Rev Mol Cell Biol* 2007 [4])

clotting platelets [56]. These platelet thrombi close the connection between the lymph sacs and the cardinal vein (◘ Fig. 1.4a). Consequently, podoplanin negative mice show bleedings within their lymphatic capillaries. Descending from the largest ducts, blood and lymphatic vessels sprout into smaller and smallest capillaries, where the direct interaction with the interstitial compartment occurs. Finally, the two vascular systems run in parallel and in close proximity, but completely separated, to each other.

**Lymph nodes**   Already during lymph sac formation, small secondary lymph sacs reorganize into lymph node-like structures. Mesenchymal cells immigrate into these sacs and build a network of lymph conduits within these cavities, the so-called lymphatic sinus. Some mesenchymal cells form the lymph node capsule and stromal structures (◘ Fig. 1.4b). Myeloid progenitor cells later immigrate from the bone marrow and thymus, the so-called primary lymphatic organs [57], while secondary are spleen, lymph nodes, lymph nodules, and tonsils [58, 59]. Only briefly before and after birth, the lymph follicles and immune germ centers develop to confer immune response [60].

**De novo lymph–/angiogenesis**   A formation of new blood or lymphatic vessels from pre-existing ones is called lymph–/angiogenesis. In fully established, resting vessels, BECs, LECs, and vSMCs express differentiation markers, and their proliferation rate is extremely low. However, they maintain their ability to dedifferentiate and reenter the cell cycle in response to environmental stimuli. Adult de novo lymph or blood vessel growth is unusual, but may occur during certain pathological conditions like tissue inflammation, wound healing, and cancer metastasis [61]. This formation of capillary networks requires a complex series of cellular events, in which endothelial cells locally degrade their basement membrane, migrate into the connective tissue, proliferate at the migrating tip cells, and elongate and organize into capillary loops [62].

---

**Box 1.1   Blood and Lymphatic Vessel Formation**
- Endothelial progenitor cells give rise to a primitive vascular labyrinth of arteries and veins.
- During subsequent angiogenesis, the vessel network expands.
- After establishment of the veins, specific lymphatic differentiation markers emerge.
- Emergence of marker molecules specifies and determines lymphatic vasculature formation.

---

## 1.2.3   Blood Vessel Morphology

■ **Blood Vessel Anatomy and Structure**

The cardiovascular system consists of the heart and blood vessels, which comprise the arteries, arterioles, capillaries, venules, and veins. Arteries deliver oxygenated blood to the capillaries where bidirectional exchange occurs between blood and surrounding tissue. Veins collect deoxygenated blood from the microvascular bed and carry it back to the heart. The blood endothelial cells show tight interendothelial junctions [9]. Arterioles and venules dispose of a basement membrane and are covered by *pericytes* (◘ Fig. 1.5a). In addition, the collecting venules have *valves* to prevent backflow of blood. In higher vertebrates, the lung and body are perfused in two separate blood circulation systems. Oxygen-poor venous blood is pumped to the pulmonary artery and to the lung capillaries, where gas-exchange occurs. Then, the oxygen-enriched blood is returned to the left part of the heart, from where it is transported by the main aorta to the organs and tissues of the organism.

**1**

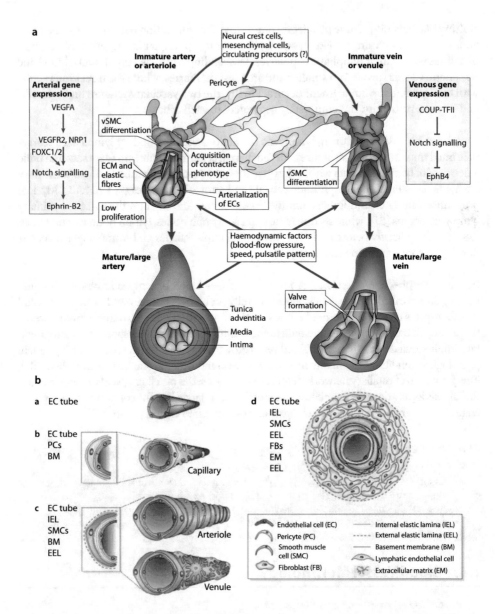

⬛ **Fig. 1.5** Morphology of blood vessels. **a** Arteriovenous differentiation and mural-cell recruitment. Arterial endothelial cells (ECs) have a characteristic spindle-like morphology and alignment in the direction of blood flow. Arteries are surrounded by extensive extracellular matrix (ECM), elastic fibers, and layers of contractile vascular smooth muscle cells (vSMCs). The wall of mature large arteries (such as the aorta) is composed of the inner tunica intima (ECs and basement membrane), the central tunica media with multiple alternating layers of matrix and vSMCs, and the outermost layer tunica adventitia, which is rich in collagen and fibroblasts. Veins have less extensive vSMC coverage and contain valves. (From: Adams & Alitalo, *Nat Rev Mol Cell Biol* 2007 [4]). **b** Wall composition of nascent versus mature blood vessels. (a) Nascent vessels consist of a tube of ECs. These mature into the specialized structures of capillaries, arteries, and veins. (b) Capillaries, the most abundant vessels in our body, consist of ECs surrounded by basement membrane and a sparse layer of pericytes embedded within the EC basement membrane. (c) Arterioles and venules have an increased coverage of mural cells compared with capillaries. (d) The walls of larger vessels consist of three specialized layers: an intima composed of endothelial cells, a media of SMCs, and an adventitia of fibroblasts, together with matrix and elastic laminae. (Adapted from: Jain, *Nat Med* 2003 [63])

■ **Structure of Arterioles, Capillaries, and Venules**

Arterial and venous parts of the smallest blood vessels can be distinguished by ultrastructural characteristics that are observed in scanning and transmission electron microscopy (SEM and TEM) (■ Fig. 1.5a): *Arterioles* have an outer diameter of 10–12 μm and a tube diameter of 4–6 μm. They are characterized by a homogeneous basement membrane forming a lamina [64, 65], which builds a continuous sheet in larger arterioles and arteries. There are seen elastic fibers, bundles, and filaments extending into the tissue, fulfilling anchoring functions. *Capillaries* are closely enveloped by *vascular smooth muscle cells* (vSMCs), which confer contractile action. Especially at the site of arteriole-capillary transition, they are forming so-called precapillary sphincters, a kind of tiny circular muscle structure that controls vessel lumen and, hence, blood flow. The finest blood capillaries form delicate structures that represent the site of substance exchange with the surrounding tissue (oxygen, nutrition). They are built by a single endothelial cell layer disposing of a thin basement membrane, and they have narrow and regular lumina (■ Fig. 1.5b). Capillary walls have a thickness of 0.1–0.3 μm, which is more prominent in the dermis (2–3 μm). The following postcapillary *venules* can be morphologically distinguished from the arterioles by a multilaminated basement membrane, pericytes, and so-called bridged fenestrations [66–68]. The finest tubes measure about 20 μm or less in diameter. They are turning into valve-containing *venules* with a diameter of 25–50 μm and later *collecting veins*, which have a diameter of 70–120 μm. Half-moon-shaped valves are inserted in order to prevent reverse blood flow. As venules became wider, elastic fibers emerge below the vSMC layer, and then become more and more prominent [69].

■ **Structure of Arteries and Veins**

Larger blood vessels are organized in three distinct layers (■ Fig. 1.5b): the inner endothelial cell layer forms the *tunica intima*, which is seen in arteries as a continuous and smooth cell lining, while veins in addition dispose of valves. They are surrounded by the *tunica media*, an elastin sheet that is covered by vSMCs, followed by the outer *tunica externa* or *tunica adventitia*, which consists of fibrous connective tissue. In arteries, the *tunica media* is more prominent than the *tunica externa*, because there are more vSMCs located as the driving forces to change vessel diameter during blood pumping.

---

**Box 1.2    Blood Capillaries**
- Morphology: regular shape, clearly rounded, interendothelial tight junctions, closed vessel walls
- Basement membrane: continuous basal lamina formed by laminated extracellular matrix deposition
- Supporting structures: pericytes (PCs) and vascular smooth muscle cells (vSMCs)
- Dimensions: arterioles 5–6 μm, venules 15–20 μm diameter, capillary wall thickness 0.1–0.3 μm
- Role: nutrient, oxygen, and material transport to and from tissue

---

## 1.2.4  Lymphatic Vessel Morphology

■ **Lymphatic Vessel Anatomy and Structure**

Lymphatic vessels are present in all vertebrates; they are found fully developed the first time in amphibians [70]. The lymphatic system not only consists of lymphatic vessels but also of the *lymphoid organs* as there are the spleen, lymph nodes, tonsils, payer's batches, and the thymus, which are essential for properly working immune response [9]. Lymphatic vessels are found throughout the body, except for the brain and central nervous system (CNS), and avascular or poorly vascularized tissues like cornea, cartilage and bone, muscle endomysium (i.e., muscle fiber sheath), and the epidermis. Instead, these dispose of interstitial channels, so-called pre-lymphatics. Conversely, particularly in the dermis, mesenterium, and breast, which are in high need of interstitial fluid drainage, lymphatic vessels density is very high [71]. Lymphatic capillaries build a very dense network, so-called lymphatic areolas. In adults, there are five different types of morphologically distinct vessel types forming the lymphatic vasculature: lymphatic capillaries, lymphatic collectors, lymph nodes, lymphatic trunks, and lymphatic ducts, having diameter ranges between 10 μm and 2 mm [11]. The lymphatic vascular system is not continuous like the blood vasculature: initial capillaries, also named peripheral or small lymphatic capillaries, emerge blind-ended within the tissue periphery. These capillaries enlarge to lymphatic vessels, followed by collecting lymphatic vessels, collecting ducts, and finally forming lymphatic trunks.

■ Structure of Initial Lymphatic Capillaries

By using ultrastructural analyses, a morphology highly distinct from that in blood capillaries is observed [72]: Lymphatic capillaries have an irregular lumen diameter ranging between 10 and 75 μm, when filled with liquid, and have approximately 100–500 μm length. A basement membrane is usually absent or discontinuous, and if present, poor and tenuously developed [73–75]. Lymphatic capillaries are not surrounded by supporting cells such as pericytes and smooth muscle cells [5]. They have irregular shape, are slightly larger than blood capillaries, and are often seen rather flattened or collapsed. They are lined by a single layer of *lymphatic endothelial cells (LECs)* that are more bulky than BECs. LECs show strong membrane expression of CD31, podoplanin, and Lyve-1. LECs are also more irregular, revealing projections reaching either into the capillary lumen or on the ablumenal side directly into the connective tissue, creating openings between the LECs. This is due to *elastin filaments* that help to tear at the LECs forming interendothelial *gaps* to the surrounding connective tissue (◘ Fig. 1.6a). These *anchoring filaments*, 4–10 nm in diameter, are essentially composed of glycoproteins emilin-1 and fibrillin [76, 77] and link the basal lamina with collagen fibers of the connective tissue [73, 74]. They are stabilizing the lymphatic capillaries against the interstitial pressure and confer the opening process during substance uptake (◘ Fig. 1.6b). The intercellular junctions in this lymphatic compartment are weak, and lymphatic capillaries have no continuous and tight interendothelial junctions. Where LECs attach at each other, they only dispose of hemidesmosome-like structures, but no tight junctions. Rather, there are openings between LECs and the surrounding connective tissue which support collection of lymph fluid. These opening gaps and structures confer permeability and enable the uptake of interstitial fluid. Inside LECs, small and smooth vesicle and larger phagocytic vesicles can be observed.

■ Structure of Collecting Lymphatics

The following larger lymphatic vessels, so-called lymphatic collectors or collecting lymphatics, show additional features: A basement membrane is becoming more prominent, the intercellular junctions are more closed and form tight regions, and the walls are covered by pericytes, vSMCs, collagen, and elastic fibers. In addition, they dispose of *bileaflet valves* [5, 9, 78], which prevent backflow of lymph fluid. Important regulators of valve development are Foxc2 and NFATc1 [79, 80], and their formation involves integrin α9, fibronectin EIIIA, and laminin α5 [81]. The task of collecting vessels is to take up interstitial material, even against low surrounding interstitial pressure. Macromolecules are retained due to size exclusion by the closed junctions, and trans-

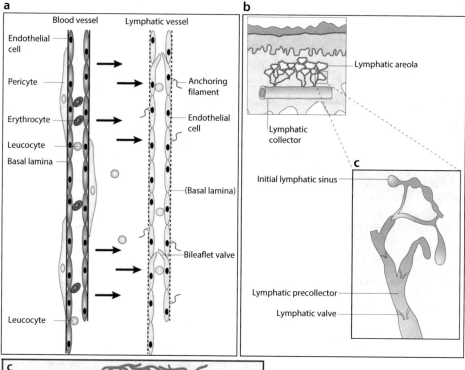

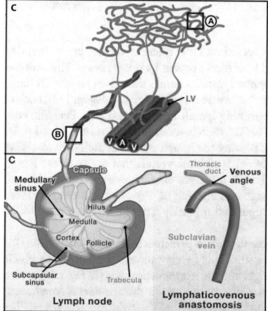

**Fig. 1.6** Morphology of lymphatic vessels. **a** Principle of fluid and substance exchange between blood and lymphatic vessels. (From: Jones et al., *Nat Rev Mol Cell Biol* 2001 [9]). **b** Representation of lymphatic capillaries and collecting vessels in human skin. (Adapted from: Oliver, *Nat Rev Immunol* 2004 [10]. **c** From the capillaries lymph moves to precollectors and on to collecting vessels, directed by changes in interstitial fluid pressure and the negative pressure within the lymphatic vascular system. Lymph finally reaches the venous system via the thoracic duct or the right lymphatic duct that connect with the subclavian veins at the venous angles. (From: Tammela + Alitalo, *Cell* 2010 [54])

**▫ Table 1.1** Structural differences of blood capillaries, lymphatic capillaries, and collecting LVs

| Feature | Blood capillaries | Lymphatic capillaries | Collecting lymphatic vessels |
|---|---|---|---|
| Vessel Lumina | Regular, narrow | Irregular, wide | Circular, wide |
| Endothelial cells | Abundant cytoplasm | Scant cytoplasm | – |
| Cell-cell junctions | Adherens junctions | Loose, valve-like overlaps | – |
| Tight junctions | Present | Absent | Present |
| Valves | Absent | Absent | Present |
| Membrane invaginations and cytoplasmic vesicles | Scant | Abundant | – |
| Basement membrane | Present | Absent | Present, but thinner |
| Anchoring filaments | Absent | Present | Absent |
| Encycling pericytes | Present | Absent | Present |
| Blood | Present | Usually absent | Usually absent |

Based on data from [72, 82–84]

cellular vesicle transport occurs, but is very slow. Only micromolecules can pass closed junctions and might reenter tissue and later blood vessels by other means. The contractility of the vSMCs is important to propulse lymph, whereas the valves prevent its backflow [54]. The region between two lymphatic valves is called *lymphangion* [52], each of which is a contractile compartment pumping lymph into the next one. Periodic contractions of these vessel segments by vSMCs, muscle tone, respiration, and blood pulse confer material transmission, which is passing the lymph nodes. Lymphatic vessels are connected to the lymph nodes via afferent lymphatic vessels that arrive at the lymph node capsule, and lymph leaves the lymph node via efferent vessels that are located centrally, providing efficient flow through and filtering (▫ Figs. 1.4b and 1.6c). In functional lymphatic vessels, which have separated regularly from blood vessels, no blood can be found intraluminally.

Collecting lymphatic vessels assemble to the five principal main *lymphatic trunks*: the lumbal, the intestinal, the bronchomediastinal, the subclavian, and the jugular lymphatic trunks. These pass the lymph into the right and the left thoracic ducts (▫ Fig. 1.6c), which are the final and largest lymphatic vessels. The right thoracic duct or *ductus lymphaticus dexter* drains lymph from the upper right body region into the right subclavian vein. The main LV is the left *thoracic duct* which drains the lymph fluid from the whole body and transports it back to the blood vasculature via the left *subclavian vein* [52, 71]. Except for these two sites where lymph enters the venous bloodstream, no connections between blood and lymphatic vascular system exist (▫ Table 1.1).

**? Questions**
1. What is a reliable and important marker for blood vasculature?
2. Why is observation and handling of lymphatic vessels hardly preserved in histology?
3. What are important specific lymphovascular markers?
4. Where do lymphatic vessels start to form?
5. What are the final connections of the lymphatic system to the blood vasculature?

---

**Box 1.3 Lymphatic Capillaries**
- Morphology: irregular shape, rather collapsed, thin-walled, interendothelial gaps, open lumina bigger than blood capillaries
- Basement membrane: tenuous and discontinuous basal lamina
- Supporting structures: anchoring fibers, no pericytes (PCs)
- Dimensions: 15–75 µm diameter when filled, usually flattened, 100–500 µm length
- Role: opening to the adjacent connective tissue, material removal from tissue.

---

## 1.3 Blood and Lymphatic Vessel Function

Blood and lymphatic vasculature serve separate functions. Blood vessels provide all regions of the organism with fluid, nutrients, and immune cells, whereas the lymphatic system accomplishes the removal of excess interstitial fluid, proteins, particles, and cells that have leaked into the tissues. The two vascular systems control important events of liquid homeostasis and vasoactive processes in higher organisms and are also key regulators in the traffic of immune and tumor cells across the vessel walls. The endothelium is the essential membrane barrier that is the interface for fluids and molecular and cellular components circulating between blood, tissue, and lymph. It is formed by blood and lymphatic endothelial cells (*BECs* and *LECs*) that build a continuous cell monolayer of varying shape and permeability and constitute the actual interface between blood, lymph, and tissue of all organs and entities of the body. BECs and LECs are polarized cells with two different liquid-facing membranes that bear compartment-specific surface molecules and secrete particular substances, which specialize them for respective functions [39].
- Arteries deliver oxygenated blood to the capillaries where exchange occurs between blood and tissues.
- Veins collect deoxygenated blood and carry it back to the heart.
- Because of the blood pressure, blood plasma continuously leaks from the capillaries.
- Tissue fluid is returned by the veins and the lymphatic vessels back to the circulation.

### 1.3.1 Blood Vessel Function

*Blood vessels* fulfill oxygen transport to all regions of the body and provide nutrients and immune cells to the organism [6]. Besides substance exchange, blood vessels are essential for body fluid homeostasis and are directly involved in the regulation of blood pressure and blood flow by secreting appropriate vasoactive mediators. They influence *hemostasis* (blood coagulation) by providing pro- or antithrombotic components on their surface and are involved in inflammatory responses. *Blood* itself provides the body with nutrients, oxygen, and cells, leads to temperature regulation, and is involved in wound healing through hemostasis. Within the bloodstream, body fluid flows rapidly and with high pressure as a plasma suspension containing erythrocytes, whereas outside the bloodstream, it circulates slowly and with low pressure as tissue fluid. The finest blood capillaries confer exchange of transported substances with the interstitial compartment through their thin and highly permeable walls.

Blood vessel volume    The blood vasculature comprises an overall intravasal volume of approximately 4–5 l, 60–70% of which is contained within the venous system. This volume overhang, under concomitant low blood flow and pressure in the veins, stabilizes the circulation. The overall *filtration rate* of the blood capillaries is roughly 20 liters of fluid per day, which is pumped by the *arteries* into the tissue and has to become reabsorbed from the interstitial space in order to maintain body fluid homeostasis. Veins contain approx. 90% of this filtrate (i.e., 18 liters per day), while the lymphatic system takes up the remaining 10% (approx. 2 liters per day). Venous activity and transport back to the heart is triggered by valve activity, skeletal muscle contractions, contraction of the diaphragm, and arterial pulsing.

- **Blood Capillary Permeability and Substance Filtration**

Capillaries and postcapillary venules are highly branched, generating a huge exchange surface. They have a diameter of 20 µm or less to permit close contact between the plasma and the interstitial space, and they have a semipermeable membrane wall. These properties enable the extravasation of fluid and substances from blood to the interstitium. Small molecules like $H_2O$, $O_2$, $CO_2$, ethanol, electrolytes, and urea are able to diffuse freely through the endothelial walls. These substances are transferred to the interstitial space by passive transport through channels and pores. Most water leakage occurs in capillaries or *postcapillary venules*, which have a semipermeable membrane wall that allows water to pass more freely than protein [85]. This is more difficult for organic molecules such as glucose, which need specific carrier proteins, and impossible for proteins, e.g., albumin, and erythrocytes. Blood endothelial surface structures are directly involved in this filtering function: In electron microscopy, it can be seen that their inner (luminal) side is covered with a *glycocalyx*, a layer of macromolecules and glycoproteins, which has a considerable thickness of 0.5–1 µm. This barrier reduces plasma flow and excludes macromolecules and red blood cells to permeate the endothelium [86, 87]. In addition, there are transendothelial mechanisms such as *pinocytosis* and *transcytosis*, which is active uptake of fluid containing solutes and proteins by specialized vesicles, to confer the specific transport of substances into the interstitial space.

- **Blood Capillary Physics of Substance Exchange**

Overall, the generation of interstitial fluid and its resorption back into the venous circulation is regulated by driving forces that are described by the *Starling equation* [88]. The physiologist Ernest *Starling* (1866–1927) understood how it is possible that the isotonic solution of extravascular fluid is drawn back into the bloodstream. Essentially, a high *hydrostatic pressure* within arterial blood vessels generated by the pulse causes water to diffuse out into the tissue and drives filtration of small molecules, but not plasma proteins, which remain in the capillaries. This leads to a difference in protein concentration (i.e., *osmotic pressure*) between blood plasma and interstitial tissue. As a result, the higher osmotic pressure generated by protein retained in the plasma tends to draw water back into the venules from the tissue [85, 89].

- ### Regulation of Blood Perfusion by Blood Flow
Essentially, the velocity and amount of blood flow will contribute to tissue perfusion and according filtration rate. Organ perfusion by blood and body fluid volume is regulated by vasoconstriction and vasodilation, which are controlled by neuronal, hormonal, and local stimuli. These blood pressure-regulating mechanisms operate at the level of the small vessels to ensure constant equilibrium of interstitial versus intravasal fluid [85]. Under physiological conditions, capillary blood flow is modulated [1] at *tissue level* by the activity of muscle and vSMCs, [2] at *vascular level* by the bloodstream, and [3] at an *excitatory level* by the nervous system. There are sensors along the vasculature that control blood pressure status and constantly monitor actual needs. Endothelial cells sense changes in hemodynamic forces and blood-borne signals and respond by immediate release of locally produced vasoactive substances [90, 91]. This is then transferred into the direct regulation of vessel diameter to balance blood perfusion.

- ### Vasoconstriction and Vasodilation: Blood Pressure Regulation
Adaptations of vessel diameter are essential to regulate blood pressure. Abnormally raised blood pressure will induce *vasoconstriction* that is characterized by a reduced vessel diameter, which leads to enhanced flow resistance and reduced blood perfusion. This can be mediated in a fast reaction by *nervus sympathicus* excitation and in a more delayed manner by humoral influence of adrenaline, noradrenaline, or angiotensin II. Finally, local factors like enhanced ATP consumption and lowered $O_2$ levels (muscle work) can be active. Conversely, low blood pressure will be counteracted by *vasodilation* or vasorelaxation that leads to enhanced vessel diameter to promote smooth muscle relaxation, which is accompanied by reduced flow resistance and enhanced blood perfusion.

- ### Hemostasis: Blood Coagulation and Thrombus Resolution
Blood endothelial cells (BECs) actively mediate *anticoagulation* in order to maintain blood fluidity and prevent vascular occlusion: They have two important anticoagulant systems, the heparan sulfate-antithrombin system and the thrombomodulin-protein C system. Under physiological conditions, these two systems inhibit activation of coagulation on BECs. Under inflammatory condition, TNF-alpha or other cytokines produced by monocytes reduce the anticoagulant properties of endothelial cells by downregulating expression of heparan sulfate and thrombomodulin [92], which might lead to thrombosis.

Injuries can lead to undesirable vessel leakage and blood loss, which has to be interrupted. Bleeding usually depends on the number of injured vessels and the rate at which blood is supplied to them, that is, on the degree of tissue perfusion (hyperemia or ischemia). However, bleeding normally retards and stops spontaneously within a few minutes by *blood coagulation*. This hemostatic mechanism is highly effective: A primary reaction leads to vessel closure by *thrombocytes* (white clot), while in a secondary reaction, the factor *thrombin* catalyzes *fibrin* formation from fibrinogen, which leads to the formation of a red thrombus. Conversely, thrombi have to be resolved over time to prevent vessel clogging. The reverse process of thrombus formation is *fibrinolysis*, the enzymatic resolution of a thrombus. Plasmin, which is derived from plasminogen, leads to fibrin degradation and clot resolution.

**1**

- **Blood Vessels in Inflammation**

Inflammation, the response against damage and danger of the tissue, essentially affects the blood vasculature, as it provides the route for its relief. In order to enable the outflow of fluid, proteins, and leukocytes to the injured tissue site, vessel permeability is increased by pro-inflammatory cytokines. At the same time, blood flow is unusually enhanced in the arterial vessel compartment, which is due to changes in both, vessel diameter and permeability. The accumulation of fluid outside the blood vessels is called *edema*, and this fluid loss results in extremely low blood flow in the venous compartment that results in an increase of blood viscosity, concentration of the blood, erythrocyte aggregation, and eventually even fluid stasis. During an inflammatory immune reaction, *leukocytes* change expression of their surface molecules. Via upregulation of *integrins* (CD11/CD18, VLA-4), they firmly stick to *adhesion molecules* (VCAM-1, ICAM-1) present on blood endothelium and emigrate across the basement membrane of small vessels to reach the site of inflammation. Although painful and unpleasant (redness, heat, swelling, pain), overall, these processes drive elimination of infections and wound healing.

- **New Blood Vessel Growth: Angiogenesis**

At homeostasis, blood endothelial cells (BECs) and vascular smooth muscle cells (vSMCs) are fully differentiated and have a very low proliferative index. Angiogenesis, the process of new blood vessel formation, is an integral part of development and wound repair [93]. In this process, fully differentiated BECs and vSMCs regain proliferative properties, divide, and newly form capillary structures. In addition, this process of neovascularization is supported by the incorporation of circulating *endothelial precursor cells* (EPCs) into growing blood vessels. Physiological angiogenesis takes place during wound healing, female reproductive cycle and pregnancy, and neovascularization, while pathological angiogenesis is found during persistent inflammatory conditions, retinopathies (macular degeneration or diabetes-associated retinopathy), and tumor growth. On the contrary, ischemic regions that are in need of blood supply might arise due to a lack of proper angiogenic activity.

**?** Questions
6. What is the approximate tissue fluid filtrate amount per day?
7. What is the interstitial fluid uptake rate of venules versus lymphatic vessels?
8. What are the three key regulators of blood flow?

Box 1.4    Blood Vasculature Functions
- Transport of tissue fluid and cells throughout the body
- Substance filtration: Exchange of nutrients, ions, oxygen, etc.
- Uptake and evacuation of tissue fluid and remnant substances by the venous system
- Regulation of the body temperature, blood pressure, wound healing
- Modulation of the immune system

## 1.3.2  Lymphatic Vessel Function

The principal function of the lymphatic system is the removal of particles, macromolecules, and excess fluid from the interstitial space and its return to the blood (□ Fig. 1.7a).

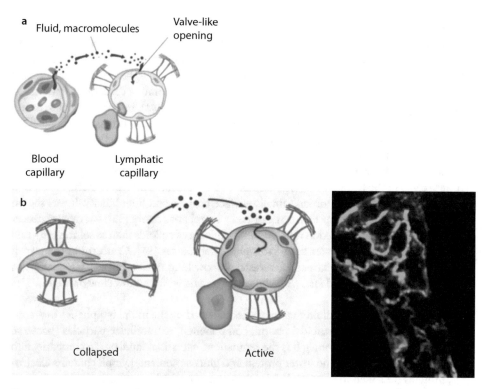

**a** Fluid, macromolecules

Valve-like opening

Blood capillary

Lymphatic capillary

**b**

Collapsed

Active

□ **Fig. 1.7** Lymphatic vessel functioning: Interstitial fluid uptake by lymphatic capillary. **a** Interstitial fluid, macromolecules, and cells enter lymphatic capillaries through interendothelial openings. Lymphatic capillaries are stabilized by anchoring filaments. (From: Alitalo et al., *Nature* 2005 [5]). **b** Mechanism of interstitial tissue fluid uptake by lymphatic capillaries: During tissue compression, filaments are relaxed and junctions close. During high interstitial pressure, filaments are strained and junctions open. (From: Tammela & Alitalo, *Cell* 2010). **c** Lymphatic capillaries are sealed when the vessel is filled: Shown is the blind end of a lymphatic capillary in a whole-mount preparation of the mouse ear skin. Lymphatic vessel hyaluronan receptor-1 (Lyve-1) immunostaining (green) shows the oak-leaf-shaped lymphatic endothelial cells (LECs). (Adapted from: Tammela & Alitalo, *Cell* 2010 [54])

Lymphatic vessels collect extravasated tissue fluid, filter it through lymph nodes, and return it to the circulation through the thoracic and lymphatic ducts to the venous system. Tissue fluid, plasma components, infectious agents such as bacteria, and extravasated white blood cells enter the lymphatic vessels through openings/intercellular gaps. Therefore, it forms a highly branched network of *blind-ended*, thin-walled lymphatic capillaries and lymphatic collectors which drain tissue fluid (lymph) from the tissues and transport it back via the thoracic duct into the venous blood circulation.

Hence, lymphatic vessels play an important part in maintaining body fluid homeostasis. In addition, lymphatic vessels are involved in absorbing lipids or dietary fats from the intestinal tract. Moreover, lymphatic vessels are an integral part of the immune system and fulfill immunological tasks like the transport of antigen-presenting cells (APCs) and white blood cells to lymph nodes and lymphoid organs (spleen, tonsils, thymus) [10, 94].

- Lymph capillaries show no basement membrane and no pericytes.
- Lymph capillaries drain almost all body sites.
- Collecting lymphatic vessels transport lymph fluid.
- Lymph is the residuum of the blood capillary filtrate.

■ **Interstitial Fluid Uptake and Lymph Formation**

Lymphatic vessels are linked to the extracellular matrix by *anchoring filaments*, which stabilize the *opening gaps/junctions* between LECs and prevent lymphatic vessel collapse in conditions of high interstitial pressure. These are the primary sites of uptake of components that are too huge for blood capillaries, which are bacteria, cells, immune cells, macromolecules, and lipid particles. Also huge particles like carbon particles (dust), infectious agents (bacteria), lipid droplets, and tumor cells are transported. Closed junctions between LECs allow for the passage of ions and small molecules (~ 40 kDa) up to approximately 500 kDa. Besides uptake into capillaries between the LECs, also transcytosis and vesicular transport have been observed. The intercellular opening gaps provide entry of high molecular weight material. Selectivity of the lymphatic uptake is conferred by restricted diffusion, steric hindrance, molecular size and dimensions, and finally surface charges and lipophilicity. It was shown that molecules with a size up to 1 μm can enter lymphatics freely [95]. Moreover, cancer drugs enter lymphatic circulation much faster as in form of colloids than as solutes [96], and lipids show higher uptake rates than hydrophilic substances [97]. Concerning the transit time, macromolecules and extracellular vesicles show long removal times. Chylomicrons are taken up and transported fast, while huge phagocytic vesicles show slow passage.

**Lymph fluid contents**    The fluid that has been collected by the initial lymphatics that contains solubilized substances (lipids, plasma components), extracellular particles (bacteria, vesicles), and cells is called *lymph*. It is the remnant of interstitial fluid, but in comparison to plasma, it has higher water and lower protein and glucose content. Lymph contains electrolytes, proteins (less than in plasma), lipids (chylomicrons), leukocytes, microorganisms and pathogens, and tumor cells. As it contains proteins and lipids, it is a milky suspension. In healthy lymphatic vessels, there are no erythrocytes present.

■ **Mechanism of Lymph Drainage**

Lymph transport is a one-way system, yet working without active pumps in the initial collectors. Lymphatic capillaries have low intraluminal pressure and very low lymph flow. Lymph formation is dependent on interstitial tissue pressure and tightness of the extracellular matrix [11]. Fluid uptake is performed during tissue relaxation and concomitant strain of the *anchoring filaments*, which provides a filling phase via the *opening junctions* (◻ Fig. 1.7b). Again, the rate of uptake can be described by the Starling equation and is mainly a result of diffusion and filtration and dependent on hydrostatic and oncotic forces, the so-called Starling forces [98]. Normally, all forces together lead to a diffusion or filtration rate of 2 liters of interstitial fluid by the lymphatic vessels per day, most of which is taken up by blind-ended lymphatic capillaries. This comprises ten percent of the blood capillary filtrate. More recently, an alternative model has been presented, where in addition to the Starling forces, interstitial pressure forces are considered. When these are taken into account, the lymphatic vessel absorption rate would increase substantially [99]. Slow backflow into the tissue prevents loss of material. However, probably there is some dilution effect of proteins during passage along the vessels. Basically, lymph fluid is transported against a higher hydrostatic pressure and protein concentration present in the interstitial space. Transport is provided by muscle contractions or by enhanced pressure through the surrounding tissue. The net flow rate of lymph is approximately 100–500 times less than that of blood [11].

- **Lymph Transport and Return to Blood**

Lymph is transported by passive and active mechanisms [100]. The *passive* lymph pump is dependent on forces acting by the surrounding tissue, e.g., skeletal muscle and cardiac contraction and respiratory or gastrointestinal peristaltic pressure. This is mainly relevant for lymph capillaries due to the lack of surrounding vascular smooth muscle cells (vSMCs). Tissue compression leads to relaxation of the *anchoring filaments*, upon which the valve-like *junctions* are closed, and forced by vessel compression, lymph fluid passes on (◘ Fig. 1.7b). From the capillaries, lymph moves to lymphatic precollectors and on to collecting ducts, which is conferred by changes in interstitial fluid pressure and by the negative pressure within the lymphatic vascular system. *Active* lymph pumps (intrinsic forces) are present in larger collecting ducts, which is the contractility of the surrounding vSMC layer and support by *lymphatic valves*, which prevent backflow. The lymphatic vessel part between two valves is called the *lymphangion*. Lymph propulsion is mediated there by regular contractions of vSMCs with an approximate frequency of 10–12 contractions per minute and supported by systemic forces like respiration [101] and blood pressure [102]. In addition, lymph transport might be extrinsically enhanced by exercise [103] and massage [104].

Finally, lymph is transported back to the blood vasculature by the left and right thoracic ducts.

- **Lymph Node Filtration**

Afferent lymphatic vessels are intercalated by a multitude of lymph nodes (LNs). Lymph is filtered when passing through them from the outer capsule (afferent lymphatic vessels) to inside, leaving on the opposite side via the efferent lymphatic vessel. During passage, foreign antigens are taken up and presented by antigen-presenting cells (APCs) to initiate a specific immune reaction. Other deposited material is cells, including circulating leukocytes and tumor cells [105]. Pollutants get caught in lymph nodes due to macrophage uptake that leads to characteristic dark coloration, the so-called anthracotic pigmentation. Such dark colored LNs can be traced prominently in the head and neck region and around the trachea, where high amounts of dirt particles are filtered out from the interstitial space.

- **Intestinal Lipid Uptake and Transport**

Another important function of lymphatic vessels is the absorption and transport of fat and free fatty acids (FFAs) from the intestine [5, 106]. *Lacteals*, specialized and blind-ended lymphatics in the center of each *villus* (= intestinal projections), are crucial for the uptake of dietary lipids packed as large lipoprotein particles, so-called chylomicrons [82]. The lymph retrieved from the intestine is designated as *chyle*, a milky, FFA, and chylomicron-rich fluid. Chylomicrons and lipoproteins are passing through many open junctions and are immediately pumped further, which prevent backflow. Both are also seen entering the lymphatic endothelial cells and lying inside them, in caveolae and vesicles [107]. It is not known whether lymphatic vessels can actively influence and regulate lipid uptake and transport into the lacteals. However, there seems to exist a close crosstalk between lymphatic vessels with the intestine on one hand and with the adipose tissue on the other regulated by gut hormones, adipokines, and/or lipids themselves. Recently, an in vitro model of the enterocyte-lacteal interface using differentiated LECs and intestinal Caco-2 cells was established [108]. By using a fluorescently labeled fatty acid, it was shown that its transport is polarized from the enterocytes to the lymphatic vessels through both, transcellular and paracellular pathways. This model highlights not only the importance of lymphatics for lipid uptake but provides an interesting tool for evaluating drug delivery from the intestine into the circulation.

- **Lymphatic Vessels and Adipose Tissue**

There is increasing evidence that there is a close relationship between lymphatic function, lipid metabolism, and adipogenesis [109, 110]: Lymph nodes and lymphatic vessels are often surrounded by adipose tissue [110, 111], and lymphatic vessel-adjacent adipocytes respond to local immune reactions by increasing lipolysis [112, 113]. Strikingly, lymph node development is necessary for development of associated lymph node fat pads and vice versa [114]. However, to what extent lymph vasculature regulates lipid metabolism and adipose tissue formation is still speculative and needs to be further clarified. Especially, knowing the regulation of uptake and transport of adipogenic factors from and to the adipose tissue or from the gut would be of eminent importance to evaluate potential contributions of lymphatics to conditions like lymphedema and obesity.

- **Interaction with Immune Cells and Immune Control Function**

Lymphatic functioning comprises important tasks in inflammation, infection, and transport of immune cells. During the acute phase of inflammation, lymphatic vessels serve as principal conduits for *antigen-presenting cells* (APCs) to reach regional lymph nodes and to educate immune cells by presenting foreign antigens. Their migration is regulated via specific chemotactic mechanisms: *Dendritic cells* (DCs) express chemokine C-C motif chemokine receptor 7 (CCR7), which binds to C-C motif chemokine ligand 21 (CCL21) secreted by LECs [115, 116]. CCR10-positive *T-lymphocytes* are especially attracted by a subpopulation of collecting lymphatic vessels expressing high levels of CCL27, but low levels of podoplanin [117]. Moreover, expression of macrophage-mannose receptor 1 (MMR-1) [118, 119] and common lymphatic endothelial and vascular endothelial receptor-1 (CLEVER-1) [120, 121] regulate lymphocyte trafficking in lymphatic vessels. Toll-like receptor 4 (TLR4) is highly expressed on the surface of LECs and the main mediator of LPS-induced activation of NFκB (nuclear factor "kappa-light-chain-enhancer" of activated B cells) during inflammation. Via this signaling pathway, various chemokines such as CCL2, CCL5, and CX3CL1 are released, which contribute to subsequent chemotaxis of *macrophages* toward sites of inflammation [122]. These release TNF-α that stimulates the expression of the leukocyte adhesion molecules intercellular adhesion molecule 1 (ICAM-1), vascular cell adhesion molecule 1 (VCAM-1), and E-selectin (SELE) on LECs, which in turn stimulate DC adhesion and transmigration [123]. Further, lymphatic vessels take part in the clearance of leukocytes from sites of inflammation [54]. However, detailed knowledge on the specific interaction mechanisms of different immune cell types with the lymphatic endothelium is still sparse.

**? Questions**
9. What is lymph fluid?
10. How are the smallest lymphatic capillaries opening and closing?
11. What prevents backflow of lymph fluid in collecting lymphatic vessels?

Box 1.5   Lymphatic Vessel Permeability and Substance Uptake
- Transport phase – closed gaps: during tissue compression and low interstitial pressure anchoring, filaments relaxed and junctions closed passage of ions and small molecules, cutoff at 500 kDa
- Filling phase – open gaps: during tissue relaxation, high interstitial pressure, and edema anchoring, filaments strained and junctions open passage of high molecular weight material (blood constituents, cells, etc.)
- Direct passage via gaps: only in severely injured lymphatic vessels

## 1.4   Blood and Lymphatic Vessel Pathology

Dysfunction or malfunction of the vascular systems contributes to the pathogenesis of many diseases [5, 63, 124, 125]. A range of distinct molecular equipments has been identified, which indicate the functional specifications of blood and lymphatic vessels [54] and have helped to elucidate pathological conditions. While the blood vessel system has been in the center of interest during recent years, the lymphatic system now also is emerging as a crucial player in diseases ranging from cancer to inflammatory processes. Its role in tumor spread, asthma, transplant rejection, or lymphedema is attracting now more attention [4]. Unphysiologic changes occur on all levels of the aforementioned functions: functional insufficiency, including altered body liquid homeostasis, chronic vessel damages, inflammatory conditions, systemic diseases such as consequences of metabolic derailments, and finally aberrant vessel growth. Because of the contribution of BEC and LEC proliferation to the pathogenesis of several diseases, including cancer and cardiovascular disease, considerable effort has been devoted to particularly elucidate the mechanisms that regulate cell cycle progression and growth in these cell types.

Box 1.6   Importance of Vascular Markers
- Insights into lymph–/angiogenesis. Understanding of vasculature development
- New imaging techniques: Visualization and understanding of diseases
- Novel lympho–/vascular markers: Improvement of diagnosis and treatment

## 1.4.1   Blood Vessels in Disease

The pathogenesis of vascular complications is complex but well studied in blood vessels. Several diseases essentially affect the smallest blood capillaries in their physiology, as these are mainly exposed to changes of blood flow and composition. Constant disturbance of these delicate homeostatic mechanisms over time leads to endothelial dysfunction, which presents with excessive matrix production, abnormal vasoconstriction, increased vessel permeability, platelet activation, low-density lipoprotein (LDL) oxidation and deposition, but also vascular smooth muscle cell proliferation and migration.

1

■ **Disturbed Blood Capillary Permeability and Edema**

Disturbed equilibrium of body liquid homeostasis leads to shifts in fluid distribution within body compartments. The accumulation of tissue fluid reduces hydrostatic pressure difference until edema appears.

Blood vessel leakage and consecutive liquid accumulation within the interstitial tissue compartment is called *edema*, and this can be caused by all components involved in body fluid circulation: blood and blood vessels, the tissue itself, and the lymphatic system. These liquid shifts arise due to altered hydrostatic or osmotic pressure gradients, enhanced permeability for fluid, and/or water-retaining capabilities of the involved compartments.

■ **Disturbed Blood Pressure**

*Hypertension (HT)* is a condition in which the blood pressure in the arteries is persistently elevated due to increased blood vessel resistance or vascular damage [126]. Long-term high blood pressure is a major risk factor for consecutive vessel diseases that are nowadays the main death cause in Western countries: these are coronary artery disease, stroke, heart failure, atrial fibrillation, peripheral vascular disease, vision loss, chronic kidney disease, and dementia [127–129]. Nearly all incidences are provoked by careless *lifestyle* habits, such as salty diet, overweight, smoking, and alcohol abuse. Only a small portion of patients suffer from a clear defined disease that is associated with high blood pressure, such as kidney diseases or endocrine dysfunctions. The disease mechanisms can be assigned to disturbances in the systems that control vasodilation and vasorelaxation, such as abnormalities of the renal renin–angiotensin [130] or of the sympathetic nervous system [131]. *Hypotension*: On the contrary, blood pressure can also be dangerously lowered. Low blood pressure can be caused by low blood volume, hormonal changes, widening of blood vessels, medicine side effects, anemia, and heart or endocrine problems. Severely low blood pressure leads to lack of oxygen and nutrients in the brain and organs, which induces the condition called *shock*.

■ **Disturbed Blood Flow and Vessel Perfusion**

*Peripheral arterial occlusive disease (PAOD)* describes a narrowing of arteries in the periphery of the body, which are not those that supply the heart or the brain. The burden of consecutive emergence of health risks, including atrial fibrillation (AF), and the associated necessary healthcare is increasing [132].

*Aneurysms* are caused by weakened blood vessel walls, which may be a result of a hereditary condition or an acquired disease. It is an unusual dilation of a vessel due to weakness of the vessel wall and the surrounding connective tissue, which later can also be the starting point for clot formation (thrombosis) and embolization. In worst case, a life-threatening vessel rupture causing lethal inner bleedings can occur.

*Raynaud syndrome* is characterized by spontaneous vasoconstriction of (vasospasms) that lead to sudden and painful paling of fingers and toes, because the organism tempts to minimize heat loss when exposed to cold. Underlying cause is an unphysiologic activation of the sympathetic system that causes exaggerated vessel constriction of the arterioles in order to store warm blood remote of the body surface within the deep veins.

*Venous diseases: Chronic venous insufficiency (CVI)* is characterized by venous valve insufficiency of superficial and deep veins, especially in the lower limbs, which leads to blood retention and increased blood pressure in the venous system [133]. This cannot be properly counterbalanced by muscle contractions that normally lead to reduction of venous pressure. Over time, the excessive blood filling causes swollen legs, pain, and skin alterations. On the long term, complications such as varicose veins, ulcers, and hemorrhoids manifest. This disease may not be confused with *deep vein thrombosis (DVT)*, in which the valves of deep veins have been damaged by a previous event of embolism or detachment of a blood clot [134].

- Disturbed Blood Coagulation

*Hemophilia* is the tendency to develop enhanced bleedings and hematomas. It is a genetic disorder that leads to altered levels of factors of the blood clotting cascade, which reduces the ability to coagulate [135]. There are two main types of hemophilia: hemophilia A, due to reduced clotting factor VIII, and hemophilia B, which is caused by reduced clotting factor IX.

*Thrombophilia,* on the contrary, is the increased disposition to have blood clotting caused by enhanced blood coagulation. This highly enhances the risk of thrombotic events, the emergence of blood clots in blood vessels such as deep vein thrombosis in the legs [136, 137].

*Thrombosis* describes the process of abnormal thrombus (= blood clot) formation within vessels that occurs as a consequence of alterations of the endothelial cell layer, reduced blood flow, or unphysiological blood composition [138]. It is also induced by vessel wall abnormalities such as deposits (plaques), fissures, and dilations (aneurysms). Thrombosis occurs during onset of inflammatory processes, particularly those in which necrosis has occurred. Conversely, inflammation is induced in thrombotic processes (e.g., the presence of inflammatory infiltrates after an event of ischemic infarct). Hence, thrombosis can be both cause and consequence of inflammation.

- Chronic Blood Capillary Damage in Atherosclerosis

Atherosclerosis is a multifactorial disease in which blood vessels chronically develop irregular material deposits within the arterial vessel walls, so-called plaques. These changes over time influence blood flow and induce aberrant blood coagulation (thrombosis) and reduction of vessel diameter. Vessel wall hardening and obstructions finally cause strokes, infarcts, and even vessel ruptures (aneurisms). Atherosclerosis has become a prevalent cause for mortality in developed countries. The exact initiative of these structural alterations is not known, but risk factors include mainly lifestyle habits such as diet (high cholesterol), high blood pressure, diabetes, smoking, obesity, and genetic predisposition (family history). The onset of the atherosclerotic process is supposed to follow the so-called *response-to-injury* cascade: Any constant unphysiological stimulus such as metabolic changes, inflammation, or trauma over time will lead to endothelial dysfunction that initiates an inflammatory reaction of the endothelial cells [139]. This leads to retention of low-density lipoprotein (LDL) particles and plaque formation, which contain fat, cholesterol, and calcium in the vessel walls. Altered lipoprotein particle composition (oxidized LDL) and subsequent calcification leads to loss of vessel wall elasticity. After plaque formation, the injury progresses toward intima fibrosis, which is characterized by vSMC proliferation and migration from the tunica media into the intima. Successive chronic activation of platelets and coagulation factors leads to microthrombus formation and vessel narrowing (*stenosis*).

- Chronic Blood Capillary Damage in Type 2 Diabetes

Type 2 diabetes (T2D) is a multifactorial disease; however, it is fundamentally associated with vascular damage, overall causing *cardiovascular disease (CVD)* [140, 141]. Vascular complications include *micro-* and *macroangiopathy* [142–144]. Persistent

**1**

metabolic changes and insulin resistance cause a sub-chronic state of inflammation that leads to frequent infections in the skin and chronic ulcers due to impaired wound healing. This causes persistent infections, dramatically delayed wound healing, neuropathy, and, ultimately, amputations. Overall, the cardiovascular complications are the main cause of increased morbidity and mortality in fully developed T2D. Adults diagnosed with diabetes have a two- to fourfold increased risk for development of cardiovascular problems [145]. All phenomena described above are also characteristic for diabetic microangiopathy of which endothelial dysfunction is thought to be the first step [146]. Macroangiopathy affects coronary, carotid, and peripheral arteries and increases the risk for myocardial infarction, stroke, and diabetic foot syndrome [147–149]. On a molecular level [150], it has been established that over time, reduced insulin-sensitivity (insulin resistance) exerts persistent hyperglycemia [151]. Chronically enhanced glucose and glycosylated protein (AGE) levels in blood exert glycotoxic effects that directly induce endothelial damage. These include enhanced oxidative stress (ROS) and inflammation (NF-κB). The alterations manifest as blood flow abnormalities (reduced eNOS, ET-1), impaired vascular relaxation (RAAS, NO production) [152], enhanced vascular permeability (enhanced VEGF), capillary occlusions (TGF-β), and vascular occlusion due to thromboembolisms (increased PAI-1) [153]. Hence, the implications of therapeutics to treat CVD that is associated with T2D are highly complex [154].

- **Diabetic Microangiopathy (DMA)**

DMA is the underlying cause of nephropathy, retinopathy, and neuropathy [155]. DMA usually begins with reversible functional disorders and ends with irreversible loss of organ function after years of disease. Diabetic *nephropathy* is a disease of the glomerulus. It begins with hyperfiltration followed by proteinuria and progressive renal failure [156]. The prevalence of diabetic microangiopathy increases with the duration of the disease. The prevalence of nephropathy increases to 50% during the first 20 years after diagnosis and levels off thereafter. Diabetic nephropathy accounts for more than 30% of all end-stage chronic renal failure patients [157]. Diabetic *retinopathy* is a disease of the retinal vasculature. Clinical signs are microaneurysms, retinal infarcts, and neovascularization finally associated with loss of vision [158]. Retinopathy is rarely seen within the first few years, but after 15 years, it is found in 25% and after 20 years in almost 100% of the patients, respectively [159].

*Blood vessel morphology of DMA*: Although diabetic complications lead to a multifaceted clinical picture, a feature observed ubiquitously in histological specimens from diabetic patients is a prominent capillary basement membrane enlargement. Essentially, an increase in type IV and VI collagen, fibronectin, and laminin protein deposition and a decrease in proteoglycan abundance occur (◘ Fig. 1.8a). This alteration of protein composition and thickness is accompanied by concomitant changes of filtering function and charge. Due to the latter, permeability to blood-borne molecules is increased. In addition, pericytes are partly lost and/or dysfunctional, which results in destabilization of the wall integrity [161, 162]. Basic membrane thickening is the most robust morphological parameter for clinically symptomatic DMA. However, the molecular mechanism behind its formation is still unknown, and there is no remedy to reverse this process.

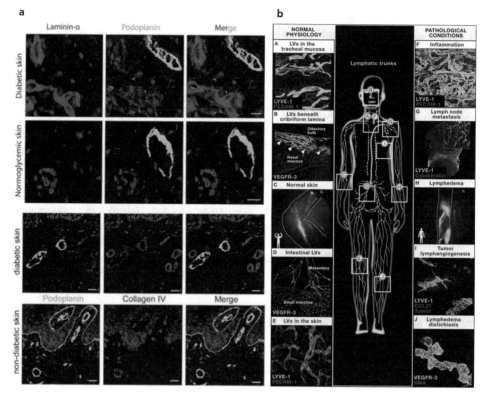

□ **Fig. 1.8** **a** Diabetic microangiopathy of blood capillaries in the human skin. Blood capillaries show enhanced expression of laminin and collagen IV (both red), but no changes of lymphatic capillaries (podoplanin, green). (From: Haemmerle et al., *Diabetes* 2013 [160]). **b** Lymphatic vessels in physiological and pathological conditions. (From: Tammela & Alitalo, *Cell* 2010 [54])

■ **Blood Vessels in Inflammation**

Blood vasculature responds to insults (heat, cold, burn, chemical injury, etc.) with vasodilation and increased permeability to plasma proteins. Plasma proteins usually leave the inflamed vessels by passing in gaps between the endothelial cells. Concomitantly, substantial amounts of plasma proteins leak into the tissue, which causes *edema* [85]. In inflammation, there is also increased proteolysis, which leads to further protein accumulation. Subsequently, inflammatory cells immigrate into the tissue. Emigration of inflammatory cells occurs with delay and exerts immune reactions. In persistent inflammatory states, there are chronic alterations of affected tissue. One special incidence is *vasculitis,* an inflammatory disorder that destroys blood vessels and affects both arteries and veins. Vasculitis is caused by leukocyte migration and resultant exaggerated vessel damage. Therapy of inflammation should aim at removing the cause of inflammation (antibiotics, antihistamines) and treatment of edema (anticoagulants, surgery). In second line, the treatment will target edema fluid and proteins to reduce blood volume and water-loading within the tissue.

1

■ **Angiogenesis: Aberrant De Novo Blood Vessel Growth**

Unphysiological blood vessel overgrowth (pathological angiogenesis) involves abnormal endothelial and vascular smooth muscle cell (vSMC) proliferation and plays important roles in the pathogenesis of vascular diseases. Because of the public health importance and economic impact of these pathological processes, elucidating the regulatory factors and molecular mechanisms that control endothelial cell and vSMC growth is currently the subject of active research. Abnormal vSMC proliferation is thought to contribute to the pathogenesis of vascular occlusive lesions, including atherosclerosis, vessel renarrowing (restenosis) after angioplasty, and graft atherosclerosis after coronary transplantation. Particularly, malignant tumors induce aberrant growth of blood vessels for oxygen and nutrient supply and, moreover, for their metastatic spread to other organs [163, 164]. Undoubtedly, certain kinds of tumors (sarcomas, colorectal and kidney cancer) mainly spread via *hematogenous metastasis*. Some mechanisms leading to tumor angiogenesis are quite well understood, one milestone being the release of pro-angiogenic growth factor VEGF into the progressive tumor environment. This has led to the establishment of anti-angiogenic treatments [165]. One prominent example is the application of *anti-VEGF antibodies* (Avastin/bevacizumab), which is also successfully used in humid macular degeneration, where exaggerated neovascularization disturbs and over time destroys vision. Conversely, states of insufficient blood vessel supply, as seen in ischemia, are treated by VEGF administration to stimulate vessel growth [166].

**? Questions**
12. What are the major lifestyle-induced blood vessel diseases?
13. Why do tumors induce blood vessel growth?

---

**Box 1.7   Effects of Hyperglycemia on Blood Endothelial Cells**
Hyperglycemia induces an endothelial increase of protein kinase C (PKC) which in turn causes:
- Reduced expression of the vasodilator nitric oxide: aberrant vasoconstriction
- Increased deposition of collagen: thickening of basement membrane, capillary occlusion
- Increased expression of the NF-κB pathway: enhanced inflammatory reactions
- Increased production of reactive oxygen species: enhanced oxidative stress

---

### 1.4.2 Lymphatic Vessels in Disease

Only during the recent two decades, the role of the lymphatic system in disease states has been recognized (◘ Fig. 1.8b): Malfunction or dysfunction of the lymphatic system contributes to the pathogenesis of chronic lymphedema, ascites, and/or aberrant adipogenesis and fibrosis, it is relevant for enhanced or impaired inflammation and immune responses, and it is involved in infectious diseases and spread of parasites [54, 167]. In cancerous diseases, the lymphatic vasculature is involved in tumor growth and metastasis. Not least, there are lymphatic tumors as there are Kaposi's sarcoma [168] and lymphangiomas [169]. Novel and exclusive lymphovascular marker molecules and the use of genetic rodent gene knockout models have enabled the identification of unphysiologic changes and deficiencies of lymphatic vessels and have helped to elucidate the pathophysiologic mechanisms behind these pathologies.

### ▪ Disturbed Lymph Transport: Lymphedema

Lymphedema arises through failure of lymph drainage, when lymphatic vessel capacity is saturated, resulting in accumulation of fluid in the tissue interstitium. This is generated by exaggerated microvascular filtration, either by increased capillary pressure or by reduced plasma osmotic pressure, which leads to enhanced amounts of interstitial fluid [99], accompanied by painful swelling. Over time, lymphedema leads to irreversible tissue fibrosis, accumulation of fat tissue and recurrent infections [170]. Lymphedema is a problem resulting from natural/primary (genetic defects, inflammation, and infection) or artificial/secondary (breast cancer surgery, radiotherapy) processes [171].

*Primary/hereditary lymphedema* (1.1 people/100.000) is caused by congenital absence or abnormalities of lymphatic vessels. Lymphedema arises either as *lymphedema praecox* (onset during puberty, Meige disease) or as *lymphedema tarda* (onset after age of 35 years), but this state comprises only 10% of the cases. A detailed geno- and phenotyping of the disease state is performed to classify the condition. There have been identified nine causal genes, which explain one-third of phenotypes. For example, Milroy disease is based on a *VEGFR3* gene mutation. Mutations in the *FOXC2* gene lead to abnormally shaped lymphatic valves, combined with enhanced smooth muscle cells coverage, which is called *lymphedema distichiasis*. This causes obstruction of lymph vessels and consecutive stasis of tissue fluid.

The concomitant tissue edema causes the filaments to dilate the vessels, which causes strongly dilated LVs due to unphysiologically extended opened junctions. In extreme cases, this might damage the junctions during the compression phase of the initial lymphatics transport cycle. Breaks in the plasma membranes destroy the effectiveness of the lymphatic pump.

*Secondary lymphedema* arises due to physical obstruction or interruption of lymphatic vessels that can be caused by malignancies, infections, tissue trauma, and excision/surgery or postradiation fibrosis. In severely injured tissue, the opening mechanisms of the lymphatic capillaries, the anchoring filaments, and opening gaps are defective. The injury has led to damage of LECs and destruction of the tight junctions or breaks in the plasma membranes of LECs. Hence, substances can directly enter the lymphatic system via gaps (infections, aggressive tumor cells). In moderate burns, there is a threefold concentration of proteins inside the lymphatics, based on higher protein concentrations in lymphatic vessels than in the surrounding tissue space, which inverses the osmotic pressure, highly overwhelming the lymphatic system. Injurious substances may also damage the junctions and cause endothelial contraction.

For *lymphedema treatment*, physical methods may be applied that comprise heat, cold, pressure, elevation, compression, massage, and exercise. As a mild technique, *manual lymph drainage* (MLD) has been developed by the Austrian surgeon Alexander von Winiwarter (1848–1917), who worked in Luttich, Belgium. The Danish couple Emil (1896–1986) and Estrid (1898–1996) *Vodder* continued and improved this procedure and established the so-called lymphology in the thirties of the twentieth century. These careful manual movements should only be practiced by specialized and trained persons.

Knowledge of LEC molecules that contribute to the uptake or passage of interstitial fluid into the lymphatic system may fundamentally help to develop therapeutic agents for treatment of this painful disease. Moreover, abnormal formations of lymphatic vessels, as seen in lymphangiomatosis and lymphangiectasis (dilatation of lymphatic vessels), may be understood better by finding specific biomarkers that might represent drugable targets.

1

■ **Lymph Fluid Accumulations**

*Chylothorax/chylous ascites* refers to the abnormal accumulation of chyle (lymph fluid) in the pleural cavity. The *cisterna chyli* is a dilated collection of lymphatic ducts located within the thoracic cavity in the abdomen that accumulates chyle before entry into the thoracic duct. Accumulation of lymph into either the chest (chylothorax) or abdomen (chylous ascites) is rare. Both can result from congenital abnormalities of the lymphatic system usually as part of a more complex lymphatic anomaly or as a consequence of physical trauma. However, the appearance of free chyle in the abdominal or thoracic cavity is a sign of lymphatic dysfunction [172]. This is seen in mouse models with mutations in PROX-1, VEGFR-3, ANGPT2, and SOX18, important lymphatic markers.

*Lymphocele* is the term for secondary types of lymph fluid accumulation. It arises within a body cavity usually after surgical intervention that involves removal of lymph nodes (lymphadenectomy), where lymphatic vessels were injured or not properly sealed. If it does not regress spontaneously, postoperative treatments such as puncture are necessary.

■ **Lymph Is Adipogenic**

Lymph stasis and/or leakage into the tissue leads to irreversible accumulation of lipids and fat formation. An indication for the involvement of dysfunctional lymphatics in tissue lipid accumulation is provided by patients with lymphedema: In late stages of the disease, they usually suffer from severe and irreversible fibrosis and overgrowth of adipose tissue in the edematous limbs [171]. Strikingly, haploinsufficiency in the Prox-1 gene, a master regulator of lymphatic development, causes adult-onset obesity in mice [173, 174]. Further, it was shown that enhanced blood cholesterol levels can influence lymphatic function. ApoE-/- mice fed a high-fat diet showed lymphatic vessel dysfunction and degeneration in [175]. Lymphatic vessel integrity was disturbed and lymph transport declined, resulting in tissue swelling. Moreover, lymphatic collectors lacked vSMC coverage and had dysfunctional valves [175]. Hence, the hypothesis that lymph leakage or failed removal of tissue fluid and adipogenic factors may cause adipocyte maturation is supported by these studies. This in turn could have therapeutic implications for treating obesity and lymphedema by promoting lymphatic integrity or preventing the release of adipogenic factors [176].

■ **Chronic Lymphatic Vessel Damage in Type 2 Diabetes**

Though diabetic microangiopathy is recognized as the prominent cause of patient morbidity, little is known about alterations of lymphatic vessels under chronic hyperglycemic conditions. In human T2D skin, on ultrastructural level, morphopathological alterations of lymphatic vessels such as deposition of basement material [177], dilations, and dislocated LECs were observed [178], indicating altered permeability and disordered lymphatic function. Patients with T2D are highly susceptible to all kinds of infections, especially skin infections [179, 180], which are also a consequence of lymphatic dysfunction. It was shown that the number of peripheral dendritic cells are reduced in T2D patients [181], pointing at restricted immune surveillance and defense response, which is in accord with impaired skin wound healing [182, 183]. On the other hand, increased lymphatic vessel density was observed, which correlated with enhanced macrophage recruitment, high abundance of pro-inflammatory cytokines, and emergence of lymphangiogenesis [160]. This kind of de novo lymphangiogenesis might be a direct consequence of increased inflammatory disposition in diabetic condition.

■ **Lymphatic Vessels in Inflammation**

Lymphatic vessels are involved in the resolution of tissue edema and wound debris and lead to tissue clearing and wound healing after an infectious process: Inflammation-associated tissue edema induces a higher hydrostatic pressure gradient into venous but also lymphatic vessels. Hence, during inflammatory processes, lymph composition is changing. Due to its task as a recycling route, the lymphatic vessel compartment consequently shows a rise in protein concentration, cellular debris, cells, and enzymes. In order to counterbalance this process, there are enhanced contractions of collecting vessels proximal to an inflammatory region, which lead to increased lymph flow. However, in worst case, coagulation of proteins or debris taken up is possible, which leads to fibrosis and, conversely, reduced lymph flow.

Lymphatic vessels can be directly affected by injurious substances, bacteria, and parasites. Eventually, *lymphangitis* can occur, which is an inflammatory process along the lymphatic vessels per se and further spread of infectious agents along lymphatics to lymph nodes and within the whole body. The detailed molecular mechanisms behind these disease states are still not clarified.

*Regeneration of lymphatic vessels* during and after inflammation comprises diverse mechanisms: initial lymphatics heal by sprouting from existing lymphatics, by rapid invasion of the healing region. They reveal a similar structure as blood capillaries, but slower growth. Concerning regeneration of collecting lymphatics, collateral lymphatic channels are formed that are able to replace obstructed lymphatics. Defective lymphatic function of lymphatics leads to an inefficient clearance of inflammatory cells and pathogens via the lymph and consequently causes an increased risk for tissue infections [171].

- **Lymphatic Vessel Infection: Filariasis**

Direct infection of the lymphatic system by *parasites* and the consecutive lymph vessel obstruction impressively show the importance of proper functioning of this system. Lymphatic filariasis is caused by infection with parasites classified as nematodes (roundworms) of the family Filarioidea. This causes severe lymphedema, creating enlarged and swollen limbs which is termed "elefantiasis." Late-stage consequences of the disease are chronic inflammation, fibrosis, adipose tissue degeneration, poor immune function, and impaired wound healing. More than 120 Mio people are affected worldwide, a huge portion of it suffering from the late effects [99].

- **Lymphangiogenesis: Aberrant De Novo LEC Proliferation**

*Increase of lymphatic vessel density in inflammation:* The inflammatory condition itself involves stimulation of lymphangiogenesis [5, 184], which can be observed during wound healing [185]. Pro-inflammatory factors such as *tumor necrosis factor alpha* (TNF-$\alpha$) induce expression of the lymphangiogenic factor VEGF-C, which is mainly produced by macrophages [186, 187]. Thereby, they are able to stimulate lymphangiogenesis [188]. These lymphatic vessels show exaggerated growth [189], as seen during chronic inflammatory diseases like psoriasis [190] or rheumatoid arthritis [191] but also in chronic intestinal disorders such as inflammatory bowel [192] and Crohn's disease [193]. Reminiscent of defective blood vessels seen in tumor angiogenesis, these newly established lymphatic vessels reveal disordered structures and function and rather enhance immune reactions. This raises the question whether suppression of lymphangiogenesis could be used to improve persistent immune reactions such as seen in transplant rejection.

*Tumor-associated lymphangiogenesis/increase of lymphatic vessel density in cancer:* Tumors can directly promote lymphangiogenesis, presumably to enable specific lymphatic metastasis toward the lymph nodes [194]. However, the specific molecular functions and target molecules altered in these conditions are relatively unknown (Tobler and Detmar 2006). Lymphangiogenic growth factors like VEGF-C and VEGF-D induce the formation

**1**

of new lymphatic vessels [195] and were shown to be upregulated in a variety of tumors [196]. Lymphatic vessel count around tumors has been identified as a prognostic factor [197], and antibodies neutralizing VEGF-C, VEGF-D, or VEGFR-3 were sufficient to block tumor growth [198].

■ **Lymphatic Metastasis: Tumor Spread**

Although still rather neglected, lymphatic vessels represent the key entry point for tumor cells when starting to disseminate within the body, as they have more delicate walls. Certain tumors (e.g., breast cancer and carcinomas such as melanoma) spread via lymphatic vessels to form metastases in regional lymph nodes or distant organs. The detection of tumor cells within lymph nodes, especially in the so-called sentinel lymph nodes, which are those residing immediately next to a tumor, is clinically used and important for tumor staging and therapy [199]. It has been established that tumors prepare lymphatic metastasis by inducing a *pre-metastatic niche* [200]. Tumors start to metastasize by induction of peritumoral lymphatic vessel growth and by invading the newly formed as well as pre-existing afferent lymphatic vessels [194, 201]. However, the mechanism on how tumor cells invade into lymphatic vessels is poorly understood. Further, chemokines and their corresponding receptors are suggested to be involved in the interaction of tumor cells with the lymphatic endothelium, for example, C-C motif chemokine ligand 21 (CCL21) and chemokine receptor 7 (CCR7) expressed on tumor cells [202]. Additionally, lipoxygenases (ALOX15) are implicated in breast cancer cell invasion of lymphatic vessels by interendothelial gap formation that leads to consecutive lymph node colonization [200]. The detailed molecular recognition mechanisms are still largely unknown, and also the prognostic significance of lymphatic vessel invasion has to be further studied.

❓ **Questions**
14. What is the difference between primary and secondary lymphedema?
15. What are key hallmarks of lymphedema?

---
**Summary: Take-Home Message**

Blood and lymphatic vasculature are two highly specialized vessel systems that are key to regulating body homeostasis. Despite the obvious enormous differences regarding their functions, until now, only a few distinct molecular markers specific for blood (BECs) or lymphatic endothelial cells (LECs) are known. As we find both systems affected in disease, e.g., cancer spread, chronic inflammation, obesity, wound healing, or type 2 diabetes, novel vascular targets are needed to detect and evaluate alterations thereof. The specific entrance of leukocytes and tumor cells into the microvasculature indicates that endothelia carry unique marker molecules enabling specific adsorption to and transport of cells inside the vessels. Understanding vessel function will be improved by detailed analysis of the molecular equipments of its two entities, also in regard to different tissue compartments (e.g., skin, eye, and lung). The endothelium represents a highly versatile and reactive cell surface that modulates its molecular equipments according to external stimuli and needs. These responses involve modulation of membrane proteins and ion channels, activation of transcription factors, cellular reorganization, and change of cell shape. They are accomplished within seconds to hours and may be mechanistically important in the pathogenesis of vascular diseases.

Box 1.8 Lymphatic Vessels in Disease
- Lymph transport: Lymphedema, ascites
- Lipid uptake and transport: Adipogenesis, obesity
- Lymphatic proliferation: Aberrant lymphangiogenesis
- Lymphatic malignancy: Kaposi's sarcoma, lymphangioma
- Matrix degradation: Wound healing, fibrosis
- Infectious agents: Infection, lymphangitis, elefantiasis
- Immune cell interaction: Chronic inflammation
- Tumor cell interaction: Metastasis

Box 1.9 Lymphangiogenesis in Inflammation and Tumor Metastasis
- Inflammation is associated with new growth of lymphatic vessels, so-called lymphangiogenesis.
- Inflammatory lymphatic vessels enhance immune reactions (arthritis, inflammatory bowel disease, dermatitis).
- Suppression of lymphangiogenesis might improve exaggerated immune reactions.

## 1.5 Answers to Questions

1. ✓ Vasculature can be specifically labeled with PECAM1 / CD31 staining.

2. ✓ Lymphatic vessel phenotype is thin-walled and fragile.

3. ✓ Most important lymphatic markers are Prox-1, Lyve-1, VEGFR-3, and podoplanin.

4. ✓ Lymphatic vessels start to form from the primary lymph sacs that derive from the jugular vein.

5. ✓ There are two final connections: (a) the main thoracic duct is connected to the left subclavian vein, and (b) the right thoracic duct is connected to the right subclavian vein.

6. ✓ There is an approximate amount of 20 liters filtration fluid per day.

7. ✓ It is assumed that the ratio of venous versus lymphatic fluid uptake is 9:1 (18:2 liters per day).

8. ✓ The main regulators are surrounding muscle activity, blood flow, and nervous excitation.

9. ✓ Lymph is the residuum of the blood capillary filtrate, a watery, milky fluid. Lymph contains electrolytes, proteins, lipids, leukocytes, microorganisms + pathogens, and tumor cells.

1

✅ 10. Lymphatic vessels are linked to the extracellular matrix by anchoring filaments. When these relax, interendothelial gaps are closed, and lymph transport is induced.

✅ 11. In collecting lymphatic vessels, bileaflet valves prevent backflow of lymph fluid.

✅ 12. These are chronic hypertension, peripheral arterial occlusive disease, atherosclerosis, and type 2 diabetes.

✅ 13. Tumors are in need of nutrients and oxygen, and they develop the tendency to spread.

✅ 14. Congenital dysfunction versus physical obstruction of lymphatic vessels.

✅ 15. Lymphedema leads to chronic inflammation, impaired wound healing, and adipose tissue generation.

## References

1. Aukland K. Distribution of body fluids: local mechanisms guarding interstitial fluid volume. J Physiol Paris. 1984;79:395–400.
2. Bianchetti MG, Simonetti GD, Bettinelli A. Body fluids and salt metabolism – part I. Ital J Pediatr. 2009;35:36. https://doi.org/10.1186/1824-7288-35-36.
3. Hill LL. Body composition, normal electrolyte concentrations, and the maintenance of normal volume, tonicity, and acid-base metabolism. Pediatr Clin N Am. 1990;37:241–56.
4. Adams RH, Alitalo K. Molecular regulation of angiogenesis and lymphangiogenesis. Nat Rev Mol Cell Biol. 2007;8:464–78.
5. Alitalo K, Tammela T, Petrova TV. Lymphangiogenesis in development and human disease. Nature. 2005;438:946–53. https://doi.org/10.1038/nature04480.
6. Wilting J, Papoutsi M, Becker J. The lymphatic vascular system: secondary or primary? Lymphology. 2004;37:98–106.
7. Aird WC. Endothelium as an organ system. Crit Care Med. 2004;32:S271–9.
8. Carmeliet P. Angiogenesis in life, disease and medicine. Nature. 2005;438:932–6.
9. Jones N, Iljin K, Dumont DJ, Alitalo K. Tie receptors: new modulators of angiogenic and lymphangiogenic responses. Nat Rev Mol Cell Biol. 2001;2:257–67.
10. Oliver G. Lymphatic vasculature development. Microlymphatic Biology. Nat Rev Immunol. New York. 2004;4:35–45. https://doi.org/10.1038/nri1258.
11. Swartz MA. The physiology of the lymphatic system. Adv Drug Deliv Rev. 2001;50:3–20.
12. Zawieja DC, von der Weid PY, Gashev AA. Microlymphatic biology. In: Handbook of physiology: microcirculation, Chap. 5. New York: Academic Press; 2006. p. 125–58.
13. Sauter B, Foedinger D, Sterniczky B, Wolff K, Rappersberger K. Immunoelectron microscopic characterization of human dermal lymphatic microvascular endothelial cells. Differential expression of CD31, CD34, and type IV collagen with lymphatic endothelial cells vs blood capillary endothelial cells in normal human skin, lymphangioma, and hemangioma in situ. J Histochem Cytochem. 1998;46:165–76.
14. Podgrabinska S, et al. Molecular characterization of lymphatic endothelial cells. Proc Natl Acad Sci U S A. 2002;99:16069–74.
15. Hirakawa S, et al. Identification of vascular lineage-specific genes by transcriptional profiling of isolated blood vascular and lymphatic endothelial cells. Am J Pathol. 2003;162:575–86.
16. Chi JT, et al. Endothelial cell diversity revealed by global expression profiling. Proc Natl Acad Sci U S A. 2003;100:10623–8.
17. Amatschek S, et al. Blood and lymphatic endothelial cell-specific differentiation programs are stringently controlled by the tissue environment. Blood. 2007;109:4777–85.
18. Myers K, Hannah P. Chapter 2. Anatomy of veins and lymphatics. In: Manual of venous and lymphatic diseases. 1st ed: CRC Press; 2018. ISBN 9781138036864.

19. West JB. Ibn al-Nafis, the pulmonary circulation, and the Islamic golden age. J Appl Physiol (1985). 2008;105:1877–80. https://doi.org/10.1152/japplphysiol.91171.2008.
20. Manual of Venous and Lymphatic Diseases. 1st edn, CRC Press; 2018.
21. Ribatti D. William Harvey and the discovery of the circulation of the blood. J Angiogenes Res. 2009;1:3. https://doi.org/10.1186/2040-2384-1-3.
22. Pearce JM. Malpighi and the discovery of capillaries. Eur Neurol. 2007;58:253–5. https://doi.org/10.1159/000107974.
23. Suy R, Thomis S, Fourneau I. The discovery of lymphatic system in the seventeenth century. Part I: the early history. Acta Chir Belg. 2016;116:260–6. https://doi.org/10.1080/00015458.2016.1176792.
24. De Aselli G. lacteibus sive lacteis venis. Quarto vasorum mesaroicum genere novo invento. Gasp. Asellii Cremonensis Anatomici Ticiensis qua sententiae anatomicae multae, nel perperam receptae illustrantur. Milan: Mediolani, apud Jo. Baptistam Bidellium; 1627.
25. Suy R, Thomis S, Fourneau I. The discovery of the lymphatic system in the seventeenth century. Part II: the discovery of Chyle vessels. Acta Chir Belg. 2016;116:329–35. https://doi.org/10.1080/00015458.2016.1195587.
26. Eriksson G. Olaus Rudbeck as scientist and professor of medicine. Sven Med Tidskr. 2004;8:39–44.
27. Ambrose CT. Immunology's first priority dispute–an account of the 17th-century Rudbeck-Bartholin feud. Cell Immunol. 2006;242:1–8. https://doi.org/10.1016/j.cellimm.2006.09.004.
28. Suy R, Thomis S, Fourneau I. The discovery of the lymphatics in the seventeenth century. Part iii: the dethroning of the liver. Acta Chir Belg. 2016;116:390–7. https://doi.org/10.1080/00015458.2016.1215952.
29. Natale G, Bocci G, Ribatti D. Scholars and scientists in the history of the lymphatic system. J Anat. 2017;231:417–29. https://doi.org/10.1111/joa.12644.
30. Suri C. The emergency of molecular and transgenic lymphology: what do we (really) know so far? Lymphology. 2006;39:1–7.
31. Sabin FR. On the origin of the lymphatic system from the veins and the development of the lymph hearts and thoracic duct in the pig. Am J Anat. 1902;1:367–89.
32. Sabin FR. On the development of the superficial lymphatics in the skin of the pig. Am J Anat. 1904;3:183–95.
33. Huntington GS, McClure CFW. The anatomy and development of the jugular lymph sac in the domestic cat (Felis domestica). Am J Anat. 1910;10:177–311.
34. Schneider M, Othman-Hassan K, Christ B, Wilting J. Lymphangioblasts in the avian wing bud. Dev Dyn. 1999;216:311–9.
35. Wigle JT, Oliver G. Prox1 function is required for the development of the murine lymphatic system. Cell. 1999;98:769–78.
36. Wilting J, Papoutsi M, Schneider M, Christ B. The lymphatic endothelium of the avian wing is of somitic origin. Dev Dyn. 2000;217:271–8.
37. Wilting J, et al. Development of the avian lymphatic system. Microsc Res Tech. 2001;55:81–91.
38. Breiteneder-Geleff S, et al. Angiosarcomas express mixed endothelial phenotypes of blood and lymphatic capillaries: podoplanin as a specific marker for lymphatic endothelium. Am J Pathol. 1999;154:385–94. https://doi.org/10.1016/S0002-9440(10)65285-6.
39. Kriehuber E, et al. Isolation and characterization of dermal lymphatic and blood endothelial cells reveal stable and functionally specialized cell lineages. J Exp Med. 2001;194:797–808.
40. Harris ES, Nelson WJ. VE-cadherin: at the front, center, and sides of endothelial cell organization and function. Curr Opin Cell Biol. 2010;22:651–8.
41. Coultas L, Chawengsaksophak K, Rossant J. Endothelial cells and VEGF in vascular development. Nature. 2005;438:937–45.
42. Middleton J, et al. A comparative study of endothelial cell markers expressed in chronically inflamed human tissues: MECA-79, Duffy antigen receptor for chemokines, von Willebrand factor, CD31, CD34, CD105 and CD146. J Pathol. 2005;206:260–8. https://doi.org/10.1002/path.1788.
43. Constans J, Conri C. Circulating markers of endothelial function in cardiovascular disease. Clin Chim Acta. 2006;368:33–47. https://doi.org/10.1016/j.cca.2005.12.030.
44. Kaipainen A, et al. Expression of the fms-like tyrosine kinase 4 gene becomes restricted to lymphatic endothelium during development. Proc Natl Acad Sci U S A. 1995;92:3566–70.
45. Petrova TV, et al. Lymphatic endothelial reprogramming of vascular endothelial cells by the Prox-1 homeobox transcription factor. EMBO J. 2002;21:4593–9.
46. Banerji S, et al. LYVE-1, a new homologue of the CD44 glycoprotein, is a lymph-specific receptor for hyaluronan. J Cell Biol. 1999;144:789–801.

47. Risau W, Flamme I. Vasculogenesis. Annu Rev Cell Dev Biol. 1995;11:73–91. https://doi.org/10.1146/annurev.cb.11.110195.000445.
48. Risau W. Mechanisms of angiogenesis. Nature. 1997;386:671–4. https://doi.org/10.1038/386671a0.
49. Choi K, Kennedy M, Kazarov A, Papadimitriou JC, Keller G. A common precursor for hematopoietic and endothelial cells. Development (Cambridge, England). 1998;125:725–32.
50. Burri PH, Hlushchuk R, Djonov V. Intussusceptive angiogenesis: its emergence, its characteristics, and its significance. Dev Dyn. 2004;231:474–88. https://doi.org/10.1002/dvdy.20184.
51. Dumont DJ, et al. Cardiovascular failure in mouse embryos deficient in VEGF receptor-3. Science. 1998;282:946–9.
52. Jeltsch M, Tammela T, Alitalo K, Wilting J. Genesis and pathogenesis of lymphatic vessels. Cell Tissue Res. 2003;314:69–84.
53. Koltowska K, Betterman KL, Harvey NL, Hogan BM. Getting out and about: the emergence and morphogenesis of the vertebrate lymphatic vasculature. Development (Cambridge, England). 2013;140:1857–70.
54. Tammela T, Alitalo K. Lymphangiogenesis: molecular mechanisms and future promise. Cell. 2010;140:460–76. https://doi.org/10.1016/j.cell.2010.01.045.
55. van der Putte SC. The early development of the lymphatic system in mouse embryos. Acta Morphol Neerl Scand. 1975;13:245–86.
56. Uhrin P, et al. Novel function for blood platelets and podoplanin in developmental separation of blood and lymphatic circulation. Blood. 2010;115:3997–4005. https://doi.org/10.1182/blood-2009-04-216069.
57. Cyster JG. Lymphoid organ development and cell migration. Immunol Rev. 2003;195:5–14.
58. Cupedo T, Mebius RE. Cellular interactions in lymph node development. J Immunol. 2005;174:21–5.
59. Randall TD, Carragher DM, Rangel-Moreno J. Development of secondary lymphoid organs. Annu Rev Immunol. 2008;26:627–50. https://doi.org/10.1146/annurev.immunol.26.021607.090257.
60. Blum KS, Pabst R. Keystones in lymph node development. J Anat. 2006;209:585–95. https://doi.org/10.1111/j.1469-7580.2006.00650.x.
61. Cueni LN, Detmar M. The lymphatic system in health and disease. Lymphat Res Biol. 2008;6:109–22.
62. Detry B, et al. Digging deeper into lymphatic vessel formation in vitro and in vivo. BMC Cell Biol. 2011;12:29.
63. Jain RK. Molecular regulation of vessel maturation. Nat Med. 2003;9:685–93. https://doi.org/10.1038/nm0603-685.
64. Braverman IM, Keh-Yen A. Ultrastructure of the human dermal microcirculation. III The vessels in the mid- and lower dermis and subcutaneous fat. J Inves Dermatol. 1981;77:297–304.
65. Braverman IM. The cutaneous microcirculation: ultrastructure and microanatomical organization. Microcirculation. 1997;4:329–40.
66. Rhodin JA. The diaphragm of capillary endothelial fenestrations. J Ultrastruct Res. 1962;6:171–85.
67. Yen A, Braverman IM. Ultrastructure of the human dermal microcirculation: the horizontal plexus of the papillary dermis. J Invest Dermatol. 1976;66:131–42.
68. Braverman IM, Yen A. Ultrastructure of the human dermal microcirculation. II The capillary loops of the dermal papillae. J Invest Dermatol. 1977;68:44–52.
69. Braverman IM. Ultrastructure and organization of the cutaneous microvasculature in normal and pathologic states. J Invest Dermatol. 1989;93:2S–9S.
70. Ny A, et al. A genetic Xenopus laevis tadpole model to study lymphangiogenesis. Nat Med. 2005;11:998–1004.
71. Fritsch H, Kuehnel W. Taschenatlas der Anatomie, Vol 2: Innere Organe. Thieme; 2013
72. Gnepp DR, Green FH. Scanning electron microscopy of collecting lymphatic vessels and their comparison to arteries and veins. Scan Electron Microsc. 1979;3:756–62.
73. Leak LV, Burke JF. Fine structure of the lymphatic capillary and the adjoining connective tissue area. Am J Anat. 1966;118:785–809.
74. Leak LV, Burke JF. Ultrastructural studies on the lymphatic anchoring filaments. J Cell Biol. 1968;36:129–49.
75. Leak LV. Electron microscopic observations on lymphatic capillaries and the structural components of the connective tissue-lymph interface. Microvasc Res. 1970;2:361–91.
76. Gerli R, Solito R, Weber E, Aglianó M. Specific adhesion molecules bind anchoring filaments and endothelial cells in human skin initial lymphatics. Lymphology. 2000;33:148–57.
77. Danussi C, et al. Emilin1 deficiency causes structural and functional defects of lymphatic vasculature. Mol Cell Biol. 2008;28:4026–39. https://doi.org/10.1128/MCB.02062-07.

78. Bazigou E, Wilson JT, Moore JE Jr. Primary and secondary lymphatic valve development: molecular, functional and mechanical insights. Microvasc Res. 2014;96:38–45. https://doi.org/10.1016/j.mvr.2014.07.008.

79. Petrova TV, et al. Defective valves and abnormal mural cell recruitment underlie lymphatic vascular failure in lymphedema distichiasis. Nat Med. 2004;10:974–81.

80. Norrmen C, et al. FOXC2 controls formation and maturation of lymphatic collecting vessels through cooperation with NFATc1. J Cell Biol. 2009;185:439–57.

81. Bazigou E, et al. Integrin-alpha9 is required for fibronectin matrix assembly during lymphatic valve morphogenesis. Dev Cell. 2009;17:175–86. https://doi.org/10.1016/j.devcel.2009.06.017.

82. Schmid-Schonbein GW. Microlymphatics and lymph flow. Physiol Rev. 1990;70:987–1028. https://doi.org/10.1152/physrev.1990.70.4.987.

83. Nathanson SD. Insights into the mechanisms of lymph node metastasis. Cancer. 2003;98:413–23. https://doi.org/10.1002/cncr.11464.

84. Tammela T, Petrova TV, Alitalo K. Molecular lymphangiogenesis: new players. Trends Cell Biol. 2005;15:434–41. https://doi.org/10.1016/j.tcb.2005.06.004.

85. Crone C. Capillary permeability: II. Physiological considerations. In: Zweifach BW, Grant L, McCluskey RI, editors. The inflammatory process, Vol. I, Chap. 3. New York: Academic Press; 1973. p. 95–119.

86. Pries AR, Secomb TW, Gaehtgens P. The endothelial surface layer. Pflugers Arch. 2000;440:653–66. https://doi.org/10.1007/s004240000307.

87. Salmon AH, Satchell SC. Endothelial glycocalyx dysfunction in disease: albuminuria and increased microvascular permeability. J Pathol. 2012;226:562–74. https://doi.org/10.1002/path.3964.

88. Levick JR. Introduction to cardiovascular physiology. London: Hodder Arnold; 2010.

89. Woodcock TE, Woodcock TM. Revised Starling equation and the glycocalyx model of transvascular fluid exchange: an improved paradigm for prescribing intravenous fluid therapy. Br J Anaesth. 2012;108:384–94. https://doi.org/10.1093/bja/aer515.

90. Pollock DM, Pollock JS. Endothelin and oxidative stress in the vascular system. Curr Vasc Pharmacol. 2005;3:365–7.

91. Sandoo A, van Zanten JJ, Metsios GS, Carroll D, Kitas GD. The endothelium and its role in regulating vascular tone. Open Cardiovasc Med J. 2010;4:302–12. https://doi.org/10.2174/1874192401004010302.

92. Vallet B, Wiel E. Endothelial cell dysfunction and coagulation. Crit Care Med. 2001;29:S36–41.

93. Hoeben A, et al. Vascular endothelial growth factor and angiogenesis. Pharmacol Rev. 2004;56:549–80. https://doi.org/10.1124/pr.56.4.3.

94. Stacker SA, et al. Lymphangiogenesis and lymphatic vessel remodelling in cancer. Nat Rev Cancer. 2014;14:159–72. https://doi.org/10.1038/nrc3677.

95. Bergqvist L, Strand SE, Persson BR. Particle sizing and biokinetics of interstitial lymphoscintigraphic agents. Semin Nucl Med. 1983;13:9–19.

96. Hagiwara A, Takahashi T, Oku N. Cancer chemotherapy administered by activated carbon particles and liposomes. Crit Rev Oncol Hematol. 1989;9:319–50.

97. Jackson AJ. Intramuscular absorption and regional lymphatic uptake of liposome-entrapped inulin. Drug Metab Dispos. 1981;9:535–40.

98. Jackson DG. Immunological functions of hyaluronan and its receptors in the lymphatics. Immunol Rev. 2009;230:216–31.

99. Mortimer PS, Rockson SG. New developments in clinical aspects of lymphatic disease. J Clin Invest. 2014;124:915–21. https://doi.org/10.1172/JCI71608.

100. Bridenbaugh EA, Gashev AA, Zawieja DC. Lymphatic muscle: a review of contractile function. Lymphat Res Biol. 2003;1:147–58.

101. Schad H, Flowaczny H, Brechtelsbauer H, Birkenfeld G. The significance of respiration for thoracic duct flow in relation to other driving forces of lymph flow. Pflugers Arch. 1978;378:121–5.

102. Parsons RJ, McMaster PD. The effect of the pulse upon the formation and flow of lymph. J Exp Med. 1938;68:353–76.

103. Olszewski W, Engeset A, Jaeger PM, Sokolowski J, Theodorsen L. Flow and composition of leg lymph in normal men during venous stasis, muscular activity and local hyperthermia. Acta Physiol Scand. 1977;99:149–55. https://doi.org/10.1111/j.1748-1716.1977.tb10365.x.

104. Mortimer PS, et al. The measurement of skin lymph flow by isotope clearance–reliability, reproducibility, injection dynamics, and the effect of massage. J Invest Dermatol. 1990;95:677–82.

105. Swartz MA, Skobe M. Lymphatic function, lymphangiogenesis, and cancer metastasis. Microsc Res Tech. 2001;55:92–9.

106. Jurisic G, Detmar M. Lymphatic endothelium in health and disease. Cell Tissue Res. 2009;335:97–108. https://doi.org/10.1007/s00441-008-0644-2.
107. Casley-Smith JR. The identification of chylomicra and lipoproteins in tissue sections and their passage into jejunal lacteals. J Cell Biol. 1962;15:259–77.
108. Dixon JB, Raghunathan S, Swartz MA. A tissue-engineered model of the intestinal lacteal for evaluating lipid transport by lymphatics. Biotechnol Bioeng. 2009;103:1224–35. https://doi.org/10.1002/bit.22337.
109. Rosen ED. The molecular control of adipogenesis, with special reference to lymphatic pathology. Ann N Y Acad Sci. 2002;979:143–58.; discussion 188–196.
110. Harvey NL. The link between lymphatic function and adipose biology. Ann N Y Acad Sci. 2008;1131:82–8.
111. Mattacks CA, Sadler D, Pond CM. The cellular structure and lipid/protein composition of adipose tissue surrounding chronically stimulated lymph nodes in rats. J Anat. 2003;202:551–61.
112. Pond CM, Mattacks CA. Interactions between adipose tissue around lymph nodes and lymphoid cells in vitro. J Lipid Res. 1995;36:2219–31.
113. Pond CM, Mattacks CA. In vivo evidence for the involvement of the adipose tissue surrounding lymph nodes in immune responses. Immunol Lett. 1998;63:159–67.
114. Benezech C, et al. Lymphotoxin-beta receptor signaling through NF-kappaB2-RelB pathway reprograms adipocyte precursors as lymph node stromal cells. Immunity. 2012;37:721–34. https://doi.org/10.1016/j.immuni.2012.06.010.
115. Ohl L, et al. CCR7 governs skin dendritic cell migration under inflammatory and steady-state conditions. Immunity. 2004;21:279–88. https://doi.org/10.1016/j.immuni.2004.06.014.
116. Forster R, Davalos-Misslitz AC, Rot A. CCR7 and its ligands: balancing immunity and tolerance. Nat Rev Immunol. 2008;8:362–71. https://doi.org/10.1038/nri2297.
117. Wick N, et al. Lymphatic precollectors contain a novel, specialized subpopulation of podoplanin low, CCL27-expressing lymphatic endothelial cells. Am J Pathol. 2008;173:1202–9. https://doi.org/10.2353/ajpath.2008.080101.
118. Irjala H, et al. Mannose receptor is a novel ligand for L-selectin and mediates lymphocyte binding to lymphatic endothelium. J Exp Med. 2001;194:1033–42.
119. Marttila-Ichihara F, et al. Macrophage mannose receptor on lymphatics controls cell trafficking. Blood. 2008;112:64–72. https://doi.org/10.1182/blood-2007-10-118984.
120. Salmi M, Koskinen K, Henttinen T, Elima K, Jalkanen S. CLEVER-1 mediates lymphocyte transmigration through vascular and lymphatic endothelium. Blood. 2004;104:3849–57. https://doi.org/10.1182/blood-2004-01-0222.
121. Karikoski M, et al. Clever-1/Stabilin-1 regulates lymphocyte migration within lymphatics and leukocyte entrance to sites of inflammation. Eur J Immunol. 2009;39:3477–87. https://doi.org/10.1002/eji.200939896.
122. Kang S, et al. Toll-like receptor 4 in lymphatic endothelial cells contributes to LPS-induced lymphangiogenesis by chemotactic recruitment of macrophages. Blood. 2009;113:2605–13. https://doi.org/10.1182/blood-2008-07-166934.
123. Johnson LA, Prevo R, Clasper S, Jackson DG. Inflammation-induced uptake and degradation of the lymphatic endothelial hyaluronan receptor LYVE-1. J Biol Chem. 2007;282:33671–80. https://doi.org/10.1074/jbc.M702889200.
124. Carmeliet P. Angiogenesis in health and disease. Nat Med. 2003;9:653–60. https://doi.org/10.1038/nm0603-653.
125. Cueni LN, Detmar M. New insights into the molecular control of the lymphatic vascular system and its role in disease. J Invest Dermatol. 2006;126:2167–77. https://doi.org/10.1038/sj.jid.5700464.
126. Oh YS, Galis ZS. Anatomy of success: the top 100 cited scientific reports focused on hypertension research. Hypertension. 2014;63:641–7. https://doi.org/10.1161/HYPERTENSIONAHA.113.02677.
127. Udani S, Lazich I, Bakris GL. Epidemiology of hypertensive kidney disease. Nat Rev Nephrol. 2011;7:11–21. https://doi.org/10.1038/nrneph.2010.154.
128. Lackland DT, Weber MA. Global burden of cardiovascular disease and stroke: hypertension at the core. Can J Cardiol. 2015;31:569–71. https://doi.org/10.1016/j.cjca.2015.01.009.
129. Hernandorena I, Duron E, Vidal JS, Hanon O. Treatment options and considerations for hypertensive patients to prevent dementia. Expert Opin Pharmacother. 2017;18:989–1000. https://doi.org/10.1080/14656566.2017.1333599.

130. Navar LG. Counterpoint: activation of the intrarenal renin-angiotensin system is the dominant contributor to systemic hypertension. J Appl Physiol (1985). 2010;109:1998–2000; discussion 2015,. https://doi.org/10.1152/japplphysiol.00182.2010a.
131. Esler M, Lambert E, Schlaich MP. Chronic activation of the sympathetic nervous system is the dominant contributor to systemic hypertension. J Appl Physiol (1985). 2010;109:1996–8; discussion 2016. https://doi.org/10.1152/japplphysiol.00182.2010.
132. Lau DH, Nattel S, Kalman JM, Sanders P. Modifiable risk factors and atrial fibrillation. Circulation. 2017;136:583–96. https://doi.org/10.1161/CIRCULATIONAHA.116.023163.
133. Eberhardt RT, Raffetto JD. Chronic venous insufficiency. Circulation. 2005;111:2398–409. https://doi.org/10.1161/01.CIR.0000164199.72440.08.
134. Kyrle PA, Rosendaal FR, Eichinger S. Risk assessment for recurrent venous thrombosis. Lancet. 2010;376:2032–9. https://doi.org/10.1016/S0140-6736(10)60962-2.
135. Peyvandi F, Garagiola I, Young G. The past and future of haemophilia: diagnosis, treatments, and its complications. Lancet. 2016;388:187–97. https://doi.org/10.1016/S0140-6736(15)01123-X.
136. Heit JA. Thrombophilia: common questions on laboratory assessment and management. Hematology Am Soc Hematol Educ Program. 2007;2007:127–35. https://doi.org/10.1182/asheducation-2007.1.127.
137. Mitchell RSKV, Abbas AK, Fausto N. In: Robbins Basic Pathology, Chap. 4. Saunders; 2007. p. 81–106.
138. Mitchell RSKV, Abbas AK, Fausto N. Hemodynamic disorders, thrombosis, and shock. In: Robbins basic pathology, Chapter 4. Philadelphia: Saunders; 2007. p. 81–106.
139. Li X, et al. Mitochondrial reactive oxygen species mediate Lysophosphatidylcholine-induced endothelial cell activation. Arterioscler Thromb Vasc Biol. 2016;36:1090–100. https://doi.org/10.1161/ATVBAHA.115.306964.
140. Mazzone T, Chait A, Plutzky J. Cardiovascular disease risk in type 2 diabetes mellitus: insights from mechanistic studies. Lancet. 2008;371:1800–9. https://doi.org/10.1016/S0140-6736(08)60768-0.
141. Defronzo RA. Insulin resistance, lipotoxicity, type 2 diabetes and atherosclerosis: the missing links. The Claude Bernard lecture 2009. Diabetologia. 2010;53:1270.
142. Stehouwer CD, Schaper NC. The pathogenesis of vascular complications of diabetes mellitus: one voice or many? Eur J Clin Investig. 1996;26:535–43.
143. Stehouwer CD, Lambert J, Donker AJ, van Hinsbergh VW. Endothelial dysfunction and pathogenesis of diabetic angiopathy. Cardiovasc Res. 1997;34:55–68.
144. Schalkwijk CG, Stehouwer CD. Vascular complications in diabetes mellitus: the role of endothelial dysfunction. Clin Sci (Lond). 2005;109:143–59.
145. Fox CS, et al. Trends in cardiovascular complications of diabetes. JAMA. 2004;292:2495–9. https://doi.org/10.1001/jama.292.20.2495.
146. De Vriese AS, Verbeuren TJ, Van de Voorde J, Lameire NH, Vanhoutte PM. Endothelial dysfunction in diabetes. Br J Pharmacol. 2000;130:963–74. https://doi.org/10.1038/sj.bjp.0703393.
147. Porta M, Bandello F. Diabetic retinopathy a clinical update. Diabetologia. 2002;45:1617–34. https://doi.org/10.1007/s00125-002-0990-7.
148. Goldberg RB. Cardiovascular disease in patients who have diabetes. Cardiol Clin. 2003;21:399–413, vii.
149. Kikkawa R, Koya D, Haneda M. Progression of diabetic nephropathy. Am J Kidney Dis. 2003;41:S19–21. https://doi.org/10.1053/ajkd.2003.50077.
150. Roberts AC, Porter KE. Cellular and molecular mechanisms of endothelial dysfunction in diabetes. Diab Vasc Dis Res. 2013;10:472–82.
151. Hadi HAR, Suwaidi JA. Endothelial dysfunction in diabetes mellitus. Vasc Health Risk Manag. 2007;3:853–76.
152. Manrique C, Lastra G, Sowers JR. New insights into insulin action and resistance in the vasculature. Ann N Y Acad Sci. 2014;1311:138–50. https://doi.org/10.1111/nyas.12395.
153. Brownlee M. Biochemistry and molecular cell biology of diabetic complications. Nature. 2001;414:813–20. https://doi.org/10.1038/414813a.
154. DeSouza C, Fonseca V. Therapeutic targets to reduce cardiovascular disease in type 2 diabetes. Nat Rev Drug Discov. 2009;8:361–7.
155. Duby JJ, Campbell RK, Setter SM, White JR, Rasmussen KA. Diabetic neuropathy: an intensive review. Am J Health Syst Pharm. 2004;61:160–73; quiz 175-166.
156. Thorp ML. Diabetic nephropathy: common questions. Am Fam Physician. 2005;72:96–9.
157. Levey AS, et al. National Kidney Foundation practice guidelines for chronic kidney disease: evaluation, classification, and stratification. Ann Intern Med. 2003;139:137–47.

158. Bek T. Diabetic retinopathy: a review of the Aarhus approach to studies on epidemiology, computerised grading, and the pathophysiology of the disease. Horm Metab Res. 2005;37(Suppl 1):35–8. https://doi.org/10.1055/s-2005-861396.

159. van Hecke MV, et al. Diabetic retinopathy is associated with mortality and cardiovascular disease incidence: the EURODIAB prospective complications study. Diabetes Care. 2005;28:1383–9.

160. Haemmerle M, et al. Enhanced lymph vessel density, remodeling, and inflammation are reflected by gene expression signatures in dermal lymphatic endothelial cells in type 2 diabetes. Diabetes. 2013;62:2509–29.

161. Williamson JR, Tilton RG, Chang K, Kilo C. Basement membrane abnormalities in diabetes mellitus: relationship to clinical microangiopathy. Diabetes Metab Rev. 1988;4:339–70.

162. Peltonen JT, Kalliomaki MA, Muona PK. Extracellular matrix of peripheral nerves in diabetes. J Peripher Nerv Syst. 1997;2:213–26.

163. Folkman J. Tumor angiogenesis: therapeutic implications. N Engl J Med. 1971;285:1182–6. https://doi.org/10.1056/NEJM197111182852108.

164. Hanahan D, Folkman J. Patterns and emerging mechanisms of the angiogenic switch during tumorigenesis. Cell. 1996;86:353–64.

165. Ribatti D. The discovery of antiangiogenic molecules: a historical review. Curr Pharm Des. 2009;15:345–52.

166. Nagy JA, Benjamin L, Zeng H, Dvorak AM, Dvorak HF. Vascular permeability, vascular hyperpermeability and angiogenesis. Angiogenesis. 2008;11:109–19. https://doi.org/10.1007/s10456-008-9099-z.

167. Alitalo K, Carmeliet P. Molecular mechanisms of lymphangiogenesis in health and disease. Cancer Cell. 2002;1:219–27.

168. Weninger W, et al. Expression of vascular endothelial growth factor receptor-3 and podoplanin suggests a lymphatic endothelial cell origin of Kaposi's sarcoma tumor cells. Lab Investig. 1999;79:243–51.

169. Wang Y, Oliver G. Current views on the function of the lymphatic vasculature in health and disease. Genes Dev. 2010;24:2115–26.

170. Alitalo K. The lymphatic vasculature in disease. Nat Med. 2011;17:1371–80. https://doi.org/10.1038/nm.2545.

171. Rockson SG. Lymphedema. Am J Med. 2001;110:288–95.

172. Rockson SG. Diagnosis and management of lymphatic vascular disease. J Am Coll Cardiol. 2008;52:799–806.

173. Nougues J, Reyne Y, Dulor JP. Differentiation of rabbit adipocyte precursors in primary culture. Int J Obes. 1988;12:321–33.

174. Harvey NL, et al. Lymphatic vascular defects promoted by Prox1 haploinsufficiency cause adult-onset obesity. Nat Genet. 2005;37:1072–81.

175. Lim HY, et al. Hypercholesterolemic mice exhibit lymphatic vessel dysfunction and degeneration. Am J Pathol. 2009;175:1328–37. https://doi.org/10.2353/ajpath.2009.080963.

176. Schneider M, Conway EM, Carmeliet P. Lymph makes you fat. Nat Genet. 2005;37:1023–4. https://doi.org/10.1038/ng1005-1023.

177. Ohkuma M. Histochemical change of the endothelial basal lamina of the diabetic lymphatic vessel. Lymphology. 1979;12:37–9.

178. Kaufmann A, Molnar B, Craciun C, Itcus A. Diabetic lymphangiopathy. An optical and electron microscopic study. Lymphology. 1980;13:202–6.

179. Shah BR, Hux JE. Quantifying the risk of infectious diseases for people with diabetes. Diabetes Care. 2003;26:510–3.

180. Muller LM, et al. Increased risk of common infections in patients with type 1 and type 2 diabetes mellitus. Clin Infect Dis. 2005;41:281–8. https://doi.org/10.1086/431587.

181. Seifarth CC, Hinkmann C, Hahn EG, Lohmann T, Harsch IA. Reduced frequency of peripheral dendritic cells in type 2 diabetes. Exp Clin Endocrinol Diabetes. 2008;116:162–6. https://doi.org/10.1055/s-2007-990278.

182. Saaristo A, et al. Lymphangiogenic gene therapy with minimal blood vascular side effects. J Exp Med. 2002;196:719–30.

183. Maruyama K, et al. Decreased macrophage number and activation lead to reduced lymphatic vessel formation and contribute to impaired diabetic wound healing. Am J Pathol. 2007;170:1178–91. https://doi.org/10.2353/ajpath.2007.060018.

184. Abouelkheir GR, Upchurch BD, Rutkowski JM. Lymphangiogenesis: fuel, smoke, or extinguisher of inflammation's fire? Exp Biol Med (Maywood). 2017;242:884–95. https://doi.org/10.1177/1535370217697385.

185. Saharinen P, Tammela T, Karkkainen MJ, Alitalo K. Lymphatic vasculature: development, molecular regulation and role in tumor metastasis and inflammation. Trends Immunol. 2004;25:387–95.
186. Baluk P, et al. Pathogenesis of persistent lymphatic vessel hyperplasia in chronic airway inflammation. J Clin Invest. 2005;115:247–57. https://doi.org/10.1172/JCI22037.
187. Baluk P, et al. TNF-alpha drives remodeling of blood vessels and lymphatics in sustained airway inflammation in mice. J Clin Invest. 2009;119:2954–64. https://doi.org/10.1172/JCI37626.
188. Maruyama K, et al. Inflammation-induced lymphangiogenesis in the cornea arises from CD11b-positive macrophages. J Clin Invest. 2005;115:2363–72. https://doi.org/10.1172/JCI23874.
189. Kerjaschki D. Lymphatic neoangiogenesis in renal transplants: a driving force of chronic rejection? J Nephrol. 2006;19:403–6.
190. Halin C, Detmar M. Chapter 1. Inflammation, angiogenesis, and lymphangiogenesis. Methods Enzymol. 2008;445:1–25. https://doi.org/10.1016/S0076-6879(08)03001-2.
191. Zhang Q, et al. Increased lymphangiogenesis in joints of mice with inflammatory arthritis. Arthritis Res Ther. 2007;9:R118. https://doi.org/10.1186/ar2326.
192. Alexander JS, Chaitanya GV, Grisham MB, Boktor M. Emerging roles of lymphatics in inflammatory bowel disease. Ann N Y Acad Sci. 2010;1207(Suppl 1):E75–85. https://doi.org/10.1111/j.1749-6632.2010.05757.x.
193. von der Weid PY, Rehal S, Ferraz JG. Role of the lymphatic system in the pathogenesis of Crohn's disease. Curr Opin Gastroenterol. 2011;27:335–41.
194. Mandriota SJ, et al. Vascular endothelial growth factor-C-mediated lymphangiogenesis promotes tumour metastasis. EMBO J. 2001;20:672–82. https://doi.org/10.1093/emboj/20.4.672.
195. Jeltsch M, et al. Hyperplasia of lymphatic vessels in VEGF-C transgenic mice. Science. 1997;276:1423–5.
196. Stacker SA, Achen MG, Jussila L, Baldwin ME, Alitalo K. Lymphangiogenesis and cancer metastasis. Nat Rev Cancer. 2002;2:573–83. https://doi.org/10.1038/nrc863.
197. Schoppmann SF, Birner P, Studer P, Breiteneder-Geleff S. Lymphatic microvessel density and lymphovascular invasion assessed by anti-podoplanin immunostaining in human breast cancer. Anticancer Res. 2001;21:2351–5.
198. Stacker SA, et al. VEGF-D promotes the metastatic spread of tumor cells via the lymphatics. Nat Med. 2001;7:186–91. https://doi.org/10.1038/84635.
199. Lyman GH, et al. American Society of Clinical Oncology guideline recommendations for sentinel lymph node biopsy in early-stage breast cancer. J Clin Oncol. 2005;23:7703–20. https://doi.org/10.1200/JCO.2005.08.001.
200. Kerjaschki D, et al. Lipoxygenase mediates invasion of intrametastatic lymphatic vessels and propagates lymph node metastasis of human mammary carcinoma xenografts in mouse. J Clin Invest. 2011;121:2000–12. https://doi.org/10.1172/JCI44751.
201. Skobe M, et al. Induction of tumor lymphangiogenesis by VEGF-C promotes breast cancer metastasis. Nat Med. 2001;7:192–8. https://doi.org/10.1038/84643.
202. Shields JD, et al. Autologous chemotaxis as a mechanism of tumor cell homing to lymphatics via interstitial flow and autocrine CCR7 signaling. Cancer Cell. 2007;11:526–38. https://doi.org/10.1016/j.ccr.2007.04.020.

# The Heart: The Engine in the Center of the Vascular System

*Svitlana Demyanets*

© Springer Nature Switzerland AG 2019
M. Geiger (ed.), *Fundamentals of Vascular Biology*, Learning Materials in Biosciences,
https://doi.org/10.1007/978-3-030-12270-6_2

**2**

**What You Will Learn in This Chapter**

In this chapter, you will learn about the anatomy and physiology of the healthy human heart. Mechanisms of the cardiovascular pathologies will be discussed separately in other chapters of this book. This chapter first will focus on the physiological processes responsible for cardiac contraction and relaxation with the emphasis on the components of the excitatory and conductivity system, differences in action potential in nodal versus non-nodal cells, and excitation-contraction coupling. This chapter will further cover the basic aspects of the heart cycle as well as the heart innervation and coronary circulation.

Next, this chapter will provide you with brief summary of characteristics and functions of cardiac cells, more specifically cardiac myocytes, fibroblasts, and heart resident macrophages. We will then discuss the current knowledge about the myocardial metabolic regulation and energy supply with the focus on glucose, fatty acid, and ketone metabolism in the healthy heart. Finally, sex-related differences in the physiology of the human heart also during aging will be highlighted.

## 2.1 Physiology of the Heart

### 2.1.1 Anatomy of the Heart

The heart is the organ composed of four cavities, four valves, large arteries, and veins. Anatomical components of the heart precisely interact in order to reach appropriate filling, ejection, contraction, and global pump task.

The two superior receiving cavities of the heart are the atria, and the two inferior pumping cavities are the ventricles. The *right atrium* (RA) obtains the deoxygenated blood from the body through the superior vena cava (SVC) and inferior *vena cava* (IVC). Just to remember, veins return blood to the heart. Blood flows from the RA into the *right ventricle* (RV) through the tricuspid or right *atrioventricular* (AV) *valve*. The inside of the RV encloses so-called trabeculae carneae, some of which mediate part of the conduction system of the heart. Blood moves from the RV through the *pulmonary semilunar valve* into a large artery called the *pulmonary trunk*, which separates into right and left pulmonary arteries and carry blood to the lungs. Just to remember, arteries bring blood away from the heart.

The *left atrium* (LA) takes oxygenated blood from the lungs through four *pulmonary veins*. Blood flows from the LA into the *left ventricle* (LV) through the bicuspid or *mitral valve* which, as its name says, has two leaflets. Blood passes from the LV through the *aortic semilunar valve* into the *aorta*. Some of the blood in the aorta distributes into the coronary arteries. The remainder of the blood flows into the thoracic and abdominal aorta, which carry blood throughout the body (to the upper or lower parts of the body, respectively) [1–3].

### 2.1.2 Excitatory and Conductivity System of the Heart

Harmonized contractile function of the heart is crucial for life. Cardiac conduction system is a collective term for a combination of diverse specific muscular tissues. Each contraction of the heart is exactly controlled by interaction between electrical signals and mechanical forces. The reason why an electrical excitation occurring in one cardiac muscle cell can spread to neighboring cells is that the heart muscle is a "syncytium." "Syncytium" means a

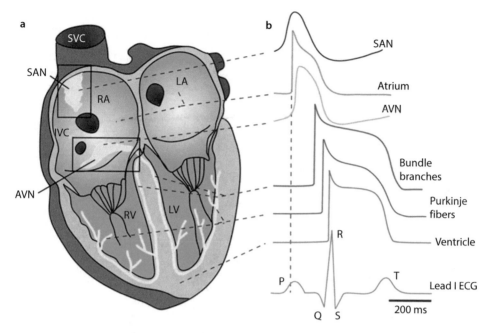

**Fig. 2.1** **a** Schematic of different regions of the human heart and the corresponding. **b** AP wave-forms with representative lead I ECG. (Reprinted from [6] with permission of the publisher. SVC superior vena cava, SAN sinoatrial node, RA right atrium, LA left atrium, IVC inferior vena cava, AVN atrioventricular node, RV right ventricle, LV left ventricle)

fusion of muscle cells interconnected by cytoplasmic bridges that is fastened inside a protein complex primarily consisted of type I fibrillar collagen. Formation and function of the cardiac conduction system underlie complex transcriptional regulation [4, 5].

The basis of rhythmical electrical activity is a system of specialized cardiac muscle fibers called autorhythmic fibers because they are self-excitable and repeatedly generated action potentials (electrical activity) that trigger heart contractions. *Autorhythmic fibers function as a pacemaker and form the conduction system.*

Cardiac excitation starts in the *sinoatrial (SA) node (SAN)*, located in the RA. It is spreading to the LA via *Bachmann's bundle.* In analogous, action potential travels to the *AV node (AVN).* The AV node is located at the apex of the triangle of Koch of the RA. The triangle of Koch is formed by the ostium of the coronary sinus, the tendon of Todaro, and the tricuspid valve. After a delay in the AV node, the potential is leaded through the *bundle of His* to the left and right branches and then to the *Purkinje fibers.* Purkinje fibers conduct the stimulus to the rest of the ventricular myocardium (■ Fig. 2.1a, [6]).

Both SAN and AVN have pacemaker activity potential. Pacemaker cells of SAN and AVN are characterized by the process of diastolic depolarization. Nevertheless, the activities of these two nodes demonstrate significant differences, for example, in firing rate, which is 20–60 beats per minute (bpm) for AVN versus 60–100 bpm of the SAN. The components of the AVN are characterized by structural, molecular, and functional heterogeneities [6–8].

The resting membrane potential of most cardiac cells, with the exception of the SAN and AVN, is approximately −90 milivolt (mV). Activation of cardiac cells is a consequence of ion movement through the cell membrane, resulting in a temporary depolarization

2

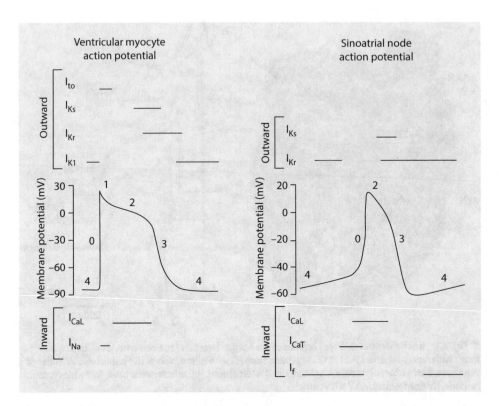

■ **Fig. 2.2**  Action potential in ventricular myocytes and SAN. (Reprinted from [9] with permission of the publisher. $I_{to}$ transient outward $K^+$ current, $I_{Ks}$ slowly activating delayed outward rectifying current, $I_{Kr}$ rapidly activating delayed outward rectifying current, $I_{K1}$ inward rectifying current, $I_{CaL}$ L-type inward $Ca^{2+}$ current, $I_{CaT}$ T-type $Ca^{2+}$ current, $I_{Na}$ $Na^+$ current, $I_f$ pacemaker current)

known as the *action potential* (*AP*). Different ionic species are recruited in the generation of the AP among the cardiac tissues, and the shape of the AP is exclusive to each tissue (■ Fig. 2.1b). Important ions for cardiac cells are potassium ($K^+$), sodium ($Na^+$), and calcium ($Ca^{2+}$). The electrochemical gradient of the respective ions regulate the direction – inward (into the cell) or outward (out of the cell) – of ion currents. The current amplitude (*I*) defines by the membrane potential (*V*) and the ion channel conductivity (*G*): $I = V \times G$ [9].

The SAN AP differs from that in ventricular muscle (■ Fig. 2.2). Nodal cells spontaneously and repeatedly depolarize to threshold, which is around −40 mV, and do not have a steady resting potential. Phase 4 in SAN is the spontaneous depolarization (=pacemaker potential). Phase 0 is the depolarization. This is followed by phase 2 (plateau) and phase 3 (repolarization) of the AP. When nodal cells are entirely repolarized at approximately −60 mV, the cycle is spontaneously repeated (■ Fig. 2.2).

Non-nodal APs are typical of atrial and ventricular myocytes and the fast-conducting Purkinje system in the ventricles. Non-nodal APs have a true resting potential (phase 4, approximately −85 mV), rapid depolarization or initial upstroke (phase 0), early repolarization (phase 1), and a prolonged plateau phase (phase 2). This is followed by phase 3 – repolarization (■ Fig. 2.2).

Thus, non-nodal cell AP differs from nodal cell AP by the presence of phase 1 (initial repolarization) and prolonged phase 2 (plateau) as well as by the differences in the phases 4 and 0 [8–10].

Coordinated contraction is allowed by the specialized organization of cardiac myocytes in muscle slides, which are connected by inserted discs containing fascia adherens, desmosomes, and gap junctions. First two components provide mechanical coupling, while gap junctions allowed dynamic coupling between cardiac myocytes and control normal heart rhythm [11].

Proper conduction of electrical and chemical (movement of ions) signal throughout the heart also depends on gap junctional channels, which are membrane structures. Gap junctions are composed of connexins. There are diverse isoforms of connexins that account for the conductance of gap junctions. Connexin 40 is the predominate isoform in the AVN and is characterized by large conductance. Other isoforms expressed at lower level in AVN are connexin 43 with medium conductance, connexin 45 with small conductance, and connexin 30.2/31.9 with ultra-small conductance [12]. Recent study characterized perinexus, the nanodomain, a tiny pocket, sodium channel-rich separation adjacent to gap junction, in human cardiac tissue. Interestingly, the authors found an association between the width of perinexus and higher risk for developing atrial fibrillation [13].

## 2.1.3 Excitation-Contraction Coupling

Excitation-contraction coupling defines the process of transforming an electrical stimulus (that is AP) to a mechanical response (that is muscle contraction). This process starts with cardiomyocyte electrical excitation, which primes the entry of calcium, and succeeding sarcomere shortening and myofibrillar contraction [14].

Excitation-contraction coupling in cardiac muscle is reliant on a process called $Ca^{2+}$-induced $Ca^{2+}$ release. $Ca^{2+}$ ions are central mediators modifying different properties in the cells. $Ca^{2+}$ stores intracellular in the sarcoplasmic reticulum (SR). Its release into the cytosol is regulated in subcellular microdomains, so-called couplons or $Ca^{2+}$ release units. As a result of suprathreshold myocyte excitation, pacemaker cell-stimulated action potential travels along T-tubules, triggers voltage-gated L-type (long-lasting) $Ca^{2+}$ channels in the cytoplasmic membrane, and causes an influx of calcium into the cytosol of cardiac myocyte. Ryanodine receptors, among others ryanodine receptor type 2 (RyR2), in the membrane of the SR detect increased intracellular calcium and transport calcium into the cytosol that potentiate a temporary increase of the cytosolic $Ca^{2+}$ concentration ($[Ca^{2+}]c$). This calcium binds to troponin C of the troponin complex (consisted of troponin C, I, and T subunits), which permits the actin and tropomyosin filaments to slide past one another, and tension starts to develop. An interaction between actin and myosin depends on the $Ca^{2+}$ concentration. When concentration of $Ca^{2+}$ is low, tropomyosin inhibits binding sites of myosin on actin. At increased $Ca^{2+}$ concentration, tropomyosin stimulates the contact of myosin and actin.

To interrupt contraction and allow the relaxation, $Ca^{2+}$ is taking away from the cytosol and forwarded to the SR, mostly by the adenosine triphosphate (ATP)-dependent *sarcoplasmic/endoplasmic reticulum $Ca^{2+}$ pump* (SERCA) and the *sarcolemmal $Na^+/Ca^{2+}$ exchanger 1* (NCX1 or SLC8A1). The replacement via $Na^+/Ca^{2+}$ exchanger usually leads to discharge of one $Ca^{2+}$ ion in exchange for three $Na^+$ ions. The action of the $Ca^{2+}$ transporters SERCA and $Na^+/Ca^{2+}$ exchanger is controlled by different kinases and phosphatases.

Myofilament $Ca^{2+}$ sensitivity is controlled by phosphorylation of troponin I, regulatory light chain of myosin, and myosin-binding protein C. When intracellular calcium concentration drops, contraction is finished. The SR reservoir stores $Ca^{2+}$ for upcoming calcium-induced calcium release [15].

Hohendanner et al. demonstrated the presence of regional differences in the subcellular control of cytosolic $Ca^{2+}$ decay between intracellular regions with slow $Ca^{2+}$ removal and fast $Ca^{2+}$ removal sites associated with mitochondria in cardiomyocytes during diastole [16].

Cardiac excitation-contraction coupling is also regulated by neurohumoral β-adrenergic as well as different paracrine signaling. The formation of cyclic adenosine monophosphate (cAMP) from ATP is mediated by adenylyl cyclase, which is in turn stimulated by the activation of the guanosintriphosphat (GTP)-binding protein α after β-adrenergic receptor stimulation by binding of epinephrine/norepinephrine. A rise in cAMP leads to protein kinase A activation and phosphorylation of different proteins. As a sequence, ion flux through the L-type $Ca^{2+}$ channel and RyR is increased. Release of nitric oxide (NO) and endothelin-1 by cardiac endothelial cells (ECs) is one of the strongest paracrine modulator of excitation-contraction coupling. Another example of the paracrine regulation is the chemokine stromal cell-derived factor, which stimulates its receptor C-X-C chemokine receptor type 4 and weakens β-adrenergic-mediated cardiomyocyte contraction [14].

### 2.1.4 Heart Cycle

Heart cycle comprise all the events related to one heartbeat. That means a heart cycle involves systole and diastole of the atria plus systole and diastole of the ventricles (◻ Table 2.1). The atria and ventricles contract and relax by turns. They power blood from regions of higher pressure to zones of lower pressure. ◻ Table 2.1 summarizes characteristic features of phases within one heart cycle in the left side of the heart including electrical signals registered by electrocardiogram (ECG) and heart sounds.

Volumes to remember:
- *End-diastolic volume* (EDV) = 110 to 130 ml
- *End-systolic volume* (ESV) = 40–60 ml
- *Stroke volume* (SV), the volume of blood ejected by the ventricle during each contraction: EDV – ESV = 60–80 ml
- *Ejection fraction* (EF) = SV/EDV ~60%

### 2.1.5 Cardiac Output

*Cardiac output* (CO) is the volume of blood expelled from the left (or right) ventricle into the aorta (or pulmonary artery) each minute.

CO equals the SV multiplied by the *heart rate* (HR):

$$CO = SV \times HR$$

CO = 70 mL/beat × 75 beats/min = 5250 mL/min = 5.25 L/min

In healthy, adult resting men, CO is 4–6 L/min. For women this value is 10–20% less.

◻ **Table 2.1** Heart cycle in the left side of the heart

| Phase | AV valve | Aortic valve | ECG | Heart sound | Characteristic |
|---|---|---|---|---|---|
| Phase 1 – atrial contraction | Opens | Closes | P wave (electrical depolarization of the atria) | | Atrium contracts → the pressure within the atrium increases → blood flows into the ventricle |
| Phase 2 – isovolumetric contraction | Closes | Remains closed | QRS complex (ventricular depolarization) | First heart sound as a result of AV valve closure | Increase in intraventricular pressure → AV valve closed → ventricular volume does not change because there is no ejection |
| Phase 3 – rapid ejection | Remains closed | Opens | | No heart sounds because the opening of healthy valve is silent | Intraventricular pressure exceeds the pressure within aorta → aortic valve opens → initial ejection of blood from the LV into the aorta |
| Phase 4 – reduced ejection | Remains closed | Remains opened | T-wave (ventricular repolarization) | | Decline in the rate of ejection |
| Phase 5 – isovolumetric relaxation | Closed | Closed | | Second heart sound as a result of aortic valve closure | All valves are closed → volume remains constant |
| Phase 6 – rapid filling | Opens | Closed | | Ventricular filling is normally silent | Ventricle continues to relax → the intraventricular pressure falls below atrial pressure → AV valve opens and ventricular filling begins |
| Phase 7 – reduced filling | Opens | Closed | | | Normally, by the end of phase 7, the ventricle is about 90% filled |

As CO depends on SV and HR, all factors influencing these parameters can influence CO. Several factors regulate SV: contractility (inotropy), preload, and afterload. *Preload –* the pressure with which the ventricle begins to contract. The preload is proportional to the

2

EDV and is determined by the duration of ventricular diastole and by the venous return. *Afterload* – the pressure against which the ventricle must to contract and is proportional to the ESV. *Contraction* of heart muscles is a consequence of muscle fibers' activation. Both SV and ESV are affected by the inotropy – SV is increased and ESV is decreased, if inotropy is increased. Inotropy is stimulated by autonomic nervous system (sympathetic activation or parasympathetic inhibition), circulating catecholamines, afterload (*Anrep effect*, an increase in afterload can lead to an increase in inotropy), and HR (*Bowditch effect*, an increase in HR can cause an increase in inotropy). HR in turn is regulated by autonomic nervous system, hormones released by the adrenal medullae (epinephrine and norepinephrine), body temperature, gender, age, and physical fitness.

## 2.1.6  Innervation of the Heart

Nervous system regulation of the heart arises in the *cardiovascular center of the medulla oblongata*. This center obtains input from a multiplicity of sensory receptors and from higher brain centers, such as the limbic system and cerebral cortex. Important sensory receptors are the *proprioreceptors* that monitor the position of limbs and muscles; the *chemoreceptors*, which monitor chemical changes in the blood; and *baroreceptors* to monitor the stretching of major arteries and veins caused by blood pressure. Important baroreceptors are positioned in the arch of the aorta and in the carotid arteries.

*Sympathetic neurons* spread from the medulla oblongata into the spinal cord. From the thoracic region of the spinal cord, sympathetic cardiac accelerator nerves extend out to the SA node, AV node, and the myocardium. Impulses in the cardiac accelerator nerves trigger the release of *noradrenaline (NA)*, which binds to β1 receptors in the heart, specifically in the SA node, AV node, and on cardiac muscle fibers in atria and ventricle. This interaction has several effects: increase of HR (via the SA node; *positive chronotropic effect*), increase of contractility (due to augmented calcium release from SR; *positive inotropic effect*), and upregulation of conduction velocity (due to the interaction with β1 receptors in AV node; *positive dromotropic effect*) (◻ Fig. 2.3).

*Parasympathetic nerve* impulses reach the heart via the vagus nerves. SA node, AV node, and atrial myocardium are innervated by vagal axons. They release *acetylcholine (Ach)*, which decreases HR. As only a few vagal fibers innervate ventricular muscles, changes in parasympathetic activity have little effect on contractility of the ventricles. Two types of muscarinic receptors, the M2 and M3, are responsible for the effects of the parasympathetic nervous system on the cardiovascular system. Parasympathetic nervous system has the following effects: decrease in HR (*negative chronotropic effect*, due to the innervation of the SA node by vagus nerve); decrease in myocardial contractility is disputable as the vagus nerve does not directly innervate ventricular myocytes, as already mentioned above; and decrease in AV node conduction velocity (*negative dromotropic effect*) (◻ Fig. 2.3).

The SAN of the heart is innervated by both sympathetic and parasympathetic nerve fibers. Under circumstances of physical or emotional stress sympathetic nerve fibers release NA. NA acts to increase the pacemaker potential and to up-regulate HR. Other effects of the stimulation of the sympathetic nervous system include constriction of the blood vessel, dilatation of the bronchiole and pupil, and inhibition of peristalsis.

Under circumstances of rest the parasympathetic fibers release ACh. ACh slows the pacemaker potential of the SA node and reduces HR. Other effects of the stimulation of

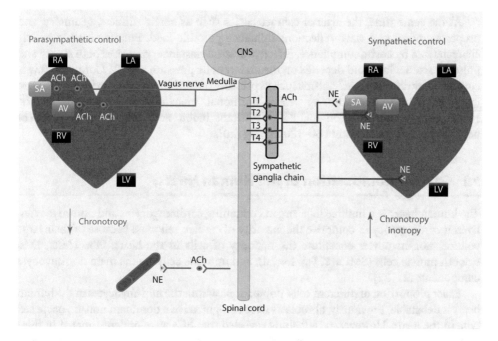

**Fig. 2.3** Autonomic nervous system regulation of the heart function. (Reprinted from [17] with the permission of the publisher. The autonomic nervous system affects the rate and force of heart contractions. CNS central nervous system, RA right atria, LA left atria, RV right ventricle, LV left ventricle, SA sinoatrial node, AV atrioventricular node, NE Norepinephrine, ACh acetylcholine)

the parasympathetic nervous system include constriction of the bronchiole and pupils, increase of peristalsis, as well as increase secretion of different glands (pancreatic, salivary, and eye) [17, 18].

### 2.1.7 Coronary Circulation

Coronary blood flow regulates the equilibrium between oxygen supply and oxygen demand. French anatomist Raymond de Vieussens was the first who described the blood vessels of the heart in 1706. Later on, in 1715, he extended his study by the description of the coronary vessels, cardiac muscle fibers, and the pericardium [19].

The left and the right coronary arteries derived at the base of the aorta from the sinuses of Valsalva. The coronary arteries bend around the heart like a crown, which give also their name from Latin. Extravascular compression during systole markedly affects coronary flow; therefore, most of the coronary flow occurs during diastole. Perfusion of arteries supplying the RV is less affected by contracting myocardium because tension developed in RV is lower than tension in LV.

Two components of arterial (hemodynamic) load – "steady"/"resistive" and pulsatile load (total arterial compliance [TAC], aortic impedance) – exist. Mean arterial pressure (MAP) is determined by CO and "steady" load, which is in turn dependent on the total peripheral resistance (TPR): $MAP = CO \times TPR$. Systemic vascular resistance is mainly dependent on the resistance of the arterioles and capillaries.

2

At the same time, the arterial characteristics such as aortic stiffness, geometry, and properties of arterial wave reflections influence pulsatile load. Pulsatile load is mostly characterized by aortic compliance. Effective arterial elastance comprises both steady and pulsatile arterial load and depends on the end-systolic pressure and SV. Coronary flow is controlled by different mechanisms such as pressure outside the vessels, perfusion pressure inside the coronary arteries, as well as hormonal, metabolic, and myogenic factors. Pressure-flow autoregulation maintains constant blood flow over a broad range of perfusion pressures of around 60–120 mmHg [20].

## 2.2  Cellular Composition of the Human Heart

The human heart is a multicellular organ containing cardiomyocytes and non-myocytes. Even if cardiomyocytes comprise the majority of cardiac cell mass because of their large volume, non-myocytes constitute the majority of cells in the heart. Fibroblasts, ECs, smooth muscle cells (SMCs, ◘ Fig. 2.4c, d), and immune cells are the main non-myocyte components [21, 22].

Exact proportion of different cells' populations within the non-myocytes in the human heart is debatable. Previously, fibroblasts were recognized as a dominant non-myocyte cell type in the heart. However, recent study revealed that ECs were underestimated in their importance, and account for the majority (>60%) of non-myocyte. Fibroblasts comprise only <20% of the non-myocyte cell in the human heart [21]. By performing immunohistochemistry, Pinto et al. showed that in human cardiac tissue, cardiomyocytes (defined as ACTN2 positive cells) account for around 30%, ECs (CD31 positive cells) for >50%, and leukocytes (CD45 positive cells) for >5%. These results were confirmed by flow cytometry analysis of single-cell preparations. In general, cardiac ECs include cells in the coronary vasculature (◘ Fig. 2.4e, f), lymphatic vessels, and endocardial and intramyocardial ECs. Within heart ECs, around 90% belongs to vascular ECs (defined as $CD102^+$ $CD105^+$ nodes) and around 5% to lymphatic ECs (defined as $podoplanin^+$ nodes) [21].

Each cell population has different characteristics and functions. However, they coordinate in a precise manner in order to achieve effective cardiac function. Cardiac cells together with *extracellular matrix* (ECM) are highly organized structure, allowing proper heart function. Cardiac development and regular cardiac function depends on coordinated interactions between these cells. Moreover, all cardiac cells are able to secrete factors that influence properties of neighboring cells [23].

As characteristics and functions of ECs and SMCs are extensively described in the separate chapters of this book, following sections in the present chapter will discuss the properties and functions of cardiac myocytes, cardiac fibroblasts, and resident immune cells (with the focus on macrophages).

### 2.2.1  Cardiac Myocytes

Cardiac myocytes comprise >70% of the myocardial tissue mass, but merely around 30% of the total cell number [21]. The function of the cardiac myocytes is directed to provide proper contraction-relaxation cycle. This task is achieved by creating contractile force and crystalline-like myofibrillar structure of adult cardiomyocytes is helpful in this process. Signals generated in the SAN enforce depolarization of cardiac myocytes. Calcium is a mediator that helps to transmit the signal into muscular contraction. Each cardiomyocyte

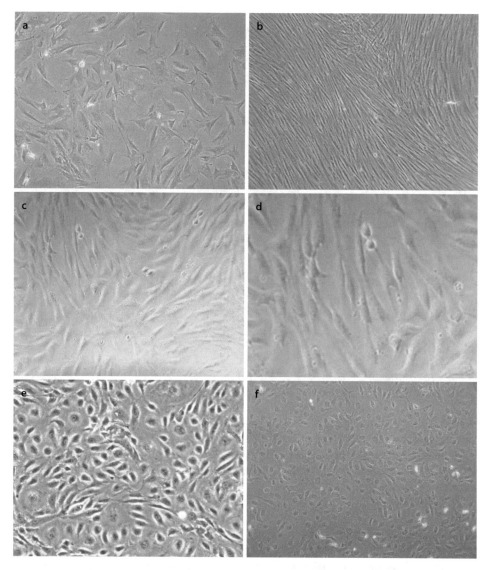

◻ **Fig. 2.4** Representative morphology of human adult cardiac myocytes (HACM, panel **a**), human adult cardiac fibroblasts (HACF, panel **b**), human coronary artery smooth muscle cells (HCASMC, panel **c**), human aortic smooth muscle cells (HASMC, panel **d**). human coronary artery endothelial cells (HCAEC, panel **e**), and human aortic endothelial cells (HAEC, panel **f**) cultured in vitro (own unpublished data)

comprises so-called sarcomeres, which are the contractile units and organized with about 2 μm interspace. In order to harmonize contraction, a lot of sarcomeres have to function coordinated in a single cell.

The human heart was traditionally seen as a postmitotic terminally differentiated organ with limited capacity for regeneration by cardiomyocytes. However, this view was challenged by the findings that new cardiac myocytes appear also in the adult heart. The technique called radiocarbon ($^{14}$C) birth dating established renewal of the heart muscle until the third decade of life [24]. However, the potential sources, generation rates, triggers, and molecular pathways for cardiac myocytes regeneration are still the matter of

**2**

■ **Table 2.2**  Cytokines and growth factors expressed and/or released by human adult cardiac myocytes and their regulation

| Factor | Constitutive expression | Regulation | References |
|---|---|---|---|
| Plasminogen activator inhibitor-1 (PAI-1) | Yes | Increased by interleukin (IL)-1α, tumor necrosis factor (TNF)-α, transforming growth factor (TGF)-β, oncostatin M (OSM) | [25] |
| Vascular endothelial growth factor (VEGF) | Yes | Increased by OSM and prostaglandin E1 | [26, 27] |
| Macrophage colony stimulating factor (M-CSF) | Yes | Increased by TNF-α | [28] |
| Stromal-derived factor 1 (SDF-1) | Yes | Increased by OSM | [29] |
| IL-33 | Yes (intracellular) | Increased by TNF-α, IL-1β, interferon (IFN)-γ | [30] |
| Brain natriuretic peptide (BNP); atrial natriuretic peptide (ANP) | Yes | Decreased in hypoxic right atrial tissue | [31] |
| Pigment epithelium-derived factor (PEDF) | Yes | Decreased by hypoxia | [32] |

debate. Among others, progenitor or stem cells as well as dedifferentiated pre-existing cardiac myocytes are discussed as potential sources.

Human adult cardiac myocytes (■ Fig. 2.4a) are able to secrete different cytokines and growth factors, as summarized in the ■ Table 2.2, and influence the neighboring cells in a paracrine manner.

## 2.2.2  Cardiac Fibroblasts

Cardiac fibroblasts are responsible for the production and degradation of the components of the ECM that provide scaffold of the heart. The importance of cardiac fibroblasts is underlined by the fact that ECM conformation is also crucial for the function and proliferation of cardiomyocytes in a healthy heart. Production of fibronectin; collagens (e.g., I, III, IV, V, VI); α1-, α2-, and α5-integrins; fibrillin; periostin; glycoproteins; proteoglycans; glycosaminoglycans; matrix metalloproteinases (MMPs); and tissue inhibitors of metalloproteinases (TIMPs), which are the components of ECM, are regulated by the cardiac fibroblasts. Collagen is the main component of the ECM and is comprised of diverse variants such as full-length, fragmented, and posttranslationally modified forms [33, 34].

However, cardiac fibroblasts have been recognized to have more functions than just maintain the structural basis of the heart. Not only cardiac myocytes but also cardiac fibroblasts take part in mechano-electrical signaling responsible for cardiac contraction. Although cardiac fibroblasts themselves are electrically non-excitable, they conduct indirect and direct influence on electrophysiological characteristics of the heart. Indirect interplay between fibroblasts and cardiac myocytes depend on ECM that acts as an electrical buffer. ECM proteins surrounding the cardiomyocytes permit the spreading of mechanical force throughout the myocardium. But also direct cardiomyocyte-fibroblast interaction could play a substantial role in gap junction. Gap junctions containing connexins provide the connections for electrical excitation. Moreover, cardiac fibroblasts express mechanosensitive channels and $K^+$ channels, which define their resting membrane potential. Additionally, cardiac fibroblasts also possess immunomodulatory properties due to the interaction with immune cells [14, 23, 35, 36].

Fibroblasts are mesenchymal-derived cells, but also the bone marrow-derived progenitor cells, perivascular cells, fibrocytes, and monocytes could differentiate into fibroblasts. Under pathological conditions, cardiac fibroblasts differentiate into another type of cells, namely, myofibroblasts, which are characterized by many futures of SMCs including ability for migration and as such invasion. Based on their morphology, fibroblasts are defined as active or inactive.

Cardiac fibroblasts (◘ Fig. 2.4b) show several unique characteristics [33, 35]:
- The absence of a basement membrane
- A prominent Golgi apparatus
- Extensive endoplasmic reticulum
- Flat, spindle- or sheet-shaped morphology with numerous filopodia

Identification of cardiac fibroblasts is often problematic because they have no single specific cell marker. In order to distinguish myofibroblasts from cardiac fibroblasts, α-smooth muscle actin should be used. Often used markers such as vimentin, the discoidin domain receptor 2 (DDR2), CD90, and Sca-1 are not exclusive to fibroblasts. Moreover, depending on the tissue origin, not all fibroblasts express these antigens. Current recommendations for the identification of the cells of fibroblast origin include the use of a combination of different markers. Several examples of the markers expressed by the fibroblasts are provided below [37]:
- Vimentin identifies fibroblasts with grate sensitivity, but it is also expressed by ECs and macrophages as well as by myofibroblasts. In order to distinguish fibroblasts from other cell types, the combinations of vimentin with CD45 (immune cells) or CD31 (ECs) are used.
- DDR2, which is a collagen receptor, is present on the surface of cardiac fibroblasts but also on myofibroblasts and other cells.
- The fibroblast-specific protein 1 (FSP-1) possesses cytosolic localization; however, it marks not only fibroblasts but also myocytes, ECs, SMCs, and immune cells.
- Cadherin-11 is a promising fibroblast marker, but it is unclear, if cadherin-11 expressed by myofibroblasts.
- Thymus cell antigen-1 (Thy-1/CD90) is expressed by fibroblasts, ECs, SMCs, a subset of $CD34^+$ bone marrow cells, and fetal liver-derived hematopoietic cells.
- Transcription factor 21 (Tcf21) is expressed in cardiac fibroblasts in the nucleus and acts as a transcription factor. Tcf21 has also limited value in identification of cardiac fibroblasts and has to be used in combination with other markers.
- Platelet-derived growth factor receptor-α (PDGFRα) in combination with Collagen1a1-GFP should detect most cardiac fibroblasts.
- Wilms tumor 1 (WT1) is postulated as a fibroblast lineage indicator, however, have to be also used in combination with other markers.

Therefore, cardiac fibroblasts sustain mechanical, electrical, and paracrine regulation within the heart, also because of their communications with cardiac myocytes, vascular cells, and immune cells.

2

### 2.2.3 Immune Cells

Immune cells are not only recruited to the heart in the response to injury but also present as resident cells within the human heart. Pinto et al. detected myeloid cells (CD11b$^+$, around 80% of leukocytes), B cells (B220$^+$, around 10%), T cells (CD3$^+$, around 3%), and lymphoid cells (CD11b$^-$B220$^-$CD3$^-$, around 7%) to be present within the population of cardiac immune cells [21]. Myeloid cells comprise monocytes, macrophages, neutrophils, eosinophils, basophils, mast cells, dendritic cells, and natural killer cells.

Cardiac monocytes and macrophages attract special attention in the recent years. Therefore, this section will discuss these cell populations within the heart with emphasis on steady-state and not diseased heart. Monocytes could enter heart tissue via lymphatic or systemic circulation. In the cardiac tissue, surviving monocytes transdifferentiate into different types of macrophages so-called M1 or M2 subtype. In the healthy heart, alternatively activated M2 macrophages with anti-inflammatory properties are dominant. Contrary, the number of extravasated monocytes in the non-diseased heart is low. Therefore, the concept of macrophage heterogeneity is also applied to the human heart [38].

Monocytes and macrophages can be also subdivided based on their expression of the chemokine C-C motif ligand 2 (CCL2). In diseased heart obtained from heart failure patients, CCR2$^-$ macrophages, CCR2$^+$ macrophages, and CCR2$^+$ monocytes with distinct gene expression patterns and functions were detected in LV myocardial tissue [39]. CCR2$^+$ monocytes and CCR2$^+$ macrophages seem to play pathological inflammatory role and are associated with poor LV systolic function in heart failure patients. CCR2$^-$ macrophages have reparative functions as they may initiate tissue repair.

Macrophages are actively involved in the regulation of cardiac homeostasis. As in other tissues and organs, heart macrophages also possess phagocytic activity, eliminate senescent and dying cells (also dying cardiomyocytes), and provide defense against infection. Macrophages located in the human heart are able to communicate with stromal cells such as cardiac myocytes, fibroblasts, and ECs and provide clearance function by eliminating unnecessary factors from surrounding cells or blood [40]. However, the functions of cardiac macrophages are not limited to the defense properties. Exciting recent data revealed that cardiac macrophages regulate electrical conduction. Hulsmans et al. described the presence of macrophages in AV node in healthy human and mouse heart, where their amount was much higher compared to the LV tissue. Macrophages situated in human AV node coupled with cardiomyocytes via connexin 43 [41]. Therefore, the ability of cardiac macrophages to interdigitate with cardiac myocytes in AV node can control the contractile properties of the human heart.

Altogether, cardiac macrophages show heterogeneous subtypes and can regulate tissue homeostasis within the healthy human heart.

## 2.3 Metabolic Control of the Myocardial Function

The process of excitation-contraction coupling that ensures myocardial contractility (as discussed in the ▶ Sect. 2.1.3. of this chapter) requires large amount of energy. To gain this energy, ATP is used. Therefore, constant formation of ATP is a vital process needed for proper cardiac function and is covered mainly by mitochondria via the process of oxidative phosphorylation (responsible for more than 90% of ATP generation), but also by anaerobic glycolysis in the cytosol. The balance between the oxidative phosphorylation and ATP hydrolysis keep ATP amount constant [42].

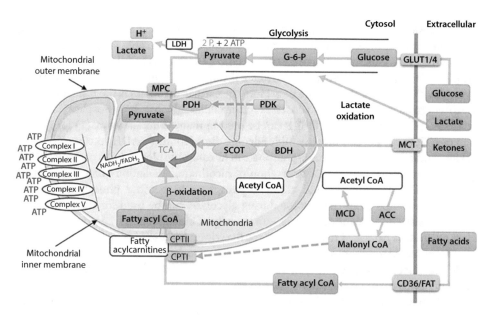

**Fig. 2.5** Energy metabolism in normal heart. (Reprinted from [43] with the permission of the publisher. MPC mitochondrial pyruvate career, PDH pyruvate dehydrogenase, PDK pyruvate dehydrogenase kinase, TCA tricarboxylic acid cycle, SCOT succinyl-CoA-3-oxaloacid CoA transferase, BDH β-hydroxybutyrate dehydrogenase, CPT carnitine palmitoyltransferase, MCD malonyl CoA dehydrogenase, ACC acetyl-CoA carboxylase, MCT monocarboxylate transporter, CD36 cluster of differentiation, FAT fatty acid translocase, GLUT glucose transporter type (1 or 4))

To cover ATP production, the healthy heart oxidizes fatty acids (responsible for 60% to 90% of ATP generation) as well as pyruvate (responsible for 10% to 40% of ATP formation). On the other hand, oxidation of lactate and glycolysis is responsible for the production of pyruvate. Under normal conditions, the human heart catches the possibility to use amino acids and ketone bodies only to a small amount, and as alternative substrates. As hormonal, substrate, and energy status of the human body is dynamically variable, cardiac myocytes need to adapt to these environmental shifting. This unique property to shift between available substrates to generate enough energy called "metabolic flexibility" and is an important characteristic feature of the normal human heart [43, 44].

*Fatty acids* are used by the heart for respiration and normally sustain the main energy source for cardiac myocytes in adult human heart. The glucose fatty acid or *Randle cycle* arranges the use of the respective substrate for the production of the energy. Free fatty acids in plasma – albumin-bound or released from very low density lipoproteins or chylomicrons – are the predominant origins of fatty acids used for beta-oxidation. Fatty acids from extracellular space reach into the cytosol of cardiac myocytes via passive diffusion or via active transport mediated by the fatty acid translocase (FAT, CD36) or fatty acid transport protein (FATP). Intracellular free fatty acids transformed to long-chain fatty acyl coenzyme A (CoA) by esterification. Long-chain fatty acyl CoA either entered the mitochondria mediated by carnitine palmitoyltransferase isomers (CPTI and CPTII) or stored in the triacylglycerol (TAG) ( Fig. 2.5). In general, fatty acid metabolism is less effective, if measured as generated ATP molecules in relation to the consumed oxygen molecules, compared to glucose oxidation.

*Glucose* is the most advantageous substrate with high produced phosphate/consumed oxygen ratio. The citric acid or *Krebs cycle* is a fundamental chemical sequence leading to the conversion of pyruvate to succinate. Although glycogen stores are not very prominent in the heart, they, together with exogenous glucose, are responsible for glucose metabolism in the heart. Glucose transporter type 1 (GLUT1), which is insulin-independent, and GLUT4, which is insulin-dependent, mediate the uptake of glucose into the cardiac myocytes. Another way for the glucose to get into the cardiac myocytes is the transmembrane glucose gradient. Hexokinase processes intracellular glucose by phosphorylation to glucose-6-phosphate (G-6-P). G-6-P can be used in many different ways: (1) for generation of ATP, pyruvate, and nicotinamide adenine dinucleotide (NADH) during glycolysis; (2) for synthesis of glucosamine-6-phosphate in hexosamine biosynthesis pathway; (3) for glycogen formation; and (4) for generation of nicotinamide adenine dinucleotide phosphate (NADPH), pentoses, and ribose 5-phosphate in pentose phosphate pathway. Pyruvate generated from the glucose during glycolysis can be used for the synthesis of lactate or entered to the mitochondria with the help of the mitochondrial pyruvate carrier. Here pyruvate is catalyzed to acetyl-CoA for the *tricarboxylic acid cycle* (■ Fig. 2.5).

*Ketone bodies* represent a minor substrate for the production of ATP compared to fatty acids and glucose. Despite this fact, ketones are more efficient for the energy production than fatty acids based on their phosphate/oxygen ratio. Fasting periods result in increased production of acetone, beta-hydroxybutyrate (βOHB), and acetoacetate in the liver. Acetone is excreted via the respiratory tract, but βOHB and acetoacetate continue to circulate in the blood. Ketone bodies entered the mitochondria of cardiac myocytes in such a way. βOHB can be oxidized into acetoacetate and finally acetyl-CoA is generated and can take part in the Krebs cycle to produce ATP (■ Fig. 2.5).

## 2.4  Male and Female Heart

Cardiac physiology displays evident sex and gender differences that were often underappreciated in the past. These differences begin already with the cardiac anatomy especially with the structure of the LV and are age-dependent even if aging is associated with physiological changes in cardiac structure and functions in both sexes. Peak systolic apical mechanics measured by the echocardiography were greater in healthy middle-aged men than in middle-aged women. Additionally, postmenopausal healthy women demonstrated lower LV relaxation compared with premenopausal women [45].

Another study also showed sex differences in age-associated changes in cardiac physiology. Thus, although LV EF increases in both male and female with age, this increase is obvious to a greater extent especially in women. Increase in LV stiffness is also more distinct in aging women as in aging men. By using magnetic resonance phase-contrast imaging, different gender-dependent and gender-independent age-related changes in myocardial motion were established. Gender-dependent changes include more pronounced decrease in long-axis contraction velocity in the systole and diastole as well as prolonged time-to-peak apical rotation in the diastole in women [46]. Therefore, with aging women are more predisposed to LV diastolic dysfunction as men. Interestingly, substantial differences in the anatomy or functional capacity of LA were not observed [47]. Female heart seems to be able better as male heart to sustain cardiac myocytes' number during aging among other due to the higher replicative capacity and longer telomeres of their cardiac stem cells. Therefore, the loss of cardiac myocytes is less pronounced in female myocardium compared with male myocardium during the process of aging [48].

Sex differences are seen not only on the anatomical but also on the cellular, molecular, and transcriptional levels [47]. LV diastolic function is determined by different structural proteins. *Titin* is one of the key proteins, and its characteristic is depending on the phosphorylation by protein kinases A and G. An interesting link to the *female hormones* provides the fact that *estrogen* is able to diminish the protein kinase A activity. The content and the type of collagen isoforms are also detrimental for the structural integrity of the human heart. Young women displayed lower levels of collagen I and III than men. However, with aging this relation is changed, and men express lower levels of collage I and III.

Sex differences in cardiac contractile function could be partially explained by distinctions in calcium handling. Calcium flow, SR $Ca^{2+}$ sparks, intracellular cAMP levels, and excitation-contraction coupling gain were lower in female cardiomyocytes [49]. Interestingly, also $K^+$-channel subunits, connexin43, and phospholamban, all involved in the regulation of conduction and repolarization, are expressed at reduced levels in female hearts [50].

Cardiac function is controlled by the substrate metabolism as discussed previously in ▶ Sect. 2.3 of this chapter. It is of note that in response to obesity, women displayed different myocardial metabolic picture as men because they have lower myocardial glucose uptake and conversion as well as higher fatty acids metabolism. Sex differences in triglyceride and acylcarnitine metabolism in the heart were also reported. Female gender was also associated with higher myocardial blood flow and oxygen consumption in that study [51].

Control of immune system and adequate inflammatory response is crucial for all aspects of the cardiac physiology and pathophysiology. It is known for a long time that women assemble stronger immunity than men [52]. This phenomenon was also obvious in human hearts of men and women, where distinct expression of genes involved in pro-inflammatory cascade was identified [47].

Biologically, sex is defined by sex chromosomes and related genes as well as sex hormones mainly estrogens and androgens. Epidemiological studies showed that premenopausal women had lower risk of developing cardiovascular events, and experimental studies proved cardio- and vasculo-protective influence of estrogen (E2), the female hormone. HR is regulated among others by estrogen receptor (ER)-α and ER-β, which are expressed in the central nervous system. Regulation of cardiac diastolic function by E2 is also accepted, and many mechanisms involved are described: E2 regulates the function of mitochondria, the energy source, NO action, and $Ca2^+$ balance [53].

---

**Take-Home Message**

- Coordinated cardiac contraction and relaxation is facilitated by organized excitatory and conductivity system including the process of excitation-contraction coupling that is reliant on $Ca^{2+}$-induced $Ca^{2+}$ release.
- Parasympathetic and sympathetic parts of the autonomic nervous system regulate the function of the heart in antagonistic manner.
- The human heart contains both cardiomyocytes and non-myocytes with endothelial cells comprising the majority of non-myocyte.
- Cardiac cells together with extracellular matrix are highly organized and interacting structure, allowing proper heart function.
- Cardiac macrophages show heterogeneity and can regulate tissue homeostasis including electrical conduction within healthy human heart.
- Substrate "metabolic flexibility" is an important characteristic feature of the normal human heart.
- Sex-related differences on anatomical, cellular, molecular, and transcriptional levels exist between male and female hearts.

2

# References

1. Tortora GJ, Derrickson BH. Principles of anatomy and physiology. 14th ed. Hoboken: John Wiley & Sons; 2013.
2. Klabunde RE. Cardiovascular physiology concepts. 2nd ed: Lippincott Williams & Wilkins; 2012. https://www.lww.com.
3. Ward JPT, Linden RWA. Physiology at a glance. 4th ed: John Wiley & Sons; 2017. https://www.wiley.com/en-at/aboutus.
4. van Eif VWW, Devalla HD, Boink GJJ, Christoffels VM. Transcriptional regulation of the cardiac conduction system. Nat Rev Cardiol. 2018; https://doi.org/10.1038/s41569-018-0031-y. [Epub ahead of print].
5. Weber KT, Sun Y, Bhattacharya SK, Ahokas RA, Gerling IC. Myofibroblast-mediated mechanisms of pathological remodelling of the heart. Nat Rev Cardiol. 2013;10(1):15–26.
6. Bartos DC, Grandi E, Ripplinger CM. Ion channels in the heart. Compr Physiol. 2015;5(3):1423–64.
7. George SA, Faye NR, Murillo-Berlioz A, Lee KB, Trachiotis GD, Efimov IR. At the atrioventricular crossroads: dual pathway electrophysiology in the atrioventricular node and its underlying heterogeneities. Arrhythm Electrophysiol Rev. 2017;6(4):179–85.
8. Mohan R, Boukens BJ, Christoffels VM. Lineages of the cardiac conduction system. J Cardiovasc Dev Dis. 2017;4(2):5.
9. Amin AS, Tan HL, Wilde AA. Cardiac ion channels in health and disease. Heart Rhythm. 2010;7(1):117–26.
10. Liu J, Laksman Z, Backx PH. The electrophysiological development of cardiomyocytes. Adv Drug Deliv Rev. 2016;96:253–73.
11. Johnson RD, Camelliti P. Role of non-myocyte gap junctions and connexin hemichannels in cardiovascular health and disease: novel therapeutic targets? Int J Mol Sci. 2018;19(3):pii: E866.
12. Hood AR, Ai X, Pogwizd SM. Regulation of cardiac gap junctions by protein phosphatases. J Mol Cell Cardiol. 2017;107:52–7.
13. Raisch TB, Yanoff MS, Larsen TR, Farooqui MA, King DR, Veeraraghavan R, Gourdie RG, Baker JW, Arnold WS, AlMahameed ST, Poelzing S. Intercalated disk extracellular nanodomain expansion in patients with atrial fibrillation. Front Physiol. 2018;9:398.
14. Mayourian J, Ceholski DK, Gonzalez DM, Cashman TJ, Sahoo S, Hajjar RJ, Costa KD. Physiologic, pathologic, and therapeutic paracrine modulation of cardiac excitation-contraction coupling. Circ Res. 2018;122(1):167–83.
15. Bertero E, Maack C. Calcium signaling and reactive oxygen species in mitochondria. Circ Res. 2018;122(10):1460–78.
16. Hohendanner F, Ljubojević S, MacQuaide N, Sacherer M, Sedej S, Biesmans L, Wakula P, Platzer D, Sokolow S, Herchuelz A, Antoons G, Sipido K, Pieske B, Heinzel FR. Intracellular dyssynchrony of diastolic cytosolic [$Ca^{2+}$] decay in ventricular cardiomyocytes in cardiac remodeling and human heart failure. Circ Res. 2013;113(5):527–38.
17. Gordan R, Gwathmey JK, Xie LH. Autonomic and endocrine control of cardiovascular function. World J Cardiol. 2015;7(4):204–14.
18. Meng L, Shivkumar K, Ajijola O. Autonomic regulation and ventricular arrhythmias. Curr Treat Options Cardiovasc Med. 2018;20(5):38.
19. Loukas M, Clarke P, Tubbs RS, Kapos T. Raymond de Vieussens. Anat Sci Int. 2007;82(4):233–6.
20. Weber T, Chirinos JA. Pulsatile arterial haemodynamics in heart failure. Eur Heart J. 2018;39:3847. https://doi.org/10.1093/eurheartj/ehy346. [Epub ahead of print].
21. Pinto AR, Ilinykh A, Ivey MJ, Kuwabara JT, D'Antoni ML, Debuque R, Chandran A, Wang L, Arora K, Rosenthal NA, Tallquist MD. Revisiting cardiac cellular composition. Circ Res. 2016;118(3):400–9.
22. Segers VFM, Brutsaert DL, De Keulenaer GW. Cardiac remodeling: endothelial cells have more to say than just NO. Front Physiol. 2018;9:382.
23. Fountoulaki K, Dagres N, Iliodromitis EK. Cellular communications in the heart. Card Fail Rev. 2015;1(2):64–8.
24. Lázár E, Sadek HA, Bergmann O. Cardiomyocyte renewal in the human heart: insights from the fallout. Eur Heart J. 2017;38(30):2333–42.
25. Macfelda K, Weiss TW, Kaun C, Breuss JM, Zorn G, Oberndorfer U, Voegele-Kadletz M, Huber-Beckmann R, Ullrich R, Binder BR, Losert UM, Maurer G, Pacher R, Huber K, Wojta J. Plasminogen activator inhibitor 1 expression is regulated by the inflammatory mediators interleukin-1alpha, tumor necrosis factor-alpha, transforming growth factor-beta and oncostatin M in human cardiac myocytes. J Mol Cell Cardiol. 2002;34(12):1681–91.

26. Weiss TW, Speidl WS, Kaun C, Rega G, Springer C, Macfelda K, Losert UM, Grant SL, Marro ML, Rhodes AD, Fuernkranz A, Bialy J, Ullrich R, Holzmann P, Pacher R, Maurer G, Huber K, Wojta J. Glycoprotein 130 ligand oncostatin-M induces expression of vascular endothelial growth factor in human adult cardiac myocytes. Cardiovasc Res. 2003;59(3):628–38.
27. Weiss TW, Mehrabi MR, Kaun C, Zorn G, Kastl SP, Speidl WS, Pfaffenberger S, Rega G, Glogar HD, Maurer G, Pacher R, Huber K, Wojta J. Prostaglandin E1 induces vascular endothelial growth factor-1 in human adult cardiac myocytes but not in human adult cardiac fibroblasts via a cAMP-dependent mechanism. J Mol Cell Cardiol. 2004;36(4):539–46.
28. Hohensinner PJ, Kaun C, Rychli K, Niessner A, Pfaffenberger S, Rega G, de Martin R, Maurer G, Ullrich R, Huber K, Wojta J. Macrophage colony stimulating factor expression in human cardiac cells is upregulated by tumor necrosis factor-alpha via an NF-kappaB dependent mechanism. J Thromb Haemost. 2007;5(12):2520–8.
29. Hohensinner PJ, Kaun C, Rychli K, Niessner A, Pfaffenberger S, Rega G, Furnkranz A, Uhrin P, Zaujec J, Afonyushkin T, Bochkov VN, Maurer G, Huber K, Wojta J. The inflammatory mediator oncostatin M induces stromal derived factor-1 in human adult cardiac cells. FASEB J. 2009;23(3):774–82.
30. Demyanets S, Kaun C, Pentz R, Krychtiuk KA, Rauscher S, Pfaffenberger S, Zuckermann A, Aliabadi A, Gröger M, Maurer G, Huber K, Wojta J. Components of the interleukin-33/ST2 system are differentially expressed and regulated in human cardiac cells and in cells of the cardiac vasculature. J Mol Cell Cardiol. 2013;60:16–26.
31. Hopkins WE, Chen Z, Fukagawa NK, Hall C, Knot HJ, LeWinter MM. Increased atrial and brain natriuretic peptides in adults with cyanotic congenital heart disease: enhanced understanding of the relationship between hypoxia and natriuretic peptide secretion. Circulation. 2004;109(23):2872–7.
32. Rychli K, Kaun C, Hohensinner PJ, Dorfner AJ, Pfaffenberger S, Niessner A, Bauer M, Dietl W, Podesser BK, Maurer G, Huber K, Wojta J. The anti-angiogenic factor PEDF is present in the human heart and is regulated by anoxia in cardiac myocytes and fibroblasts. J Cell Mol Med. 2010;14(1–2):198–205.
33. Rog-Zielinska EA, Norris RA, Kohl P, Markwald R. The living scar–cardiac fibroblasts and the injured heart. Trends Mol Med. 2016;22(2):99–114.
34. Frangogiannis NG. Fibroblasts and the extracellular matrix in right ventricular disease. Cardiovasc Res. 2017;113(12):1453–64.
35. Klesen A, Jakob D, Emig R, Kohl P, Ravens U, Peyronnet R. Cardiac fibroblasts : active players in (atrial) electrophysiology? Herzschrittmacherther Elektrophysiol. 2018;29(1):62–9.
36. Furtado MB, Hasham M. Properties and immune function of cardiac fibroblasts. Adv Exp Med Biol. 2017;1003:35–70.
37. Tarbit E, Singh I, Peart JN, Rose'Meyer RB. Biomarkers for the identification of cardiac fibroblast and myofibroblast cells. Heart Fail Rev. 2018;24:1. https://doi.org/10.1007/s10741-018-9720-1. [Epub ahead of print].
38. Nahrendorf M, Swirski FK. Monocyte and macrophage heterogeneity in the heart. Circ Res. 2013;112(12):1624–33.
39. Bajpai G, Schneider C, Wong N, Bredemeyer A, Hulsmans M, Nahrendorf M, Epelman S, Kreisel D, Liu Y, Itoh A, Shankar TS, Selzman CH, Drakos SG, Lavine KJ. The human heart contains distinct macrophage subsets with divergent origins and functions. Nat Med. 2018;24(8):1234–45.
40. Ma Y, Mouton AJ, Lindsey ML. Cardiac macrophage biology in the steady-state heart, the aging heart, and following myocardial infarction. Transl Res. 2018;191:15–28.
41. Hulsmans M, Clauss S, Xiao L, Aguirre AD, King KR, Hanley A, Hucker WJ, Wülfers EM, Seemann G, Courties G, Iwamoto Y, Sun Y, Savol AJ, Sager HB, Lavine KJ, Fishbein GA, Capen DE, Da Silva N, Miquerol L, Wakimoto H, Seidman CE, Seidman JG, Sadreyev RI, Naxerova K, Mitchell RN, Brown D, Libby P, Weissleder R, Swirski FK, Kohl P, Vinegoni C, Milan DJ, Ellinor PT, Nahrendorf M. Macrophages facilitate electrical conduction in the heart. Cell. 2017;169(3):510–22.
42. Stanley WC, Recchia FA, Lopaschuk GD. Myocardial substrate metabolism in the normal and failing heart. Physiol Rev. 2005;85(3):1093–129.
43. Karwi QG, Uddin GM, Ho KL, Lopaschuk GD. Loss of metabolic flexibility in the failing heart. Front Cardiovasc Med. 2018;5:68.
44. Bertero E, Maack C. Metabolic remodelling in heart failure. Nat Rev Cardiol. 2018;15:457. https://doi.org/10.1038/s41569-018-0044-6. [Epub ahead of print].
45. Nio AQX, Stöhr EJ, Shave RE. Age-related differences in left ventricular structure and function between healthy men and women. Climacteric. 2017;20(5):476–83.
46. Föll D, Jung B, Schilli E, Staehle F, Geibel A, Hennig J, Bode C, Markl M. Magnetic resonance tissue phase mapping of myocardial motion: new insight in age and gender. Circ Cardiovasc Imaging. 2010;3(1):54–64.

2

47. Beale AL, Meyer P, Marwick TH, Lam CS, Kaye DM. Sex differences in cardiovascular pathophysiology: why women are overrepresented in heart failure with preserved ejection fraction. Circulation. 2018;138(2):198–205.
48. Kajstura J, Gurusamy N, Ogórek B, Goichberg P, Clavo-Rondon C, Hosoda T, D'Amario D, Bardelli S, Beltrami AP, Cesselli D, Bussani R, del Monte F, Quaini F, Rota M, Beltrami CA, Buchholz BA, Leri A, Anversa P. Myocyte turnover in the aging human heart. Circ Res. 2010;107(11):1374–86.
49. Parks RJ, Ray G, Bienvenu LA, Rose RA, Howlett SE. Sex differences in SR Ca(2+) release in murine ventricular myocytes are regulated by the cAMP/PKA pathway. J Mol Cell Cardiol. 2014;75:162–73.
50. Gaborit N, Varro A, Le Bouter S, Szuts V, Escande D, Nattel S, Demolombe S. Gender-related differences in ion-channel and transporter subunit expression in non-diseased human hearts. J Mol Cell Cardiol. 2010;49(4):639–46.
51. Peterson LR, Soto PF, Herrero P, Mohammed BS, Avidan MS, Schechtman KB, Dence C, Gropler RJ. Impact of gender on the myocardial metabolic response to obesity. JACC Cardiovasc Imaging. 2008;1(4):424–33.
52. Klein SL, Flanagan KL. Sex differences in immune responses. Nat Rev Immunol. 2016;16(10):626–38.
53. Li S, Gupte AA. The role of estrogen in cardiac metabolism and diastolic function. Methodist Debakey Cardiovasc J. 2017;13(1):4–8.

# Regulation of Tissue Perfusion and Exchange of Solutes, Macromolecules, and Water Between Blood Vessels and the Interstitial Space

*Manuel Salzmann, Diethart Schmid, and Margarethe Geiger*

© Springer Nature Switzerland AG 2019
M. Geiger (ed.), *Fundamentals of Vascular Biology*, Learning Materials in Biosciences,
https://doi.org/10.1007/978-3-030-12270-6_3

**What You Will Learn in this Chapter**

In this chapter you will learn how the human body regulates the supply of blood and its components to different organs and tissues in such a way that every tissue gets exactly what is needed for its function in different situations. In general, blood supply can be regulated by increasing or decreasing the total cardiac output as well as by locally increasing or decreasing the tissue perfusion. Blood transports substances to the sites where they are needed (e.g., nutrients and oxygen to tissues) or removes substances from tissues for disposal (e.g., uric acid and carbon dioxide). At these sites substances therefore have to cross the vascular barrier to reach the tissue cells or the blood, respectively. In general, this happens at the smallest blood vessels, the capillaries. You will learn how blood supply to these capillaries is locally regulated and by which mechanisms the different blood components can be transported though the vessel wall. You will also see that the described mechanisms may differ between different organs/tissues, and you will be provided with examples on how they can be adapted according to actual needs. Furthermore, you will learn that in some organs, the circulatory system fulfills special functions relevant for the whole organism (e.g., urine formation in the kidney).

## 3.1 Introduction

The morphological and functional characteristics of the heart and the blood vessels have been described in detail in ▶ Chaps. 1 and 2, respectively. This chapter deals with the regulation of tissue perfusion, in other words with mechanisms how the body distributes blood to different organs/tissues to meet their actual needs. The needs of the organs for oxygen and nutrients are not constant but depend, e.g., on the activity of the organ. A typical example for an organ that needs much more energy and therefore much more blood supply when active, is the skeletal muscle. At physical exercise blood supply to the muscles increases dramatically. This cannot be accomplished simply by increasing the total cardiac output but also needs redistribution of blood from organs with lower needs to the working skeletal muscles.

Blood vessels of the different organs and tissues in the body are arranged in parallel. In general, this allows regulation of *regional perfusion* in an organ−/tissue-specific manner independent from the rest of the body. To understand how the body distributes blood to different parts of the body, a few physical principles have to be considered (see ▶ Sect. 3.2). It should also be mentioned that blood vessels do not have the same characteristics in all organs, e.g., they may be differently equipped with receptors for substances mediating vasodilation or vasoconstriction. This may be related to differences in the embryonic origin of vascular smooth muscle cells as outlined in ▶ Chap. 5 by J. Wojta. The basic principles of the regulation of *blood vessel diameter* are discussed in ▶ Sect. 3.3. Also the *permeability of the capillary wall* is not the same throughout the body. For example, capillaries in the brain are very tight, whereas capillaries in the liver have wide "pores," which allow even proteins to get out of the blood vessels and to reach the liver cells. The permeability of the capillaries and the transport of gases, fluid, solutes, and macromolecules through the capillary wall will be discussed in ▶ Sect. 3.4. In ▶ Sect. 3.5 we describe examples of some organs, in which perfusion fulfills special functions in addition to providing oxygen and nutrients to the tissues.

We would like to emphasize that the purpose of this chapter is to give an introduction to basic principles in the physiology of blood vessels. It therefore simplifies very complex mechanism with the aim to make readers who don't have a physiological background aware of these principles. For deeper understanding and knowledge, readers are referred to current textbooks of physiology [1–3].

3

## 3.2 Physics: Some Basic Principles Relevant for the Circulatory System

Although blood vessels and their content (blood) are more complex than ordinary pipes filled with water, some general physical principles apply to blood vessels as well. Some of these principles and their relevance for the regulation of local blood supply are discussed in this section. As shown in ▢ Fig. 3.1, the blood flow through a vessel is proportional to the *pressure gradient (ΔP)* along the vessel and indirectly proportional to the *resistance (R)*. Since the resistance is inversely proportional to the 4th power of the radius, also the blood flow increases in proportion to the 4th power of the vessel radius. This implies that minor changes in vessel diameter result in large changes of perfusion through this vessel.

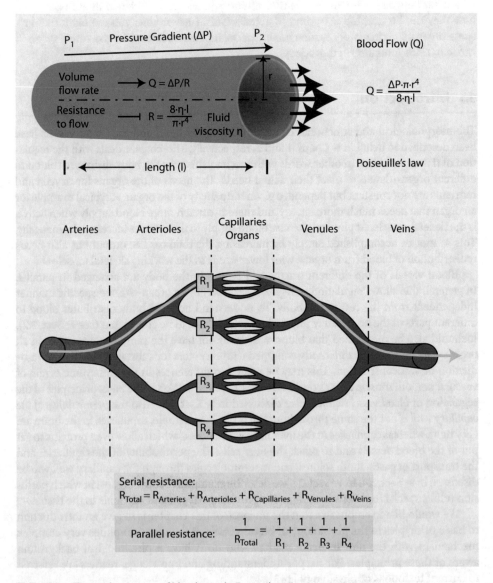

▢ **Fig. 3.1**   Flow and resistance of blood vessels. For explanations see text

By decreasing or increasing the diameter of blood vessels, the body can redistribute blood to those organs/tissues that are in current need. Decreasing or increasing vessel diameter is mediated by contraction or relaxation of vascular smooth muscles, respectively.

> **Box 3.1**
> The blood flow is proportional to the fourth power of the vessel radius. This means that small changes of the vessel diameter have a dramatic effect on the blood flow

As can be seen from ◻ Fig. 3.1 (upper part), the resistance that a blood vessel opposes to blood flow depends on its radius, its length, and on the viscosity of the blood inside the vessel. Therefore, also the blood flow through a vessel depends on these parameters. This holds for a single vessel, but what about the whole network of blood vessels in the body and the resistance of this network? When blood vessels are connected in series, their total resistance is the sum of the resistances of each individual vessel as shown in ◻ Fig. 3.1 (lower part, equation underlaid in light green). When vessels are arranged in parallel – as are the vessels supplying blood to the different organs – the total resistance is lower than the lowest individual resistance of a single vessel as shown in ◻ Fig. 3.1 (lower part, equation underlaid in light blue). This equation also implies that the more vessels are involved in forming a parallel network, the lower the total resistance of the network. In the body the microvasculature forms a large network of parallel vessels.

Another parameter to consider is the *flow velocity* in different parts of the circulatory system. Since the flow (volume/time) has to be the same at all cross sections of the vascular tree (e.g., the flow in aorta has to be the same as the flow in the collectivity of all perfused capillaries), the flow velocity decreases with increasing total cross-sectional area of the vascular bed. This fact is expressed in the so-called continuity equation: $Q_{total} = A_1 * v_1 = A_2 * v_2 = A_3 * v_3$ and so on, where $A_{1,2,3}$ are the cross-sectional areas and $V_{1,2,3}$ are the corresponding flow velocities at the different cross sections. Low flow velocity in capillaries offers optimal conditions for the exchange of substances between blood and tissues (see also ▶ Sect. 3.4).

**❓ Question 1**
Which increase in blood flow can you expect when the radius of a blood vessel is increased by 20%?

## 3.3 Regulation of Microvascular Perfusion/Blood Vessel Diameter

The *microvasculature* comprises arterioles (1st to 4th order), capillaries, and venules (1st to 4th order). The capillaries and small venules are the sites where gases, nutrients, water, signaling molecules, and others can be exchanged between blood and tissue cells. In some organs the microvasculature fulfills additional special functions such as urine formation in the kidney or gas exchange in the lungs (see ▶ Sect. 3.5). The cardiac output is distributed by the arterial blood vessels to the different organs. The regulation of blood supply according to the needs of the organs is a function of the microvasculature and is accomplished by smooth muscle cells of small arteries and arterioles. Smooth muscle cells of these vessels always exhibit a certain stage of active tension/contraction ("basal tone"). This tone can be increased or decreased by different stimuli, which thereby modulate the diameter of the vessel. As compared to arterial blood vessels, veins have very little basal tone, and therefore there is very little influence of *vasodilators* on the diameter of these vessels. Both, contraction and relaxation of *vascular smooth muscle cells* can be induced by a variety of stimuli. These stimuli can be neuronal, humoral (through circulating mediators), or mechanical. The common mechanism of vascular smooth muscle contraction is either an increase in intracellular, cytosolic $Ca^{++}$, which binds to calmodulin and triggers the activation of the myosin light-chain kinase (MLCK) and phosphorylation of the myosin light chain, or a direct activation of proteins mediating phosphorylation of the myosin light chain. Phosphorylation of the myosin light chain is necessary for its ATPase activity; and cleavage of ATP is necessary for the contraction itself. A few intracellular second messengers mediate contraction or relaxation of vascular smooth muscle cells depending on their concentrations. These are:

— Changes in the intracellular $Ca^{++}$-concentration ($\uparrow$, contraction; $\downarrow$, relaxation)
— Changes in intracellular cAMP-concentration($\uparrow$, relaxation; $\downarrow$, contraction)
— Changes in intracellular cGMP-concentration ($\uparrow$, relaxation; $\downarrow$, contraction)

The intracellular pathways triggered by the changes of these second messengers are described in detail by J. Wojta in ▶ Chap. 5. Some stimuli that activate one of these intracellular second messenger pathways and thereby cause *vasoconstriction* or *vasodilation* are summarized in ◻ Table 3.1.

As outlined in ◻ Fig. 3.1 (*Poiseuille's law*) a small change in diameter causes a dramatic change in the resistance and therefore in the blood flow to the organ/tissue served by these vessels. Although all blood vessels (except capillaries) have a layer of smooth muscle cells, smooth muscle cells of different arterial vessels don't behave in the same way. There are tonic vessels (e.g., aorta, efferent arterioles in the kidney) and vessels with a phasic phenotype (e.g., small resistance vessels, afferent arterioles in the kidney). In addition to these differences, there are also differences in the magnitude of response to vasoconstrictors/vasodilators.

Small arteries and arterioles are able to regulate their diameter and thereby the blood flow by themselves (*autoregulation*). This autoregulation can occur as a response of the vascular smooth muscle cells to intraluminal pressure changes (*myogenic autoregulation*) [4] or as a response to metabolic changes in the tissue surrounding the vessels (*metabolic autoregulation*) [5]. The susceptibility for autoregulatory stimuli and the strength of the response are not the same for every vascular bed.

> **Box 3.2**
> Note: Arteries and especially veins do not behave like rigid pipes. Their wall can be stretched when the pressure inside increases. On the other hand, vascular smooth muscle cells especially in arteries can react to increased pressure by contraction thereby decreasing the diameter and keeping the flow constant despite an increased pressure gradient

### 3.3.1 Myogenic Autoregulation

According to Poiseuille's law, the flow through a blood vessel depends on the pressure gradient across the vessel and its diameter. This means that the flow would increase, when the pressure gradient increases (and vice versa). The flow also increases, when

☐ **Table 3.1** Mediators causing contraction or relaxation of vascular smooth muscle cells

| Vasoconstrictors (in alphabetical order) | | | |
|---|---|---|---|
| Increased intracellular $Ca^{++}$ | | Decreased intracellular cAMP | |
| ADH (vasopressin) | | Catecholamines | (via alpha-2-adrenorecep-tors in some vessels) |
| Angiotensin II | | Neuropeptide Y (NPY) | |
| Catecholamines | (via alpha-1-adrenore-ceptors in most vessels) | | |
| Endothelin | | | |
| Serotonin[a] (if not acting on endothelial cells) | | | |
| Stretch | (via stretch-activated cation channels) | | |
| Thromboxane $A_2$ | | | |
| Vasodilators (in alphabetical order) | | | |
| Decreased intracellular $Ca^{++}$ | Increased cAMP | | Increased cGMP |
| Adenosine | Adenosine | | Atrial natriuretic peptide (ANP) |
| Atrial natriuretic peptide (ANP) | Catecholamines (epinephrine) | (via beta-2-adrenoreceptors in many vessels) | NO (NO is generated by various stimuli such as acetylcholine, bradyki-nin, serotonin[a] or shear stress in endothelial cells and diffuses to smooth muscle cells) |
| Prostacyclin $(PGI_2)$, prosta-glandin $E_2$ $(PGE_2)$ | Histamine | | |
| | Vasoactive intestinal peptide (VIP) | | |

[a]Serotonin can act both as a vasodilator and as a vasoconstrictor. This depends on the condition of the endothelial cells. In healthy vessels serotonin leads to a vasodilation because of a NO release by the endothelial cells. However, if endothelial cells are injured, the constricting effect on the smooth muscles cells outweighs the vasodilating effect

the mean hydrostatic mean pressure rises because of stretching of the vessel wall. With blood vessels this can be observed only during the first few minutes of pressure change. Thereafter the blood flow returns to the initial values. The reason is that smooth muscle cells contract. The result is a decrease in vessel diameter. This autoregulation of vascular diameter is an intrinsic property of vascular smooth muscle cells and is independent of autonomous innervation or the endothelium. This so-called *Bayliss effect* [named after the English physiologist William Bayliss (1860–1924)] guarantees maintenance of constant tissue perfusion at varying systemic blood pressure levels. In addition, it also prevents overstretching of vessels due to the influence of hydro-static pressure, especially in the arteries of legs and feet while standing. The molecular mechanism responsible for this effect includes stretch-induced smooth muscle cell depolarization, opening of voltage-gated $Ca^{++}$-channels, and increase in the intracellular $Ca^{++}$ concentration [6].

### 3.3.2 Metabolic Autoregulation

It can be observed in many tissues that an increased rate of energy metabolism of an organ/tissue is accompanied by increased perfusion (= increased blood flow). The best known example is the working skeletal muscle. There is a clear association between oxygen consumption and blood flow. Following are some metabolites/metabolic changes that have been shown to cause vasodilation: decrease in oxygen and intracellular ATP, increase in $CO_2$, [$H^+$], and extracellular [$K^+$]. To assure increased tissue perfusion, it is important that vasodilation does not only occur directly at the site where the metabolites accumulate but also upstream of this site. This upstream vasodilation is mediated by direct conduction along the vessel wall. Electronic conduction via the gap junctions of the smooth muscle cells seems to play a role in this upstream spread of vasodilation, but also the increased flow, which causes the release of vasodilators like NO and $H_2S$ from endothelial cells of upstream vessels, has been made responsible.

### 3.3.3 Autoregulation by Red Blood Cells

While it sounds reasonable that the demand for oxygen ($O_2$) is one of the major regulators of tissue perfusion, it is much less clear how decreased $O_2$ increases blood supply to tissues. Recently, there is increasing evidence that *red blood cells*, the $O_2$-transport vehicles in blood, most likely also act as $O_2$-*sensors*. In red blood cells, $O_2$ is reversibly bound to *hemoglobin*. Decrease in the $O_2$ concentration in peripheral tissues results in the release of $O_2$ from hemoglobin. Released $O_2$ freely diffuses through the plasma and the vascular wall to the tissues following its concentration gradient. Diffusion is not restricted to capillaries but also occurs in upstream vessels. As a consequence hemoglobin in red blood cells in the periphery of blood vessels is less saturated than hemoglobin in red blood cells in the central blood stream (◘ Fig. 3.2, left side). For hemodynamic reasons and under conditions of laminar flow, the concentration and velocity of red blood cells are higher in the center of the vessels than adjacent to the vessel wall. At bifurcations these phenomena (i.e., less red blood cells with lower velocity and lower oxygen saturation in the vicinity of the vessel wall) result in lower hematocrit and lower $O_2$-saturation in the vessel with the lower diameter (◘ Fig. 3.2, left side). Therefore vasoconstriction and vasodilation not only regulate the amount of blood supply to an organ/tissue but also the $O_2$-concentration in the delivered blood [7]. Decreased saturation of hemoglobin causes the release of ATP from red blood cells, which – together with its derivatives ADP and AMP – stimulates endothelial cells to produce and release NO [7, 8] (◘ Fig. 3.2, right side). In red blood cells, NO can also be released from N-nitrosohemoglobin when the oxygen saturation of hemoglobin is decreased [9]. As outlined below and as shown in ◘ Table 3.1, NO is a potent vasodilator. Therefore red blood cells themselves can increase the oxygen supply to areas of increased demand by increasing the perfusion of hypoxic tissues.

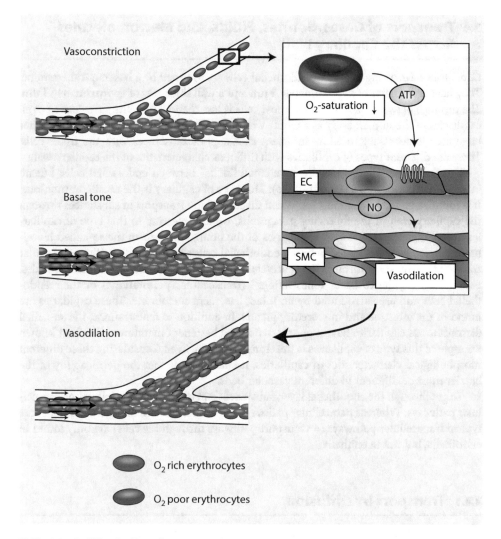

☐ **Fig 3.2**   Red blood cells mediate oxygen-dependent regulation of vascular diameter. For explanations see text. EC endothelial cells, SMC smooth muscle cells, the length of the arrows inside the blood vessels symbolizes the flow velocity profile

### 3.3.4  Regulation of Blood Vessel Diameter by Endothelial Cells

The vascular *endothelium*, the innermost layer of the blood vessels, is an important source of mediators for vascular smooth muscle cell contraction and relaxation: Some substances mediate vasodilation by releasing NO from endothelial cells, although their direct action on smooth muscle cells would cause contraction. For example, bradykinin, serotonin, ATP, and acetylcholine act in such a way. In addition to NO, there are also other substances released from endothelial cells that mediate relaxation of vascular smooth muscle cells and thereby vasodilation (e.g., prostacyclin). On the other hand, endothelial cells also secrete endothelin, which is a very potent vasoconstrictor.

## 3.4  Transport of Gases, Solutes, Fluids, and Macromolecules Across the Capillary Wall

*Capillaries* have a single layer of endothelial cells surrounded by a basement membrane. They have a diameter of approximately 5 µm and a wall thickness of approximately 1 µm. The distance between capillaries and most cells is less than 10 µm. The surface area of all capillaries of the human body as a whole would cover approximately 100 m². This rather huge area is necessary to allow sufficient exchange of substances with the tissue cells. There are different types of capillaries with different characteristics of the capillary wall:

Continuous capillaries usually have small "clefts" between endothelial cells, but no intracellular perforations (fenestrations). This type of capillary is the most common one. It is found, e.g., in skeletal muscles. When discussing the transport of substances through the capillary wall in the following paragraphs, we are referring to this type of capillaries unless indicated otherwise. Capillaries of the brain, which form the so-called blood-brain barrier, have no "clefts" between endothelial cells and do not allow the paracellular transport (= transport through the space/clefts between cells) of water-soluble molecules. Fenestrated capillaries have "fenestrations" (transcellular perforations) in their endothelial cells and are surrounded by an intact basement membrane. These capillaries are found in the intestine and in secretory glands. In addition to fenestrations, the so-called discontinuous capillaries have a poorly structured basement membrane. The best known example of this type of capillaries is the sinusoids in the liver. Considering these different morphological characteristics of capillaries, it became clear that the permeability of the barrier must be different in different vascular beds.

To get through the endothelial layer, substances can use either transcellular or paracellular pathways. Whereas paracellular pathways are similar in epithelia and endothelia, the typical transcellular pathways (= transport pathways through the cells) are only found in endothelia, but not in epithelia.

### 3.4.1  Transport by Diffusion

#### 3.4.1.1  Transport of Gases and Lipid-Soluble Substances

Gases, such as oxygen ($O_2$), carbon dioxide ($CO_2$), or nitric oxide (NO), and other lipid-soluble substances can use the whole surface of the capillary wall for diffusion, based on the fact that these molecules are apolar ($O_2$, $CO_2$) or almost apolar (NO) and small. There is general agreement that they diffuse easily through phospholipid membranes. Recently there is evidence that diffusion is not the only means of transport for gases through cellular membranes. At least in some epithelial cells and in red blood cells, gases also use so-called gas channels. Aquaporins and Rhesus proteins have been shown to function as channels for $CO_2$ and ammonia ($NH_3$), respectively (for review see [10, 11]).

Water-soluble substances have to use water-filled pores for crossing the capillary wall. These are mainly small paracellular clefts (~3 nm) between adjacent endothelial cells. The surface area covered by these water permeable pores is orders of magnitude lower than the surface area available for the diffusion of gases. Nevertheless small water-soluble molecules can diffuse easily through theses pores, and these molecules equilibrate rapidly between plasma and the interstitial space. Larger molecules (~10 nm or more) can be transported through so-called large pores by transcytosis through the endothelial cells.

## 3.4.2 Transport of Water and Small Solutes by Convection (Filtration and Reabsorption)

Water crosses the capillary wall mainly by convection. Water can use transcellular (via aquaporin 1) and paracellular routes. The concept of *fluid exchange (filtration and reabsorption)* at the capillary wall dates back to the late nineteenth century, when Ernest H. Starling developed the so-called Starling principle (Eq. 3.1, ◼ Fig. 3.3).

$$Q = K_f \left[ \left( P_c - P_i \right) - \sigma \left( \pi_p - \pi_i \right) \right]$$
(3.1)

$Q$: net flow of fluid

$K_f$: capillary filtration coefficient (characterizes permeability of the filtration membrane)

$P_c$: hydrostatic pressure in the capillary

$P_i$: hydrostatic pressure in the interstitial fluid

$\pi_p$: colloid osmotic pressure of plasma in the capillary

$\pi_i$: colloid osmotic pressure of interstitial fluid

$\sigma$: reflection coefficient (ability of membrane to prevent extravasation of solute particles)

Most physiology textbooks give the following values for fluid filtration and reabsorption in capillaries: filtration at the arterial limb of capillaries, ~ 20 l/day; reabsorption at the venous limb of capillaries, ~16–18 l/day; and back transport of fluid via the lymphatic system, ~2–4 l/day.

Until nowadays this Starling principle has been taught to students and is still current knowledge in many Physiology textbooks. However, in more recent experimental studies, it has been shown that under steady-state conditions, no reabsorption occurred upon lowering the hydrostatic pressure in the capillaries below the colloid osmotic pressure [12, 13]. To be in agreement with the estimated lymph flow (2–4 l/day), this would implicate that the total body filtration in "standard" capillaries must be quite below the originally estimated 20 l/day. The Starling principle takes into account hydrostatic pressures and colloid

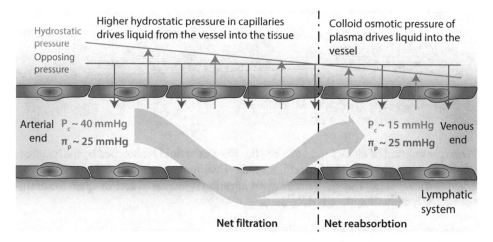

◼ **Fig. 3.3** Starling's concept of filtration and reabsorption at capillaries. According to this model, filtration occurs at the arterial end of the capillary. Most of the fluid is reabsorbed at the venous end of the capillary, and ~10% of the filtered fluid is transported back via the lymphatics. $P_c$ hydrostatic pressure in capillary, $\pi_p$ colloid osmotic pressure of plasma mainly brought about by plasma proteins

osmotic (=oncotic) pressures in the capillaries and in the interstitial fluid (as such, in total). What has not been considered in the Starling model is the *glycocalyx* covering the endothelial cells. There is now convincing evidence that this glycocalyx represents the true *filtration barrier*. The glycocalyx at the blood exposed side of endothelial cells is composed of proteoglycans and glycoproteins, which carry negatively charged glycosaminoglycans. Plasma components (e.g., proteins) are also bound to these structures. Together they form the actual filtration barrier to the "*subglycocalyx space.*" The tight junctions between the endothelial cells are the outside border of the subglycocalyx space. Therefore the endothelial filtration barrier is situated completely at the apical (blood) side of the endothelial layer. Therefore the "true" hydrostatic and colloid osmotic pressures outside the capillaries are the hydrostatic and colloid osmotic pressures of the subglycocalyx space (compartment) [14–16]. In a revised model of microvascular fluid exchange (◘ Fig. 3.4), it is assumed that reabsorption of fluid in the venous end of capillaries only occurs for a limited time and in situations such as hemorrhage, when the hydrostatic pressure in the capillaries is very low (decreased). This assumption has been explained as follows: Any reabsorption of fluid from the subglycocalyx space results in a lowering of hydrostatic pressure and in an increase of colloid osmotic pressure in the subglycocalyx space. This terminates reabsorption and promotes filtration. Recent experimental data suggest that most of the filtered fluid, which seems to be much less than assumed according to Starling's equation, is not reabsorbed by the capillaries but removed from the interstitium and returned to the vasculature mainly via the lymphatic vessels (see ▶ Chap. 1) [12, 13]. Therefore lymphatics play an essential role for the interstitial fluid homeostasis. The situation is different for capillary beds, where large amounts of fluid are regularly absorbed (e.g., in the gastrointestinal tract and in the kidney). Here large amounts of fluid are transported from epithelial cells to the interstitial space causing increased hydrostatic pressure in the interstitial space and favoring absorption of fluid by the capillaries.

**Question 2**
After a major hemorrhage activation of the sympathetic nervous system occurs, what is the effect on the fluid transport in capillaries?

**Box 3.3**
The glycocalyx of the endothelial cells together with blood components bound to this glycocalyx forms the filtration barrier. A pressure gradient (hydrostatic – colloid osmotic) between the intravascular space and the subglycocalyx space drives fluid filtration. The subglycocalyx space is the space between the glycocalyx and the structures forming the tight junctions between endothelial cells

When the amount of filtered fluid exceeds the amount of fluid reabsorbed by the lymphatics, fluid accumulates in the interstitial space. This fluid accumulation is called edema. Edema formation can therefore occur when filtration is increased and/or when the drainage via the lymphatic system is decreased.

**Box 3.4**
An edema is the accumulation of excess fluid in the interstitial space. Edema formation occurs when the amount of fluid filtered from the blood vessels exceeds the amount of fluid returned to the blood vessels by the lymphatics

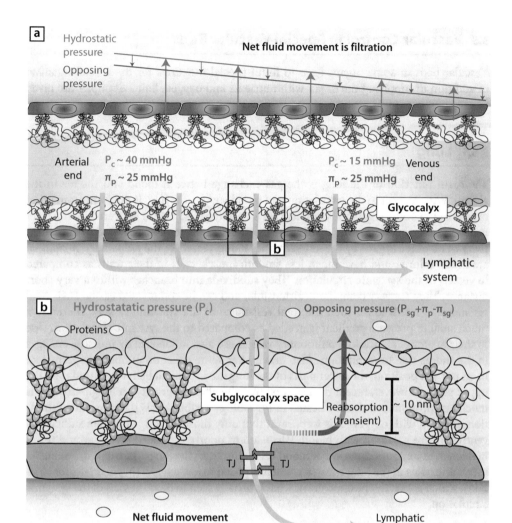

**Fig. 3.4** Revised concept of filtration and reabsorption at capillaries. Panel **a**: Filtration is less than assumed in the Starling model, and most of the filtered fluid is cleared away by the lymphatic system. Panel **b**: The "true" filtration barrier is the glycocalyx, which is located on the luminal surface of the endothelial cells and therefore completely inside the endothelial barrier. $P_c$ hydrostatic pressure in capillary, $P_{sg}$ hydrostatic pressure in subglycocalyx space, $\pi_p$ colloid osmotic pressure of plasma, $\pi_{sg}$ colloid osmotic pressure in subglycocalyx space, TJ thight junction with clefts allowing transport of fluid and small solutes

**? Question 3**

Edema is the accumulation of fluid in the interstitial space. What can lead to edema formation? Explain at least three mechanisms.

## 3.5  Vascular Control in Special Vascular Beds

Vascular beds in some organs have to fulfill special functions for the whole organism in addition to supplying the organ with nutrients and oxygen. Some examples are given below.

### 3.5.1  Lung

The main function of the lung is the gas exchange between blood and the air in the alveoli. With respect to the circulation and blood supply, the lung is different from other organs of the body. It gets the complete cardiac output (~5 l/min) delivered by the right ventricle of the heart, in other words the same amount of blood as delivered to the collectivity of all other organs by the left ventricle. Blood vessels of the *pulmonary circulation* are much thinner and have much less smooth muscle cells in their walls as compared to vessels of the systemic circulation. They subdivide into branches within a very short distance. They are very compliant (distensible) and can be compressed easily. Their total resistance is much lower than the total resistance of the systemic vessels. Therefore a much smaller pressure gradient is needed as compared to the systemic circulation. Due to their anatomical location, pulmonary vessels are also influenced by pressure changes in the thorax due to respiration. Also gravity has some effect on pulmonary perfusion: in an upright position, the lower parts of the lung are in general better perfused than the upper parts. However, recent data suggest that the effect of gravity on lung perfusion is much less pronounced than indicated in Physiology textbooks [17]. For optimal saturation of blood with oxygen, it is important that only those alveoli, which are sufficiently ventilated, are perfused with blood. Perfusion of poorly ventilated alveoli would result in decreased oxygen saturation of blood returning from the lungs. To fulfill this requirement, arterioles in the lung constrict, when exposed to low oxygen (*hypoxic vasoconstriction*), which is exactly the opposite of what is usually seen in blood vessels of the systemic circulation.

### 3.5.2  Kidney

The kidney gets approximately 20% of the cardiac output, i.e., ~1.2 l blood per min. Urine formation is achieved by filtration of a large amount of plasma in the capillaries of the glomerula and subsequent unregulated and regulated reabsorption of most of the solutes and water back to the circulatory system. The kidney has actually two capillary networks arranged in series: the first forms the *glomerular capillaries* and the second the *peritubular capillaries*. Blood supply to the capillaries as well as the pressure in the glomerular capillaries is kept constant over a wide range of systemic blood pressure. This is accomplished by special properties of the blood vessels: the inflow as well as the outflow from the glomerular capillaries occurs via arterioles. The diameter of these arterioles can be adjusted independently from each other. The glomerular capillaries themselves are very permeable to water and solutes smaller than albumin allowing filtration of ~180 l/day. Due to the high filtration rate, the colloid osmotic pressure in the post-glomerular vessels and in the second capillary network is higher than at the venous end of other capillaries. This facilitates the reabsorption of filtered fluid from the renal tubules into these capillaries.

### 3.5.3 **Skin**

The human skin plays an essential role in the regulation of the body temperature, since it is the major site of heat dissipation. With the circulatory system, heat is transported from the inner parts to the surface of the body. From the body's surface, excess heat can be emitted by radiation, conduction, and convection. Therefore the perfusion of the skin is mainly regulated by the requirements of body temperature control.

> **Take-Home Messages**
>
> — Blood supply to organs/tissues is regulated according to current (metabolic) needs.
> — Blood flow through a vessel is directly proportional to the pressure gradient along the vessel and indirectly proportional to the resistance opposed by the vessel. This resistance is indirectly proportional to the forth power of the vessel radius. Therefore small changes in vessel diameter cause dramatic changes in blood flow.
> — Blood supply to different organs/tissues is locally regulated by increasing or decreasing the diameter of blood vessels. This is achieved by relaxation or contraction of vascular smooth muscle cells, respectively.
> — Regulation of vascular smooth muscle contraction/relaxation can be triggered by neuronal stimuli from the autonomous nervous system (not described in this chapter), by mechanical stimuli (stretch), by local metabolites, by circulating mediators, and by the oxygen content of red blood cells.
> — Capillaries and small venules are the sites where substances can be exchanged between blood and interstitial tissue. Depending on the nature of the substance, passage through the vascular wall can be accomplished by diffusion and/or by convection (i.e., filtration and absorption).
> — Capillaries in different organs/vascular beds may have different permeability; e.g., capillaries (sinusoids) in the liver are very leaky, whereas capillaries in the brain are very tight.
> — In "standard" capillaries fluid and small solutes leave the vasculature by pressure-dependent filtration; fluid is returned to the blood vessels mainly via the lymphatic vessels. In "special" capillary beds (e.g., in the intestine or in the kidney), fluid is directly (re)absorbed from the interstitial space into the capillaries.
> — Blood vessels in some organs fulfill additional special functions, e.g., urine formation in the kidney, heat dissipation in the skin, gas exchange in the lung.

## 3.6 **Answers to Questions**

✓ Question 1: Ca. 100%

✓ Question 2: Extreme activation of the sympathetic nervous system causes strong vasoconstriction in the arterioles. This decreases the hydrostatic pressure in the capillaries, and fluid can be transiently reabsorbed from the interstitial space.

✅ Question 3: Edema formation occurs when the amount of fluid filtered in the capillaries exceeds the amount of fluid reabsorbed by blood and lymphatic vessels. Mechanisms: (i) increased filtration because of increased capillary pressure and/or decreased intravascular colloid osmotic pressure (e.g., plasma protein deficiency); (ii) decreased reabsorption due to lymphatic dysfunction or lymphatic overload; (iii) increased vascular permeability (e.g., in inflammation).

**3**

## References

1.  Boron WF, Boulpaep EL. Medical physiology. Philadelphia: Elsevier; 2017.
2.  Barrett KE, Barman SM, Boitano S, Brooks HL. Ganong's review of medical physiology. New York: McGraw-Hill Education; 2016.
3.  Raff H, Levitzky M. Medical physiology: a systems approach. New York: McGraw-Hill Medical; 2011.
4.  Jensen LJ, Nielsen MS, Salomonsson M, Sorensen CM. T-type Ca(2+) channels and autoregulation of local blood flow. Channels (Austin). 2017;11(3):183–95.
5.  Reglin B, Pries AR. Metabolic control of microvascular networks: oxygen sensing and beyond. J Vasc Res. 2014;51(5):376–92.
6.  Clifford PS. Local control of blood flow. Adv Physiol Educ. 2011;35(1):5–15.
7.  Ellsworth ML, Ellis CG, Goldman D, Stephenson AH, Dietrich HH, Sprague RS. Erythrocytes: oxygen sensors and modulators of vascular tone. Physiology (Bethesda). 2009;24:107–16.
8.  Ellsworth ML, Ellis CG, Sprague RS. Role of erythrocyte-released ATP in the regulation of microvascular oxygen supply in skeletal muscle. Acta Physiol (Oxf). 2016;216(3):265–76.
9.  Stamler JS, Jia L, Eu JP, McMahon TJ, Demchenko IT, Bonaventura J, Gernert K, Piantadosi CA. Blood flow regulation by S-nitrosohemoglobin in the physiological oxygen gradient. Science. 1997;276(5321):2034–7.
10. Boron WF. Sharpey-Schafer lecture: gas channels. Exp Physiol. 2010;95(12):1107–30.
11. Cooper GJ, Occhipinti R, Boron WF. CrossTalk proposal: physiological CO2 exchange can depend on membrane channels. J Physiol. 2015;593(23):5025–8.
12. Michel CC, Phillips ME. Steady-state fluid filtration at different capillary pressures in perfused frog mesenteric capillaries. J Physiol. 1987;388:421–35.
13. Zhang X, Adamson RH, Curry FE, Weinbaum S. Transient regulation of transport by pericytes in venular microvessels via trapped microdomains. Proc Natl Acad Sci U S A. 2008;105(4):1374–9.
14. Levick JR, Michel CC. Microvascular fluid exchange and the revised Starling principle. Cardiovasc Res. 2010;87(2):198–210.
15. Jacob M, Chappell D. Reappraising Starling: the physiology of the microcirculation. Curr Opin Crit Care. 2013;19(4):282–9.
16. Thind GS, Zanders S, Baker JK. Recent advances in the understanding of endothelial barrier function and fluid therapy. Postgrad Med J. 2018;94(1111):289–95.
17. Galvin I, Drummond GB, Nirmalan M. Distribution of blood flow and ventilation in the lung: gravity is not the only factor. Br J Anaesth. 2007;98(4):420–8.

# Endothelial Cells: Function and Dysfunction

*Rainer de Martin*

© Springer Nature Switzerland AG 2019
M. Geiger (ed.), *Fundamentals of Vascular Biology*, Learning Materials in Biosciences,
https://doi.org/10.1007/978-3-030-12270-6_4

**4**

**What You Will Learn in This Chapter**

About the role of nitric oxide (NO) that is generated enzymatically in EC and how it acts on smooth muscle cells (SMC) to cause vasorelaxation.

A general overview over acute and chronic inflammation, how the endothelium reacts to inflammatory stimuli, and how leukocytes transmigrate through the endothelium.

How the large number of genes with specific functions are (up)regulated in EC during the inflammatory response and how the reaction is terminated/resolved.

Due to its strategic localization on the interphase between the bloodstreem and the underlying tissue, it is evident that everything that needs to pass between these two compartments has to cross the endothelial cell barrier. This includes oxygen, nutrients, proteins, molecules of different size, and also cells of the immune system. It has been increasingly recognized that the endothelium, although only a monolayer of cells, is not just a passive inner lining of the blood vessels but controls very actively the passage of many of these agents. During the inflammatory response, the endothelium changes its properties fundamentally from a previously quiescent to an activated phenotype by synthesizing a wide variety of cytokines, chemokines, and adhesion molecules that mediate the attachment and transmigration of leukocytes. The same holds true for another important function, namely, the maintenance of an anticoagulant surface that is changed to procoagulant when necessary. A third main function is the regulation of the vascular tone through the generation of nitric oxide (NO) and other endothelium-derived vasoactive agents, which act on the underlying smooth muscle cells.

This chapter will focus on these functions, leaving apart others that are equally important such as proliferation, migration, death and survival, renewal through stem cells, angiogenesis and vasculogenesis, metastasis, homing of lymphocytes, or the heterogeneity between endothelial cells (EC) in arterial, venous, and lymphatic vessels, in different organs (e.g., brain, lung, kidney) and in tumors; these are covered by other chapters within this book.

## 4.1  Endothelial Function and Dysfunction: Nitric Oxide (NO) and the Control of the Vascular Tone

Although the endothelium fulfills many different functions (as described above), endothelial dysfunction is often and historically used in a more narrow sense to describe its disability to promote relaxation of smooth muscle cells. Several endothelium-derived factors can regulate the vascular tone, the most prominent being NO, the molecule of the year 1992 and the topic of the Nobel Prize in 1998. NO is generated from L-arginine that is oxidized to citrulline by the enzyme endothelial nitric oxide synthase (eNOS). EC also express an inducible form of NOS (iNOS) in response to pro-inflammatory stimulation [1]. NO as a diffusible gas can then act on SMC where it activates guanylate cyclase resulting in increased cGMP levels that further lead to activation of cGMP-dependent kinases, decrease of intracellular calcium, and activation of potassium channels, all resulting in SMC relaxation (◘ Fig. 4.1). Furthermore, it can s-nitrosylate cysteine residues in proteins, e.g., NF-κB, leading to dampening of their activity [2]. It is conceivable that NO would also act on target proteins at its site of generation, the EC. eNOS is activated by shear stress (in its laminar form), adenosine, serotonin, bradykinin, and also hypoxia (via VEGF induction). However, eNOS can switch under certain circumstances to produce reactive oxygen species (ROS), which have the opposite effects and result in EC activation

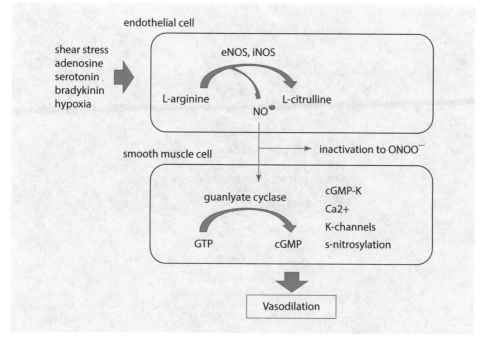

◻ **Fig. 4.1** Generation and functions of NO: In response to stimulation by different agents, nitric oxide synthases (NOS) are activated and generate NO through oxidation of L-arginine to L-citrulline. NO diffuses to SMC and causes relaxation through different pathways. Part may be inactivated by conversion to, e.g., peroxonitrite (ONOO⁻)

instead of silencing. This phenomenon is termed eNOS uncoupling [3]. Amyl nitrite and nitroglycerin are NO donors and have been used as drugs for short-term vasorelaxation.

Other vasoactive factors produced by the endothelium include prostacyclin (a vasodilator generated by the enzyme cyclooxygenase), and the vasoconstrictors endothelin, prostanoids, and angiotensin-converting enzyme (ACE) that cleaves angiotensin-I to -II. ACE inhibitors are widely used as drugs for the treatment of high blood pressure.

## 4.2 Inflammation

### 4.2.1 Acute Inflammation

The acute inflammatory response is essentially a beneficial reaction of the organism to various kinds of noxae. These can be either microbial (bacterial or viral) but also physical, chemical, or mechanical stimuli, e.g., UV or γ-irradiation, burn, frost bite, trauma, tissue necrosis, these latter being termed sterile inflammation. Inflammation has already been described by the ancient Romans who have introduced the four cardinal signs, namely, calor, rubor, tumor, and dolor (heat, redness, swelling, pain), that reflect mainly vascular effects (Aulus Cornelius Celsus, *De Medicina*, AD 25). ◻ Box 4.1 summarizes the vascular effects that happen during inflammation. Later, Virchow has added a fifth sign, functio laesa (loss of function). However, the response is not limited to the endothelium but includes a wide variety of other cells such as neutrophils, macrophages, mast cells, platelets, fibroblasts, soluble mediators secreted by these cells, and other structures like the

**4**

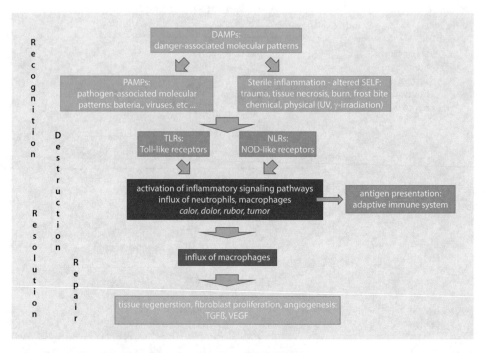

**Fig. 4.2**   Inflammation as the first line of defense. For details, see text

extracellular matrix (□ Fig. 4.2). The activation phase involves neutrophils that are first attracted to fight microbia through the generation of reactive oxygen species (ROS), followed by macrophages that engulf them and present antigens, a prerequisite for the adaptive immune reaction. At a later stage, they also remove neutrophils, and a repair process (resolution phase) begins that involves fibroblast proliferation and the expression of IL-10 and of TGFß for wound healing through promotion of, e.g., collagen synthesis, and the secretion of VEGF for the generation of new blood vessels [4].

---

**Box 4.1: Summary of Vascular Events During Inflammation**

**Vasodilation**
- Relaxation of the smooth muscle cells results in increased blood flow into the capillaries.
- Increased tissue perfusion causes redness (*rubor*) and warmth (*calor*). Warmth is also generated by increased metabolic activity of the involved cells.

**Vascular permeability**
- Capillaries become leaky, allowing more fluid (blood plasma) to exude into the connective tissue spaces.
- The fluid buildup is causing edema and is visible as swelling (*tumor*).

**Pain and/or itching**
- (*dolor*) is caused by the action of chemical agents (e.g., prostaglandins) released by different cells.

**Emigration of leukocytes**
- Expression of adhesion molecules on EC and secretion of chemoattractants promote the adhesion and transmigration of leukocytes into the inflamed tissue. A combination of vasodilation with thickening of the blood (due to fluid leaking out of the vessels) causes a slowing of the flow rate and helps leukocytes to stick to the sides of the vessels.

## 4.2.2 Chronic Inflammation

It is conceivable that such a powerful response as the inflammatory needs to be tightly controlled, since its destructive power may easily turn against structures of the organism. This is essentially what happens in chronic inflammation. Chronic inflammation is a hallmark of several diseases including those of the skin (psoriasis, atopic dermatitis), the intestine (Crohn's disease, ulcerative colitis), the joints (rheumatoid arthritis), the lung (asthma), the central nervous system (Alzheimer, multiple sclerosis), and last but not least the blood vessels (arteriosclerosis), as well as others. Unfortunately, it is very poorly understood how acute inflammation may turn into a chronic state and, above all, that chronic inflammation is more refractory to treatment as compared to its acute form. One simple explanation is that the initiating stimulus persists, is either not efficiently removed, or is constantly re-generated. The other possibility is that the resolution phase is not or inappropriately initiated or conducted, i.e., mechanisms of downregulation are failing. Such mechanisms exist in the endothelium and will be described below.

## 4.2.3 Endothelial-Leukocyte Interactions During Inflammation

At a site of infection, foreign structures such as bacterial lipopolysaccharide (LPS) or altered self are recognized by a variety of different receptors, e.g., the toll-like receptor (TLR) or nucleotide oligomerization domain (NOD) family, depending on the nature of the noxae. Danger signals are generated (interleukins, TNF or, in the case of sterile inflammation, high-mobility group binding protein B1 (HMGB1), a nonhistone DNA-binding protein released by necrotic cells) that act on nearby EC. In other settings such as hyperlipidemia or diabetes, EC can be directly activated by these or other stimuli, e.g., oxidized lipids, advanced glycation end products (AGE), or also by turbulent flow at certain areas of the vascular bed, e.g., bifurcations. EC respond by the expression of a plethora of cytokines, chemokines, adhesion molecules, and pro-inflammatory molecules (but also of others; see below) that lead to the attraction, binding, and transmigration of leukocytes into the underlying tissue and to their migration to the site of infection.

The first step in leukocyte-endothelial interaction is the so-called rolling, where leukocytes (e.g., neutrophils) adhere only lightly to the endothelium, detaching and reattaching to result in the name-giving phenomenon. An electron microscopic view of leukocytes attaching to the endothelium is shown in ◘ Fig. 4.3. It can be nicely observed through intravital microscopy, and a view of corresponding videos on YouTube or else is highly recommended (▶ http://www.youtube.com/watch?v=QCXpqT4_3bQ&feature=related). This initial low-affinity adhesion is regulated by two members of the selectin family of cell adhesion molecules, P-selectin and E-selectin. Whereas E-selectin is synthesized de novo and requires gene expression, P-selectin is released from Weibel-Palade bodies and is therefore available within minutes (but is also resynthesized). Selectins bind to their counter receptors PSGL-1 (P-selectin glycoprotein ligand) and ESL-1 (E-selectin ligand), respectively, on leukocytes through their N-terminal lectin homology domain. This domain recognizes specific carbohydrate structures (sialyl Lewis X) on their counter receptors.

This initial binding has effects on the neutrophil: engagement of the counter receptors as well as presentation of chemokines by EC evokes a structural change of integrins on leukocytes [6]. Integrins represent a large family, are composed of two non-covalently bound chains (α and ß), and link the extracellular matrix to the cytoskeleton. In this context, the ß2 integrins LFA-1 (lymphocyte function-associated antigen 1) and MAC-1

4

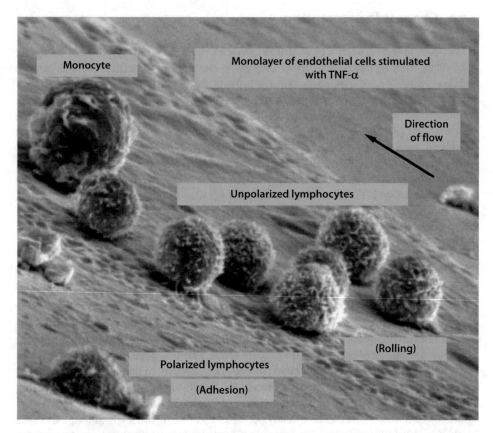

**Fig. 4.3** Lymphocytes bind to the surface of a high endothelial venule. (Reproduced by permission from [5]; © Sociedad Española de Cardiología (2009))

(macrophage 1 antigen) on neutrophils and macrophages, respectively, and the ß1 integrin VLA-4 (very late antigen) on leukocytes are operative. These structural changes of the integrins (inside-out signaling) lead to an increase in their affinity that enables binding to their counter receptors on EC, which are in this case ICAM-1, ICAM-2, and VCAM-1. These are members of the immunoglobulin family that are characterized by several repeats of their respective Ig-loop structures. ICAM-1 and VCAM-1 are inducibly expressed, whereas ICAM-2 is constitutively present on EC. These receptor pairs mediate the firm adhesion between leukocytes and EC. The different phases of rolling, adhesion, and transmigration are depicted in ◘ Fig. 4.4. Again, viewing a video animation is illustrative (e.g., ▶ https://www.youtube.com/watch?v=LB9FYAo7SJU).

The third step is the transmigration of the leukocyte through the endothelium. This is a complex multistep process that involves the sequential association, dissociation, and interaction of cell surface receptors but also intracellular proteins, as it is conceivable that such a process will require reorganization of the cytoskeleton of both the leukocyte and the endothelial cell. Again, it should be emphasized that diapedesis is from the view of the endothelium not a passive process where leukocytes kind of digest their way through the monolayer but that EC actively contribute.

The very first steps toward transmigration actually take place already during the process of rolling (◘ Fig. 4.5), since E- and P-selectin are coupled to cortactin, one of the

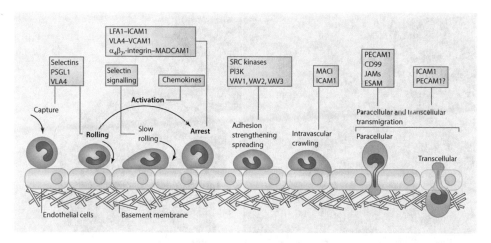

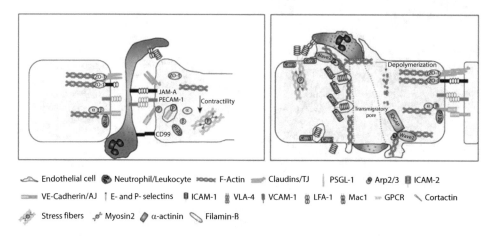

several actin-binding proteins (ABPs). ABPs have important functions by connecting to actin and thereby influencing cytoskeletal changes. Upon firm adhesion, more ABPs are recruited that bind also to the intracellular domains of ICAM-1 and VCAM-1. These include filamin B, α-actinin, and ERM proteins (ezrin/radixin/moesin) [8]. At this stage EC can extend membrane structures that engulf leukocytes that adhere and lead to the formation of so-called docking structures, cups, or domes [9–11] (□ Fig. 4.6). Likewise, neutrophils search the EC surface for "soft spots" that contain less actin filaments using "invadosome-like" protrusions [12]. Consecutively, signaling cascades (Ca2+, src, small GTPases Rho, Rac) are activated, leading to cytoskeletal remodeling. Other ABPs such as the cytoskeletal linker ZO-1 and α-catenin that connect tight junctions and adherens junctions to the cytoskeleton are disassembled, resulting in opening of these junctions,

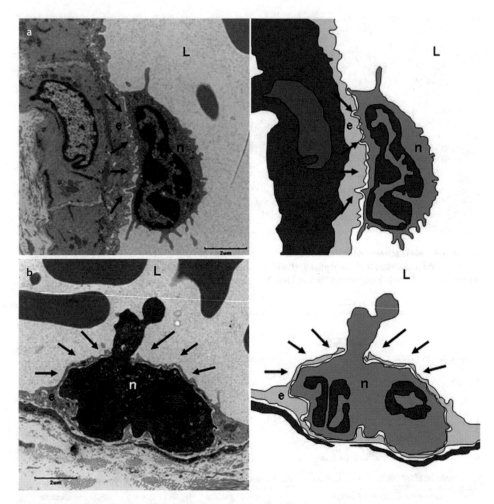

**Fig. 4.6** Endothelial "dome" formation through engulfment of adherent leukocytes by EC. Electron micrographs (left) and schematic representation (right). **a**, Crawling neutrophil with podosome-like protrusions (arrows) invaginating the endothelium. **b**, "dome"-like structures (arrows) in vivo may be formed from docking structures or transmigratory cups. Abbreviations: e endothelial cell, n neutrophil, L lumen, arrows depict EC that that engulf the neutrophil. (Reproduced by permission from [14]; ©The American Association of Immunologists, Inc. (2008))

including disassembly of the VE-cadherin-catenin complex. This is a prerequisite for several homo- and heterotypic interactions between leukocyte and EC surface receptors, including JAM-A, CD99, and PECAM-1 (platelet-endothelial cell adhesion molecule, CD31). The latter is expressed not only on these cells but also on macrophages, Kupffer cells, granulocytes, lymphocytes (T cells, B cells, and NK cells), megakaryocytes, and osteoclasts; like ICAM-1 and VCAM-1, it is a member of the immunoglobulin family. PECAM-1 signals to activate leukocyte integrins that share a common ß1 chain but also proteases (NE, neutrophil elastase; MMP-9, matrix metalloproteinase). These molecular interactions between leukocytes and EC and the formation of actomyosin stress fibers are thought to pull the leukocyte through the gap between the EC. Last but not least, leukocytes have to find also gaps between pericytes and to digest the basal membrane using the abovementioned proteases to move further into the tissue where they follow a chemotactic gradient toward the site of inflammation. Mechanisms to prevent vascular leakage need to be operative during transmigration and involve the expression of LSP-1 (lymphocyte-specific protein 1), a protein expressed on EC during dome formation [13].

However, in contrast to this pericellular diapedesis, leukocytes can also use an alternative route, namely, passage through the EC (transcellular diapedesis). Here, the initial steps such as formation of the transmigratory cup are similar; however, a transmigratory pore is then formed due to ICAM-1 being internalized within a ring of caveolin-1-enriched caveolae. Transmigrating cells extend protrusions into this pore and are transcytosed toward the basal side. It is estimated that this route is used only to a minor extent, but this may vary depending on pathophysiological settings such as barrier strength, which can in turn be influenced by histamine or TNF-induced stress fiber formation [15]; also, organotypic differences in EC may play a role. For further reading, see [8].

## 4.3 Molecular Mechanisms of Inflammation

Researchers have from very early on studied the gene expression repertoire of EC stimulated with TNF, IL-1, LPS, or other pro-inflammatory mediators and analyzed mRNA levels using differential screening, differential display, microarrays, or RNASeq depending on technical progress. As a result, several hundred genes (depending on where the threshold for induction is set) were found to be up- but also downregulated [16]. These include primarily the pro-inflammatory cytokines, chemokines, and their receptors (IL-1, IL-6, IL-8, CXC, and CCL families), as well as cell adhesion molecules (E-selectin, ICAM-1, VCAM-1) that mediate the chemoattraction and adhesion of leukocytes as described above. Also, genes related to coagulation such as tissue factor and plasminogen activator inhibitor (PAI-1) are found that relate to the switch from anti- to procoagulant properties during inflammation. However, also genes that mediate proliferation or the inhibition of apoptosis (programmed cell death) were discovered. During proliferation and migration, EC loosen their contact to other cells and the extracellular matrix (ECM) and are prone to a special form of cell death (anoikis, the Greek word for homelessness). Anti-apoptotic genes (IAP, inhibitors of apoptosis gene family) protect EC during this process. Proliferative genes include the immediate-early genes c-jun and c-fos; the growth factors GM-CSF, G-CSF, and to a lesser extent vascular endothelial growth factor (VEGF); as well as matrix metalloproteases that are required for ECM degradation. Also, SOX18, a transcription factor that is downregulated in this setting, falls into this category. SOX18 controls lymphatic vessel development but also vascular permeability and the expression of guidance molecules that are necessary for directed vessel growth. It might be that during inflammation, a less directed growth of vessels takes place, e.g., in solid tumors. Moreover, genes encoding other transcription factors are expressed, e.g., members of the nuclear hormone (NR4A) and early growth response (EGR) families, which evoke a secondary wave of gene expression. Taken together, upon pro-inflammatory stimulation, EC express not only pro-inflammatory genes as can be anticipated but also others that reflect distinct functions that are also necessary during this process, including proliferation, apoptosis, survival, cell and tissue dynamics (migration), and metabolism. A selection of relevant genes is given in ◘ Table 4.1.

**□ Table 4.1**    A selection of genes and associated functions (mostly) upregulated upon pro-inflammatory stimulation of EC

| | |
|---|---|
| Interleukins | IL-1, IL-6, IL-8, IL-18 |
| Chemokines and receptors | CXCL3, CX3CL1, CXCL2, CCL8, CXCL1, CXCL6, CCL5, CXCL5, CCL4, CCRL2 |
| Adhesion molecules | E-selectin, ICAM-1, VCAM-1 |
| Procoagulant molecules | Tissue factor, PAI-1 (plasminogen activator inhibitor) |
| Anti-apoptotic | IAP gene family (XIAP, cIAP1, cIAP-2) |
| Proliferation | GM-CSF, G-CSF; c-jun, c-fos, VEGF, matrix metalloproteinases |
| Transcription factors | Nr4a1/Nur77, Nr4a2/Nurr1, Nr4a3/Nor1; EGR1, −2, −3, ATF3, Sox18 |
| Cell and tissue dynamics | RND1/Rho6, tenascin, laminin C2 |
| Metabolism | PTGS2 (prostaglandin-endoperoxide synthase), SNARK/NUAK2 (stress-activated kinase involved in tolerance to glucose starvation), INSIG1 (insulin-induced gene 1), KYNU (kynureninase; catabolism of Trp and the biosynthesis of NAD cofactors), solute carrier family members |

### 4.3.1   Regulation of Inflammatory Gene Expression Through NF-κB

Transcription factors (TF) control gene expression through binding to specific recognition sites in the promoter region of genes. Through the years, NF-κB has emerged as a TF of central importance for inflammatory gene expression in EC, as the vast majority of these genes is dependent on this transcription factor. Although originally found in B cells (the name stands for "nuclear factor binding to the kappa-light-chain-enhancer of activated B cells"), it was soon found to be operative in many other cell types, particularly in cells of the immune system. In most cell types, NF-κB is an inducible factor that becomes active only upon demand and is shut down afterward. As such it is well suited to mediate transient pro-inflammatory gene expression. NF-κB consists of five family members that form homo- or heterodimers, RelA, RelB, c-Rel, NF-κB1 (p50), and NF-κB2 (p52), whereby the latter two are synthesized as inactive precursors (p105 and p100, respectively) that require proteolytic cleavage of their C-terminal domains for activation. Also, they do not contain a transactivation domain and act as repressors when forming homodimers but are active upon hetero-dimerization with, e.g., RelA; the most common combination is this RelA-NF-κB1 heterodimer. Inhibitory proteins of the IκB family, most notably IκBα, can also be considered part of the family.

#### 4.3.1.1  The Classical Pathway

The classical (canonical) pathway of NF-κB allows for very rapid activation, as the TF is preformed in the cytoplasm but prevented from translocation into the nucleus through complex formation with IκBα. Activation of the pathway is initiated through engagement

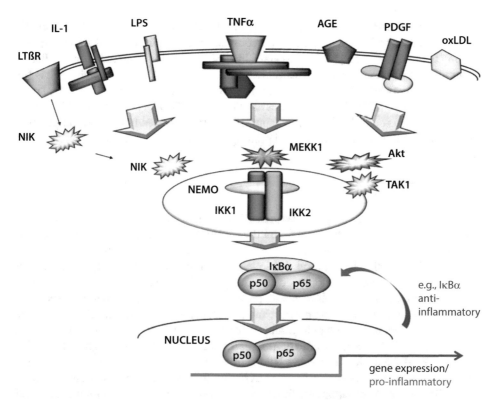

□ **Fig. 4.7** Basic scheme of the classical (canonical) NF-κB signaling pathway. In EC, engagement of different receptors leads through receptor-specific adapter proteins to MAP 3-type kinases (NIK, MEKK1, Akt, TAK1) which in turn activate the signalosome (consisting of IKK-1, IKK-2, and the regulatory subunit IKKγ/NEMO. IKKs further phosphorylate IκBα, thus targeting it for ubiquitination and proteasomal degradation. Thereby the NLS of the NF-κB RelA(p65)/p50 heterodimer is exposed, and the TF translocates to the nucleus where it mediates the expression of pro- but also anti-inflammatory genes

of receptors for, e.g., IL-1, TNF, or LPS, oxidized lipids, or advanced glycation end products (AGE), and proceeds through receptor-specific adapter molecules to a common structure, the signalosome. This consists of the two IκB kinases IKK-1 and IKK-2 (also designated IKK-α and IKK-ß) and a third subunit, the non-catalytical IKKγ (NEMO). The IκB kinases, mainly IKK2, phosphorylates IκBα on N-terminal serine residues, which is a signal for its K48-linked ubiquitination and degradation via the proteasome. Liberated from IκBα, the nuclear translocation signal of RelA is exposed, recognized by the translocation machinery, and the TF is transported to the nucleus where it can bind to its respective elements in the promoter region of pro-inflammatory genes (□ Fig. 4.7). Although this controlled nuclear translocation is the main step in the regulation of NF-κB, additional levels may apply through posttranslational modifications (phosphorylation, acetylation, s-nitrosylation) that modulate transactivation, DNA binding, or interaction with other factors such as histone deacetylases or the peroxisome proliferator-activated receptors (PPARs).

### 4.3.1.2 The Alternative Pathway

In contrast, the alternative (or noncanonical) pathway originates at receptors, e.g., CD40 or lymphotoxin-ß receptor, proceeds again through receptor-specific adaptors, but leads via NIK to IKK1. The regulatory step in this pathway is the stabilization of NIK, which is recruited by TRAF3 to a complex composed of TRAF2, TRAF3, and cIAP1 and cIAP2 (cellular inhibitor of apoptosis). The latter constitutively ubiquitinate NIK in resting cells leading to its degradation [17, 18]. Receptor ligation results in the degradation of TRAF3, accumulation of NIK, and subsequently IKK1 activation. IKK1 then phosphorylates the NF-κB2 precursor p100 leading to ubiquitin-mediated degradation of its C-terminal part, resulting in active NF-κB2 that associates usually with RelB to form the active TF. Of note, this process takes considerably longer (approx. 4 h) as compared to the classical pathway and would result in a more delayed and prolonged NF-κB activation. For further reading, see [19, 20].

### 4.3.1.3 Feedback Mechanisms for Pro-inflammatory Gene Expression

Termination of the inflammatory response is an important issue, since the prolonged maintenance of an "emergency state" and the continuous production of powerful mediators thereby would be detrimental to the organism. Again, this appears to be an active process and not a simple cessation due to, e.g., consumption of necessary molecules. In line with the complex activation process, multiple and partially redundant mechanisms are operative on different levels to ensure proper shut down the inflammatory response.

In the NF-κB signaling pathway, a main endogenous feedback mechanism is the NF-κB-dependent expression of its inhibitor IκBα. Newly synthesized IκBα, which occurs in EC within 45 min. after stimulation, replaces IκBα that had been degraded but also translocates to the nucleus to remove NF-κB from its DNA-binding site, and the complex can re-shuttle to the cytoplasm [21, 22]. A20, another NF-κB-dependent gene with an originally described anti-apoptotic function, has an additional role in ubiquitination and can interrupt the signaling pathway in two ways: First, it functions as a deubiquitinase to remove K63-linked (activating) ubiquitin chains from target proteins, e.g., RIP1, TRAF6, and IKKγ, and second, it acts as an ubiquitin-conjugating enzyme to attach K48-linked chains (in the case of RIP) leading to its degradation [23]. A third inducible gene is tristetraprolin (TTP, Zfp36) that mediates mRNA degradation through binding to AUUUA-rich elements in the 3'-untranslated region of certain mRNAs but also inhibits NF-κB at the level of nuclear translocation [24]. Last but not least, IKK2 activity, which is controlled by phosphorylation of serine residues in its activation loop, is at a later stage downregulated by hyperphosphorylation of its C-terminal HLH domain, thus inactivating a central component of the NF-κB signaling pathway [25].

However, several feedback mechanisms that target other pathways exist on the level of gene expression: a survey of IL-1-induced genes in EC revealed that several have demonstrated potential negative regulatory function (☐ Table 4.2). They illustrate that in addition to NF-κB, several other signaling pathways are activated (and required to be shut down). These include the MAP kinases that are dephosphorylated by DUSPs (dual-specificity phosphatases), the NF-AT pathway (nuclear factor of activated T cells) through inhibition of calcineurin by DSCR1, JAK-STAT through SOCS-1, and G-protein-coupled receptor signaling through RGS7. Moreover, they act on different levels of the signaling cascade, ranging from receptors to nuclear TFs. Examples are genes encoding EHD1 and ARTS1 that control endosomal sorting (recycling or degradation) and ectodomain shedding of receptors, thereby regulating their activity; TTP that was already mentioned above, which also regulates in its function as an mRNA destabilizing enzyme the levels of GM-CSF, COX2, and TNF mRNAs; and the expression of TFs like ATF3 and members of the NR4A family of nuclear hormone receptors that interfere with NF-κB function [26] (☐ Fig. 4.8).

▣ **Table 4.2** Negative feedback regulators. Genes with demonstrated or potential negative regulatory function that are induced by inflammatory stimulation in EC within 6 h are listed

| | |
|---|---|
| IκBα | Inhibitor of NF-κB |
| A20 | Inhibitor of NF-κB, inhibitor of apoptosis |
| Zfp36/TTP | Destabilization of specific ARE-containing mRNAs |
| DSCR1 | Inhibitor of NF-AT activation |
| C8FW (TRIB1) | Homologue of SINK and SHIK (involved in NF-κB p65 transactivation) |
| Etr101 | Inhibitor of several signaling pathways |
| DUSP1 | Dual-specificity phosphatase, inhibits MAPKs |
| DUSP5 | Dual-specificity phosphatase, inhibits MAPKs |
| SOCS1 | Suppressor of cytokine signaling (JAK-STAT, TLR pathways) |
| RGS7 | Regulator of G-protein signaling |
| ARTS-1 | Aminopeptidase, promotes TNFR1 ectodomain shedding |
| EHD1 | Endosomal sorting, recycling of growth factor receptors (?) |
| TNFRSF11B | Osteoprotegerin, soluble decoy receptor for RANKL |
| OATP-C | Na-independent transport of prostaglandin E2, thromboxane B2, leukotriene C3, leukotriene E4 (clearance of pro-inflammatory mediators?) |
| TCF8 | Transcriptional repressor in T cells |
| GG2–1 | Negative regulator of T cell signaling |
| TBX3 | Transcriptional repressor (in development) |
| ADAMTS1 | Disintegrin, metalloproteinase, anti-angiogenic properties, associated with inflammatory processes |
| TNFSF15 | Inhibits EC proliferation |
| Jag1 | Activates notch signaling, thereby represses EC proliferation via Rb |

Reproduced by permission from [26]; ©Georg Thieme Verlag KG (2007)

Finally, it should be mentioned that a mechanism of feedback regulation occurs on the level of prostaglandins (PGs): the synthesis of these essentially pro-inflammatory mediators such as PGE has been found to change at a later stage to anti-inflammatory cyclopentenone PGs (15dPGJ2) due to a switch in COX2 [27]; also, lipoxins and products of omega-3-polyunsaturated fatty acids (resolvins and protectins) are generated. As addressed in the introductory part, due to the involvement of many different cell types, also resolution of inflammation is not restricted to EC, and further reading is encouraged [28].

These observations, and the fact that activation-dependent genes with negative regulatory function are already expressed at a very early state, essentially concomitant with activation, have led to the concept of "the beginning programs the end" [4], i.e., that the stage for resolution is already set at an early stage of the activation process.

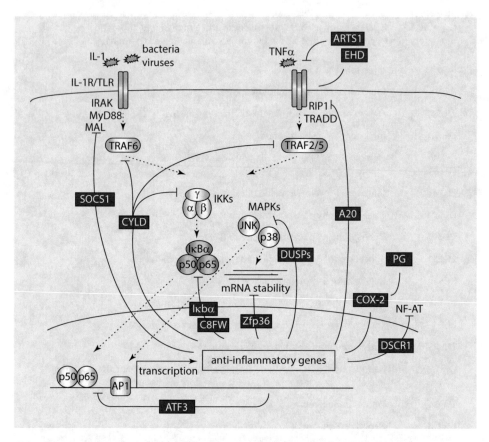

**Fig. 4.8** Endogenous autoregulatory feedback mechanisms in the endothelium. Genes with demonstrated or potential negative regulatory function that are expressed upon stimulation of human umbilical vein endothelial cells with IL-1 within 6 h are shown. See Table 4.2 for description. (Reproduced by permission from [26]; ©Georg Thieme Verlag KG (2007))

Take-Home Messages

- Nitric oxide is one of the main vaso-relaxing factors.
- It is generated in EC from L-arginine by two enzymes, the constitutively expressed eNOS and the inducible iNOS.
- It acts on SMC leading to vasorelaxation.
- Inflammation is a beneficial response that leads either directly or indirectly (through activation of the adaptive immune system) to elimination of the bacterial, viral, or other noxae; however, under poorly understood circumstances, it can turn into a chronic state that can lead to the destruction of body structures.
- Transmigration of leukocytes through the vessel wall is a key event during inflammation and involves a multitude of cell adhesion molecules, chemoattractants, and cytoskeletal changes; we distinguish pericellular and transcellular transmigration.
- Upon inflammatory stimulation, EC express genes encoding, e.g., cell adhesion molecules, cytokines and chemokines, proliferation, migration, survival, metabolism, and coagulation.

- NF-κB is a family of transcription factors of central importance for the expression of these genes; its activation is regulated mainly by nuclear translocation and follows a classical and an alternative pathway.
- Feedback mechanisms exist on multiple levels ranging from receptors and adapter proteins to signaling molecules and transcriptional regulators with negative regulatory function in order to shut down the inflammatory reaction at a later stage.

# References

1. Beck KF, Eberhardt W, Frank S, Huwiler A, Messmer UK, Muhl H, Pfeilschifter J. Inducible NO synthase: role in cellular signalling. J Exp Biol. 1999;202:645–53.
2. Stamler JS, Lamas S, Fang FC. Nitrosylation. the prototypic redox-based signaling mechanism. Cell. 2001;106:675–83.
3. Karbach S, Wenzel P, Waisman A, Munzel T, Daiber A. eNOS uncoupling in cardiovascular diseases–the role of oxidative stress and inflammation. Curr Pharm Des. 2014;20:3579–94.
4. Henson PM. Dampening inflammation. Nat Immunol. 2005;6:1179–81.
5. Barreiro O, Sanchez-Madrid F. Molecular basis of leukocyte-endothelium interactions during the inflammatory response. Rev Esp Cardiol. 2009;62:552–62.
6. McDowall A, Leitinger B, Stanley P, Bates PA, Randi AM, Hogg N. The I domain of integrin leukocyte function-associated antigen-1 is involved in a conformational change leading to high affinity binding to ligand intercellular adhesion molecule 1 (ICAM-1). J Biol Chem. 1998;273:27396–403.
7. Ley K, Laudanna C, Cybulsky MI, Nourshargh S. Getting to the site of inflammation: the leukocyte adhesion cascade updated. Nat Rev Immunol. 2007;7:678–89.
8. Schnoor M. Endothelial actin-binding proteins and actin dynamics in leukocyte transendothelial migration. J Immunol. 2015;194:3535–41.
9. Barreiro O, Yanez-Mo M, Serrador JM, Montoya MC, Vicente-Manzanares M, Tejedor R, Furthmayr H, Sanchez-Madrid F. Dynamic interaction of VCAM-1 and ICAM-1 with moesin and ezrin in a novel endothelial docking structure for adherent leukocytes. J Cell Biol. 2002;157:1233–45.
10. Carman CV, Springer TA. A transmigratory cup in leukocyte diapedesis both through individual vascular endothelial cells and between them. J Cell Biol. 2004;167:377–88.
11. Phillipson M, Kaur J, Colarusso P, Ballantyne CM, Kubes P. Endothelial domes encapsulate adherent neutrophils and minimize increases in vascular permeability in paracellular and transcellular emigration. PLoS One. 2008;3:e1649.
12. Carman CV. Mechanisms for transcellular diapedesis: probing and pathfinding by 'invadosome-like protrusions'. J Cell Sci. 2009;122:3025–35.
13. Petri B, Kaur J, Long EM, Li H, Parsons SA, Butz S, Phillipson M, Vestweber D, Patel KD, Robbins SM, Kubes P. Endothelial LSP1 is involved in endothelial dome formation, minimizing vascular permeability changes during neutrophil transmigration in vivo. Blood. 2011;117:942–52.
14. Petri B, Phillipson M, Kubes P. The physiology of leukocyte recruitment: an in vivo perspective. J Immunol. 2008;180:6439–46.
15. Martinelli R, Zeiger AS, Whitfield M, Sciuto TE, Dvorak A, Van Vliet KJ, Greenwood J, Carman CV. Probing the biomechanical contribution of the endothelium to lymphocyte migration: diapedesis by the path of least resistance. J Cell Sci. 2014;127:3720–34.
16. Mayer H, Bilban M, Kurtev V, Gruber F, Wagner O, Binder BR, de Martin R. Deciphering regulatory patterns of inflammatory gene expression from interleukin-1-stimulated human endothelial cells. Arterioscler Thromb Vasc Biol. 2004;24:1192–8.
17. Vallabhapurapu S, Matsuzawa A, Zhang W, Tseng PH, Keats JJ, Wang H, Vignali DA, Bergsagel PL, Karin M. Nonredundant and complementary functions of TRAF2 and TRAF3 in a ubiquitination cascade that activates NIK-dependent alternative NF-kappaB signaling. Nat Immunol. 2008;9:1364–70.
18. Zarnegar BJ, Wang Y, Mahoney DJ, Dempsey PW, Cheung HH, He J, Shiba T, Yang X, Yeh WC, Mak TW, Korneluk RG, Cheng G. Noncanonical NF-kappaB activation requires coordinated assembly of a regulatory complex of the adaptors cIAP1, cIAP2, TRAF2 and TRAF3 and the kinase NIK. Nat Immunol. 2008;9:1371–8.

19. de Martin R, Hoeth M, Hofer-Warbinek R, Schmid JA. The transcription factor NF-kappa B and the regulation of vascular cell function. Arterioscler Thromb Vasc Biol. 2000;20:E83–8.
20. Zhang Q, Lenardo MJ, Baltimore D. 30 Years of NF-kappaB: a blossoming of relevance to human pathobiology. Cell. 2017;168:37–57.
21. Arenzana-Seisdedos F, Thompson J, Rodriguez MS, Bachelerie F, Thomas D, Hay RT. Inducible nuclear expression of newly synthesized I kappa B alpha negatively regulates DNA-binding and transcriptional activities of NF-kappa B. Mol Cell Biol. 1995;15:2689–96.
22. de Martin R, Vanhove B, Cheng Q, Hofer E, Csizmadia V, Winkler H, Bach FH. Cytokine-inducible expression in endothelial cells of an I kappa B alpha-like gene is regulated by NF kappa B. EMBO J. 1993;12:2773–9.
23. Beyaert R, Heyninck K, Van Huffel S. A20 and A20-binding proteins as cellular inhibitors of nuclear factor-kappa B-dependent gene expression and apoptosis. Biochem Pharmacol. 2000;60:1143–51.
24. Schichl YM, Resch U, Hofer-Warbinek R, de Martin R. Tristetraprolin impairs NF-kappaB/p65 nuclear translocation. J Biol Chem. 2009;284:29571–81.
25. Rothwarf DM, Karin M. The NF-kappa B activation pathway: a paradigm in information transfer from membrane to nucleus. Sci STKE. 1999;1999:RE1.
26. Winsauer G, de Martin R. Resolution of inflammation: intracellular feedback loops in the endothelium. Thromb Haemost. 2007;97:364–9.
27. Rossi A, Kapahi P, Natoli G, Takahashi T, Chen Y, Karin M, Santoro MG. Anti-inflammatory cyclopentenone prostaglandins are direct inhibitors of IkappaB kinase. Nature. 2000;403:103–8.
28. Gilroy D, De Maeyer R. New insights into the resolution of inflammation. Semin Immunol. 2015;27:161–8.

4

# Vascular Smooth Muscle Cells: Regulation of Vasoconstriction and Vasodilation

*Johann Wojta*

© Springer Nature Switzerland AG 2019
M. Geiger (ed.), *Fundamentals of Vascular Biology*, Learning Materials in Biosciences,
https://doi.org/10.1007/978-3-030-12270-6_5

**What You Will Learn from This Chapter**

In this chapter you will learn about the diverse embryonic origin of vascular smooth muscle cell and their high degree of plasticity.

The essential contribution of vascular smooth muscle cells to maintaining the structural integrity of blood vessels and to regulating *vasodilation* and *vasoconstriction* will be discussed in detail. In particular you will learn how intracellular, cytosolic $Ca^{2+}$ levels modulate contraction and relaxation of vascular smooth muscle cells and which modulators and intracellular pathways are involved in regulating these cytosolic $Ca^{2+}$ levels.

You will also learn about the impact of dysfunctional vascular smooth muscle cells on the development of cardiovascular pathologies. Dysregulation of *vascular tone* by such dysfunctional vascular smooth muscle cells might cause *hypertension*, whereas a *phenotypic switch* from differentiated contractile vascular smooth muscle cells to a highly proliferative dedifferentiated phenotype might contribute to cardiovascular pathologies such as *atherosclerosis* and *restenosis* after stent implantation.



## 5.1 Introduction

Vascular smooth muscle cells, as integral part of the wall of blood vessels, are responsible for maintaining stability of the blood vessels. They are the predominant cell type in the tunica media, whereas the tunica intima is formed by a single layer of endothelial cells, and the tunica externa consists mainly of connective tissue (◘ Fig. 5.1).

Besides this role in determining vascular structure, the main function of vascular smooth muscle cells is the regulation of vascular tone and resistance and consequently the modulation of blood pressure through vasoconstriction and vasodilation. On a cellular level, vasoconstriction and vasodilation are dependent on the contraction and relaxation of vascular smooth muscle cells in the vascular wall, which on a molecular level depends on the concentration of cytosolic $Ca^{2+}$ whereby an increase in cytosolic $Ca^{2+}$ leads to contraction and a decrease results in relaxation of the respective smooth muscle cell. Influx of $Ca^{2+}$ into the cytosol and removal of $Ca^{2+}$ out of this cellular compartment are finely tuned and regulated by a large variety of biomolecules and intricate intracellular pathways. Dysfunction or dysregulation of such pathways and consequently impaired vasoconstriction or vasodilation might lead to vascular pathologies such as hypertension, which itself is a risk factor for the development of yet another vascular pathology, namely, atherosclerosis [1]. Under certain pathological conditions causing vascular injury, vascular smooth muscle cells can switch from a contractile phenotype to a highly proliferative migratory phenotype that is involved in the development of atherosclerotic lesions and in the phenomenon of restenosis after stent implantation [2].

In the following sections, the *developmental origin* of smooth muscle cells, the regulation of smooth muscle cell contraction and relaxation and the link between dysfunctional vascular smooth muscle cells and vascular pathologies will be discussed in detail.

## 5.2 Origin and Differentiation of Vascular Smooth Muscle Cells

Due to their different functional roles including vasodilation and vasoconstriction, proliferation in response to injury and matrix deposition in order to maintain the structure of the respective blood vessel, it is not surprising that vascular smooth muscle cells show

**Structure of the blood vessel**

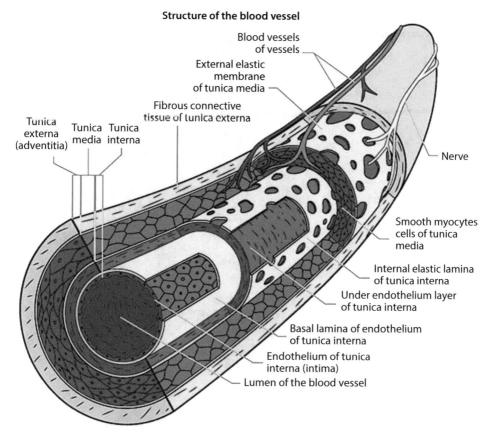

Blood vessels
of vessels

External elastic
membrane
of tunica media

Fibrous connective
tissue of tunica externa

Tunica
externa
(adventitia)

Tunica
media

Tunica
interna

Nerve

Smooth myocytes
cells of tunica
media

Internal elastic lamina
of tunica interna

Under endothelium layer
of tunica interna

Basal lamina of endothelium
of tunica interna

Endothelium of tunica
interna (intima)

Lumen of the blood vessel

☐ **Fig. 5.1**   Structure of the vessel wall

a high degree of plasticity and diversity in their developmental origin. Contrary to long held belief, vascular smooth muscle cells do not originate from a single type of precursor cell that is recruited to newly formed blood vessels from the surrounding mesenchyme. Recent evidence, however, suggest that during embryogenesis progenitors that ultimately develop into mature vascular smooth muscle cells are recruited from various regions of the embryo. During embryogenesis vascular smooth muscle cells develop from various mesodermal tissues such as the somatic or paraxial mesoderm, the lateral plate mesoderm and the splanchnic mesoderm but also from the ectoderm such as the neural crest (☐ Fig. 5.2) [3, 4].

This different embryonic origin of vascular smooth muscle cells results in a mosaic pattern of differentiation in the vascular tree. In particular evidence gathered from various animal models showed that vascular smooth muscle cells in the wall of the branchial arch originate from precursors developing in the neural crest of the embryo [6, 7]. The vascular smooth muscle forming the wall of the dorsal aorta are derived from cells that migrate from somites to that part of the aorta, whereas the smooth muscle cells of the abdominal aorta and the main arteries supplying the lower limbs are derived from cells of the splanchnic mesoderm [5, 8]. The smooth muscle cells of the coronary arteries originate from the proepicardium. However, it was shown in mice recently that postnatal coronary arteries

5

**◘ Fig. 5.2** Developmental fate map of vascular smooth muscle cells [5]

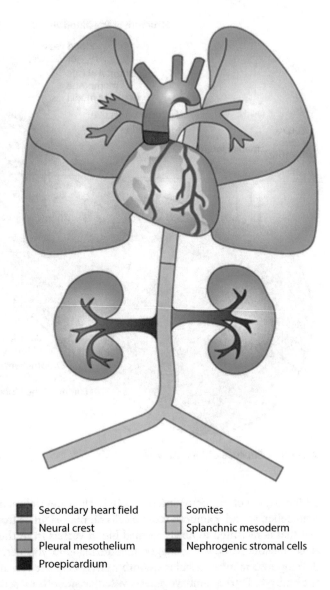

■ Secondary heart field    ☐ Somites
■ Neural crest             ☐ Splanchnic mesoderm
☐ Pleural mesothelium      ■ Nephrogenic stromal cells
■ Proepicardium

also develop de novo [9]. Finally, vascular smooth muscle cells in the blood vessel of the kidney originate from nephrogenic stromal cells during embryonic development, whereas vascular smooth muscle cells of the blood vessels in the lung develop from precursor cells from the pleural mesothelium [5, 9].

## 5.3    Regulation of Vascular Tone

Contraction and relaxation of vascular smooth muscle cells and consequently vasoconstriction and vasodilation are on a cellular level brought about by interactions of the thick filaments composed of myosin and the thin filaments composed of actin, tropomyosin, caldesmon (CaD) and calponin (CaP). Their interaction is mainly regulated by changes in the concentration of cytosolic $Ca^{2+}$, whereby an increase to 1 µM $Ca^{2+}$ leads to contraction and a decrease to 0.1 µM results in relaxation of the smooth muscle cells. To initiate contraction, $Ca^{2+}$ entering the cytosol binds to calmodulin (CaM) and induces a conformational change in this protein so that CaM can bind 4 $Ca^{2+}$. This $Ca^{2+}$-CaM complex activates myosin light chain kinase (MLCK) which subsequently phosphorylates the regulatory 20kD regulatory light chain of myosin ($MLC_{20}$) on serine 19. This enables myosin adenosine triphosphatase (ATPase) to be activated by actin and cyclic binding of the head of myosin to actin and tilting of the myosin head which ultimately transforms the chemical energy gained from cleavage of adenosine triphosphate (ATP) to adenosine diphosphate (ADP) into movement and mechanical force. When the concentration of $Ca^{2+}$ falls below 1 µM, $Ca^{2+}$ dissociates from CaM and the latter from MLCK which results in inactivation of MLCK. Now myosin phosphatase (MP), which in its activity is independent of $Ca^{2+}$, dephosphorylates $MLC_{20}$. This results in inactivation of the myosin ATPase and in relaxation of the smooth muscle cell (◘ Fig. 5.3) [10].

Interestingly, in contrast to skeletal muscle, smooth muscle cells can maintain high contractile force at low expense of ATP. This so-called "latch-bridge" phenomenon was thought to be brought about by dephosphorylation of myosin generating a slow cycling of

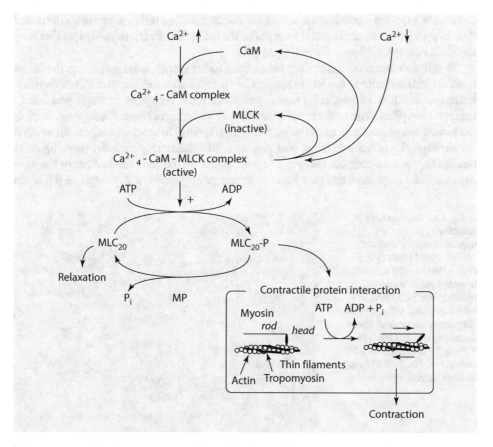

◘ **Fig. 5.3** Role of $Ca^{2+}$ in the regulation of contraction of vascular smooth muscle cells [11]. Abbreviations: ADP adenosine diphosphate, ATP adenosine triphosphate, CaM calmodulin, $MLC_{20}$ 20kD regulatory light chain of myosin, MLCK myosin light chain kinase, MP myosin phosphatase

myosin-actin interactions to maintain contraction force [12]. However, the exact molecular basis of the "latch-bridge" mechanism remains to be fully understood and might involve enhanced myofilament force sensitivity even in the absence of increased cytosolic $Ca^{2+}$ or myosin light chain (MLC) phosphorylation.

The concentration of $Ca^{2+}$ in the extracellular space is between 1 and 2 mM and thus several orders of magnitude higher than cytosolic $Ca^{2+}$ in vascular smooth muscle which reaches 0.1 µM in the resting state. $Ca^{2+}$ is highly water soluble, and therefore its diffusion through the lipid bilayer of the cell membrane into the cytosol is negligible despite a steep concentration gradient. In case of activation, excitable $Ca^{2+}$ channels in the plasma membrane open and allow influx of $Ca^{2+}$ into the cytosol where a concentration >1 µM has to be reached to initiate contraction. In the resting state, however, the concentration of $Ca^{2+}$ is kept well below this level by plasma membrane calcium ATPase (PMCA) and a $Na^+$/$Ca^{2+}$ exchanger (NCX) in the plasma membrane of the smooth muscle cell. Furthermore, sarcoplasmic/endoplasmic reticulum ATPase (SERCA) actively transports $Ca^{2+}$ into the sarcoplasmic reticulum (SR) and through its connections to the SR also into the nucleus. There is also passive transport of $Ca^{2+}$ into the lumen of the mitochondria. However, the role of the mitochondria in $Ca^{2+}$ homeostasis is considered to be minor in comparison to that of the SR (■ Fig. 5.4). It should be emphasized, however, that under physiological conditions vascular smooth muscle cells are not in a resting but in a partially contracted state in order to maintain vascular tone under the influence of various mediators that keep the intracellular $Ca^{2+}$ levels above 0.1 µM [11].

To initiate vasoconstriction $Ca^{2+}$ influx from either intracellular stores, e.g. the SR, or from the extracellular space has to occur. $Ca^{2+}$ influx follows either membrane depolarization or binding of particular agonist to membrane receptors through pharmacochemical coupling. The majority of mediators of pharmacochemical coupling, such as the potent vasoconstrictors angiotensin II (AngII), norepinephrine, endothelin or ATP, act through the phosphatidylinositol pathway. To initiate this cascade, these agonists bind to $G_q$-protein-coupled receptors in the plasma membrane. Interaction of the agonists with the respective receptor leads to exchange of guanosine diphosphate (GDP) to

■ **Fig. 5.4** Mechanisms reducing cytosolic $Ca^{2+}$ concentrations in vascular smooth muscle cells [11]. Abbreviations: ADP adenosine diphosphate, ATP adenosine triphosphate, NCX $Na^+$/$Ca^{2+}$ exchanger, MTC mitochondrium, PM plasma membrane, PMCA plasma membrane calcium ATPase, SERCA sarcoplasmic/endoplasmic calcium ATPase, SR sarcoplasmic reticulum

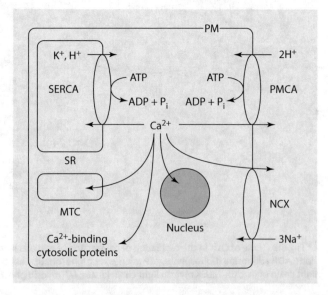

guanosine triphosphate (GTP) on the $\alpha$-subunit of the heterotrimeric $G_q$-protein and subsequent dissociation of the $\alpha$-subunit from the $\beta$- and $\gamma$-subunits. The $\alpha$-subunit then activates phospholipase C (PLC) which cleaves membrane-bound phosphatidylinositol-biphosphate ($PIP_2$) into diacylglycerol (DAG) and inositol-triphosphate ($IP_3$). The latter diffuses into the cytosol and induces release of $Ca^{2+}$ from the SR. DAG on the other hand activates protein kinase C (PKC) which phosphorylates CaP thereby allowing more actin to bind to myosin and also phosphorylates C-kinase potentiated protein phosphatase-1 inhibitor (CPI-17) which in turn inhibits MP. Thus, under these conditions, MP is not active and therefore unable to dephosphorylate the light chain of myosin to initiate relaxation [11].

It should be noted that the initial phase of rapid, phasic contraction of vascular smooth muscle cells is thought to be the result of the release of $Ca^{2+}$ from the SR in response to activation of the phosphatidylinositol pathway. A following tonic increase of $Ca^{2+}$ influx into the cytosol is believed to occur through agonist-activated $Ca^{2+}$ channels in the plasma membrane [11].

### 5.3.1 $Ca^{2+}$ Release from Intracellular Stores

The SR, which is an intracellular system of tubules, makes up to 7.5% of the volume of a vascular smooth muscle cell [13]. $Ca^{2+}$ efflux of this intracellular store is induced by $IP_3$ as described above but also by $Ca^{2+}$ itself through the interaction of the former with a specific receptor for $IP_3$, $IP_3$ receptor ($IP_3R$), and of the latter with the ryanodine receptor (RyR). Both receptors are located in the membrane of the SR and are second messenger-operated $Ca^{2+}$ channels (SMOCC) which open in response to interaction with their respective ligands [11].

The $IP_3R$ has binding domains for $IP_3$ and $Ca^{2+}$ and an aqueous pore for $Ca^{2+}$. Interestingly the $Ca^{2+}$ binding domain on the $IP_3R$ regulates a biphasic effect on the $IP_3$-induced $Ca^{2+}$ release from the SR: up to an increase in cytosolic $Ca^{2+}$ to 300 nM, a positive feedback is initiated enhancing $Ca^{2+}$ efflux from the SR, whereas cytosolic $Ca^{2+}$ concentrations of >300 nM lead to a negative feedback inhibiting $Ca^{2+}$ efflux from the SR [14].

Similar to the $IP_3R$, the RyR is a SMOCC with two known physiological agonists, namely, $Ca^{2+}$ and cyclic ADP ribose (cADP ribose). The RyR is also sensitive to caffeine and, hence the name, to the plant alkaloid ryanodine [14]. $Ca^{2+}$ efflux from the SR through activation of the RyR is suggested to occur once the local concentration of $Ca^{2+}$ near the SR rises above 3 $\mu$M. It is further thought that this initial increase in $Ca^{2+}$ concentration close to the SR is caused by activation of the $IP_3R$. It should also be noted that this $Ca^{2+}$-induced $Ca^{2+}$-release from the SR through activation of the RyR is other than in cardiac and skeletal muscle independent of $Ca^{2+}$ influx from the extracellular space (◻ Fig. 5.5) [13].

### 5.3.2 $Ca^{2+}$ Influx from the Extracellular Space

As diffusion of $Ca^{2+}$ through the lipid bilayer of the cell membrane is minimal, $Ca^{2+}$ influx large enough to activate the contraction machinery has to rely on various $Ca^{2+}$ channels located in the plasma membrane of vascular smooth muscle cells such as voltage-operated

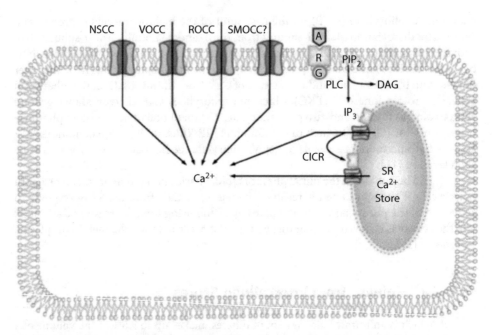

**◻ Fig. 5.5** $Ca^{2+}$ influx into the cytosol of vascular smooth muscle cells from the sarcoplasmic reticulum and from the extracellular space, modified from [13]. Abbreviations: A agonist, CICR calcium-induced calcium release, DAG diacylglycerol, G G-protein, $IP_3$ inositol-triphosphate, $PIP_2$ phosphatidylinositol-biphosphate, PLC phospholipase C, NSCC nonselective cation channel, R receptor, ROCC receptor-operated calcium channel, SMOCC second messenger-operated calcium channel, SR sarcoplasmic reticulum, VOCC voltage-operated calcium channel

calcium channels (VOCC), receptor-operated calcium channels (ROCC) and store-operated calcium channels (SOCC). SMOCC such as the $IP_3R$ and the RyR are considered a subgroup of ROCC as they are indirectly activated by diffusible second messengers.

VOCC respond to changes in the membrane potential. In vascular smooth muscle cells, the resting membrane potential ranges from −40 to −55 mV in vivo, which is within the activation range of VOCC. Thus, it is believed that $Ca^{2+}$ influx through these VOCC is crucial to maintain vasomotor tone [15]. Three types of VOCC have been identified in vascular smooth muscle cells, namely, "long-lasting", L-type channels, "transient", T-type channels and "resting", R-type channels.

The threshold for the L-type channels lies at approximately −40 mV, and full activation is achieved at a membrane potential of approximately 0 mV. The L-type channels are inactivated after 300–600 ms. They consist of an $\alpha_1$-, an $\alpha_2$-, a β-, a γ- and a δ-subunit, whereby the $\alpha_1$-subunit has specific sites sensitive for phosphorylation by, e.g. PKC, a voltage sensor and gates for activation and deactivation [16]. Deactivation of the L-type channels depends on voltage and the concentration of $Ca^{2+}$. L-type channels are the most numerous VOCC in vascular smooth muscle cells and are thus considered to be the most important channels for $Ca^{2+}$ influx from the extracellular space into the cytosol. They might also play a role in replenishing intracellular stores, e.g. in the SR. The T-type channels are activated at a membrane potential of around −30 mV and are inactivated voltage-dependently quickly after 20–60 ms [17]. It has been suggested that influx of extracellular $Ca^{2+}$ through these channels induces release of $Ca^{2+}$ from the SR via the RyR. The R-type channels are thought to be responsible for a continuous slow

$Ca^{2+}$ influx into the cytosol of vascular smooth muscle cells. However, the significance of their contribution to contraction of vascular smooth muscle cells has not yet been clarified.

Several mediators of vasoconstriction such as AngII, norepinephrine, endothelin, serotonin, or ATP have been shown to activate nonselective cation channels (NSCC) and by doing so are thought to contribute to membrane depolarization and thus activation of VOCC [16]. Although the existence of SMOCC sensitive to $IP_3$ or $Ca^{2+}$ in the plasma membrane of vascular smooth muscle cells has been proposed, evidence for this notion is limited [16].

SOCC in the plasma membrane of vascular smooth muscle cells are responsible for the so-called capacitative $Ca^{2+}$ entry from the extracellular space to replenish $Ca^{2+}$ when intracellular stores in the SR are depleted. However, the exact molecular mechanisms regulating this process are still unclear [18].

Vascular smooth muscle cells are capable of fine-tuning peripheral blood flow through a process called autoregulation: They respond to an increase in intravascular pressure with an increase in contractile force, whereas a decrease in intravascular pressure induces smooth muscle cell relaxation. It has been shown that stretch induces influx of $Ca^{2+}$ from the extracellular space in the cytosol through activation of VOCC and mechanosensitive NSCC in the cell membrane and enhances $Ca^{2+}$ sensitivity of myofilaments (◘ Fig. 5.5) [16].

### 5.3.3 Removal of $Ca^{2+}$ from the Cytosol

Two $Ca^{2+}$-ATPases are responsible for pumping $Ca^{2+}$ from the cytosol of vascular smooth muscle cells into the extracellular space and into the SR, respectively, namely, PMCA, located in the plasma membrane of the cell, and SERCA present in the membrane of the SR [19].

PMCA is an electroneutral pump as it pumps one $Ca^{2+}$ into the extracellular space in exchange for 2 $H^+$ pumped into the cytosol. The cytosolic part of PMCA has binding sites for CaM and for PKC, cyclic adenosine monophosphate (cAMP)-dependent protein kinase (PKA) and cyclic guanosine monophosphate (cGMP)-dependent protein kinase (PKG) [19]. PMCA is activated directly by CaM binding to its binding site, whereas the pump is inhibited in the absence of CaM. In addition, phosphorylation of the CaM binding site by PKA, PKC or PKG also results in activation of PMCA (◘ Fig. 5.4) [19].

SERCA is also electroneutral as it pumps 1 $H^+$ and 1 $K^+$ into the cytosol in exchange for each $Ca^{2+}$ moved into the SR [20]. SERCA does not have a CaM binding site and thus is not activated by CaM. Instead phospholamban, a transmembrane protein showing considerable sequence homology with CaM and located in membrane of the SR, is regulating the activity of SERCA. When not phosphorylated, phospholamban inhibits SERCA, whereas once phosphorylated by PKA, PKC or PKG, phospholamban activates SERCA [14]. The phosphorylation of phospholamban by PKG is thought to contribute to vasodilation in response of modulators such as NO, epinephrine or atrial natriuretic peptide (ANP) which increase cytosolic levels of cGMP (◘ Fig. 5.4) [19].

As already mentioned above, vascular smooth muscle cells also express an NCX which is located in the plasma membrane that transports 1 $Ca^{2+}$ out of the cytosol in exchange for 3 $Na^+$. It is driven by the steep inward directed concentration gradient for $Na^+$ that is maintained by the $Na^+/K^+$ATPase. However, the physiological relevance of NCX for regulating cytosolic $Ca^{2+}$ is still unclear (◘ Fig. 5.4) [14].

### 5.3.4 Regulation of Vascular Tone by Modulating Ca²⁺ Sensitivity of Myofilaments

In addition to heterotrimeric G-proteins, such as the $G_q$-protein described above, also small monomeric G-proteins exist. Ras homolog gene family member A (RhoA) is such a small GTPase that is involved in modulating $Ca^{2+}$ sensitivity of myofilaments in vascular smooth muscle cells. Thereby RhoA activates rho-associated protein kinase (ROCK) which in turn phosphorylates the myosin phosphatase target subunit (MYPT) of MP. Once MYPT is phosphorylated by ROCK, MP is rendered inactive. Consequently, $MLC_{20}$ phosphorylation and secondary to that contraction of the vascular smooth muscle cell are enhanced. ROCK also phosphorylates CPI-17, which once phosphorylated becomes active and inactivates MP by inhibiting its catalytic subunit, termed catalytic subunit of type 1 phosphatase δ (PP1Cδ) [21, 22].

It should be emphasized that arachidonic acid, which is cleaved off phospholipids of the plasma membrane by phospholipase $A_2$ ($PLA_2$) under, e.g. inflammatory conditions, can impact upon vasoconstriction by activating ROCK independently of RhoA [23].

As described above, several potent vasoconstrictors such as AngII, endothelin or norepinephrine by binding to a $G_q$-protein-coupled receptor activate PLC which in turn cleaves $PIP_2$ into $IP_3$ and DAG. DAG then activates PKC which phosphorylates and activates the inhibitor of MP, CPI-17. In addition, PKC phosphorylates caldesmon and CaP thereby allowing more binding interactions between myosin and actin [24].

### 5.3.5 Regulation of Vascular Tone by Cyclic Nucleotides

The cyclic nucleotides cGMP and cAMP are considered to be the major mediators of relaxation of vascular smooth muscle cells. Increased cytosolic levels of cAMP and in particular of cGMP therefore are critically involved in vasodilation.

cGMP levels are increased by the activation of either membrane-bound guanylyl cyclase by *natriuretic peptides* such as ANP, brain natriuretic peptide (BNP) or C natriuretic peptide (CNP) or by the activation of soluble guanylyl cyclase by NO or CO [25]. cGMP subsequently activates PKG which phosphorylates phospholamban. Phosphorylated phospholamban activates SERCA which pumps $Ca^{2+}$ from the cytosol into the SR as described above. PKG increases $Ca^{2+}$ efflux into the extracellular space by phosphorylating and activating PMCA. Furthermore, PKG blocks the release of $Ca^{2+}$ from the SR through phosphorylation of $IP_3R$ and inhibition of $IP_3$ synthesis and activates K⁺ channels in the plasma membrane leading to hyperpolarization. It also blocks VOCC indirectly through phosphorylation of protein phosphatase 2A which in turn dephosphorylates and inactivates VOCC thereby blocking $Ca^{2+}$ entrance into the cytosol [25]. PKG furthermore reduces $Ca^{2+}$ sensitivity of myofilaments by activating MLP which in turn dephosphorylates and thus inactivates $MLC_{20}$.

Receptor-bound adenylyl cyclase is activated in the membrane of vascular smooth muscle cells by agonists such as adrenomedullin, prostacyclin ($PGI_2$), calcitonin generelated peptide (CGRP) and epinephrine. Subsequently activated adenylyl cyclase results in increased cAMP levels in the cytosol of these cells. Such increased cAMP levels in turn activate PKA. PKA activates K⁺ channels and blocks VOCC in the plasma membrane, activates PMCA and SERCA and blocks $Ca^{2+}$ release from the SR through mechanisms

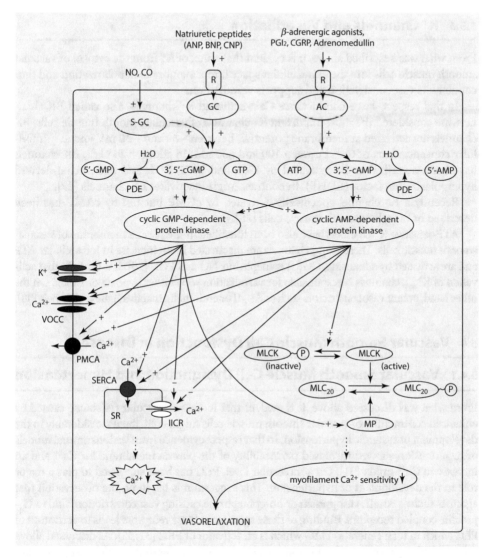

◘ **Fig. 5.6** Cyclic nucleotides and vasorelaxation [26]. AC adenylyl cyclase, AMP adenosine monophosphate, ANP atrial natriuretic peptide, ATP adenosine triphosphate, BNP brain natriuretic peptide, cAMP cyclic adenosine monophosphate, cGMP cyclic guanosine monophosphate, CGRP calcitonin gene-related peptide, CNP C-type natriuretic peptide, GC guanylyl cyclase, GMP guanosine monophosphate, GTP guanosine triphosphate, IICR $IP_3$-induced calcium release, $PGI_2$ prostacyclin, $MLC_{20}$ 20kD regulatory light chain of myosin, MLCK myosin light chain kinase, MP myosin phosphatase, PDE phosphodiesterase, PMCA plasma membrane calcium ATPase, R receptor, SERCA sarcoplasmic/endoplasmic calcium ATPase, SGC soluble guanylyl cyclase, SR sarcoplasmic reticulum, VOCC voltage-operated calcium channel

described above for PKG. In addition, PKA phosphorylates MLCK thereby reducing its affinity for CaM [25].

Finally, cross-activation between PKG and PKA in both directions may enhance the effects described above (◘ Fig. 5.6) [26].

### 5.3.6  K⁺ Channels and Vasodilation

From what was described above, it is evident that efflux of K⁺ from the cytosol of vascular smooth muscle cells into the extracellular space limits membrane depolarization and thus counteracts vasoconstriction and supports vasodilation.

In that respect, large-conductance $Ca^{2+}$-activated K⁺ channels, also called BK channels, are considered the most important K⁺ channels of vascular smooth muscle cells. BK channels are activated at membrane potentials between −60 and −30 mV and at intracellular concentrations of $Ca^{2+}$ between 100 and 600 nM. In addition to $Ca^{2+}$, BK channels have been described to be activated by PKA and PKG, $PGI_2$ and endothelial-derived hyperpolarization factor (EDHF). In contrast AngII inactivates BK channels [27].

Recently a K⁺ channel specifically activated by cGMP but not by cAMP has been described in vascular smooth muscle cells [28].

ATP-sensitive K⁺ channels have also been identified in the plasma membrane of vascular smooth muscle cells. These $K^+_{ATP}$ channels are inactivated by an increase in intracellular ATP and are activated by adenosine, CGRP, epinephrine, NO and $PGI_2$ [29]. It is thought that activation of $K^+_{ATP}$ channels is responsible for vasodilation seen under hypoxia or shock. On the other hand, potent vasoconstrictors such as AngII or endothelin inactivate these channels [30].

## 5.4  Vascular Smooth Muscle Cell Dysfunction in Disease

### 5.4.1  Vascular Smooth Muscle Cell Dysfunction and Hypertension

From what was discussed above, it is evident that increased vascular resistance caused by enhanced contractility of vascular smooth muscle cells might contribute considerably to the development of systemic hypertension. In that respect evidence from various animal models of hypertension showed increased permeability of the plasma membrane for $Ca^{2+}$ and an increase in $Ca^{2+}$ influx [31]. On a molecular level, PKC has been suggested to play a major role in the development of hypertension. This suggestion is based on the observation that agonists such as AngII, vasopressin or norepinephrine causing vasoconstriction bind to $G_q$-protein-coupled receptors. Binding of these agonists to these receptors leads to activation of PLC which in turn generates DAG which is an activator of PKC [32, 33]. As discussed above PKC phosphorylates CPI-17 which in turn inhibits MP. Furthermore, PKC phosphorylates CaP thereby allowing more actin to bind to myosin [13]. The notion that PKC is involved in the pathogenesis of hypertension is further supported by the finding that overexpression of a particular isoform of PKC, namely, α-PKC, in a mouse model results in hypertension [34]. Thus, inhibition of PKC might represent a therapeutic strategy for the treatment of hypertension. Unfortunately, however, the first generation of PKC inhibitors lacks specificity.

By inhibiting MP and by increasing $Ca^{2+}$ sensitivity, the RhoA/ROCK pathway represents yet another pathway that seems to be critically involved in the pathogenesis of systemic hypertension [13]. Tissue samples obtained from arteries of hypertensive animals showed increased RhoA/ROCK activation [35]. On the other hand, the ROCK inhibitor Y-27632 normalized arterial pressure in animal models of hypertension [36]. Furthermore stretch-induced activation of mitogen-activated protein kinase (MAPK) of vascular smooth muscle cells was inhibited by the inhibition of ROCK [37]. In hypertensive rats blocking the type I receptor for AngII reduced RhoA/ROCK activity, whereas long-time infusion of AngII increased the activity of these biomolecules [38, 39].

## 5.4.2 Vascular Smooth Muscle Cell Proliferation, Phenotypic Switching and Progenitor Cells in Cardiovascular Disease

As already discussed above, vascular smooth muscle cells exhibit a considerable degree of plasticity. This plasticity is also reflected by their ability to switch between two functionally different phenotypes [40]. The quiescent phenotype of *differentiated vascular smooth muscle cells* is responsible for vasoconstriction and vasodilation and is characterized by the expression of contractile proteins such as smooth muscle α-actin, smooth muscle myosin heavy chain, CaP and smooth muscle protein 22α (SM22α) [5]. The expression of these contractile proteins is downregulated in the dedifferentiated phenotype of vascular smooth muscle cells, also referred to as synthetic phenotype characterized. The phenotypic switch from the contractile, differentiated phenotype to the synthetic, dedifferentiated phenotype occurs in response to vascular injury [41]. The synthetic phenotype of vascular smooth muscle cells is characterized by increased proliferation, migratory behaviour and the production of matrix proteins [42]. Thus, *dedifferentiated vascular smooth muscle cells* are involved in *neointima formation* and therefore are found in atherosclerotic lesions and are responsible for restenosis after percutaneous transluminal angioplasty and stent implantation [2]. However, recent evidence obtained through lineage tracing suggests that only a small proportion of vascular smooth muscle cells present in atherosclerotic lesions originate from mature vascular smooth muscle cells of the tunica media [43]. Currently it is thought that the majority of vascular smooth muscle cells in atherosclerotic lesions may originate from various sources such as resident progenitor cells of the vasculature, transdifferentiated endothelial cells, fibroblasts from the adventitia or bone marrow haematopoietic cells [44–47].

---

┌─ **Take-Home Message** ─────────────────────────────────────────

- During embryonic development vascular smooth muscle cells originate from different progenitor cells that are derived from various tissues.
- Contraction and relaxation of vascular smooth muscle cells, and thus vasoconstriction and vasodilation, are brought about by change intracellular, cytosolic $Ca^{2+}$ levels, whereby an increase in intracellular $Ca^{2+}$ leads to vasoconstriction and a decrease to vasodilation.
- $Ca^{2+}$ influx into and efflux from the cytosol of vascular smooth muscle cells from and to the extracellular space and from and to the sarcoplasmic reticulum occurs through a wide variety of channels or pumps and exchangers, respectively, whose function is finely tuned by a host of agonists activating intricate intracellular pathways.
- Vascular smooth muscle cells show a high degree of plasticity; e.g. in case of vascular injury, they can switch from a differentiated contractile to a dedifferentiated synthetic phenotype.
- Vascular smooth muscle cells are involved in the development of various cardiovascular pathologies such as hypertension, atherosclerosis and restenosis after stent implantation.

# References

1. Willerson JT, Ridker PM. Inflammation as a cardiovascular risk factor. Circulation. 2004;109:II2–10.
2. Torrado J, Buckley L, Duran A, Trujillo P, Toldo S, Valle Raleigh J, Abbate A, Biondi-Zoccai G, Guzman LA. Restenosis, stent thrombosis, and bleeding complications: navigating between scylla and charybdis. J Am Coll Cardiol. 2018;71:1676–95.
3. Wasteson P, Johansson BR, Jukkola T, Breuer S, Akyurek LM, Partanen J, Lindahl P. Developmental origin of smooth muscle cells in the descending aorta in mice. Development. 2008;135:1823–32.
4. Jiang X, Rowitch DH, Soriano P, McMahon AP, Sucov HM. Fate of the mammalian cardiac neural crest. Development. 2000;127:1607–16.
5. Wang G, Jacquet L, Karamariti E, Xu Q. Origin and differentiation of vascular smooth muscle cells. J Physiol. 2015;593:3013–30.
6. Voiculescu O, Papanayotou C, Stern CD. Spatially and temporally controlled electroporation of early chick embryos. Nat Protoc. 2008;3:419–26.
7. Nakamura T, Colbert MC, Robbins J. Neural crest cells retain multipotential characteristics in the developing valves and label the cardiac conduction system. Circ Res. 2006;98:1547–54.
8. Pouget C, Gautier R, Teillet MA, Jaffredo T. Somite-derived cells replace ventral aortic hemangioblasts and provide aortic smooth muscle cells of the trunk. Development. 2006;133:1013–22.
9. Tian X, Hu T, Zhang H, He L, Huang X, Liu Q, Yu W, He L, Yang Z, Yan Y, Yang X, Zhong TP, Pu WT, Zhou B. Vessel formation. De novo formation of a distinct coronary vascular population in neonatal heart. Science. 2014;345:90–4.
10. Somlyo AP, Somlyo AV. Signal transduction and regulation in smooth muscle. Nature. 1994;372:231–6.
11. Akata T. Cellular and molecular mechanisms regulating vascular tone. Part 1: basic mechanisms controlling cytosolic Ca2+ concentration and the Ca2+-dependent regulation of vascular tone. J Anesth. 2007;21:220–31.
12. Hai CM, Murphy RA. Ca2+, crossbridge phosphorylation, and contraction. Annu Rev Physiol. 1989;51:285–98.
13. Khalil RA. Regulation of vascular smooth muscle function. San Rafael (CA): Morgan & Claypool Life Sciences; 2010.
14. Marin J, Encabo A, Briones A, Garcia-Cohen EC, Alonso MJ. Mechanisms involved in the cellular calcium homeostasis in vascular smooth muscle: calcium pumps. Life Sci. 1999;64:279–303.
15. Gollasch M, Nelson MT. Voltage-dependent Ca2+ channels in arterial smooth muscle cells. Kidney Blood Press Res. 1997;20:355–71.
16. Hughes AD. Calcium channels in vascular smooth muscle cells. J Vasc Res. 1995;32:353–70.
17. McDonald TF, Pelzer S, Trautwein W, Pelzer DJ. Regulation and modulation of calcium channels in cardiac, skeletal, and smooth muscle cells. Physiol Rev. 1994;74:365–507.
18. Gibson A, McFadzean I, Wallace P, Wayman CP. Capacitative Ca2+ entry and the regulation of smooth muscle tone. Trends Pharmacol Sci. 1998;19:266–9.
19. Gonzalez JM, Jost LJ, Rouse D, Suki WN. Plasma membrane and sarcoplasmic reticulum Ca-ATPase and smooth muscle. Miner Electrolyte Metab. 1996;22:345–8.
20. Grover AK, Khan I. Calcium pump isoforms: diversity, selectivity and plasticity. Review article. Cell Calcium. 1992;13:9–17.
21. Koyama M, Ito M, Feng J, Seko T, Shiraki K, Takase K, Hartshorne DJ, Nakano T. Phosphorylation of CPI-17, an inhibitory phosphoprotein of smooth muscle myosin phosphatase, by Rho-kinase. FEBS Lett. 2000;475:197–200.
22. Hartshorne DJ. Myosin phosphatase: subunits and interactions. Acta Physiol Scand. 1998;164:483–93.
23. Parmentier JH, Muthalif MM, Saeed AE, Malik KU. Phospholipase D activation by norepinephrine is mediated by 12(s)-, 15(s)-, and 20-hydroxyeicosatetraenoic acids generated by stimulation of cytosolic phospholipase a2. tyrosine phosphorylation of phospholipase d2 in response to norepinephrine. J Biol Chem. 2001;276:15704–11.
24. Horowitz A, Menice CB, Laporte R, Morgan KG. Mechanisms of smooth muscle contraction. Physiol Rev. 1996;76:967–1003.
25. Vaandrager AB, de Jonge HR. Signalling by cGMP-dependent protein kinases. Mol Cell Biochem. 1996;157:23–30.
26. Akata T. Cellular and molecular mechanisms regulating vascular tone. Part 2: regulatory mechanisms modulating Ca2+ mobilization and/or myofilament Ca2+ sensitivity in vascular smooth muscle cells. J Anesth. 2007;21:232–42.

27. Waldron GJ, Cole WC. Activation of vascular smooth muscle K+ channels by endothelium-derived relaxing factors. Clin Exp Pharmacol Physiol. 1999;26:180–4.
28. Yao X, Segal AS, Welling P, Zhang X, McNicholas CM, Engel D, Boulpaep EL, Desir GV. Primary structure and functional expression of a cGMP-gated potassium channel. Proc Natl Acad Sci U S A. 1995;92:11711–5.
29. Standen NB, Quayle JM. K+ channel modulation in arterial smooth muscle. Acta Physiol Scand. 1998;164:549–57.
30. Brayden JE. Potassium channels in vascular smooth muscle. Clin Exp Pharmacol Physiol. 1996;23: 1069–76.
31. Sugiyama T, Yoshizumi M, Takaku F, Urabe H, Tsukakoshi M, Kasuya T, Yazaki Y. The elevation of the cytoplasmic calcium ions in vascular smooth muscle cells in SHR--measurement of the free calcium ions in single living cells by lasermicrofluorospectrometry. Biochem Biophys Res Commun. 1986;141:340–5.
32. Griendling KK, Rittenhouse SE, Brock TA, Ekstein LS, Gimbrone MA Jr, Alexander RW. Sustained diacylglycerol formation from inositol phospholipids in angiotensin II-stimulated vascular smooth muscle cells. J Biol Chem. 1986;261:5901–6.
33. Takagi Y, Hirata Y, Takata S, Yoshimi H, Fukuda Y, Fujita T, Hidaka H. Effects of protein kinase inhibitors on growth factor-stimulated DNA synthesis in cultured rat vascular smooth muscle cells. Atherosclerosis. 1988;74:227–30.
34. Liou YM, Morgan KG. Redistribution of protein kinase C isoforms in association with vascular hypertrophy of rat aorta. Am J Phys. 1994;267:C980–9.
35. Seko T, Ito M, Kureishi Y, Okamoto R, Moriki N, Onishi K, Isaka N, Hartshorne DJ, Nakano T. Activation of RhoA and inhibition of myosin phosphatase as important components in hypertension in vascular smooth muscle. Circ Res. 2003;92:411–8.
36. Uehata M, Ishizaki T, Satoh H, Ono T, Kawahara T, Morishita T, Tamakawa H, Yamagami K, Inui J, Maekawa M, Narumiya S. Calcium sensitization of smooth muscle mediated by a Rho-associated protein kinase in hypertension. Nature. 1997;389:990–4.
37. Numaguchi K, Eguchi S, Yamakawa T, Motley ED, Inagami T. Mechanotransduction of rat aortic vascular smooth muscle cells requires RhoA and intact actin filaments. Circ Res. 1999;85:5–11.
38. Kataoka C, Egashira K, Inoue S, Takemoto M, Ni W, Koyanagi M, Kitamoto S, Usui M, Kaibuchi K, Shimokawa H, Takeshita A. Important role of Rho-kinase in the pathogenesis of cardiovascular inflammation and remodeling induced by long-term blockade of nitric oxide synthesis in rats. Hypertension. 2002;39:245–50.
39. Higashi M, Shimokawa H, Hattori T, Hiroki J, Mukai Y, Morikawa K, Ichiki T, Takahashi S, Takeshita A. Long-term inhibition of Rho-kinase suppresses angiotensin II-induced cardiovascular hypertrophy in rats in vivo: effect on endothelial NAD(P)H oxidase system. Circ Res. 2003;93:767–75.
40. Salmon M, Gomez D, Greene E, Shankman L, Owens GK. Cooperative binding of KLF4, pELK-1, and HDAC2 to a G/C repressor element in the SM22alpha promoter mediates transcriptional silencing during SMC phenotypic switching in vivo. Circ Res. 2012;111:685–96.
41. Herring BP, Hoggatt AM, Burlak C, Offermanns S. Previously differentiated medial vascular smooth muscle cells contribute to neointima formation following vascular injury. Vasc Cell. 2014;6:21.
42. Yoshida T, Kaestner KH, Owens GK. Conditional deletion of Kruppel-like factor 4 delays downregulation of smooth muscle cell differentiation markers but accelerates neointimal formation following vascular injury. Circ Res. 2008;102:1548–57.
43. Feil S, Fehrenbacher B, Lukowski R, Essmann F, Schulze-Osthoff K, Schaller M, Feil R. Transdifferentiation of vascular smooth muscle cells to macrophage-like cells during atherogenesis. Circ Res. 2014;115:662–7.
44. Chade AR, Zhu XY, Grande JP, Krier JD, Lerman A, Lerman LO. Simvastatin abates development of renal fibrosis in experimental renovascular disease. J Hypertens. 2008;26:1651–60.
45. Sata M, Saiura A, Kunisato A, Tojo A, Okada S, Tokuhisa T, Hirai H, Makuuchi M, Hirata Y, Nagai R. Hematopoietic stem cells differentiate into vascular cells that participate in the pathogenesis of atherosclerosis. Nat Med. 2002;8:403–9.
46. Sartore S, Chiavegato A, Faggin E, Franch R, Puato M, Ausoni S, Pauletto P. Contribution of adventitial fibroblasts to neointima formation and vascular remodeling: from innocent bystander to active participant. Circ Res. 2001;89:1111–21.
47. Zengin E, Chalajour F, Gehling UM, Ito WD, Treede H, Lauke H, Weil J, Reichenspurner H, Kilic N, Ergun S. Vascular wall resident progenitor cells: a source for postnatal vasculogenesis. Development. 2006;133:1543–51.

# Embryonic Development of the Cardiovascular System

*Wolfgang J. Weninger and Stefan H. Geyer*

**What You Will Learn in This Chapter**

This chapter briefly recapitulates milestones in the morphogenesis of the cardiovascular system. Its focus rests on the development and remodelling of the heart and great arteries.

## 6.1  Researching Cardiovascular Morphogenesis and Remodelling

For obvious reasons descriptive and experimental research cannot be performed on human embryos. Hence, various biomedical models are used for studying the genetic and biomechanic mechanisms orchestrating the development and remodelling of the cardiovascular system. The chick, frog and zebrafish are mainly employed for unravelling basic genetic decisions and for analysing the influence of biomechanic factors on heart remodelling [6, 14, 17, 21, 22]. Rodents, especially the mouse, are chiefly used for examining the genetic orchestration of normal and pathologic developmental events. They have short reproduction times, and the advancement of molecular tools has reached a level that allows for specific deletion of every single gene of the mouse genome and targeted disruption of gene and gene product function in specific tissues and at specific time points [1, 2, 4, 5, 8, 13, 19].

The advent of sophisticated imaging methods permitting three-dimensional (3D) visualisation of early to late embryos, foetuses and infants of biomedical models in high detail [12, 15, 16, 20, 23, 24] heavily boosted researching the morphogenetic events in cardiovascular development. This chapter will primarily make use of virtual 3D models produced with the high-resolution episcopic microscopy (HREM) technique [11, 18, 25] and traditional wax models produced by Ziegler at the beginning of the twentieth century.

## 6.2  Early Blood Vessels and Primitive Circulation

In mammals, the formation of the vascular system starts with a process called *vasculogenesis*. Small clusters of haemangioblasts appear in the extraembryonic mesoderm of the yolk sac, body stalk and chorion. They form isolated blood islands and differentiate into endothelial cells surrounding primitive blood cells. Finally, these islands coalesce and form a primitive capillary plexus. In humans this process happens between days 13 and 15 of *intrauterine* development (i.e. end of 4th week of pregnancy) and in the mouse, at approximately embryonic day (E) 8. Around days 15–17, vasculogenesis starts also in the mesoderm inside the embryo, and a plexus of primitive intraembryonal blood vessels forms.

Once a primitive vascular system is shaped, it expands by a process named *angiogenesis*. New, endothelial lined channels sprout from existing vessels, or already established larger channels split after invasion of additional cell material. For details we refer to the chapter "Cellular and Molecular Mechanisms of Vasculogenesis, Angiogenesis, and Lymphangiogenesis".

In human embryos of developmental day 21, the then tubular heart connects to the intra- and extraembryonic vessels. Since it is already pumping, this establishes a primitive unidirectional circulation. First large arteries, the left and right primitive aortae, pass as continuation of the outflow of the tubular heart lateral to the pharynx and turn caudally to descend along the forming intestine. Each dorsal aorta gives rise to segmental arteries, vitelline branches and an umbilical artery. The venous channels are collected by anterior and posterior cardinal veins that enter the cardiac inflow. The components of the primitive circulatory system undergo dramatic remodelling during the rest of the embryonic period until the foetal circulation is established. The following chapters aim at describing the morphological milestones of these processes in detail.

## 6.3 Development of the Heart

Even at the beginning of the 2nd week of human *intrauterine* development, cells destined to become heart cells can be identified. At the end of the 3rd week, the embryos have developed a pumping primitive heart then a plump, cranio-caudally arranged tube. This primitive heart tube undergoes complex morphological remodelling in the embryonic period until it has largely achieved its foetal appearance in week 8 of *intrauterine* development. Immediately after birth changes occur in heart morphology to adapt the foetal heart to postnatal requirements.

Correct heart formation and remodelling is essential for sufficient supply of oxygen and nutrients to all the developing organs of the embryo and foetus. Hence an estimated 10% of miscarriages are considered to be a direct result of disrupted embryonic heart development.

### 6.3.1 Heart Progenitor Cells and First and Second Heart Field

Around the time of implantation into the endometrium, the embryo is composed of two single-cell layers – a dorsal epiblast and a ventral hypoblast. In these early embryos, cell-tracing methods identify cardiac progenitor cells in the epiblast anteriorly to the primitive streak. During gastrulation, which produces the three germ layers ecto-, meso- and endoderm, these cells and their daughter cells invaginate during mesoderm formation through the primitive streak and end up as parts of the ventral portion of the lateral plate mesoderm, the so-called splanchnic mesoderm.

In the three-layered embryo, the region occupied by the heart progenitor cells is referred to as heart field. As part of the mesoderm, the cells are sandwiched between ento- and ectoderm and, in the 3rd to 4th week of gestation, are arranged in a horseshoe-shaped, crescent-like structure, the *cardiac crescent*. In mice the two branches of the cardiac crescent are connected anteriorly to the developing neural folds. In humans they are not fused. The lateral area of the branches and the forefront of the cardiac crescent (the toe of the horseshoe) are named as the *first heart field* and the medial part of the branches as the *second heart field*.

### 6.3.2 Formation of the Linear Heart Tube

On the left and right sides of the embryo, the splanchnic mesoderm of the first heart field forms antero-posteriorly (cranio-caudally) oriented tube-like structures comprised of endothelial (endocardial) cells. Cells of the second heart field are not involved in this process. In the 3rd week, when the most anterior parts of the embryo bend ventrally and form the head fold, these tube-like structures are passively shifted caudally and to the ventral midline. Here they fuse and form a single, very plump primitive heart tube immediately ventral to the developing head and foregut. Caudally (posteriorly), two venous channels drain into this tube. Cranially (anteriorly), two primitive ventral aortae leave the

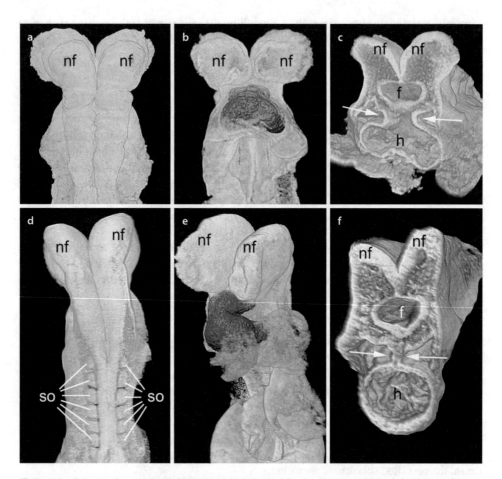

**Fig. 6.1** Primitive heart tube. Volume-rendered 3D models of early **a–c** and late **d–f** mouse embryos of embryonic day (E) 8 and surface-rendered models of their heart tube (red). Models were created out of "high-resolution episcopic microscopy" (HREM) volume data. Anterior (cranial) parts of the embryo from dorsal **a, d** and ventral **b, e**. Note the plump primitive heart tube and its slender and looped appearance in the later stage. **c, f**: Axial sections through the heart region of volume-rendered models. View from anterior. The mesocardium (between arrows) connects the heart tube to the dorsal body wall near the foregut (f). h heart, nf neural fold, so somite

tube ( Fig. 6.1). The paired arms of the inflow are destined to become the precursors of the atria, the linear part of the tube transforms to eventually become the left ventricle and the two vessels leaving the outflow are destined to form transitory connections to two primitive aortae developing in the dorsal and caudal parts of the embryo.

The primitive heart tube rests in a cavity, which is termed as the pericardial cavity, although there is no proper pericardium formed yet. It forms at a later time point by migration of cells from the proepicardial organ (see below).

The dorsal aspects of the heart tube are connected to the dorsal wall of the embryo along its foregut region by a meso-like structure called mesocardium ( Fig. 6.1). Near the attachment of the mesocardium to the body wall and at the same time laterally to the foregut, the second heart field had been shifted to during head fold formation. From here, cells migrate towards the heart tube and are incorporated into their wall.

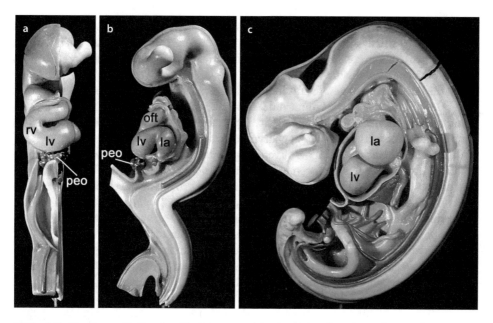

**Fig. 6.2** Looping and ballooning. Ziegler wax models of a chick embryo during looping from ventral **a** and left **b**. Note the heart loop and the proepicardial organ (peo). **c** Ballooning of the left atrium (la) and left ventricle (lv). oft outflow tract, rv right ventricle

The lumen of the primitive heart tube is lined with endocardial cells. Myocardial cells appear as if incompletely surrounding the endocardial tube almost as a mantel-like structure. This myocardial mantel layer borders against the pericardial cavity until the epicardium is formed. It does not exist where the mesocardium connects to the heart tube. After forming the mantel layer, the myocardial cells start secreting a complex proteoglycan- and glycosaminoglycan-rich matrix into the space between endocardial and myocardial cells. This matrix, sandwiched between endo- and myocardium, originally does not contain cells and is called *cardiac jelly*.

### 6.3.3 Looping

After the linear heart tube has formed, cells of the second heart field migrate to it and integrate themselves into its wall. This cell integration quickly leads to a dramatic increase of the length of the heart tube, which is much faster than the cranio-caudal growth of the whole embryo. Therefore, the straight heart tube is forced to assemble a curved shape, and the heart tube starts forming a slender loop protruding into the pericardial cavity. The vertex of the loop points to the right (■ Figs. 6.1 and 6.2).

Elongation of the heart tube is accompanied by disintegration of the mesocardium leaving only the cranial and caudal ends of the heart fixed to the embryonic body tissues. The parts in between the cranial and caudal fixation freely protrude into the pericardial cavity. Consequently, from this time of development, cells, which still migrate from the second heart field and cardiac neural crest towards the heart, can only become associated to the outflow and inflow sections of the heart tube. The cells added to the outflow section form large parts of the later right ventricle and cardiac outflow. The cells added to the inflow section contribute to the formation of the atria and inflow. Hence, severe heart malformations chiefly affecting the right ventricle and cardiac outflow are often a direct consequence of abnormal migration of cardiac neural crest and/or second heart field cells to the cranial parts of the tubular heart.

In the late looping stages, the future compartments are arranged in a sequential order along the still tubular heart. From caudal to cranial, they are the inflow tract, atrial segment, atrioventricular canal, left ventricle, right ventricle and outflow tract. The outflow tract is connected to the developing pharyngeal arch arteries (◘ Fig. 6.2).

The straight heart has its inflow or venous pole caudally and its outflow or arterial pole cranially. In principle this arrangement remains throughout life, but due to heart elongation and the associated looping and twisting and the growth of the embryo, the relative position of the two poles appears as if having changed after *looping*, with the venous pole being shifted into a dorso-cranial and the arterial pole into an antero-cranial position.

Looping is highly conserved in vertebrates. It is the first feature that makes left-right asymmetry obvious and is governed by molecular signalling cascades, which have been started during gastrulation. Perturbation of these cascades leads to abnormal or left-sided looping and problems in the situs of asymmetric organs. Such laterality defects often are associated with congenital heart defects including serious malformations [17].

### 6.3.4  Ballooning

The atria and ventricles are formed as pouches that appear at characteristic sites along the heart tube. The pouches grow and expand like balloons, wherefore the whole process of cardiac chamber formation is referred to as *ballooning*, the atria balloon laterally out of the atrial segment of the heart tube and the ventricle balloon in a proximal to distal sequence out of the tube near its vertex (◘ Fig. 6.2).

#### 6.3.4.1  Formation of the Ventricles

During the late phases of heart looping, the mitotic rate locally increases at the outer curvature of the left and right ventricle segments of the cardiac loop. This results in the formation of small thickenings and then of pouches arranged in a proximal to distal sequence along the tube. While the pouches become steadily bigger and start ballooning, the region between them does not significantly change its dimension. As seen from outside, this results in the indentation of an anterior and posterior groove between the ballooning ventricles, the later anterior and posterior interventricular sulcus.

As seen from the heart tube lumen, the ballooning process leads to a local expansion of the lumen at the level of the left and right ventricle segments. The diameter of the lumen in between these two segments does not increase dramatically. Hence the adjacent walls of the two balloons and the material in between form a septum-like structure between the two ventricle cavities. This material is the forerunner of the later muscle part of the *interventricular septum*.

Simultaneously with ballooning of the ventricles, the ventricle myocardium develops two layers: an outer compact tissue layer, with a thickness of 3–4 cardiomyocytes, and an inner layer forming myocardial protrusions into the ventricle cavity that are covered by endocardium. These are called trabeculae carneae. In the early heart, the trabecular layer creates the main contractile force of the developing heart. In addition, the extensions of the ventricle cavities between the trabeculae secure a sufficiently large surface to guarantee smooth supply of oxygen and nutrients to the myocardium until the coronary vasculature is established.

### 6.3.4.2 Formation of the Atria

In both lateral walls of the singular atrial segment of the heart tube, the mitotic activity increases. This results in the formation of left- and right-sided thickenings and then pouches, which steadily expand until they finally form the appendices of the left and right atrium. These are the forerunner of the later auricula cordis. Inside the appendices myocardial protrusions develop and form pectinate muscles.

The proper atrial spaces are derived from the atrial segment of the heart loop that is located between the atrial appendices and from structures that become secondarily associated with the left and right atrium. The right atrium integrates parts of the sinus venosus, which is the cavity that receives the large systemic veins. The left atrium integrates the vastly expanded space of an originally single pulmonary vein, which receives all four lung veins.

### 6.3.4.3 Peculiarities of Heart Looping

In contrast to the ventricles, which balloon sequentially from the large curvature of the looping heart, the atrial appendices balloon from its left and right sides. Hence sidedness of the atria is affected by left/right decisions during gastrulation. Consequently, abnormal left/right definition result in abnormal atrium morphology.

An interesting aspect is that elongation of the heart tube during looping is mainly caused by adding cells stemming from the second heart field and cardiac neural crest. In contrast the ballooning process is triggered and maintained by mitoses of cells, which are already part of the cardiac wall.

## 6.3.5 Epi- and Pericardium

Around the time when heart looping takes place, a vesicular structure formed by mesothelial cells of the septum transversum appears at the base of the sinus venosus. This vesicle is called the *proepicardial organ*. From this organ cells migrate along the cardiac surface and form the epicardium. Cells also migrate along the inside of the pericardial cavity and form the serous layer of the pericardium. A subset of the epicardial cells undergoes *epithelial-mesenchymal transformation*, invades the subepicardial space and forms the subepicardial mesenchyme. These cells also contribute to the formation of the endothelium and smooth muscle cells of the future coronary vessels as well as the forming atrioventricular cushions and valves.

## 6.3.6 Chamber Separation

### 6.3.6.1 Atrioventricular Junction

In the looped heart tube, the atrioventricular junction is a relatively long, canal-like segment. A subpopulation of endothelial cells lining this segment undergo endothelial-mesenchymal transformation and invade the cardiac jelly. They stimulate adjacent myocardial cells to produce additional extracellular matrix, which increases the amount of cardiac jelly in the atrioventricular canal and produces two thickenings, the atrioventricular *endocardial cushions*. Because of their position in respect to the orientation of the looped heart tube, these cushions are referred to as superior and inferior atrioventricular cushion. Their size increases until they fuse in the midline, dividing the atrioventricular canal into a left and right portion, the later left and right atrioventricular ostium.

In addition to separating atria and ventricles, the cushion material forms relative thick extensions that protrude into the ventricle cavities and become connected to the trabeculae of the myocardium. After reducing their thickness, these protrusions remain as leaflets of the atrioventricular valves and as chorda tendinea.

### 6.3.6.2 Atria

In the 5th week of human development, the common space between the left and right atrial appendices starts to become separated into a left and right atrium. In the mouse this happens around E10.5.

After the inflow of the heart has become shifted dorso-cranially during the looping process, muscle tissue starts growing in the midline of the dorso-cranial wall of the atrium into the atrial cavity. This forms the *septum primum*, which at first is crescent-shaped. The space between the free edge of the crescent and the superior atrioventricular cushion (between the left and right atrioventricular ostia) is referred to as *foramen primum*. At the caudal edge of the septum, at the level of the atrioventricular junction, fibrous tissue termed as *spina vestibuli* grows also from dorsal into the atrium. By growing of the septum and spina, the foramen primum becomes gradually smaller and is finally closed with spina vestibuli tissue sandwiched between the atrioventricular cushions and the septum [3]. While the foramen primum closes, parts of the dorsal portion of the septum primum disintegrate, and a foramen opens in the dorsal part, the *foramen secundum*, a forerunner of the *foramen ovale*.

In humans, in the 8th week of intrauterine development (mouse E14), an invagination of the dorso-cranial atrial wall forms a plump ridge right to the septum primum. The ridge is named as *septum secundum* and only extends for a short distance, but far enough to sufficiently cover the foramen secundum in the septum primum (◻ Fig. 6.3).

### 6.3.6.3 Ventricles

The *interventricular septum* comprises a membranous and a muscular part. First the muscular part is formed by ballooning of the ventricles. Then, during the rest of the embryonic period, its free edge is gradually shifted towards the atrioventricular junction.

The space between the atrioventricular cushions and the free edge of the muscular septum is termed *foramen interventriculare*. It permits blood exchange between the left and right ventricle and persists until the end of the 8th week of gestation (E14.5 in the mouse), which is the beginning of the foetal period. Around this time the muscular portion of the ventricular septum then shifted close to the atrioventricular junction and fuses with a small extension of the atrioventricular cushions, which transforms into the fibrous part of the interventricular septum.

### 6.3.6.4 Outflow Tract

The cardiac *outflow tract* (conotruncus) has two segments, the proximal *conus arteriosus* and the distal *truncus arteriosus*. The truncus arteriosus continues into the extracardiac *aortic sac*.

During heart looping, the endocardium and the cardiac jelly of the conotruncus form two ridges, twisting like a spiral from proximal to distal. The spiralling arrangement is triggered by haemodynamic forces. Although at this time point the ventricles are connected by an intervertebral foramen, each ventricle pumps a separate stream of blood into the conotruncus, and these blood streams spiral.

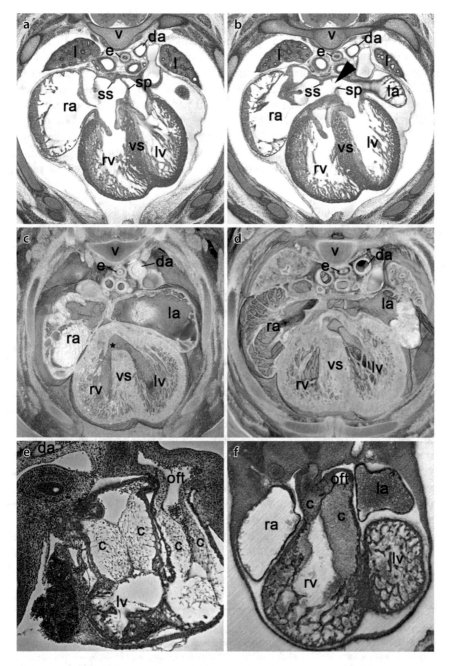

☐ **Fig. 6.3** Separation of chambers and outflow tract. **a, b** Atrium septation in an E14.5 mouse embryo. Axial HREM sections through the heart showing septum primum (sp), septum secundum (ss) and foramen secundum (ovale; arrowhead). **c, d** Septation of ventricles. Axially sectioned volume-rendered models based on high-resolution episcopic microscopy data viewed from cranial. Early **c** and late E14.5 mouse embryo **d**. Note the interventricular foramen in **c** (asterisk). **e, f** Architecture of atrioventricular and outflow tract cushions in a mouse embryo of E10.5 **e** and E12 **f**. c cushion, da descending aorta, e oesophagus, l lung, la left atrium, lv left ventricle, oft outflow tract, ra right atrium, rv right ventricle, sp septum primum, ss septum secundum, v vertebra, vs ventricular septum

Once the ridges are formed, extracardiac cells derived from the neural crest invade the truncal portion of the ridges and trigger the formation of mesenchyme. This causes the ridges to thicken until they fuse, thereby separating the distal outflow tract into two channels, the future aorta and pulmonary trunk. The conal part of the ridges does not receive neural crest cells. Here, cellular material is contributed by endothelial-mesenchymal transformation, which also leads to growing and fusion of the ridges and ultimately to separation of the cardiac outflow.

Nearby the conal parts of the ridges, two small intercalated cushions deriving their mesenchyme mainly by endothelial-mesenchymal transformation are formed. These cushions form the posterior aortic and the anterior pulmonic leaflet of the semilunar valves.

## 6.4  Great Intrathoracic Arteries

Three-week-old embryos have paired dorsal aortae, which descend from the neck to the later lumbar region left and right to the gut-forming entoderm. The lumbar parts of the two aortae fuse to form a single dorsal vessel. Their cranial segments are connected to the heart by primitive aortic arches, which obliterate and vanish after the 1st *pharyngeal arch arteries* are formed (see below) (◻ Fig. 6.4).

Between the 3rd and 6th week of development in humans (E8.5–E13 in mice), the thoracic segments of the dorsal aortae and the ventrally situated aortic sac form buds that grow together and fuse to from six "pairs" of arteries. These arteries appear in a more or less cranio-caudal temporary sequence on both sides of the pharynx and are formed in the mesenchyme of the pharyngeal arches (◻ Fig. 6.4), wherefore they are named as pharyngeal or aortic arch arteries. Immediately after formation, they start to become remodelled. Certain segments are enlarged; others are reduced in diameter or entirely vanish.

At no time point, all six pairs simultaneously exist. For example, most segments of the first two pairs have already largely vanished by the time point the 3rd to 6th are built. Remodelling is largely finished with the start of the foetal period. An exception is the ductus Botalli, which is a derivative of the left 6th pharyngeal arch artery that exists throughout the entire foetal period and obliterates postnatally.

Human embryos only form four pharyngeal (branchial) arches. The material of the 5th and 6th is condensed as the so-called ultimobranchial body. Hence the 5th and 6th pair of pharyngeal arch arteries do not develop inside proper pharyngeal arch mesenchyme but

◻ **Fig. 6.4**    Pharyngeal arch arteries. **a–c** Ziegler wax models of chick embryos. View from right of subsequent developmental stages. Right primitive aortic arch (rpaa) and right pharyngeal arch arteries 1 (r1)–4 (r4). **d–i.** Surface-rendered 3D models of the pharyngeal arch arteries of mouse embryos between embryonic day (E)12 **d–f** and E12.5 **g–i** derived from HREM volume data. View from right **d, g**, ventral **e, h** and left **f, i**. Note the not yet fully separated ascending aorta (aa) and pulmonary trunk (pt) in **d–f**. bt brachiocephalic trunk, da descending aorta, h heart, lc left common carotid artery, ldoa left dorsal aorta, ls left subclavian artery, l3 left 3rd pharyngeal arch artery, l4 left 4th pharyngeal arch artery, l5 left 5th pharyngeal arch artery, l6 left 6th pharyngeal arch artery, pa pulmonary artery, rav right anterior cardinal vein, rc right common carotid artery, rdoa right dorsal aorta, rpv right posterior cardinal vein, rs right subclavian artery, r1 right 1st pharyngeal arch artery, r2 right 2nd pharyngeal arch artery, r3 right 3rd pharyngeal arch artery, r4 right 4th pharyngeal arch artery, r6 right 6th pharyngeal arch artery, 3 3rd pharyngeal arch artery, 4 4th pharyngeal arch artery, 5 5th pharyngeal arch artery, 6 6th pharyngeal arch artery

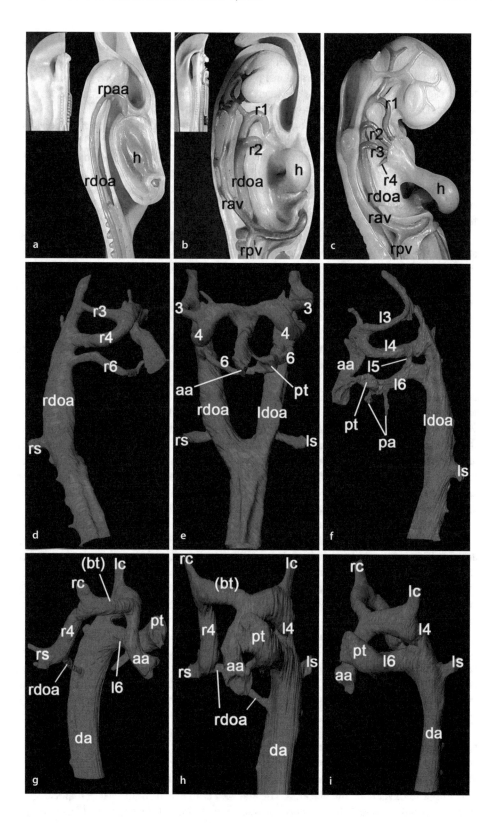

in the mesenchyme laterally to the ultimobranchial body. This has a peculiar effect on the 5th and 6th pair. The 5th pair forms after the 6th pair is already established and shows up as a connection between the 4th and 6th pharyngeal arch arteries, which runs in parallel to the segment of the dorsal aorta between the 4th and 6th pharyngeal arch arteries with a branch sometimes connecting to this segment [7, 9, 10] (◻ Fig. 6.4).

Segments of the first and second pair of pharyngeal arch arteries are remodelled to become segments of head arteries, while the 3rd to 6th are remodelled into the great intra-thoracic arteries. From the first pharyngeal arch artery, material only remains as segments of the *maxillary artery* and the *external carotid artery*. Material from the distal portion of the 2nd pharyngeal arch artery forms the *stapedial artery,* which is the main source of blood supply to the orbit in rodents, but a transient embryonic vessel in humans. Hence, from the foetal stages onwards, there are no derivatives of the second pharyngeal arch arteries.

The 3rd to 6th pairs of pharyngeal arch arteries remodel to form the unpaired aortic arch and the great intrathoracic arteries of the foetal cardiovascular system. The remodel-ling process is highly complex and influenced by haemodynamic forces and cardiac neural crest cells. At the time point the pharyngeal arch arteries start forming, the latter migrate in the mesenchyme of the forming pharyngeal arches towards the outflow tract of the heart. Some remain in the mesenchyme and become incorporated into the walls of the pharyngeal arch arteries.

Material of pharyngeal arch arteries three to six contributes to the following foetal blood vessels: Segments of the third pharyngeal arch arteries remain as parts of the *common carotid* and proximal parts of *internal carotid arteries*. Segments of the left 4th pharyngeal arch artery remain as the segment of the aortic arch between left common carotid and left subcla-vian artery (the proximal part of the aortic arch develops from the aortic sac and primitive ventral aorta). The right 4th pharyngeal arch artery remains as the proximalmost segment of the right subclavian artery. The 5th pair of pharyngeal arch arteries exists only transiently and vanishes entirely. The left 6th pharyngeal arch artery contributes to the distal pulmonary trunk and persists as *ductus arteriosus (Botalli)*. The latter is essential for foetal circulation and obliterates postnatally. The proximal part of the right 6th pharyngeal arch artery persists as the most proximal part of the right pulmonary artery (◻ Table 6.1 and ◻ Fig. 6.5).

The segments of the left and right dorsal aorta in between the pharyngeal arch arteries obliterate and vanish, except for the segment between the left 4th and 6th pharyngeal arch arteries. This segment seems to remain as a very small part of the aortic arch proximal to the origin of the left subclavian artery. The segment of the right dorsal aorta between the 6th pharyngeal arch artery and the bifurcation where the two dorsal aortae form a single vessel first elongates and then obliterates and vanishes immediately before the foetal cir-culation is established.

Abnormal remodelling, that is, abnormal regression or persistence of single pharyngeal arch arteries, is quite frequent. Examples are double lumen aortic arch as persistence of the left 5th pharyngeal arch artery, type B interrupted aortic arch as a result of total regression of the left 4th pharyngeal arch artery and retro-oesophageal right subclavian artery, as an example of abnormally total regression of the right 4th pharyngeal arch artery (◻ Fig. 6.5).

## 6.5  Foetal Circulation

The foetus swims in the amnion and lacks the ability to eat or breath. It therefore receives nutrients and oxygen via the placenta. The placenta is an organ comprised of maternal and foetal material, in which foetal blood vessels directly border to maternal blood spaces.

**▢ Table 6.1** Derivatives of pharyngeal arch arteries (PAA)

| PAA | Derivatives |
| --- | --- |
| 1st | Segments of maxillary artery |
| | Ventral part of external carotid artery |
| 2nd | Stapedial artery |
| 3rd | Distal segment of common carotid artery |
| | Proximal segment of internal carotid artery |
| Right 4th | Proximal segment of right subclavian artery |
| Left 4th | Aortic arch between left carotid and left subclavian artery |
| 5th | – |
| Right 6th | Proximal pulmonary artery |
| Left 6th | Distal pulmonary trunk |
| | Ductus arteriosus Botalli |

Small substances, up to a molecular weight of 1000 Da, can diffuse through the so-called placental barrier from the maternal into the foetal blood and vice versa.

From the placenta the umbilical vein brings blood back to the foetus. It runs in the umbilical cord, enters the foetal body through the navel and continues in the free lower rim of the ventral mesogastrium towards the liver hilus. Here it drains into the left branch of the portal vein. In contrast to the foetus, embryos have a left and right umbilical vein. But most segments of the right umbilical vein obliterate during embryogenesis leaving only a thin vessel, the paraumbilical vein that connects the subcutaneous veins of the body wall with the liver vasculature. Even this remnant of the right umbilical vein often fully obliterates during the foetal period.

From the left branch of the portal vein, a thick blood channel, the *ductus venosus* (Arantii), arises and connects to the vena cava inferior. The vena cava inferior pierces the diaphragm, enters the pericardial cavity and drains into the right atrium from caudal. Thus, the vast majority of the oxygenated blood entering the portal vein circumvents the liver vasculature via the ductus venosus and almost directly reaches the foetal heart. Only small amounts of oxygenated blood enter the liver. Since the foetal heart pumps the foetal circulation and the blood flow is from the placenta to the heart, all these vessels are veins.

In the right atrium, at the ostium of the vena cava inferior, a valvelike structure, the valva venae cavae inferioris (Eustachii) directs the blood stream directly towards the foramen ovale. On its way it passes to the almost canal-like space left to the septum secundum and the free edge of the septum primum, thereby bulging the septum primum leftwards into the cavity of the left atrium. Thus, the oxygenated blood enters the left atrium through the foramen ovale (former foramen secundum) in a caudo-cranial direction. Once in the left atrium, the relatively high pressure in the lung veins prevents the blood from draining into these veins, and it takes its way through the left atrioventricular ostium into the left ventricle.

The left ventricle pumps blood into the aorta. Through the coronary arteries, it supplies oxygen and nutrients to the heart and through the big arteries of the aortic arch to both

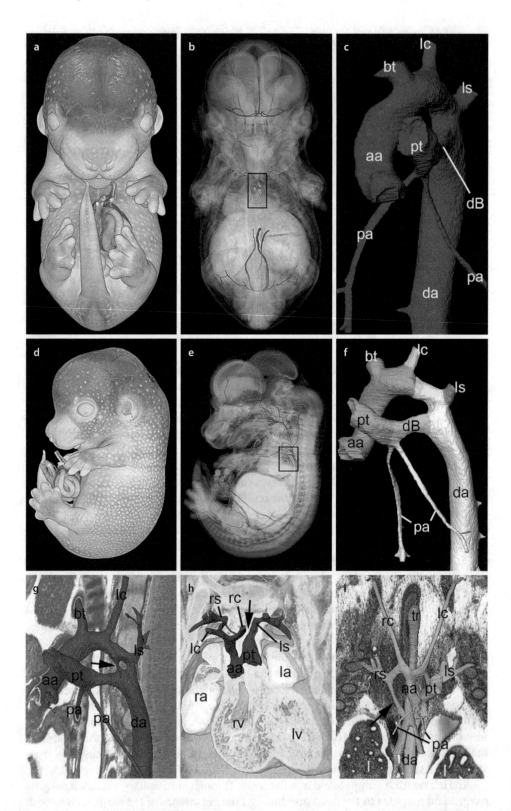

6

upper limbs and the head and neck. The rest of the oxygenated blood enters the descending aorta and mixes with the low-oxygen blood coming through the ductus arteriosus.

The venous blood from the upper body comes back to the heart via the veins collected by the superior vena cava. It enters the right atrium from cranial. Inside the atrium the blood stream passes ventrally to the stream of blood from the vena cava inferior almost without mixing. Through the right atrioventricular ostium, it then enters the right ventricle, which pumps it into the truncus pulmonalis. Since the lung is not yet inflated, the high pressure inside the lung vasculature forces the majority of the low-oxygen blood to take its way through the ductus arteriosus into the descending aorta. Thus the first segment of the descending aorta receives oxygenated blood through the aortic arch and low-oxygen (venous) blood via the ductus arteriosus.

The descending aorta gives rise to intercostal arteries, transits the diaphragm and supplies blood to the abdominal and retroperitoneal organs. It bifurcates into the common iliac arteries. Each common iliac artery in turn bifurcates into an internal and external iliac artery, which supply blood to the pelvic organs and lower limbs. The first branch of the left and right internal iliac artery is the umbilical artery. The two umbilical arteries ascend towards the navel and, running inside the umbilical cord, take the low-oxygen (venous) blood back to the placenta.

Blood from the lower body half is collected by vessels draining into the vena cava inferior or to a small amount into the azygos vein, which drains into the vena cava superior. Venous blood from the gut drains into the portal vein and either enters the liver vasculature or to a very small amount is taken into the ductus venosus, together with the oxygenated blood coming via the umbilical vein.

- Sites at Which Arterious and Venous Blood Mix

The foetal circulation has some sites where blood with high and low oxygen levels mix up:

1. Left branch of portal vein
   The portal vein and thus also its left branch receive blood from the gut. But the left branch also receives the umbilical vein and opposite to its junction gives rise to the ductus venosus. Hence, small amounts of oxygenated blood coming with the umbilical vein and low-oxygen blood coming from the gut mix up at this crossroad.
2. Connection inferior vena cava/ductus venosus
   The vena cava inferior receives venous blood from the lower body. This becomes mixed into the oxygenated blood draining into the vena cava inferior through the ductus venosus.

□ **Fig. 6.5** Foetal arrangement of great intrathoracal arteries (E14.5 mouse foetus). **a–f** Volume-rendered and surface-rendered 3D models reconstructed from "high-resolution episcopic microscopy" (HREM). **c** and **f** enlargement of arteries in the boxed region of **b** and **e**. View from ventral **a–c** and from left **d–f**. Colour coding in **f** indicates the derivation of the segments of the great arteries from the respective pharyngeal arch arteries. Aortic sac (red), left 3rd pharyngeal arch artery (orange), left 4th pharyngeal arch artery (yellow), left subclavian artery (violet) and left 6th pharyngeal arch artery (blue). **g–h** Examples of malformations caused by abnormal remodelling of the pharyngeal arch arteries. **g** Double lumen aortic arch. Arrow points to remnant of 5th pharyngeal arch artery. **h** Interrupted aortic arch type B. Arrow points to position of missing segment of left 4th pharyngeal arch artery. **I.** Persisting right dorsal aorta (arrow). aa ascending aorta, bt brachiocephalic trunk, da descending aorta, dB ductus arteriosus Botalli, I lung, la left atrium, lc left common carotid artery, ls left subclavian artery, lv left ventricle, pa pulmonary artery, pt pulmonary trunk, ra right atrium, rc right common carotid artery, rs right subclavian artery, rv right ventricle, tr trachea

3. Connection liver veins/inferior vena cava

Between the connection with the ductus venosus and it draining into right atrium, the vena cava inferior receives the liver veins. They mix low-oxygen blood draining from the liver into the oxygenated blood of the vena cava inferior.

4. Right atrium

The anterior space of the right atrium receives low-oxygen blood via the superior vena cava and the coronary sinus. The posterior part receives oxygenated blood via the inferior vena cava, which is directed towards the foramen ovale by the valva Eustachii. Most of the time, the two blood streams pass through the atrium, one ventrally and one dorsally, without much interference. But during atrium contraction, small amounts of oxygen-rich and oxygen-poor blood get mixed.

5. Left atrium

The left atrium receives oxygenated blood through the foramen ovale. Low-oxygen blood enters from the four lung veins. Since the lungs are not inflated yet, this is only a small amount.

6. Connection ductus arteriosus/descending aorta

From the aorta oxygenated blood drains into the descending aorta. Low-oxygen blood comes via the ductus arteriosus.

---

**Take-Home Message**

Detailed descriptions of cardiovascular morphogenesis and profound knowledge of the genetic and biomechanic factors triggering and orchestrating each single step and event are essential for understanding cardiovascular malformations and diseases.

---

# References

1. Abu-Issa R, Smyth G, Smoak I, Yamamura K, Meyers EN. Fgf8 is required for pharyngeal arch and cardiovascular development in the mouse. Development. 2002;129(19):4613–25.
2. Andelfinger G. Genetic factors in congenital heart malformation. Clin Genet. 2008;73(6):516–27.
3. Anderson RH, Webb S, Brown NA, Lamers W, Moorman A. Development of the heart: (2) Septation of the atriums and ventricles. Heart. 2003;89(8):949–58.
4. Anderson GA, Udan RS, Dickinson ME, Henkelman RM. Cardiovascular patterning as determined by hemodynamic forces and blood vessel genetics. PLoS One. 2015;10(9):e0137175. https://doi.org/10.1371/journal.pone.0137175.
5. Ayadi A, Birling MC, Bottomley J, Bussell J, Fuchs H, Fray M, Gailus-Durner V, Greenaway S, Houghton R, Karp N, Leblanc S, Lengger C, Maier H, Mallon AM, Marschall S, Melvin D, Morgan H, Pavlovic G, Ryder E, Skarnes WC, Selloum M, Ramirez-Solis R, Sorg T, Teboul L, Vasseur L, Walling A, Weaver T, Wells S, White JK, Bradley A, Adams DJ, Steel KP, Hrabe de Angelis M, Brown SD, Herault Y. Mouse large-scale phenotyping initiatives: overview of the European Mouse Disease Clinic (EUMODIC) and of the Wellcome Trust Sanger Institute Mouse Genetics Project. Mamm Genome. 2012;23(9–10):600–10. https://doi.org/10.1007/s00335-012-9418-y.
6. Baldessari D, Mione M. How to create the vascular tree? (Latest) help from the zebrafish. Pharmacol Ther. 2008;118(2):206–30. https://doi.org/10.1016/j.pharmthera.2008.02.010.
7. Bamforth SD, Chaudhry B, Bennett M, Wilson R, Mohun TJ, Van Mierop LH, Henderson DJ, Anderson RH. Clarification of the identity of the mammalian fifth pharyngeal arch artery. Clin Anat. 2013;26(2):173–82. https://doi.org/10.1002/ca.22101.
8. Brown CB, Wenning JM, Lu MM, Epstein DJ, Meyers EN, Epstein JA. Cre-mediated excision of Fgf8 in the Tbx1 expression domain reveals a critical role for Fgf8 in cardiovascular development in the mouse. Dev Biol. 2004;267(1):190–202.

9. Geyer SH, Weninger WJ. Some mice feature 5th pharyngeal arch arteries and double-lumen aortic arch malformations. Cells Tissues Organs. 2012;196:90–8. https://doi.org/10.1159/000330789.. 000330789 [pii].

10. Geyer SH, Weninger WJ. Metric characterization of the aortic arch of early mouse fetuses and of a fetus featuring a double lumen aortic arch malformation. Ann Anat. 2013;195(2):175–82. https://doi.org/10.1016/j.aanat.2012.09.001.. S0940-9602(12)00139-2 [pii].

11. Geyer SH, Reissig LF, Husemann M, Hofle C, Wilson R, Prin F, Szumska D, Galli A, Adams DJ, White J, Mohun TJ, Weninger WJ. Morphology, topology and dimensions of the heart and arteries of genetically normal and mutant mouse embryos at stages S21-S23. J Anat. 2017;231(4):600–14. https://doi.org/10.1111/joa.12663.

12. Handschuh S, Beisser CJ, Ruthensteiner B, Metscher BD. Microscopic dual-energy CT (microDECT): a flexible tool for multichannel ex vivo 3D imaging of biological specimens. J Microsc. 2017. https://doi.org/10.1111/jmi.12543.

13. Kelly RG, Buckingham ME, Moorman AF. Heart fields and cardiac morphogenesis. Cold Spring Harb Perspect Med. 2014;4(10):a015750. https://doi.org/10.1101/cshperspect.a015750.

14. Linask KK, Yu X, Chen Y, Han MD. Directionality of heart looping: effects of Pitx2c misexpression on flectin asymmetry and midline structures. Dev Biol. 2002;246(2):407–17. https://doi.org/10.1006/dbio.2002.0661.. S0012160602906615 [pii].

15. Liu X, Tobita K, Francis RJ, Lo CW. Imaging techniques for visualizing and phenotyping congenital heart defects in murine models. Birth Defects Res C Embryo Today. 2013;99(2):93–105. https://doi.org/10.1002/bdrc.21037.

16. Liu M, Maurer B, Hermann B, Zabihian B, Sandrian MG, Unterhuber A, Baumann B, Zhang EZ, Beard PC, Weninger WJ, Drexler W. Dual modality optical coherence and whole-body photoacoustic tomography imaging of chick embryos in multiple development stages. Biomed Opt Express. 2014;5(9): 3150–9. https://doi.org/10.1364/boe.5.003150.

17. Männer J. The anatomy of cardiac looping: a step towards the understanding of the morphogenesis of several forms of congenital cardiac malformations. Clin Anat. 2009;22(1):21–35. https://doi.org/10.1002/ca.20652.

18. Mohun TJ, Weninger WJ. Imaging heart development using high-resolution episcopic microscopy. Curr Opin Genet Dev. 2011;21(5):573–8. https://doi.org/10.1016/j.gde.2011.07.004.. S0959-437X(11)00111-0 [pii].

19. Moon AM. Mouse models for investigating the developmental basis of human birth defects. Pediatr Res. 2006;59(6):749–55. https://doi.org/10.1203/01.pdr.0000218420.00525.98.

20. Norris FC, Wong MD, Greene ND, Scambler PJ, Weaver T, Weninger WJ, Mohun TJ, Henkelman RM, Lythgoe MF. A coming of age: advanced imaging technologies for characterising the developing mouse. Trends Genet. 2013;29(12):700–11. https://doi.org/10.1016/j.tig.2013.08.004.

21. Ribeiro I, Kawakami Y, Buscher D, Raya A, Rodriguez-Leon J, Morita M, Rodriguez Esteban C, Izpisua Belmonte JC. Tbx2 and Tbx3 regulate the dynamics of cell proliferation during heart remodeling. PLoS One. 2007;2(4):e398. https://doi.org/10.1371/journal.pone.0000398.

22. Swift MR, Weinstein BM. Arterial-venous specification during development. Circ Res. 2009;104(5): 576–88. https://doi.org/10.1161/CIRCRESAHA.108.188805.

23. Tobita K, Liu X, Lo CW. Imaging modalities to assess structural birth defects in mutant mouse models. Birth Defects Res C Embryo Today. 2010;90(3):176–84. https://doi.org/10.1002/bdrc.20187.

24. Weninger WJ, Geyer SH, Mohun TJ, Rasskin-Gutman D, Matsui T, Ribeiro I, Costa Lda F, Izpisua-Belmonte JC, Müller GB. High-resolution episcopic microscopy: a rapid technique for high detailed 3D analysis of gene activity in the context of tissue architecture and morphology. Anat Embryol. 2006;211(3):213–21.

25. Weninger WJ, Geyer SH, Martineau A, Galli A, Adams DJ, Wilson R, Mohun TJ. Phenotyping structural abnormalities in mouse embryos using high-resolution episcopic microscopy. Dis Model Mech. 2014;7(10):1143–52. https://doi.org/10.1242/dmm.016337.. 7/10/1143 [pii].

# Cellular and Molecular Mechanisms of Vasculogenesis, Angiogenesis, and Lymphangiogenesis

*Pavel Uhrin*

© Springer Nature Switzerland AG 2019
M. Geiger (ed.), *Fundamentals of Vascular Biology*, Learning Materials in Biosciences,
https://doi.org/10.1007/978-3-030-12270-6_7

**What You Will Learn in This Chapter**

In vertebrates, two specialized vascular systems facilitate effective fluid circulation: the blood vessels and the lymphatic system. These two tubular networks are closely linked and contribute to the homeostasis of the organism at cellular and tissue level. In this chapter you will learn about cellular and molecular mechanisms of blood and lymphatic vascular system formation. This chapter also illustrates conditions and disease states when the mechanisms of blood and lymphatic vascular system formation are not properly working.

## 7.1 Cellular and Molecular Mechanisms of Blood Vascular System Generation

### 7.1.1 Introduction



The essential components of the vascular system are the heart, blood, and blood vessels. Functioning of the blood vascular system affects numerous physiological and pathological processes. This system plays, e.g., an important role in the delivery of oxygen and nutrients to the tissues, in hormone transport, in the removal of carbon dioxide and degradation products, in blood coagulation, in warm regulation, and in immune function. In pathological conditions, the vascular system contributes to tumor growth and dissemination of cancer cells.

### 7.1.2 Vasculogenesis, Angiogenesis, and Arteriogenesis

The blood vessel system is generated during the embryonic development by processes referred to as vasculogenesis and angiogenesis (◘ Fig. 7.1).

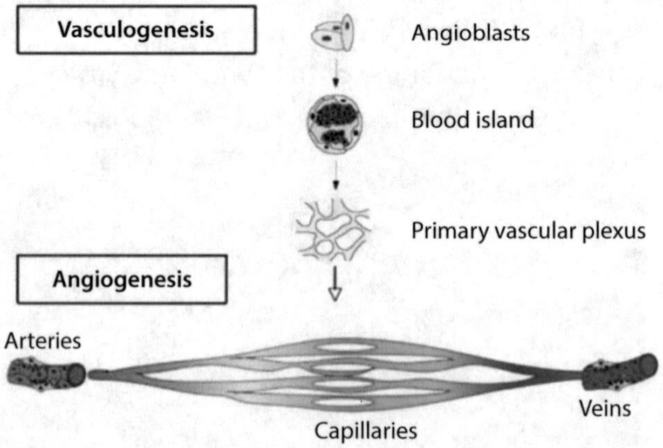


◘ **Fig. 7.1** Generation of the blood vascular system during the embryonic development. Blood vascular system is generated from precursor cells designated as angioblasts. Angioblasts together with blood cells constitute blood islands and form a primary vascular plexus before the onset of heartbeat. Veins and arteries are generated by expanding of the primitive vascular plexus in a process of angiogenesis, involving vessel remodeling and maturation. (Figure modified, and reproduced with permission, from Ref. [9] p. 32, Springer Verlag)

*Vasculogenesis* represents the de novo assembly of blood vessels from endothelial progenitor cells called angioblasts [1]. Angioblasts derive from mesoderm, and they share a common origin with blood cells. The formation of angioblasts is induced by fibroblast growth factors (FGFs). Although vasculogenesis is operative mainly during the embryonic development, it also contributes to blood vessel generation and remodeling during postnatal life, especially in ischemic, malignant, and inflamed tissues, by involvement of multipotent adult progenitor cells and other cell types [2].

*Angiogenesis* involves sprouting and subsequent stabilization of these sprouts by mural cells – pericytes in medium-sized vessels and vascular smooth muscle cells (vSMCs) in large vessels. Angiogenesis also encompasses vessel remodeling by collateral growth to a more complex and sophisticated vascular network [3].

*Arteriogenesis* denotes formation of arteries during the embryonic development. Endothelial progenitor cells may differentiate either into vein or arterial endothelial cells, and this process is influenced, e.g., by proteins of the Notch family [4, 5], as well as ephrin family members, specifically ephrinB2 and ephrinB4 [6–8].

### 7.1.3 Pro-angiogenic and Anti-angiogenic Factors Affecting Blood Vessel Formation and Regression

During vasculogenesis, vascular endothelial growth factor VEGF-A, together with its cognate VEGF receptor 2 (VEGFR2), fosters growth of endothelial cells and their survival [1, 10]. Further stimuli to this process provide cell adhesion molecules such as VE-cadherin and PECAM-1 (CD-31), as well as transcription factors (e.g., ets-1) that participate in subsequent events of endothelial cell differentiation, apoptosis, and angiogenesis [1]. However, VEGF-A does not solely act on endothelial cells, but it also affects other cell types. For example, VEGF-A acts as a chemoattractant on monocytes [11]. Other growth factors affecting angiogenesis include placental growth factor (PlGF), a homolog of VEGF-A, and angiopoietin 1 and angiopoietin 2 that bind to their receptor Tie-2. While angiopoietin 1 is important for the recruitment of vSMCs and pericytes, angiopoietin 2 acts antagonistically on angiopoietin 1 [12]. Basic fibroblast growth factor (bFGF) stimulates proliferation and differentiation of numerous cell types that are necessary for blood vessel formation, and it increases formation of vascular capillaries [13]. Platelet-derived growth factor (PDGF), and especially its isoform BB, is expressed particularly strongly in tip cells of angiogenic sprouts and in the endothelium of growing arteries, where it actively induces recruitment of pericytes and vSMCs [14, 15]. Hypoxia-inducible factor-1$\alpha$ (HIF-1$\alpha$), rapidly synthesized under hypoxic condition, robustly increases synthesis of many pro-angiogenic factors, including VEGF-A [16]. For maintaining the integrity of established blood vessels, proper blood flow and thrombocytes are very important [17–20].

The established blood vessels may also regress. Although such regression occurs physiologically when the nascent vasculature consists of too many vessels, increased regression of blood vessels may be a hallmark and contributory factor to many pathological conditions. It can be induced, e.g., by anti-angiogenic factors secreted by cells of the connective tissue or by immune cells. Thrombospondin 1, a large glycoprotein affecting adhesion, and endostatin, a C-terminal fragment derived from type XVIII collagen, suppress proliferation and survival of vascular cells [21, 22]. Angiostatin, another anti-angiogenic factor, is a cleavage product of a coagulation protein plasmin [23]. Interferons may exert their anti-angiogenic effects by diminishing bFGF levels [24, 25].

### 7.1.4 Process of Angiogenesis Affects Numerous Physiological and Pathological Processes

Numerous animal and human studies demonstrated that angiogenesis plays an important role in wound healing and menstruation as well as in disease states, such as cardiovascular disease, stroke, psoriasis, age-related blindness, and vasculitis, and the list of known conditions where angiogenesis plays an important role has been constantly growing [2]. Augmented angiogenesis occurs in the retina of diabetic patients [26], in the intestine of patients with inflammatory bowel disease [27], in the bones and joints of patients with arthritis or synovitis [28, 29], and in the lung of patients with pulmonary hypertension or asthma [30]. Furthermore, insufficient angiogenesis or vessel regression may also affect organ function. Such condition may cause heart and brain ischemia due to an impairment in collateral blood vessel growth, e.g., in diabetic patients [31, 32]. Collateral blood vessel growth is also impaired in patients with atherosclerosis [33], and it decreases with age, as seen e.g., in older experimental animals upon the exposure to limb ischemia [34] or arterial injury [35]. Reduced angiogenesis also contributes to preeclampsia [36] and osteoporosis [37].



The elaboration of a technique of gene knockout, based on the use of embryonic stem cells and gene inactivation via homologous recombination, has allowed to elucidate the role of different genes in blood vessel formation. Remarkably, in many cases it was found that functionality of blood vessels may be severely impeded by mutations in single genes. For example, disruption of only one allele of VEGF-A led to embryonic lethality of mice due to insufficient vascular development [38]. Likewise, inactivation of VEGFR2 was early embryonic lethal due to defects in the vasculature [39]. Mice deficient for PDGF-B and PDGFR-β continued to develop only until embryonic days E16–E19, at which time massive hemorrhage and edema occurred. In these animals, the lack of pericyte and vSMC was observed already at the onset of angiogenic sprouting at around E10 [14, 40, 41]. In addition, clinical studies revealed low expression levels of VEGF-A and neuropilin-1 (co-receptor of VEGF-A and semaphorin) in patients with DiGeorge syndrome, characterized by congenital heart problems and other symptoms [42]. Mutations in Tie-2 gene were found to cause venous malformations [43], and mutations in Notch-3 gene resulted in the hereditary arteriopathy and increased susceptibility to stroke [44].

Furthermore, angiogenesis has been shown to play a crucial role in tumor development, supporting primary tumor growth, and later on fostering its metastatic spreading [45]. Specifically, tumor vascularization was found very important for the initiation of tumor growth beyond certain size, as advocated in a theory of "angiogenic switch" by Judah Folkman [46]. In the initial avascular phase of tumor growth, when the tumor is still small, normal levels of interstitial nutrients are sufficient for tumor survival. However, such avascular phase ends after the tumor has reached a size of 1–2 mm. At that stage, the hypoxia in the inner mass of the tumor induces production of HIF-1α that robustly stimulates VEGF-A secretion, leading to sprouting of existing vessels, thus supplying the tumor with blood and nourishment (◘ Fig. 7.2).

In addition to secretion of HIF-1α and VEGF-A, tumors have developed remarkable ways to further stimulate their growth. For example, they release chemotactic cytokines, thus attracting immune cells that secrete angiogenic factors, further facilitating tumor vascularization. Furthermore, tumors not only utilize the existing blood vessel system for their modification by angiogenesis, but they also generate new blood vessels by vasculogenesis or use the mechanisms of "vascular mimicry." In the former process, tumor cells act as pluripotent stem cells that are able to differentiate into endothelial cells and contribute to the

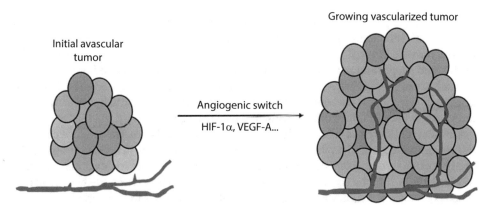

● **Fig. 7.2** Upon reaching a certain size, the hypoxic conditions within the tumor stimulate the production of HIF-1α leading to subsequent VEGF-A release from tumor cells, thus fostering sprouting of existing vessels. The generated blood vessels promote tumor growth

formation of new vessels. The latter process denotes the generation of microvascular channels by aggressive and deregulated tumor cells. All these mechanisms promote tumor vascularization and subsequent tumor dissemination to distant locations [47].

In agreement with the assumption that angiogenesis in tumors can be inhibited, and this would keep them small and clinically not relevant, drugs targeting VEGF or VEGF receptors have been developed. Avastin (bevacizumab), an antibody against VEGF-A, and ramucirumab, an antibody against VEGFR2, as well as other VEGF pathway inhibitors (including small molecule tyrosine kinase inhibitors sunitinib, sorafenib, and pazopanib), are in clinical use in combination with immunotherapy or cytostatic drugs for treatment of a variety of cancer types. However, latest findings argue that the effectiveness of the current anti-angiogenic therapies focused on targeting VEGF in cancer might be limited due to intrinsic tumor resistance or developed drug resistance [48]. Therefore researchers are developing alternative ways of tumor inhibition that are based, e.g., on pharmacological suppression of angiogenesis using metabolic targeting. As sprouting of endothelial cells is utterly glycolysis-dependent [49], transient suppression of glycolysis, e.g., by inhibiting the glycolytic activator PFKFB3, seems to be promising for defeating pathological angiogenesis without causing systemic effects [50, 51].

## 7.2 Cellular and Molecular Mechanisms of Lymphatic Vascular System Generation

### 7.2.1 Introduction

Although the blood vascular system has been recognized for centuries, the lymphatic system was initially described only in 1627. At that time Gaspare Aselli observed in dogs fed with lipid-rich meal structures that he designates as the "lacteae venae," i.e., milky veins. The lymphatic system is composed of a vascular network of thin-walled capillaries and larger lymphatic collecting vessels and the thoracic duct. In addition, the lymphatic system also comprises lymphoid organs such as the lymph nodes, tonsils, Peyer's patches, the spleen, and the thymus, which all play an important role in the immune response.

Lymphatic capillaries, in contrast to blood vessel capillaries, do not have a continuous basement membrane, and this enables them to continuously receive interstitial fluid containing macromolecules, cells, and lipids. Larger collecting lymphatic vessels are covered by vSMCs and supporting pericytes and have a basement membrane. In addition, they have luminal valves, and this helps them to prevent backflow of the lymph [52, 53]. The lymphatic system pervades almost the entire body, with the exception of some partially avascular structures such as the epidermis, cornea, nails, and hair [54]. Until recently, the central nervous system was considered as an organ devoid of lymphatic vasculature. However, recent work showed the presence of a lymphatic vessel network in the dura mater of the mouse brain [55]. The lymphatic system enters the blood circulation through the thoracic duct into the right subclavian vein [53]. Lately, also other communication routes between the lymphatic and venous systems – specifically in the axillary and subiliac regions – were reported [56].

## 7.2.2  Lymphangiogenesis and Factors and Conditions Affecting Lymphatic Vessel Formation

*Lymphangiogenesis*, the generation of the lymphatic system, starts shortly after the vascular system has been formed by vasculogenesis and angiogenesis. In mice lymphangiogenesis begins in ~ E9.0–9.5 [54] and in human in the 6th to 7th embryonic week [57].

Two models of lymphangiogenesis were presented in the beginning of the twentieth century. At around 1902 Florence Sabin injected ink into the skin of pig embryos, which allowed her to visualize lymph sacs and lymph vessels. She concluded that lymphatics originate from embryonic veins [58]. At Sabin's times Huntington and McClure proposed that mesenchymal cells are responsible for the generation of lymph sacs and lymphatic vasculature [59]. Although contemporary studies using, e.g., a detailed "line tracing approach" in mouse knockout models fully supported Sabin's hypothesis [60], it was also shown that mesenchymal cells to a certain extent contribute to the generation of lymph sacs and lymphatic vasculature [61].

Discovery of specific markers of the lymphatic endothelium, e.g., Lyve-1 and podoplanin [62, 63], as well as the extensive use of mouse knockout models in the last two to three decades, enabled researchers to elucidate the mechanism of lymphangiogenesis. It was shown that the development of the lymphatic system is governed by the polarized expression of a master control gene of lymphatic development, the Prox1 gene [64]. Subsequent formation of lymph sacs is achieved via migration and sprouting of cells derived from the cardinal vein, and this process is under the control of VEGF-C [65]. The lymphatic system further develops "centrifugally" under the influence of angiopoietin-2, ephrinB2, FOXC2, and podoplanin genes, as reviewed [57]. In addition, integrin-$\alpha$9 is responsible for the proper lymphatic valve morphogenesis [66].

During the embryonic development, the lymphatic system gets separated from the blood vessel system, and until recently it was not clear how this process takes place [57]. The "nonseparation phenotype" of the blood vessels and lymphatic system was firstly described in a pivotal study of a group of Mark Kahn, who reanalyzed previously generated mice deficient in SLP-76, Syk, or phospholipase-C$\gamma$2 genes [67]. The cause for this "nonseparation" has remained, however, enigmatic. At the Medical University of Vienna, we identified the "nonseparation phenotype" upon the disruption of podoplanin gene in mice. We showed that podoplanin and activated platelets, both present on sprouting lymph sacks (◻ Fig. 7.3), are critically responsible for the separation of the blood and lymphatic circulation during the embryonic development [68, 69]. This "platelet hypothesis" was corroborated soon by

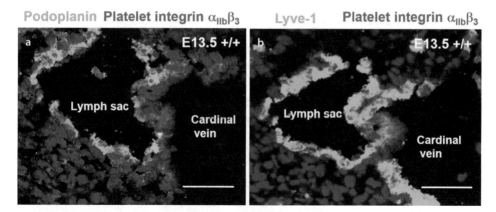

□ **Fig. 7.3** Platelet aggregation driven by podoplanin is linked to separation of the lymph sacs from cardinal veins. In the mouse, the lymphatic system starts to develop from the cardinal vein at around ~ E9.0–9.5. This process of lymphangiogenesis is initiated by polarized expression of Prox1 and Lyve-1 on the cardinal vein and later on followed by formation of the lymphatic sac under the control of VEGF-C. Pictures show the presence of platelet thrombi (labeled for integrin $\alpha_{IIb}\beta_3$ in red) on sprouting podoplanin- **a** and Lyve-1- **b** positive cells, facilitating the developmental separation of lymphatic and blood vessels, as previously reported [69]. Scale bars in panels equal 50 μm

other studies, as the "nonseparation phenotype" was found also upon disruption of the homeodomain transcription factor Meis1 in mice completely lacking platelets [70]. Further proof for the critical role of podoplanin and platelet aggregation/activation in the developmental separation of blood and lymphatic system was provided by the group of Mark Kahn. The authors showed that disruption of the CLEC-2, podoplanin's receptor expressed mainly on the platelets, induced in CLEC-2-deficient mice the "nonseparation phenotype," thus recognizing platelets as the cell type in which SLP-76 signaling is required to regulate lymphatic vascular development [71].

### 7.2.3 Proper Functioning of the Lymphatic System Affects Numerous Physiological and Pathological Processes

The physiological role of the lymphatic system is to return proteins and fluids that have seeped into the extracellular space from the blood vessels (extravasation) back into the blood circulation. The lymphatic system functions as a conduit for lymphocytes and antigen-presenting cells that are part of the lymph fluid [72]. In addition, this system enables the absorption of triglycerides from the intestine [53]. Lymphatic system is also important for obesity, inflammation, and regulation of salt storage during hypertension [73]. The failure of the lymph transport leads to *lymphedema* – an accumulation of lymphatic fluid in the tissue, a harmful, disabling, and occasionally life-threatening disease [74]. Lymphedemas can be classified as primary and secondary.

Primary lymphedemas are caused by gene mutations. In patients with Milroy disease characterized by the absence or reduction of lymphatic vessels seen at birth [75], mutations in vascular endothelial growth factor receptor 3 (VEGFR3) have been identified [76–78]. In persons with lymphedema-distichiasis syndrome characterized by distichiasis (a double row of eyelashes) at birth, and bilateral lower limb lymphedema at puberty, mutations in the transcription factor FOXC2 were found [79, 80]. In these patients, in spite of the normal number of lymphatic vessels, their lymphatic draining is not functioning,

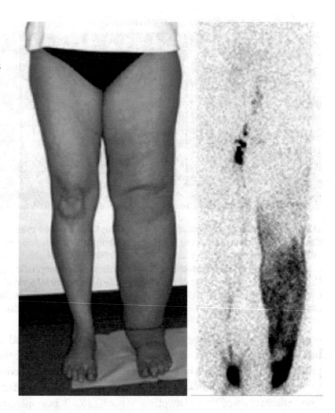

**Fig. 7.4**   A 46-year-old woman after surgery and radiation for uterine cancer. Lymphedema in her left leg was confirmed by the lymphoscinti-gram, which shows marked dermal backflow. (Figure reproduced, with permission, from Ref. [92], p. 14, Springer Verlag. Photos courtesy Emily Iker)

probably because of the lack of the lymphatic valves and impaired coverage of lymphatic capillaries with mural cells [81]. In human affected with hypotrichosis-lymphedema-tel-angiectasia (a syndrome characterized by hair loss and underdeveloped eyebrows and eyelashes, lymphedema, and dilated blood vessels on the skin), mutations in the transcription factor SOX18 were detected [82, 83]. Another cause for primary lymphedemas can be mutations in integrin-α9 gene detected in human fetuses with severe chylothorax [84].

Secondary lymphedemas represent worldwide most cases of lymphatic dysfunction, and they are instigated by some kind of external damage to the lymphatic vasculature. Secondary lymphedema can be caused by infections (e.g., filariasis), surgery, or radio-therapy. Filariasis itself is the most common secondary lymphedema affecting more than 100 million people, especially in tropical areas [85]. Filariasis is the result of a parasitic infection by mosquito-borne parasitic worms (*Wuchereria bancrofti* or *Brugia malayi*) which are located in the lymphatic system, where an inflammatory response stimulates the production of VEGF-A, VEGF-C, and VEGF-D. It often causes hyperplasia, obstruction, and extensive damage to the lymphatic vessel, finally leading to a chronic lymph-edema of the lower limbs or genitals, which leads to permanent disability [86–88].

In the industrial world, secondary lymphedemas are primarily initiated by lymph node dissection or radiation therapy [88]. Impaired lymphatic transport causing edemas occurs in 15–20% of cases after breast cancer surgery [89], but it may be caused also by surgery and radiation of other cancer types (Fig. 7.4). Regrettably, treatment of lymph-edema is still mainly restricted to conservative therapies such as compression garments, massage, manual drainage, liposuction, and modifications of the diet (primarily focused on minimizing the consumption of long-chain fatty acids) [88, 90, 91].

A study in mice showed that delivery of VEGF-C/VEGF-D via adenovirus stimulates regeneration of collecting lymphatic vessels, including the formation of intraluminal valves, after the excision of lymph nodes and the adjacent collecting lymphatic vessels [93]. These results suggested that VEGF-C/VEGF-D delivery might provide a basis for therapy of lymphedema, especially in cases of injury and restoration of primary lymph-edemas but also in other conditions [94–96]. At present, a phase I multicenter clinical study in patients with early breast cancer-related upper extremity lymphedema is currently in progress. It should assess the safety and efficacy of using a VEGF-C adenoviral vector in combination with vascularized lymph node transplantation [97].

Lymphangiogenesis also plays an important role under pathological conditions in cancer. Kaposi sarcomas in AIDS patients, as well as angiosarcomas, express both lymphatic and blood vascular endothelial markers [62]. Lymphangiogenesis also facilitates primary tumor growth and cancer cell dissemination, and the presence of cancer cells in the tumor-adjacent "sentinel" lymph node has been shown to correlate with survival of patients [98, 99]. Recently, by directly injecting cancer cells into the lymphatic vessels of mice and following up their migration into the lymph nodes, researchers demonstrated that blood vessels of the lymph nodes, designated as "high endothelial venules," serve as an exit route for rapid systemic dissemination of cancer cells [100]. Currently it remains to be determined if such form of tumor cell spreading also occurs in human cancer patients.

---
### Take-Home Message

- Blood vessel formation is controlled by a complex interplay of pro-angiogenic and anti-angiogenic factors. Vasculogenesis represents the de novo assembly of blood vessels by endothelial progenitor cells. Angiogenesis involves sprouting and subsequent stabilization of these sprouts by pericytes in medium-sized and vSMCs in large vessels. Arteriogenesis denotes formation of arteries.
- The status of blood vessels affects many conditions, including cardiovascular disease, diabetes, stroke, age-related blindness, psoriasis, osteoporosis, as well as wound healing and menstruation. Angiogenesis also contributes to primary tumor formation and metastasizing.
- The lymphatic system starts to develop shortly after the vascular system has been formed by vasculogenesis and angiogenesis, and this process is referred to as "lymphangiogenesis."
- The physiological role of the lymphatic system is to return proteins and fluids that have seeped into the extracellular space from the blood vessels (extravasation) back into the blood circulation. The lymphatic system functions as a conduit for lymphocytes and antigen-presenting cells that are part of the lymph fluid. In addition, this system enables the absorption of triglycerides from the intestine. The lymphatic system is also important for obesity, inflammation, and regulation of salt storage during hypertension. In pathological conditions, it also contributes to dissemination of cancer cells.
- Primary lymphedemas are caused by genetic mutations. Secondary lymphedemas represent worldwide most cases of lymphatic dysfunction and include, e.g., filariasis as well as lymphedemas caused by lymph node dissection or radiation therapy. Treatment of lymphedema is still mainly restricted to conservative therapies such as manual emptying, massage, compression garments, liposuction, and diet modification, and new treatment strategies are being developed.

# References

1. Risau W, Flamme I. Vasculogenesis. Annu Rev Cell Dev Biol. 1995;11:73–91.
2. Carmeliet P. Angiogenesis in health and disease. Nat Med. 2003;9:653–60.
3. Risau W. Mechanisms of angiogenesis. Nature. 1997;386:671–4.
4. Lawson ND, Scheer N, Pham VN, Kim CH, Chitnis AB, Campos-Ortega JA, Weinstein BM. Notch signaling is required for arterial-venous differentiation during embryonic vascular development. Development. 2001;128:3675–83.
5. Fischer A, Schumacher N, Maier M, Sendtner M, Gessler M. The Notch target genes Hey1 and Hey2 are required for embryonic vascular development. Genes Dev. 2004;18:901–11.
6. Wang HU, Chen ZF, Anderson DJ. Molecular distinction and angiogenic interaction between embryonic arteries and veins revealed by ephrin-B2 and its receptor Eph-B4. Cell. 1998;93:741–53.
7. Adams RH, Wilkinson GA, Weiss C, Diella F, Gale NW, Deutsch U, Risau W, Klein R. Roles of ephrinB ligands and EphB receptors in cardiovascular development: demarcation of arterial/venous domains, vascular morphogenesis, and sprouting angiogenesis. Genes Dev. 1999;13:295–306.
8. Gerety SS, Wang HU, Chen ZF, Anderson DJ. Symmetrical mutant phenotypes of the receptor EphB4 and its specific transmembrane ligand ephrin-B2 in cardiovascular development. Mol Cell. 1999;4:403–14.
9. Marmé D, Fusenig N, editors. Tumor angiogenesis: basic mechanisms and cancer therapy. Berlin/Heidelberg: Springer; 2008.
10. Ferrara N, Davis-Smyth T. The biology of vascular endothelial growth factor. Endocr Rev. 1997;18:4–25.
11. Clauss M, Gerlach M, Gerlach H, Brett J, Wang F, Familletti PC, Pan YC, Olander JV, Connolly DT, Stern D. Vascular permeability factor: a tumor-derived polypeptide that induces endothelial cell and monocyte procoagulant activity, and promotes monocyte migration. J Exp Med. 1990;172:1535–45.
12. Tait CR, Jones PF. Angiopoietins in tumours: the angiogenic switch. J Pathol. 2004;204:1–10.
13. Stegmann TJ. FGF-1: a human growth factor in the induction of neoangiogenesis. Expert Opin Investig Drugs. 1998;7:2011–5.
14. Hellstrom M, Kalen M, Lindahl P, Abramsson A, Betsholtz C. Role of PDGF-B and PDGFR-beta in recruitment of vascular smooth muscle cells and pericytes during embryonic blood vessel formation in the mouse. Development. 1999;126:3047.
15. Gerhardt H, Golding M, Fruttiger M, Ruhrberg C, Lundkvist A, Abramsson A, Jeltsch M, Mitchell C, Alitalo K, Shima D, Betsholtz C. VEGF guides angiogenic sprouting utilizing endothelial tip cell filopodia. J Cell Biol. 2003;161:1163.
16. Pugh CW, Ratcliffe PJ. Regulation of angiogenesis by hypoxia: role of the HIF system. Nat Med. 2003;9:677–84.
17. Murakami M, Simons M. Regulation of vascular integrity. J Mol Med (Berlin, Germany). 2009;87:571–82.
18. Hahn C, Schwartz MA. Mechanotransduction in vascular physiology and atherogenesis. Nat Rev Mol Cell Biol. 2009;10:53–62.
19. Ho-Tin-Noe B, Boulaftali Y, Camerer E. Platelets and vascular integrity: how platelets prevent bleeding in inflammation. Blood. 2018;131:277–88.
20. Ho-Tin-Noe B, Demers M, Wagner DD. How platelets safeguard vascular integrity. J Thromb Haemost. 2011;9(Suppl 1):56–65.
21. O'Reilly MS, Boehm T, Shing Y, Fukai N, Vasios G, Lane WS, Flynn E, Birkhead JR, Olsen BR, Folkman J. Endostatin: an endogenous inhibitor of angiogenesis and tumor growth. Cell. 1997;88:277–85.
22. Jimenez B, Volpert OV, Crawford SE, Febbraio M, Silverstein RL, Bouck N. Signals leading to apoptosis-dependent inhibition of neovascularization by thrombospondin-1. Nat Med. 2000;6:41–8.
23. O'Reilly MS. Angiostatin: an endogenous inhibitor of angiogenesis and of tumor growth. EXS. 1997;79:273–94.
24. Singh R, Bucana C, Llansa N, Sanchez R, Fidler I. Cell density-dependent modulation of basic fibroblast growth factor expression by human interferon-beta. Int J Oncol. 1996;8:649–56.
25. Sasamura H, Takahashi A, Miyao N, Yanase M, Masumori N, Kitamura H, Itoh N, Tsukamoto T. Inhibitory effect on expression of angiogenic factors by antiangiogenic agents in renal cell carcinoma. Br J Cancer. 2002;86:768–73.
26. Capitao M, Soares R. Angiogenesis and inflammation crosstalk in diabetic retinopathy. J Cell Biochem. 2016;117:2443–53.

27. Scaldaferri F, Vetrano S, Sans M, Arena V, Straface G, Stigliano E, Repici A, Sturm A, Malesci A, Panes J, et al. VEGF-A links angiogenesis and inflammation in inflammatory bowel disease pathogenesis. Gastroenterology. 2009;136:585–595.e585.
28. Paleolog EM. Angiogenesis in rheumatoid arthritis. Arthritis Res. 2002;4(Suppl 3):S81–90.
29. Elshabrawy HA, Chen Z, Volin MV, Ravella S, Virupannavar S, Shahrara S. The pathogenic role of angiogenesis in rheumatoid arthritis. Angiogenesis. 2015;18:433–48.
30. Knox AJ, Stocks J, Sutcliffe A. Angiogenesis and vascular endothelial growth factor in COPD. Thorax. 2005;60:88–9.
31. Rivard A, Silver M, Chen D, Kearney M, Magner M, Annex B, Peters K, Isner JM. Rescue of diabetes-related impairment of angiogenesis by intramuscular gene therapy with adeno-VEGF. Am J Pathol. 1999;154:355–63.
32. Waltenberger J. Impaired collateral vessel development in diabetes: potential cellular mechanisms and therapeutic implications. Cardiovasc Res. 2001;49:554–60.
33. Van Belle E, Rivard A, Chen D, Silver M, Bunting S, Ferrara N, Symes JF, Bauters C, Isner JM. Hypercholesterolemia attenuates angiogenesis but does not preclude augmentation by angiogenic cytokines. Circulation. 1997;96:2667–74.
34. Rivard A, Fabre JE, Silver M, Chen D, Murohara T, Kearney M, Magner M, Asahara T, Isner JM. Age-dependent impairment of angiogenesis. Circulation. 1999;99:111–20.
35. Gennaro G, Menard C, Michaud SE, Rivard A. Age-dependent impairment of reendothelialization after arterial injury: role of vascular endothelial growth factor. Circulation. 2003;107:230–3.
36. Chaiworapongsa T, Chaemsaithong P, Yeo L, Romero R. Pre-eclampsia part 1: current understanding of its pathophysiology. Nat Rev Nephrol. 2014;10:466–80.
37. Filipowska J, Tomaszewski KA, Niedzwiedzki L, Walocha JA, Niedzwiedzki T. The role of vasculature in bone development, regeneration and proper systemic functioning. Angiogenesis. 2017;20:291–302.
38. Carmeliet P, Ferreira V, Breier G, Pollefeyt S, Kieckens L, Gertsenstein M, Fahrig M, Vandenhoeck A, Harpal K, Eberhardt C, et al. Abnormal blood vessel development and lethality in embryos lacking a single VEGF allele. Nature. 1996;380:435–9.
39. Shalaby F, Rossant J, Yamaguchi TP, Gertsenstein M, Wu XF, Breitman ML, Schuh AC. Failure of blood-island formation and vasculogenesis in Flk-1-deficient mice. Nature. 1995;376:62–6.
40. Soriano P. Abnormal kidney development and hematological disorders in PDGF beta-receptor mutant mice. Genes Dev. 1994;8:1888–96.
41. Hellström M, Gerhardt H, Kalén M, Li X, Eriksson U, Wolburg H, Betsholtz C. Lack of pericytes leads to endothelial hyperplasia and abnormal vascular morphogenesis. J Cell Biol. 2001;153:543.
42. Stalmans I, Lambrechts D, De Smet F, Jansen S, Wang J, Maity S, Kneer P, von der Ohe M, Swillen A, Maes C, et al. VEGF: a modifier of the del22q11 (DiGeorge) syndrome? Nat Med. 2003;9:173–82.
43. Vikkula M, Boon LM, Carraway KL 3rd, Calvert JT, Diamonti AJ, Goumnerov B, Pasyk KA, Marchuk DA, Warman ML, Cantley LC, et al. Vascular dysmorphogenesis caused by an activating mutation in the receptor tyrosine kinase TIE2. Cell. 1996;87:1181–90.
44. Kalimo H, Ruchoux MM, Viitanen M, Kalaria RN. CADASIL: a common form of hereditary arteriopathy causing brain infarcts and dementia. Brain Pathol. 2002;12:371–84.
45. Bergers G, Benjamin LE. Tumorigenesis and the angiogenic switch. Nat Rev Cancer. 2003;3:401–10.
46. Folkman J. Tumor angiogenesis: therapeutic implications. N Engl J Med. 1971;285:1182–6.
47. Rafii S, Lyden D, Benezra R, Hattori K, Heissig B. Vascular and haematopoietic stem cells: novel targets for anti-angiogenesis therapy? Nat Rev Cancer. 2002;2:826–35.
48. Itatani Y, Kawada K, Yamamoto T, Sakai Y. Resistance to anti-angiogenic therapy in cancer-alterations to anti-VEGF pathway. Int J Mol Sci. 2018;19:1232.
49. De Bock K, Georgiadou M, Schoors S, Kuchnio A, Wong BW, Cantelmo AR, Quaegebeur A, Ghesquiere B, Cauwenberghs S, Eelen G, et al. Role of PFKFB3-driven glycolysis in vessel sprouting. Cell. 2013;154:651–63.
50. Schoors S, De Bock K, Cantelmo AR, Georgiadou M, Ghesquiere B, Cauwenberghs S, Kuchnio A, Wong BW, Quaegebeur A, Goveia J, et al. Partial and transient reduction of glycolysis by PFKFB3 blockade reduces pathological angiogenesis. Cell Metab. 2014;19:37–48.
51. Cantelmo AR, Conradi LC, Brajic A, Goveia J, Kalucka J, Pircher A, Chaturvedi P, Hol J, Thienpont B, Teuwen LA, et al. Inhibition of the glycolytic activator PFKFB3 in endothelium induces tumor vessel normalization, impairs metastasis, and improves chemotherapy. Cancer Cell. 2016;30:968–85.

52. Baluk P, Fuxe J, Hashizume H, Romano T, Lashnits E, Butz S, Vestweber D, Corada M, Molendini C, Dejana E, McDonald DM. Functionally specialized junctions between endothelial cells of lymphatic vessels. J Exp Med. 2007;204:2349–62.
53. Oliver G, Detmar M. The rediscovery of the lymphatic system: old and new insights into the development and biological function of the lymphatic vasculature. Genes Dev. 2002;16:773–83.
54. Oliver G. Lymphatic vasculature development. Nat Rev Immunol. 2004;4:35–45.
55. Aspelund A, Antila S, Proulx ST, Karlsen TV, Karaman S, Detmar M, Wiig H, Alitalo K. A dural lymphatic vascular system that drains brain interstitial fluid and macromolecules. J Exp Med. 2015;212:991–9.
56. Shao L, Takeda K, Kato S, Mori S, Kodama T. Communication between lymphatic and venous systems in mice. J Immunol Methods. 2015;424:100–5.
57. Alitalo K, Tammela T, Petrova TV. Lymphangiogenesis in development and human disease. Nature. 2005;438:946–53.
58. Sabin FR. On the origin of the lymphatic system from the veins, and the development of the lymph hearts and thoracic duct in the pig. Am J Anat. 1902;1:367–89.
59. Huntington GS, McClure CFW. The anatomy and development of the jugular lymph sac in the domestic cat (Felis domestica). Am J Anat. 1910;10:177–311.
60. Srinivasan RS, Dillard ME, Lagutin OV, Lin FJ, Tsai S, Tsai MJ, Samokhvalov IM, Oliver G. Lineage tracing demonstrates the venous origin of the mammalian lymphatic vasculature. Genes Dev. 2007;21:2422–32.
61. Buttler K, Ezaki T, Wilting J. Proliferating mesodermal cells in murine embryos exhibiting macrophage and lymphendothelial characteristics. BMC Dev Biol. 2008;8:43.
62. Breiteneder-Geleff S, Soleiman A, Kowalski H, Horvat R, Amann G, Kriehuber E, Diem K, Weninger W, Tschachler E, Alitalo K, Kerjaschki D. Angiosarcomas express mixed endothelial phenotypes of blood and lymphatic capillaries: podoplanin as a specific marker for lymphatic endothelium. Am J Pathol. 1999;154:385–94.
63. Banerji S, Ni J, Wang SX, Clasper S, Su J, Tammi R, Jones M, Jackson DG. LYVE-1, a new homologue of the CD44 glycoprotein, is a lymph-specific receptor for hyaluronan. J Cell Biol. 1999;144:789–801.
64. Hong YK, Harvey N, Noh YH, Schacht V, Hirakawa S, Detmar M, Oliver G. Prox1 is a master control gene in the program specifying lymphatic endothelial cell fate. Dev Dyn. 2002;225:351–7.
65. Karkkainen MJ, Haiko P, Sainio K, Partanen J, Taipale J, Petrova TV, Jeltsch M, Jackson DG, Talikka M, Rauvala H, et al. Vascular endothelial growth factor C is required for sprouting of the first lymphatic vessels from embryonic veins. Nat Immunol. 2004;5:74–80.
66. Bazigou E, Xie S, Chen C, Weston A, Miura N, Sorokin L, Adams R, Muro AF, Sheppard D, Makinen T. Integrin-alpha9 is required for fibronectin matrix assembly during lymphatic valve morphogenesis. Dev Cell. 2009;17:175–86.
67. Abtahian F, Guerriero A, Sebzda E, Lu MM, Zhou R, Mocsai A, Myers EE, Huang B, Jackson DG, Ferrari VA, et al. Regulation of blood and lymphatic vascular separation by signaling proteins SLP-76 and Syk. Science. 2003;299:247–51.
68. Uhrin P, Zaujec J, Bauer M, Breuss J, Alitalo K, Stockinger H, Kerjaschki D, Binder BR. Podoplanin-induced platelet aggregation mediates separation of blood an lymphatic vessels. Vasc Pharmacol. 2006;45:190.
69. Uhrin P, Zaujec J, Breuss JM, Olcaydu D, Chrenek P, Stockinger H, Fuertbauer E, Moser M, Haiko P, Fassler R, et al. Novel function for blood platelets and podoplanin in developmental separation of blood and lymphatic circulation. Blood. 2010;115:3997–4005.
70. Carramolino L, Fuentes J, Garcia-Andres C, Azcoitia V, Riethmacher D, Torres M. Platelets play an essential role in separating the blood and lymphatic vasculatures during embryonic angiogenesis. Circ Res. 2010;106:1197–201.
71. Bertozzi CC, Schmaier AA, Mericko P, Hess PR, Zou Z, Chen M, Chen CY, Xu B, Lu MM, Zhou D, et al. Platelets regulate lymphatic vascular development through CLEC-2-SLP-76 signaling. Blood. 2010;116:661–70.
72. Randolph GJ, Angeli V, Swartz MA. Dendritic-cell trafficking to lymph nodes through lymphatic vessels. Nat Rev Immunol. 2005;5:617–28.
73. Kerjaschki D. The lymphatic vasculature revisited. J Clin Invest. 2014;124:874–7.
74. Wang Y, Oliver G. Current views on the function of the lymphatic vasculature in health and disease. Genes Dev. 2010;24:2115–26.
75. Milroy WF. An undescribed variety of hereditary oedema. N Y Med J. 1892;56:505–8.
76. Ferrell RE, Levinson KL, Esman JH, Kimak MA, Lawrence EC, Barmada MM, Finegold DN. Hereditary lymphedema: evidence for linkage and genetic heterogeneity. Hum Mol Genet. 1998;7:2073–8.

7

77. Irrthum A, Karkkainen MJ, Devriendt K, Alitalo K, Vikkula M. Congenital hereditary lymphedema caused by a mutation that inactivates VEGFR3 tyrosine kinase. Am J Hum Genet. 2000;67:295–301.
78. Karkkainen MJ, Ferrell RE, Lawrence EC, Kimak MA, Levinson KL, McTigue MA, Alitalo K, Finegold DN. Missense mutations interfere with VEGFR-3 signalling in primary lymphoedema. Nat Genet. 2000;25:153–9.
79. Fang J, Dagenais SL, Erickson RP, Arlt MF, Glynn MW, Gorski JL, Seaver LH, Glover TW. Mutations in FOXC2 (MFH-1), a forkhead family transcription factor, are responsible for the hereditary lymphedema-distichiasis syndrome. Am J Hum Genet. 2000;67:1382 8.
80. Finegold DN, Kimak MA, Lawrence EC, Levinson KL, Cherniske EM, Pober BR, Dunlap JW, Ferrell RE. Truncating mutations in FOXC2 cause multiple lymphedema syndromes. Hum Mol Genet. 2001;10:1185–9.
81. Petrova TV, Karpanen T, Norrmen C, Mellor R, Tamakoshi T, Finegold D, Ferrell R, Kerjaschki D, Mortimer P, Yla-Herttuala S, et al. Defective valves and abnormal mural cell recruitment underlie lymphatic vascular failure in lymphedema distichiasis. Nat Med. 2004;10:974–81.
82. Irrthum A, Devriendt K, Chitayat D, Matthijs G, Glade C, Steijlen PM, Fryns JP, Van Steensel MA, Vikkula M. Mutations in the transcription factor gene SOX18 underlie recessive and dominant forms of hypotrichosis-lymphedema-telangiectasia. Am J Hum Genet. 2003;72:1470–8.
83. Bastaki F, Mohamed M, Nair P, Saif F, Tawfiq N, Al-Ali MT, Brandau O, Hamzeh AR. A novel SOX18 mutation uncovered in Jordanian patient with hypotrichosis-lymphedema-telangiectasia syndrome by whole exome sequencing. Mol Cell Probes. 2016;30:18–21.
84. Ma GC, Liu CS, Chang SP, Yeh KT, Ke YY, Chen TH, Wang BB, Kuo SJ, Shih JC, Chen M. A recurrent ITGA9 missense mutation in human fetuses with severe chylothorax: possible correlation with poor response to fetal therapy. Prenat Diagn. 2008;28:1057–63.
85. Wynd S, Melrose WD, Durrheim DN, Carron J, Gyapong M. Understanding the community impact of lymphatic filariasis: a review of the sociocultural literature. Bull World Health Organ. 2007;85:493–8.
86. Pfarr KM, Debrah AY, Specht S, Hoerauf A. Filariasis and lymphoedema. Parasite Immunol. 2009;31:664–72.
87. Taylor MJ, Hoerauf A, Bockarie M. Lymphatic filariasis and onchocerciasis. Lancet. 2010;376:1175–85.
88. Rockson SG. Lymphedema. Am J Med. 2001;110:288–95.
89. Vignes S, Arrault M, Bonhomme S, Spielmann M. Upper limb lymphedema revealing breast cancer. Rev Med Interne. 2007;28:631–4.
90. Brorson H. Liposuction in arm lymphedema treatment. Scand J Surg. 2003;92:287–95.
91. Rockson SG. Lymphedema. Vasc Med. 2016;21:77–81.
92. Tretbar LL, Morgan CL, Lee BB, Simonian SJ, Blondeau B, editors. Lymphedema: diagnosis and treatment. London: Springer; 2008.
93. Tammela T, Saaristo A, Holopainen T, Lyytikka J, Kotronen A, Pitkonen M, Abo-Ramadan U, Yla-Herttuala S, Petrova TV, Alitalo K. Therapeutic differentiation and maturation of lymphatic vessels after lymph node dissection and transplantation. Nat Med. 2007;13:1458–66.
94. Zhou Q, Guo R, Wood R, Boyce BF, Liang Q, Wang YJ, Schwarz EM, Xing L. Vascular endothelial growth factor C attenuates joint damage in chronic inflammatory arthritis by accelerating local lymphatic drainage in mice. Arthritis Rheum. 2011;63:2318–28.
95. Bouta EM, Bell RD, Rahimi H, Xing L, Wood RW, Bingham CO 3rd, Ritchlin CT, Schwarz EM. Targeting lymphatic function as a novel therapeutic intervention for rheumatoid arthritis. Nat Rev Rheumatol. 2018;14:94–106.
96. D'Alessio S, Correale C, Tacconi C, Gandelli A, Pietrogrande G, Vetrano S, Genua M, Arena V, Spinelli A, Peyrin-Biroulet L, et al. VEGF-C-dependent stimulation of lymphatic function ameliorates experimental inflammatory bowel disease. J Clin Invest. 2014;124:3863–78.
97. Schaverien MV, Aldrich MB. New and emerging treatments for lymphedema. Semin Plast Surg. 2018;32:48–52.
98. Alitalo A, Detmar M. Interaction of tumor cells and lymphatic vessels in cancer progression. Oncogene. 2012;31:4499–508.
99. Ma Q, Dieterich LC, Detmar M. Multiple roles of lymphatic vessels in tumor progression. Curr Opin Immunol. 2018;53:7–12.
100. Brown M, Assen FP, Leithner A, Abe J, Schachner H, Asfour G, Bago-Horvath Z, Stein JV, Uhrin P, Sixt M, Kerjaschki D. Lymph node blood vessels provide exit routes for metastatic tumor cell dissemination in mice. Science. 2018;359:1408–11.

# Mechanisms of Hemostasis: Contributions of Platelets, Coagulation Factors, and the Vessel Wall

*Marion Mussbacher, Julia B. Kral-Pointner, Manuel Salzmann, Waltraud C. Schrottmaier, and Alice Assinger*

© Springer Nature Switzerland AG 2019
M. Geiger (ed.), *Fundamentals of Vascular Biology*, Learning Materials in Biosciences,
https://doi.org/10.1007/978-3-030-12270-6_8

**What You Will Learn in This Chapter**

**Hemostasis** is a physiological process that allows rapid, localized, and highly regulated closure of an injured blood vessel while maintaining normal blood flow. It involves platelets (**primary hemostasis**), coagulation factors (**secondary hemostasis**), as well as components of the vessel wall. Upon vessel injury, circulating platelets rapidly adhere to the exposed subendothelial matrix, leading to platelet activation and concomitant release of secondary feedback molecules to stimulate and recruit additional platelets from the circulation. Direct platelet-platelet binding produces a platelet plug (**white thrombus**). In parallel, a cascade of coagulation factors is induced, which leads to cleavage and cross-linking of soluble fibrinogen into insoluble fibrin, forming a tight network transforming the initial, unstable platelet thrombus to a stable thrombus (**red thrombus**). Numerous positive feedback loops as well as **anticoagulant mechanisms** guarantee locally restricted coagulation at defined cellular surfaces and prevent excessive blood loss while ensuring vascular integrity and unrestricted blood flow. During wound healing processes, the thrombus is dissolved by the **fibrinolytic system**, which degrades fibrin and re-establishes physiological blood flow.

## 8

## 8.1 Principles of Primary Hemostasis

### 8.1.1 Platelets: Cells with Unique Structure

Human platelets are small, anucleate cells with a discoid shape and a life span of 7–10 days. With 150–400,000 cells/μl, platelets constitute the second most abundant circulating blood cell, though approximately one third of all platelets are stored in the spleen [1, 2]. Platelets are shed into the blood stream from their progenitor cells in the bone marrow. These large, polyploid **megakaryocytes** respond to thrombopoietin (TPO) as a major regulatory factor to form long cytoplasmic expansions (**proplatelets**), which cross through the bone marrow sinusoids into the circulation, where they fission into platelets. Thus, **thrombopoiesis** can yield up to 10,000 platelets per megakaryocyte.

Platelets contain three different types of granules: α-granules, dense granules, and lysosomes. The most abundant platelet vesicles, numbering 50–80 per platelet, are **α-granules** which may contain up to 1000 different proteins [3], including modulators of coagulation (e.g., FV, FVII, FVIII, FIX, FXIII, fibrinogen), adhesion and membrane proteins (e.g., von Willebrand factor (vWF), thrombospondin, CD40 ligand, P-selectin), antimicrobial peptides (e.g., β-thromboglobulin (β-TG), neutrophil-activating peptide 2 (NAP-2)), chemokines (e.g., platelet factor 4 (PF4/CXCL4), interleukin (IL)-1β, IL-8), as well as multifunctional growth factors (e.g., vascular endothelial growth factor (VEGF), platelet-derived growth factor (PDGF), transforming growth factor-β (TGF-β)) [4, 5]. Due to their lack of nucleus, platelets are generally unable to de novo synthesize proteins with only few reported exceptions such as IL-1β, B-cell lymphoma 3-encoded protein (Bcl-3), and tissue factor (TF) [6–8] which are spliced and translated from pre-existing mRNA. Consequently, proteins in platelet α-granules are prepackaged and derive from megakaryocytes or were taken up from plasma [9–11]. **Dense granules** store small molecules, including ADP, ATP, $Ca^{2+}$, polyphosphates, and bioactive amines such as serotonin. **Lysosomes** are rare platelet vesicles and contain metabolic enzymes such as acid proteases (e.g., cathepsins) and glycohydrolases [4, 12].

Upon activation platelets **degranulate**, releasing the soluble content of α- and dense granules via exocytosis, including the secondary feedback molecule adenosine diphosphate (**ADP**). Additionally, granular membrane proteins are translocated to the cell surface from where they can be cleaved off into the circulation to function as endocrine mediators [4, 13]. Alternatively, exposed membrane proteins (e.g., P-selectin) enable platelets to bind to other cells including other platelets, endothelial cells, or leukocytes. In resting platelets, a marginal ring of **microtubules** maintains the discoid shape. However, during activation platelets undergo a drastic **shape change** and form filopodia and lamellipodia to increase their size, allowing them to form aggregates and cover vascular injuries more easily. This shape change is accompanied by profound cytoskeletal reorganization: The marginal microtubule coil is compressed into the center of the cell and disintegrates, while individual microtubules are formed to allow the formation of cellular protrusions [14].

In addition to specific granules, platelets contain two unique membrane systems: The platelet **dense tubular system (DTS)**, a remnant of megakaryocyte smooth endoplasmic reticulum, functions as internal $Ca^{2+}$ storage and is involved in the synthesis of the secondary feedback molecule **thromboxane $A_2$ (TXA$_2$)** [15]. Furthermore, a unique invaginated membrane network, the **open canalicular system (OCS)**, permeates throughout the platelet like a sponge. Due to its surface connection, it has major functions in uptake and redistribution of substances from the plasma into organelles [16, 17]. Similarly, upon stimulation platelet granules release their content into the OCS [18]. Further, the OCS acts as membrane and adhesion receptor reservoir for the extensive surface expansion during platelet activation [19].

## 8.1.2 Platelet Function in Primary Hemostasis

Hemostasis is a tightly controlled process to ensure vascular integrity upon injury and to prevent excessive platelet activation and thrombus formation which would compromise blood flow. Under resting conditions, endothelial cells constantly release nitric oxide (NO) and prostacyclin (PGI$_2$) and express thrombomodulin (TM) which prevents platelet activation by increasing intra-platelet cyclic guanosine or adenosine monophosphate (cGMP or cAMP) levels or by scavenging thrombin, respectively [20, 21]. These resting platelets **patrol** the circulation. Upon endothelial injury, these protective molecules are lost, and the subendothelial matrix is exposed to the blood flow. Thus, primary hemostasis is initiated and can be structured in three distinct phases, platelet **adhesion**, **activation**, and **aggregation**, as illustrated in ◻ Fig. 8.1.

### 8.1.2.1 Adhesion

Although **vWF** is constantly present in the plasma, its globular form has negligible affinity for its receptor glycoprotein (GP) complex **GPIb-V-IX** on platelets. Immobilization of vWF on subendothelial collagen subjects vWF to high shear stress and induces its conformational unfolding, strongly increasing its receptor affinity [22]. Initial platelet adhesion via vWF/GPIb-V-IX is fleeting but induces platelet rolling and decreases their velocity to allow binding to **collagen** via **GPVI and GPIa/IIa (integrin $\alpha_2\beta_1$)**. While GPIa/IIa mainly mediates platelet firm adhesion to the injured vessel, GPVI primarily causes collagen-induced platelet activation [23, 24].

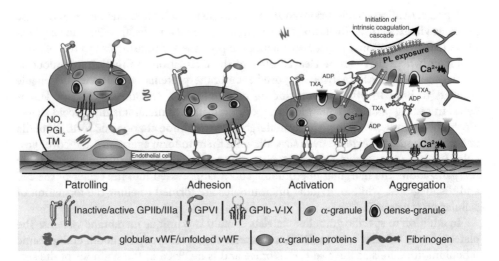

□ **Fig. 8.1** Phases of platelet function in primary hemostasis. Platelets perpetually patrol the vessel lining for endothelial injury to ensure vascular integrity. To keep circulating platelets in a resting state, the intact endothelium secretes nitric oxide (NO) and prostacyclin (PGI$_2$) and express thrombomodulin (TM) which prevents platelet activation. Upon vascular injury, the subendothelial matrix is exposed which induces initial platelet rolling via fleeting von Willebrand factor (vWF)-glycoprotein Ib-V-IX (GPIb-V-IX) binding, followed by stable adhesion via collagen/GPIa/IIa binding. Interactions of vWF/GPIb-V-IX and collagen/GPVI induce complex signaling, increasing intracellular Ca$^{2+}$ levels and resulting in platelet activation with concomitant degranulation of α- and dense granules, surface exposure of negatively charged phospholipids (PL), and activation of the fibrinogen receptor GPIIb/IIIa. Release of the secondary feedback molecules adenosine diphosphate (ADP) and thromboxane A$_2$ (TXA$_2$) recruits and activates further platelets, creating a positive feedback loop. Activated platelets interact with each other to form platelet aggregates which are stabilized by fibrinogen cross-linking to form a white thrombus. Further, PL exposure initiates coagulation, thereby promoting conversion of fibrinogen into a stable fibrin mesh and further stabilizes the platelet plug to form a red thrombus

### 8.1.2.2 Activation

Upon firm adhesion, vWF/GPIb-V-IX and collagen/GPVI interactions induce a complex series of **signaling cascades**, resulting in platelet activation. This entails vast cytoskeletal reorganization and platelet **shape change**, **degranulation** of α- and dense granules, and **integrin activation** [23, 25, 26]. Degranulation also exposes negatively charged phospholipids from the inner membrane leaflet to the surface (membrane flip-flop) to assist coagulation via supporting prothrombinase complex assembly [27]. The most abundant integrin on platelets is the fibrinogen receptor **GPIIb/IIIa ($\alpha_{IIb}\beta_3$)**, which has a low ligand affinity in resting platelets. In activated platelets, signaling events lead to binding of cytoskeletal proteins to GPIIb/IIIa, triggering a conformational change to its active form which recognizes proteins containing an Arg-Gly-Asp (RGD) motif, such as fibrinogen, fibronectin, and vitronectin [28–30]. This process is called "*inside-out signaling.*" Active GPIIb/IIIa readily binds to **fibrinogen** which is perpetually circulating in the blood but also present in the subendothelial matrix or released from α-granules of activated platelets. Fibrinogen binding triggers signaling events back into the platelet. This so-called "*outside-in signaling*" further propagates platelet activation by enhancing platelet spreading and degranulation [31].

Efficient and persistent platelet activation is highly dependent on **secondary feedback molecules ADP and TXA$_2$** to amplify the physiological responses by activating nearby platelets via inside-out signaling. Pre-stored ADP is released from platelet dense granules upon stimulation. Additionally, activated platelets use membrane phospholipids to synthesize TXA$_2$ which is released independently of degranulation. Together, ADP and TXA$_2$ recruit resting circulating platelets to sites of vascular injury or endothelial activation [22, 32]. Platelet activation, release of feedback molecules, and recruitment of circulating platelets create a positive feedback loop, producing a growing platelet plug to cover the site of vascular injury. The importance of these secondary feedback molecules is underlined by the fact that common **antiplatelet drugs** inhibit the generation of TXA$_2$ by targeting its synthesizing enzyme cyclooxygenase-1 (COX-1) (**acetylsalicylic acid**), or they block the binding of ADP to its receptor (**P2Y$_{12}$ receptor antagonists**, e.g., clopidogrel) [33].

### 8.1.2.3 Aggregation

Direct **platelet-platelet interactions** can occur via newly exposed or activated surface receptors such as P-selectin; however, these initial aggregates are unstable. Binding of multiple platelets to individual fibrinogen fibers via active GPIIb/IIIa enables **fibrinogen to cross-link adjacent activated platelets**, thereby stabilizing the platelet aggregate. Thus, primary hemostasis results in the formation of a firm but reversible platelet plug (**white thrombus**). Antiplatelet **GPIIb/IIIa inhibitors** (e.g., abciximab) use the dependency of stable plug formation on fibrinogen binding, thereby preventing thrombosis.

Concomitant activation of the coagulation cascade results in thrombin production which not only induces further platelet activation but also cleaves fibrinogen to fibrin. **Fibrin** polymerizes to large fibers to form a netlike structure that catches circulating erythrocytes and further stabilizes the platelet plug, resulting in an irreversible fibrin-rich **red thrombus** (see ▶ Sect. 8.2 "Principles of Secondary Hemostasis").

The growing thrombus is a hurdle to the laminar blood flow and restricts the vessel lumen. Locally increased shear forces promote platelet activation and coagulation which exacerbates thrombus growth. As a countermeasure, outside-in signaling in platelets induces **clot retraction**, a tightening and thus shrinking of the thrombus mediated by cytoskeletal actin and myosin [31]. This function not only reduces thrombus-induced shear stress but also pulls the wound edges closer together, which supports wound healing [34].

## 8.1.3 Molecular Signaling Events During Platelet Activation

Platelet signaling comprises of a highly complex, interwoven network involving three major pathways that are induced downstream of most major platelet agonists: **phospholipase C** (PLC), **phosphoinositide 3-kinase** (PI3K)/ **protein kinase B (Akt)**, and **protein kinase C** (PKC).

Binding of **vWF** to **GPIb-V-IX** recruits the adapter molecules 14-3-3ζ, Src kinase, Rac1, and PI3Kβ which triggers downstream activation of the major effector enzyme **PLCγ** [35]. The **collagen** receptor **GPVI** is coupled to a Fc receptor γ chain (FcRγ) and perpetually associated with the Src family kinases (SFK) Fyn and Lyn. Upon ligand binding, Fyn and Lyn phosphorylate the immunoreceptor tyrosine-based activation motif (ITAM) in FcRγ to bind and activate the tyrosine kinase Syk [26, 36, 37]. Syk and recruited PI3Kα or **PI3Kβ** orchestrate a complex signaling cascade, also leading to the activation of **PLC isoform γ** [38–40].

A number of platelet agonists signal via G protein-coupled receptors (GPCR), such as ADP and thrombin. Further, GPCR-mediated signaling is triggered by **TXA$_2$** binding to TXA$_2$/prostaglandin H2 receptor (TP) as well as by the agonist **serotonin** (5-HT) binding to 5-HT$_{2A}$ receptor [41–43]. Platelets express two **ADP** receptors, **P2Y$_1$ and P2Y$_{12}$** [43]. Engagement of P2Y$_1$ results in initial platelet aggregation, whereas P2Y$_{12}$ sustains and amplifies platelet activation, highlighting P2Y$_{12}$ as target for antiplatelet drugs [44, 45]. ADP engages **PLC isoform β** as well as PI3Kβ and **PI3Kγ** to increase intracellular Ca$^{2+}$ levels, degranulation, and integrin activation [40, 46]. The protease **thrombin** is generated as penultimate step of the coagulation cascade, acting as a potent platelet agonist. Via proteolytic cleavage of their extracellular N-terminus, thrombin irreversibly activates **protease-activated receptors 1 and 4 (PAR-1 and PAR-4)** on human platelets [47, 48], which activate downstream **PLCβ** [43].

**PLCγ and PLCβ**, induced by glycoprotein receptors (GPIb-V-XI, GPVI) or GPCRs (P2Y$_1$, PARs, TP and 5-HT$_{2A}$), respectively, cleave membrane phosphatidylinositol-4,5-bisphosphate (PIP$_2$) into the second messengers inositol-1,4,5-triphosphate (IP$_3$) and diacylglycerol (DAG) which trigger the **release of Ca$^{2+}$** from the DTS and **activation of PKC**, respectively. These steps lead to degranulation (thereby to feedback activation), cytoskeletal reorganization, flipping of phospholipids to the outer membrane leaflet, and integrin activation, as well as synthesis of TXA$_2$ by phospholipase A$_2$ (**PLA$_2$**) which converts PIP$_2$ into arachidonic acid, the substrate for COX-1 [49–51]. Furthermore, PLCβ and Ca$^{2+}$ may lead to downstream activation of SFKs to further induce the PI3Kβ pathway [52–55].

Platelet **PI3K** binds to adapter proteins downstream of glycoprotein or integrin receptors (PI3Kα and β) or to GPCRs (PI3Kγ) and is thus activated by all major agonists. PI3K phosphorylates PIP$_2$ to the second messenger phosphatidylinositol-3,4,5-triphosphate (**PIP$_3$**) [56, 57]. This induces recruitment and activation or **protein kinase B/Akt**, a central signaling hub in platelet inside-out signaling, and mediates degranulation and integrin activation [58]. An overview of the central signaling events is given in ◻ Fig. 8.2.

**Box 8.1   The Evolution of Primary Hemostasis**
Platelets and megakaryocytes are unique in realms of animal species and solely exist in mammals. Evolutionary platelets derive from so-called hemocytes/hematocytes present in invertebrates which fulfill both hemostatic and immune functions. Secretion of clotting proteins and the ability to aggregate are already present in these amoeboid-like cells [59, 60]. Specialized hemostatic cells only emerged in vertebrates. Here, the term **"platelet"** is reserved to specifically describe the anucleate mammalian cell in comparison with the nonmammalian nucleated **"thrombocyte"**. In lower vertebrates, diploid thrombocytes differentiate directly from progenitor cells in the bone marrow. In comparison, platelet production via megakaryocyte budding or rupture allows rapid production of high platelet numbers from one progenitor cell. Further, nonmammalian thrombocytes do not store nucleotides and do not express ADP receptors, indicating their lack of ADP feedback to promote activation [61]. Notably, structural and functional differences exist also between mammalian species. As such, bovine, equine, and camel platelets lack an OCS. Murine platelets, extensively used in preclinical studies, are structurally similar to human platelets but show differences in receptor expression and associated functions [19, 48, 62].

In line with their evolutionary origin from a primordial multifunctional cell, platelets retain **immunologic capacity** in addition to their primer and specialized function in hemostasis. This is emphasized by their expression of various immune receptors including **toll-like receptors and Fc receptors** that enable platelets to directly recognize and respond to invading pathogens [63–70]. By releasing antimicrobial peptides or internalizing invading pathogens, platelets themselves participate in combating infectious agents and prevent their dissemination [5, 71, 72]. Further, activated platelets rapidly secrete cytokines or bind to cells of the innate immune system to modulate their activity. Thereby, platelets stimulate leukocyte activation and recruitment to sites of infection or inflammation as well as pro-inflammatory effector functions such as neutrophil extracellular trap formation, phagocytosis, oxidative burst, and release of pro-inflammatory cyto- and chemokines.

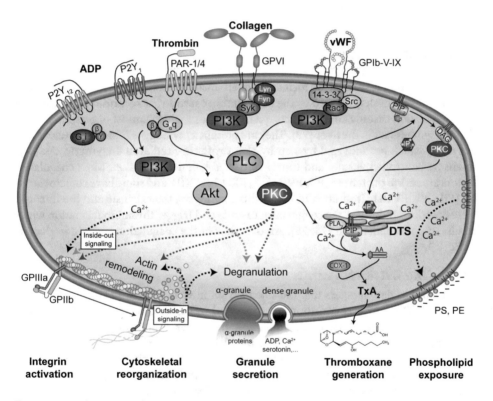

◧ **Fig. 8.2** Molecular signaling events during platelet activation. The platelet agonists thrombin and adenosine diphosphate (ADP) activate G protein-coupled receptors protease-activated receptor 1/4 (PAR-1/4) and P2Y$_1$ or P2Y$_{12}$, respectively, to induce phospholipase Cβ (PLCβ) or phosphoinositide 3-kinase (PI3K). In contrast, collagen/glycoprotein VI (GPVI) and vWF/GPIb-V-IX signaling involves multiple adaptor proteins and binding of kinases (e.g., Syk and Src) to recruit PI3Kβ, which leads to downstream activation of PLCγ. Both PLC isoforms catalyze the conversion of membrane phosphatidylinositol-4,5-bisphosphate (PIP$_2$) into inositol-1,4,5-triphosphate (IP$_3$) and diacylglycerol (DAG), which induces Ca$^{2+}$ release from the dense tubular system (DTS) and activates protein kinase C (PKC), respectively. Further, PI3K activates the central signaling hub, Akt. PKC, Akt, and the calcium burst set off complex downstream events including both inside-out and outside-in signaling that results in GPIIb/IIIa activation, cytoskeletal reorganization due to actin remodeling, and degranulation of α- and dense granules, thereby eliciting cellular responses such as fibrinogen binding, shape change, and release of granule content including feedback ADP. Moreover, PKC promotes PLA$_2$- and cyclooxygenase 1 (COX-1)-mediated conversion of PIP$_2$ into arachidonic acid (AA) and finally into feedback thromboxane A$_2$ (TXA$_2$)

## 8.2　Principles of Secondary Hemostasis

For centuries exposure of blood to air was thought to be the initiator of coagulation after injury. Only in the mid-nineteenth century, the concept that blood contains factors with the ability to develop a blood clot was proposed. However, it took until the 1960s to understand the mechanism of the coagulation cascade, where enzymes and their cofactors are stepwise cleaved into subsequent proteins. The so-called cascade model, which was studied under cell-free, static conditions, was developed. Today, we know that the in vivo situation is far more complex and involves cellular surfaces, which allow for controlled and site-specific activation of coagulation factors to prevent clot formation within the vasculature to maintain normal blood flow.

### 8.2.1　Cascade Model

Platelet activation results in the formation of an unstable thrombus, which is then stabilized by a fibrin mesh generated by the **coagulation cascade** (or **secondary hemostasis**). The coagulation cascade comprises coagulation factors, which consist of plasma proteins that are synthesized in the **liver** [73]. They are present as zymogens (inactive precursors of enzymes) which get activated by step-by-step **protease-mediated enzymatic cleavage** to ensure controlled, locally and timely restricted blood coagulation. Most coagulation factors are **serine proteases** (FII, FVII, FIX, FX, FXI, FXII), and some have cofactor activity (FIII, FV, FVIII). Traditionally, two distinct pathways, the extrinsic and the intrinsic, were described to initiate the coagulation cascade and trigger thrombin formation which results in fibrin generation [74, 75]. An overview of the coagulation cascade is given in ◘ Fig. 8.3.

#### 8.2.1.1　The Extrinsic Pathway

All evidence to date suggests that the sole relevant in vivo initiator of coagulation is **tissue factor (TF)**, which is constitutively expressed in subendothelial tissue (e.g., vascular smooth muscle cells), generating a hemostatic barrier. Injury-induced vessel rupture brings extrinsic TF in contact with circulating FVII, which represents the classical activation pathway of the coagulation cascade. TF binding to FVII activates FVII (FVIIa), and, subsequently, the TF:FVIIa complex (also called **extrinsic tenase complex**) activates FX to FXa in the presence of phospholipids and $Ca^{2+}$, leading to the common pathway. The extrinsic tenase complex further activates FIX, thereby cross-linking the extrinsic and the intrinsic pathway, which is called **Josso-loop** [74, 76] (see ◘ Fig. 8.3, Extrinsic pathway).

#### 8.2.1.2　The Intrinsic Pathway

All components of the intrinsic cascade are present in the circulation, and activation starts with a conformational change of FXII to FXIIa which is initiated by physical contact to **negatively charged surfaces** (e.g., glass surface) or the contact activation system (see ▶ Box 8.2). Subsequently, FXIIa catalyzes the formation of FXIa from FXI. FXIa then activates FIX to FIXa, which binds the cofactor FVIIIa, forming the **intrinsic tenase complex** (FIXa:FVIIIa), which again induces the common pathway [74] (see ◘ Fig. 8.3, Intrinsic pathway).

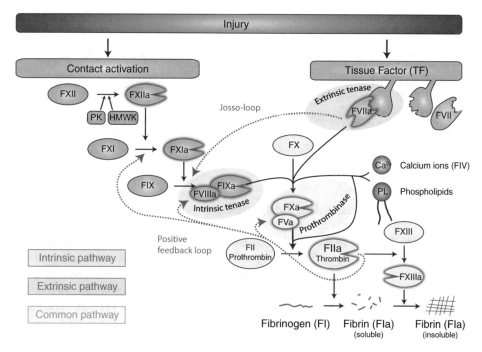

□ **Fig. 8.3** Overview of the coagulation cascade model. The coagulation cascade can be divided into three different pathways: the intrinsic, extrinsic, and common pathway. The intrinsic pathway is initiated by the contact activation system or negatively charged surfaces leading to the activation of coagulation factor (F)XII in a plasma kallikrein (PK)- and high-molecular-weight kininogen (HMWK)-dependent manner. Activated FXII (FXIIa) activates FXI, which activates FIXa to form complex with FVIIIa, phospholipids, and $Ca^{2+}$ (intrinsic tenase complex). The extrinsic pathway starts by exposure of tissue factor (TF) to the circulation leading to FVII activation. FVIIa forms a complex with TF, phospholipids, and $Ca^{2+}$ (extrinsic tenase complex). The common pathway is induced by the activation of FX by either the intrinsic or extrinsic tenase complex. Together with its cofactor FVa, FXa cleaves prothrombin (FII) to thrombin (FIIa), which catalyzes the cleavage of fibrinogen (FI) to fibrin (FIa). Moreover, thrombin activates FXI, FVIII and FV promoting a positive feedback loop. Further, FXIII is activated by thrombin and crosslinks fibrin to an insoluble fibrin network. However, the intrinsic and extrinsic pathways are connected by the so-called Josso loop, which describes the activation of FIX by the extrinsic tenase complex

### 8.2.1.3 Common Pathway

The common pathway starts with the activation of FX by either the extrinsic or the intrinsic tenase complex. Activated FX binds to thrombin-activated FV, forming the **prothrombinase complex**, which activates prothrombin (FII) to thrombin (FIIa). Thrombin in turn cleaves **fibrinogen (FI) to fibrin (FIa)**, generating an insoluble mesh that is integrated into the platelet plug [77]. Finally, both FXIIa and thrombin are able to activate the **transglutaminase FXIII**, which crosslinks the fibrin mesh into a stable, insoluble matrix [76] (see □ Fig. 8.3, Common pathway). Thrombin further activates FXI, FVIII, and FV, thereby generating a positive feedback loop, which amplifies thrombin generation [76, 78]. Although the **extrinsic pathway** is more prominent for **initiating** the coagulation cascade in vivo, the **intrinsic pathway** is essential for **propagating** thrombin generation.

**Box 8.2   Contact Activation System**
The contact activation system comprises factors thought to be involved in the in vivo activation of the intrinsic pathway. It starts with autoactivation of FXII to FXIIa upon contact to negatively charged surfaces. In the presence of high-molecular-weight kininogen (HMWK), FXIIa further converts prekallikrein into plasma kallikrein (PK). PK in turn generates FXIIa from FXII, which is called trans-activation, leading to an amplification loop and an accelerated response [79]. However, FXII autoactivation can also occur on neutral or positively charged surfaces at slower rates. Various biological substances are potent in mediating FXII autoactivation, such as heparan sulfate, polyphosphates, and vessel wall collagen [79, 80].

## 8.2.2  Cell-Based Model of Coagulation

In contrast to the cascade model, the cell-based model emphasizes the involvement of **cellular surfaces**, especially TF-bearing cells and platelets, which allow efficient and site-specific activation of coagulation factors. The cell-based model represents a significant improvement of our understanding of coagulation in vivo and divides coagulation in three phases, **initiation, amplification, and propagation** [81], as depicted in ◘ Fig. 8.4.

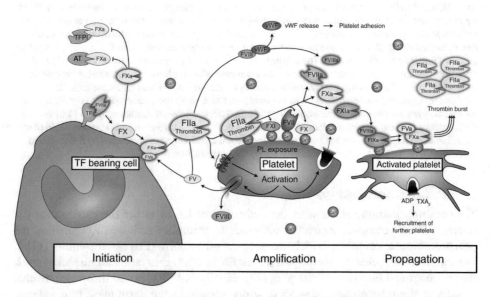

◘ **Fig. 8.4**   The cell-based model of coagulation. The cell-based model of coagulation describes the proteolytic steps of hemostasis in the context of cellular surfaces. First, tissue factor (TF)-bearing cells induce the activation of FVII followed by FXa formation resulting in small amounts of thrombin (initiation phase). Next, thrombin induces a positive feedback loop; liberates vWF, which facilitates platelet adhesion to collagen; and activates platelets via protease-activated receptors (PARs). Activated platelets release granule-stored FV, FVIII, and $Ca^{2+}$ and expose negatively charged phospholipids (PS) allowing clustering of coagulation factors (amplification phase). The procoagulant surface of activated platelets then allows intrinsic tenase and prothrombinase complex assembly resulting in a thrombin burst (propagation phase)

### 8.2.2.1 Initiation

Upon vessel injury blood components get exposed to **TF-bearing cells** (e.g., smooth muscle cells, fibroblasts). Exposed TF rapidly binds to FVIIa, the only coagulation factor circulating also in its active form (~1%) [82], forming the TF:FVIIa complex. The cell membrane-associated TF:FVIIa complex activates further FVII molecules, creating a positive feedback loop. Additionally, FXa and FIXa are generated, and their activity results in small amounts of thrombin (**thrombin spark**) [81]. While membrane-associated FXa fulfills its pro-coagulatory function, dissociated FXa diffuses away and gets inactivated (by TF pathway inhibitor (TFPI) and antithrombin (AT); see ◻ Fig. 8.4), thereby limiting FX activity to TF-bearing cells. In contrast, FIXa is only moderately inhibited by antithrombin and therefore capable of forming the intrinsic tenase complex when getting in contact with, e.g., activated platelets upon dissociation. FXa slowly activates FV, leading to the prothrombinase complex (◻ Fig. 8.4).

### 8.2.2.2 Amplification

Small amounts of thrombin generated during the initiation phase dissociate from TF-bearing cells and activate platelets. Activated platelets in turn release procoagulant mediators (FV, FVIII, and fibrinogen), undergo shape change, and expose negatively charged phosphatidylserine and phosphatidylethanolamine on their surface. The cluster of these negatively charged phospholipids allows interaction with γ-carboxyglutamic acid residues of **vitamin K-dependent coagulation factors (FVII, FIX, FX, FII)**, which are bridged by $Ca^{2+}$ [83]. Thereby, vitamin K-dependent coagulations factors are brought in close proximity to each other at the site of injury. Thrombin further liberates FVIII from vWF and activates it along with FXI and FV, thus amplifying the coagulatory response [81] (see ◻ Fig. 8.4).

### 8.2.2.3 Propagation

Activated platelets aggregate and recruit further platelets to the site of injury, mediated by the release of lipid mediators and granule content. On the negatively charged platelet surface, the formed **intrinsic tenase complex FIXa:FVIIIa** rapidly produces large amounts of FXa. Under physiological conditions, the intrinsic tenase complex is much more potent in generating larger amounts of FXa than the extrinsic tenase complex [81]. Finally, the prothrombinase complex composed of FVa:FXa assembles and generates a **thrombin burst**, which, on the one hand, further promotes coagulation via feedback mechanisms and, on the other hand, cleaves fibrinogen to fibrin (◻ Fig. 8.4). Within this process, thrombin first removes fibrinopeptide A from fibrinogen, allowing fibrinogen polymerization to longitudinal fibrin strands based on non-covalent bonds. Second, thrombin cleaves off fibrinopeptide B, leading to structural changes, which are necessary for generating an insoluble fibrin matrix [84]. Furthermore, fibrin strands are crosslinked by thrombin-activated FXIIIa, resulting in an insoluble clot. Moreover, by binding to thrombomodulin on endothelial cell surface, thrombin activates **thrombin activatable fibrinolysis inhibitor (TAFI)**, which cleaves terminal lysine residues of fibrin, thereby protecting fibrin from fibrinolytic cleavage [85].

### 8.2.2.4 Termination

As a final result, a stable clot is formed and coagulatory processes have to be terminated to prevent overwhelming coagulation. Binding of thrombin to **thrombomodulin** induces a negative feedback loop by **activating protein C** (aPC), which, together with its cofactor protein S, binds to negatively charged surfaces via $Ca^{2+}$ and in turn inactivates various FVa and FVIIIa. Moreover, aPC prevents spreading of pro-coagulatory events as it represents a potent inhibiting cofactor on endothelial cell surfaces [81].

### 8.2.3 Regulation of Coagulation

Thrombin is a central player in both primary and secondary hemostasis. To restrict thrombin formation to the site of injury and prevent its spreading to surrounding areas, three central mechanisms have been developed: *antithrombin (AT), tissue factor pathway inhibitor (TFPI), and the protein C pathway* (see ◘ Fig. 8.5 for details).

> **Box 8.3   Coagulopathies**
> Coagulopathies are bleeding disorders that affect clot formation either by prolongation of bleeding time or by increasing susceptibility to thrombotic events.
>
> **Hemophilia A and B** are X-chromosomal recessive coagulopathies caused by deficiency or lack of FVIII and FIX, respectively. Depending on the residual activity of the coagulation factors, patients may present with easy bruising, inadequate clotting of mild injury, or, in severe cases, spontaneous hemorrhages. As therapy, recombinant or plasma-derived FVIII or FIX concentrates are administered.
>
> **Von Willebrand disease (vWD)** is the most common inherent bleeding disorder with a prevalence of 0.1–1% in the total population [86]. Nevertheless, just one out of ten patients suffers from symptoms like mild spontaneous bleedings such as nose bleedings or menorrhagia. However, less common variants can also cause life-threatening bleedings [87, 88]. The vWD is characterized by quantitative (decreased synthesis/secretion or increased clearance) and/or qualitative (reduced affinity to vWF binding partners, impaired dimer assembly, enhanced susceptibility to cleavage by ADAMTS-13) deficiency of vWF [86]. Moreover, due to the FVIII-stabilizing role of vWF, hemophilia-like bleedings are observed in severe cases of vWD. Treatment options for vWD patients include desmopressin, a synthetic analogue of human antidiuretic hormone, which promotes release of stored vWF from endothelial cells, vWF/FVIII concentrates, recombinant vWF, as well as fibrinolysis inhibitors and humoral treatment.
>
> **Acquired coagulopathies** may arise from anticoagulant therapy (e.g., heparin, vitamin K antagonists), decreased production (liver dysfunction), depletion of coagulation factors (e.g., disseminated intravascular coagulation (DIC)) or from premature coagulation factor degradation due to production of autoantibodies (e.g., acquired hemophilia A: autoantibodies against FVIII [89]).
>
> Missense mutations and small deletions in fibrinogen chains cause **dysfibrinogenemia**, which may lead to impaired binding of tPA to fibrin (e.g., fibrinogen Dusard) as well as abnormal fibrin assembly and fibrinolysis [90]. Paradoxically, dysfibrinogenemia is often associated with thrombosis as physiologically thrombin is sequestered by binding to fibrinogen [91].
>
> **Protein C and protein S deficiencies** are autosomal, dominant disorders associated with recurrent venous thromboembolism [92]. Hereditary **aPC resistance** can be caused by a point mutation in the factor FV gene (factor V Leiden), which abolishes the aPC-cleavage sites in FVa and thereby prevents its degradation [93]. **Antithrombin III deficiency** can be either inherited or adapted, and 65% of inherited patients develop thrombosis, predominately deep vein thrombosis [94].

### 8.2.4 Fibrinolysis

Resolution of the fibrin clot, restoration of blood flow after vessel injury, and prevention of unnecessary intravascular fibrin generation are mediated by the highly regulated enzymatic fibrinolytic system. Fibrin clot architectures, especially fiber size and arrangement, are important determinates for fibrinolysis as fragile clots are more prone to increased

Mechanisms of Hemostasis: Contributions of Platelets, Coagulation…

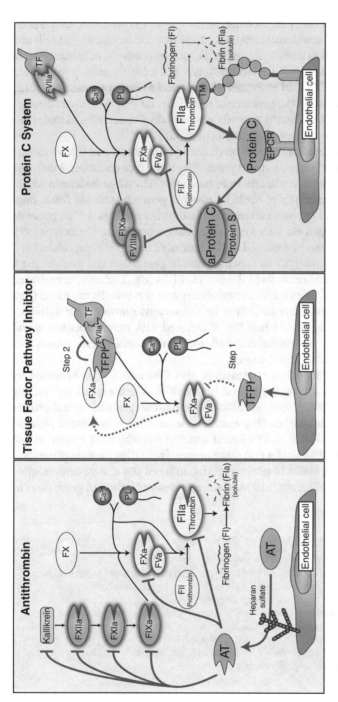

◻ **Fig. 8.5** Main regulators of coagulation. **a** Antithrombin (AT) is a serine protease inhibitor (serpin) present in the plasma with the main function to bind and inactivate FIIa and FXa and additionally FIXa, FXIa, FXIIa, and kallikrein. Its inhibitory activity can be increased up to 1000-fold upon binding of glycosaminoglycans like heparins or heparin sulfates, which are present on (microvascular) endothelial cells and trigger a conformational change within AT [117]. In addition to its role in anticoagulation, AT stimulates endothelial synthesis of PGI₂, which is a potent inhibitor of platelet aggregation and exerts anti-inflammatory properties [1ʹ8]. **b** Tissue factor pathway inhibitor (TFPI) is mainly produced by endothelial cells and inhibits coagulation by inactivation of FXa and the FVIIa/TF complex [119]. Inhibition is mediated via a two-step process, including binding and deactivation of FXa (step 1) as prerequisite to allow subsequent inhibition of the FVIIa/TF complex (step 2) [120]. Furthermore, the major fibrinolytic enzyme plasmin was shown to degrade endothelium-bound TFPI, counteracting anticoagulation [121]. **c** The protein C pathway consists of four main key elements – protein C, thrombomodulin (TM), endothelial protein C receptor (EPCR), and protein S – and inhibits the activity of FVIII and FV by cleavage forming FVIIIi and FVi, respectively. The proenzyme protein C is cleaved by thrombin bound to TM on the endothelial cell surface to activated protein C (aPC), which is a serine protease. This process is augmented by EPCR, which not only presents protein C to the thrombin-TM activation complex but also downregulates endothelial inflammatory cytokine production [122]. Once aPC is dissociated from EPCR, it binds to its cofactors protein S and phospholipids to inactivate FV and FVIII

fibrinolysis and bleeding as firm clots, which are more stable and tend to cause thrombosis [95, 96]. Generation of the major fibrinolytic protease **plasmin** from its plasma-circulating zymogen **plasminogen** is catalyzed by either of the two serine proteases, **tissue plasminogen activator (tPA)** and **urokinase (uPA),** which are released by endothelial cells and urinary epithelial cells as well as monocytes/macrophages, respectively. In comparison to uPA, which mainly acts in extravascular locations, tPA has a higher affinity to plasmin and its activity is more than 500-fold increased when bound to fibrin, promoting locally restricted formation of plasmin in the presence of the fibrin clot [97]. Moreover, through a positive feedback mechanism, plasmin converts single-chain tPA and uPA to their two-chain polypeptide counterparts, increasing their activity [98].

Fibrin is the major plasmin substrate and binds both plasminogen and tPA on its surface, thereby facilitating their interaction and promoting its own degradation. Moreover, binding of plasmin to fibrin protects plasmin from inactivation by **α2-antiplasmin** within fractions of seconds [99]. Proteolysis of fibrin by plasmin generates **soluble fibrin degradation products (FDP)** and exposes carboxy-terminal lysine residues, which promote binding of tPA and plasminogen via their kringle domains, augmenting fibrinolysis [99]. However, these binding sites can be removed by the actions of the carboxypeptidase **TAFI**, which thereby counteracts fibrinolysis by slowing plasmin generation and establishing a link between coagulation and fibrinolysis [85]. Multiple FDPs (e.g., D-dimer) are released during fibrinolysis, which also exert immunomodulatory and chemotactic functions (e.g., fibrinopeptide B). Due to immediate inhibition by the **serpins plasminogen activator inhibitor-1 (PAI-1) and PAI-2**, plasma half-life of tPA and uPA ranges between 4 and 8 min. PAI-1 is synthesized by endothelial cells and stored in platelets, and the presence of inflammatory cytokines increases PAI-1 release.

Similar to the cell-based model of coagulation, also fibrinolysis is orchestrated by endothelial cells, monocytes, macrophages, and neutrophils, which express cell surface receptors with fibrinolytic activity, serve as cofactors for plasmin generation, and protect from circulating fibrinolysis inhibitors. One example for membrane-associated plasmin generation is mediated by S100A10 and its ligand annexin A2, which are located at cell surfaces. S100A10/annexin A2 binds tPA and plasminogen, facilitating plasmin generation at cell membranes. Moreover, S100A10 prevents PAI-1-induced tPA and α2-antiplasmin-mediated plasmin inhibition. Thereby, S100A10 strongly promote plasmin generation in the absence of fibrin [100].

> **Box 8.4   Laboratory Tests of Hemostasis**
> To test alterations of hemostasis, blood needs to be collected in tubes containing anticoagulants, e.g., citrate. Via centrifugation blood is separated into cellular and noncellular components (plasma), which is then analyzed for hemostatic functions.
>
> For the evaluation of **coagulation defects** and fibrinogen disorders, prothrombin time (PT)/international normalized ratio (INR), activated partial thromboplastin time (aPTT), thrombin time, and Clauss fibrinogen assay/reptilase assays are applied in the clinics [101]. These plasma tests help to differentiate between hemophilia A/B, vitamin K deficiencies, and liver dysfunction and help to properly adjust anticoagulation therapy.

For testing plasma **fibrinolysis**, the clot lysis time (CLT) assay has been developed which combines addition of plasma with $Ca^{2+}$ and phospholipid vesicles and subsequent lysis of the clot by exogenous tPA [98]. Moreover, rotational thromboelastometry (ROTEM) allows to measures clot strength, amplitude, and maximal clot firmness in whole blood over time.

Qualitative and quantitative measurement of **platelet dysfunction** is essential to assess increased risk of hemorrhage or thrombosis. While multiple assays can be used, the gold standard to test platelet function is light transmission aggregometry (LTA), which measures the capacity of a platelet suspension to absorb light. As platelets bind to each other, the suspension becomes clearer and more light passes through, allowing real-time evaluation of platelet aggregation. The capability of platelets to aggregate can also be determined in whole blood by multiple electrode aggregometry (MEA) or using a platelet function analyzer (PFA).

## 8.3 Contribution of the Vessel Wall

Endothelial cells are necessary to preserve blood fluidity, prevent thrombus formation, and control extravasation of fluids, as well as platelets and blood cells. In concert with smooth muscle cells, endothelial cells regulate tissue perfusion and local blood supply. Due to physical and cellular differences between veins and arteries (e.g., shear stress, oxygen content, morphology), platelet activation is more likely to occur in arteries, while fibrinous thrombi more frequently occur in veins.

The endothelial surface is covered by a layer of glycoproteins and proteoglycans consisting of one or several negatively charged, sulfated **glycosaminoglycan chains** (GAG), like heparan sulfate and dermatan sulfate [102]. GAGs exert antithrombotic properties via high-affinity binding of AT and induce a conformational change within the reactive center loop of AT, increasing its ability to degrade FIIa and FXa. Heparin is a naturally occurring GAG consisting of variable chains of glucuronic acid and glucosamine and is produced by basophil granulocytes and mast cells in the blood and tissue, respectively. Besides AT, GAGs also bind to heparin cofactor II as well as TM and TFPI. Moreover, negatively charged GAGs and secreted **NO and PGI$_2$** prevent adhesion and aggregation of platelets to the endothelium. This process is supported by degradation of platelet-activating ADP by **endothelial ectonucleotidase** (CD39) [103]. In contrast, the endothelial glycocalyx can be catabolized by heparinase released from platelet granules [104].

Large amounts of thrombin can stimulate endothelial cells via PAR-1, to generate NO and PGI$_2$ as well as to release preformed **Weibel-Palade bodies (WPB)**, which store vWF, P-selectin, and angiopoietin-1 to promote platelet binding as well as leukocyte recruitment. Upon release from WPB, vWF gets unfolded into ultra-large vWF multimers on the surface of endothelial cells, where it is cut into smaller multimers by the action of a disintegrin-like and metalloprotease with thrombospondin type 1 repeats-13 (**ADAMTS-13**) [105], which exhibits enhanced cleavage activity when bound to the endothelium. Additionally, **tPA and endothelin-1** can be released by endothelial cells, promoting fibrinolysis and vasoconstriction, respectively. Continuous and thrombin-triggered release of tPA by endothelial cells ensures presence of tPA within the clot, effectively increasing fibrinolysis in comparison to addition of tPA from the outside to an existing clot [106].

On the other hand, low concentrations of thrombin lead to binding of TM and activation of protein C. This process can be limited by shedding of proteins from the endothelial surface (e.g., endothelial protein C receptor (EPCR) or TM) [107, 108]. Furthermore, binding of thrombin to TM not only prevents its clotting potential but also facilitates degradation of thrombin by plasma protease inhibitors [109]. Beside its important role in inhibition of thrombin generation, aPC additionally prevents activity of PAI-1 and exerts anti-inflammatory actions via inhibition of the NF-κB signaling pathways. Moreover, TM-bound thrombin can catalyze the inactivation of pro-urokinase and activates TAFI, which both dampen fibrinolysis.

Moreover, endothelial cells control initiation of coagulation by two different strategies: Firstly, endothelial cells express **TFPI**, which binds and inactivates TF/FXa/FVIIa activity [110]. On the other hand, endothelial cells secrete protein disulfide isomerase (PDI), which is thought to be necessary for TF decryption (conversion of Cys186 and Cys209 into a disulfide bond, resulting in a conformational change) [111]. EPCR binds protein C and aPC with equal affinity on the surface of endothelial cells adjacent to the TM-thrombin complex, effectively inactivating FVa and FVIIa.

**8**

---

**Box 8.5   Hemostasis During Sepsis**

Not only injury but also infections activate platelets and the coagulation cascade. On the one hand, the fibrin mesh ensnares bacteria to limit pathogen dissemination, but on the other hand, uncontrolled activation of platelets or the coagulation cascade leads to serious complications [113]. Occasionally, systemic infections, independent of pathogens, can cause serious physiological and pathological malfunctions of the host, which are summarized as sepsis or septic shock [114]. According to the current third international consensus definition, sepsis is defined as "life-threatening organ dysfunction caused by a dysregulated host response to infection" [115]. The major problems in septic patients are inappropriate inflammation and coagulopathy including expression of blood-borne **TF on monocytes**, inadequate function of anticoagulatory pathways (AT, aPC, TFPI) due to dysfunctional endothelium, and an excess of PAI-1 and TAFI, suppressing fibrinolysis [114]. Thus, many sepsis patients suffer from elevated levels of platelet activation as well as systemic activation of the coagulation cascade, causing fibrin and thrombus formation within the vasculature, a condition known as **disseminated intravascular coagulation (DIC)**. This increases the risk of pulmonary embolism, multi-organ failure, and thereby death. However, DIC also causes **depletion of coagulation factors** and platelets, resulting in severe hemorrhages. Taken together, septic patients suffer from a combination of hypercoagulability, lack of coagulation factors, and **thrombocytopenia** leading to life-threatening complications during systemic infection [116].

Due to the uncontrolled activation of the coagulation cascade, a plethora of anticoagulants have been tested as treatment option for sepsis. Although some patients benefited from their application, the overall improvement of patient outcome was minimal. Since inflammatory and coagulatory pathways are tightly interwoven leading to mutual amplification, novel treatment strategies aim at preventing an amplified coagulation response. Ideally, this would reduce the risk of thromboembolism together with averting the consumption of coagulations factors [116].

Take-Home Message

- Hemostasis, a physiological process that stops bleeding at the site of injury while maintaining normal blood flow elsewhere in the circulation, involves platelets (primary hemostasis), coagulation factors (secondary hemostasis), and the vessel wall.
- Platelets mediate primary hemostasis by adhering to subendothelial matrix at sites of endothelial injuries, forming platelet aggregates to cover the injury and re-establish vascular integrity.
- Platelet function heavily depends on positive feedback via ADP and $TXA_2$. Anti-platelet therapy therefore targets these feedback mechanisms to prevent thrombotic events.
- In addition to their hemostatic functions, platelets fine-tune immunity by direct or indirect effects on pathogens and/or leukocytes.
- Platelets do not contain a nucleus; thus, all proteins necessary for platelet function have to be prepackaged or taken up from plasma. Their anucleate nature also makes them ideal model cells to study non-genomic protein functions, e.g., structural roles of transcription factors.
- Secondary hemostasis compromises a cascade of zymogens, which are incrementally proteolytically activated leading to the generation of thrombin, the central protease of coagulation.
- The coagulation cascade can be classified in the extrinsic (induced by tissue factor) and intrinsic pathway (induced by negatively charged surfaces), which both convey to the common pathway.
- The cell-based model of coagulation combines both pathways and cellular components and emphasizes that first small amounts of thrombin are generated (thrombin spark; too little for stable clot formation), which then lead to a thrombin burst via a propagation phase, which is necessary for clot formation.
- To resolve the clot, the fibrinolytic enzyme plasmin is activated from its zymogen plasminogen by tissue plasminogen activator (tPA). Plasmin catalyzes proteolysis of fibrin to soluble fibrin degradation products (e.g., D-dimers).
- Resting endothelial cells store vWF in preformed Weibel-Palade bodies and prevent platelet activation by constant release of NO and $PGI_2$.
- The major inhibitors of coagulation are antithrombin, tissue factor pathway inhibitor, and the protein C pathway, consisting of protein C, thrombomodulin, endothelial protein C receptor, and protein S.
- The endothelial surface is covered by negatively charged glycosaminoglycans (e.g., heparan sulfate), which increase the anticoagulant activity of antithrombin.
- Hemophilia A and B are caused by deficiency or lack of FVIII and FIX, respectively.
- Sepsis is a syndrome caused by a systemic infection leading to serious organ dysfunction. Sepsis can cause disseminated intravascular coagulation (DIC) leading to thromboembolism; however, due to consumption coagulopathy, it also increases the bleeding risk.

## 8.4  Questions and Answers

**?** 1.  Thrombin is a key player in hemostasis. Name at least three different cell types/pathways, which are regulated by thrombin.

**✓** Thrombin cleaves fibrinogen to fibrin (common pathway of secondary hemostasis). Thrombin is a strong activator of platelets via binding to protease-activated receptors (PARs). Thrombin can bind to endothelial thrombomodulin and activate the protein C system, leading to inactivation of FV and FVIII. Moreover, TM-bound thrombin stimulates activation of TAFI (thrombin activatable fibrinolysis inhibitor) and thereby inhibits fibrinolysis.

**?** 2.  What is the difference between a white and a red thrombus?

**✓** Primary hemostasis results in the formation of a firm, but reversible platelet plug (white thrombus), whereas concomitant activation of secondary hemostasis causes polymerization of fibrin to an irreversible fibrin-rich red thrombus which ensnares circulating erythrocytes.

**?** 3.  What are the different platelet granule types, and why are they especially important for platelets?

**✓** Platelets contain α-granules (filled with coagulation factors, adhesion and membrane proteins, antimicrobial proteins, chemokines, and growth factors), dense granules (filled with ADP, ATP, polyphosphates), and lysosomes (filled with metabolic enzymes). As platelets do not have a nucleus and only show limited ability for de novo protein synthesis, packaging of platelet content by megakaryocytes and selective uptake from blood plasma is crucial for platelet activity.

**?** 4.  Explain outside-in and inside-out signaling of platelets?

**✓** These signaling pathways are important feedback mechanisms to control platelet activation. Inside-out signaling refers to the conformational change of the fibrinogen receptor GPIIb/IIIa upon platelet activation, allowing high-affinity binding to fibrinogen. Outside-in signaling involves signaling events back into the platelet upon fibrinogen binding. Both processes are necessary for platelet activation and degranulation.

**?** 5.  What are the extrinsic and intrinsic tenase complexes?

**✓** In general, "tenase" refers to FX-activating enzyme complexes which originate from either the extrinsic or intrinsic pathway of coagulation. The extrinsic tenase complex consists of TF and FVIIa and the intrinsic tenase complex of FIXa and FVIIIa. Both complexes cause the activation of the common pathway in the presence of $Ca^{2+}$ and phospholipids.

**?** 6.  Why is calcium essential for hemostasis?

**✓** Divalent positively charged calcium ions allow binding of negatively charged γ-carboxyglutamic acid residues of vitamin K-dependent coagulation factors (FVII, FIX,

FX, FII) as well as protein S and protein C to negatively charged phosphatidylserines present on cellular surfaces (e.g., platelets) via bridging their charges. This mechanism allows close proximity of coagulation factors to each other, thereby drastically increasing their enzymatic activity. Blocking of calcium by chelating agents such as EDTA and citrate is a potent mechanism used in the clinics to prevent coagulation during plasma generation.

**?** 7.  What are the two important functions of vWF?

**✓** vWF promotes platelet adhesion via binding to GPIb-V-IX on platelets, promoting platelet rolling, and decreasing their velocity to enable platelet binding to collagen.

Circulating vWF is a carrier for FVIII in the plasma and is necessary of FVIII stabilization. Thus, von Willebrand disease patients suffer from hemophilia-like bleedings due to lack of FVIII.

**?** 8.  What are the first events after vessel injury?

**✓** To prevent excessive blood loss upon vessel damage, smooth muscle cells locally constrict and minimize the vessel diameter to limit blood flow. Moreover, circulating platelets adhere to the exposed subendothelial matrix, leading to platelet activation and formation of a reversible platelet plug. In parallel the coagulation cascade gets activated via TF binding to FVIIa.

**?** 9.  What is the difference between FVa, FV, and FVi?

**✓** The little "a" stands for "activated" and means that the precursor is already activated by enzymatic cleavage. In contrast, the little "i" stands for "inactivated", indicating that the coagulation factor is inhibited or cleaved.

**?** 10.  Which anticoagulant can be used for functional testing of platelets and/or coagulation?

**✓** Anticoagulation by calcium-chelating agents like citrate inhibits both platelet activation and coagulation which, however, can be reversed upon addition of calcium.

**?** 11.  How does platelet-rich plasma therapy work?

**✓** Injection of autologous platelet-rich plasma represents a new treatment option for sports-related injuries such as tendinopathies or ligamentous injuries. Due to the high platelet concentration, platelet-derived growth factors such as VEGF, EGF, PDGF, and TGF-$\beta$ reach high levels in a locally focused area, where they support wound healing processes and angiogenesis.

**?** 12.  Explain how septic patients may present with both hypercoagulability and increased bleeding.

**✓** Systemic infection may result in excessive inflammatory host responses, conferring increased activation of both primary and secondary hemostasis throughout the body.

This results in development of microthrombi and clots all over the circulation (disseminated intravascular coagulopathy), thereby consuming platelets and coagulation factors and impairing their ability to maintain vascular integrity.

? 13. Explain the difference between plasma and serum.

✓ Plasma and serum are the cell-free fraction of whole blood which can be produced by centrifugation. In plasma, coagulation is prevented by anticoagulants, platelet activation is curtailed, and clotting factors remain present and functional. In contrast, serum is the generated following induction of the clotting cascade which activates platelets and consumes coagulation factors. Therefore, plasma and serum exhibit relevant differences in their protein content and thus usage.

? 14. How do you treat hemophilia A, and which possibilities can you think of to treat patients who developed neutralizing antibodies (inhibitors)?

✓ Patients suffering from hemophilia A are usually treated with factor replacement therapy (plasma-derived FVIII concentrates or recombinant FVIII) or with agents increasing FVIII levels (e.g., desmopressin). However, some patients develop neutralizing antibodies (inhibitors). Thus, new therapeutic strategies have to be developed. One possibility is the bypass therapy. For example, activated prothrombin complex concentrates (containing FVII/VIIa, FII/IIa, FIX/IXa, and FX/X) and recombinant FVIIa could be administered to bypass the lack of FVIII in acute cases.

? 15. Which mechanisms guarantee site-specific clot formation?

✓ To prevent systemic activation of the clotting cascade, pro-coagulatory mechanisms are restricted to sites of vessel injury due to constitutively generated anticoagulatory pathways.

Coagulation factors interact with negatively charged phospholipids (PS) on the surface of activated platelets via $Ca^{2+}$, bringing them in close proximity to each other and to the wound. This membrane interaction is important as membrane-associated FXa acts procoagulatory, whereas dissociated FXa gets rapidly inactivated by tissue factor pathway inhibitor (TFPI) and antithrombin. TFPI, further, inactivates the FVIIa/TF complex. Antithrombin, in the presence of glycosaminoglycan on intact endothelial cells, efficiently inactivates mainly thrombin and FXa but also FIXa, FXIa, FXIIa, and kallikrein. Via the protein C pathway generated activated protein C (aPC) inactivates FV and FVIII. Therefore, spreading of activated coagulation factors is prevented. Moreover, intact endothelium releases NO and $PGI_2$ as well as expresses negatively charged GAG to inhibit platelet activation.

? 16. How can uncontrolled activation of coagulation lead to a bleeding risk?

✓ Occasionally, septic patients suffer from disseminated intravascular coagulation leading, on the one hand, to intravascular thrombus formation and thereby to embolism and, on the other hand, to depletion of coagulation factors increasing the risk for bleedings.

# References

1. Ghoshal K, Bhattacharyya M. Overview of platelet physiology: its hemostatic and nonhemostatic role in disease pathogenesis. Sci World J. 2014;2014:781857.
2. Machlus KR, Italiano JE Jr. The incredible journey: from megakaryocyte development to platelet formation. J Cell Biol. 2013;201(6):785–96.
3. Fong KP, Barry C, Tran AN, et al. Deciphering the human platelet sheddome. Blood. 2011;117(1): e15–26.
4. Rendu F, Brohard-Bohn B. The platelet release reaction: granules' constituents, secretion and functions. Platelets. 2001;12(5):261–73.
5. Yeaman MR. Platelets: at the nexus of antimicrobial defence. Nat Rev Microbiol. 2014;12(6):426–37.
6. Denis MM, Tolley ND, Bunting M, et al. Escaping the nuclear confines: signal-dependent pre-mRNA splicing in anucleate platelets. Cell. 2005;122(3):379–91.
7. Weyrich AS, Dixon DA, Pabla R, et al. Signal-dependent translation of a regulatory protein, Bcl-3, in activated human platelets. Proc Natl Acad Sci U S A. 1998;95(10):5556–61.
8. Schwertz H, Tolley ND, Foulks JM, et al. Signal-dependent splicing of tissue factor pre-mRNA modulates the thrombogenicity of human platelets. J Exp Med. 2006;203(11):2433–40.
9. Sharda A, Flaumenhaft R. The life cycle of platelet granules. F1000Res. 2018;7:236.
10. Banerjee M, Whiteheart SW. The ins and outs of endocytic trafficking in platelet functions. Curr Opin Hematol. 2017;24(5):467–74.
11. Harrison P, Wilbourn B, Debili N, et al. Uptake of plasma fibrinogen into the alpha granules of human megakaryocytes and platelets. J Clin Invest. 1989;84(4):1320–4.
12. Ambrosio AL, Di Pietro SM. Storage pool diseases illuminate platelet dense granule biogenesis. Platelets. 2017;28(2):138–46.
13. Semenov AV, Romanov YA, Loktionova SA, et al. Production of soluble P-selectin by platelets and endothelial cells. Biochemistry (Mosc). 1999;64(11):1326–35.
14. Patel-Hett S, Richardson JL, Schulze H, et al. Visualization of microtubule growth in living platelets reveals a dynamic marginal band with multiple microtubules. Blood. 2008;111(9):4605–16.
15. Teijeiro RG, Sotelo Silveira JR, Sotelo JR, Benech JC. Calcium efflux from platelet vesicles of the dense tubular system. Analysis of the possible contribution of the Ca2+ pump. Mol Cell Biochem. 1999;199(1–2):7–14.
16. White JG. The transfer of thorium particles from plasma to platelets and platelet granules. Am J Pathol. 1968;53(4):567–75.
17. Escolar G, Lopez-Vilchez I, Diaz-Ricart M, White JG, Galan AM. Internalization of tissue factor by platelets. Thromb Res. 2008;122(Suppl 1):S37–41.
18. White JG, Krumwiede M. Further studies of the secretory pathway in thrombin-stimulated human platelets. Blood. 1987;69(4):1196–203.
19. Selvadurai MV, Hamilton JR. Structure and function of the open canalicular system - the platelet's specialized internal membrane network. Platelets. 2018;29(4):319–25.
20. Wu KK, Thiagarajan P. Role of endothelium in thrombosis and hemostasis. Annu Rev Med. 1996;47: 315–31.
21. Geraldo RB, Sathler PC, Lourenco AL, et al. Platelets: still a therapeutical target for haemostatic disorders. Int J Mol Sci. 2014;15(10):17901–19.
22. Jackson SP. Arterial thrombosis--insidious, unpredictable and deadly. Nat Med. 2011;17(11):1423–36.
23. Jennings LK. Mechanisms of platelet activation: need for new strategies to protect against platelet-mediated atherothrombosis. Thromb Haemost. 2009;102(2):248–57.
24. Cimmino G, Golino P. Platelet biology and receptor pathways. J Cardiovasc Transl Res. 2013;6(3): 299–309.
25. Randriamboavonjy V. In: Kerrigan S, Moran N, editors. Mechanisms involved in diabetes-associated platelet hyperactivation, the non-thrombotic role of platelets in health and disease: IntechOpen; 2015. https://doi.org/10.5772/60539. Available from: https://www.intechopen.com/books/the-non-thromboticrole-of-platelets-in-health-and-disease/mechanisms-involved-in-diabetes-associated-platelet-hyperactivation.
26. Watson SP, Auger JM, McCarty OJ, Pearce AC. GPVI and integrin alphaIIb beta3 signaling in platelets. J Thromb Haemost. 2005;3(8):1752–62.
27. Crook M. Platelet prothrombinase in health and disease. Blood Coagul Fibrinolysis. 1990;1(2):167–74.

28. Tadokoro S, Shattil SJ, Eto K, et al. Talin binding to integrin beta tails: a final common step in integrin activation. Science. 2003;302(5642):103–6.

29. Moser M, Nieswandt B, Ussar S, Pozgajova M, Fassler R. Kindlin-3 is essential for integrin activation and platelet aggregation. Nat Med. 2008;14(3):325–30.

30. Perutelli P, Mori PG. The human platelet membrane glycoprotein IIb/IIIa complex: a multi functional adhesion receptor. Haematologica. 1992;77(2):162–8.

31. Shattil SJ, Newman PJ. Integrins: dynamic scaffolds for adhesion and signaling in platelets. Blood. 2004;104(6):1606–15.

32. Haley KM, Recht M, McCarty OJ. Neonatal platelets: mediators of primary hemostasis in the developing hemostatic system. Pediatr Res. 2014;76(3):230–7.

33. Schrottmaier WC, Kral JB, Badrnya S, Assinger A. Aspirin and P2Y12 inhibitors in platelet-mediated activation of neutrophils and monocytes. Thromb Haemost. 2015;114(3):478–89.

34. Sorrentino S, Studt JD, Medalia O, Tanuj Sapra K. Roll, adhere, spread and contract: structural mechanics of platelet function. Eur J Cell Biol. 2015;94(3–4):129–38.

35. Bryckaert M, Rosa JP, Denis CV, Lenting PJ. Of von Willebrand factor and platelets. Cell Mol Life Sci. 2015;72(2):307–26.

36. Suzuki-Inoue K, Tulasne D, Shen Y, et al. Association of Fyn and Lyn with the proline-rich domain of glycoprotein VI regulates intracellular signaling. J Biol Chem. 2002;277(24):21561–6.

37. Ezumi Y, Shindoh K, Tsuji M, Takayama H. Physical and functional association of the Src family kinases Fyn and Lyn with the collagen receptor glycoprotein VI-Fc receptor gamma chain complex on human platelets. J Exp Med. 1998;188(2):267–76.

38. Carrim N, Walsh TG, Consonni A, Torti M, Berndt MC, Metharom P. Role of focal adhesion tyrosine kinases in GPVI-dependent platelet activation and reactive oxygen species formation. PLoS One. 2014;9(11):e113679.

39. Kim S, Mangin P, Dangelmaier C, et al. Role of phosphoinositide 3-kinase beta in glycoprotein VI-mediated Akt activation in platelets. J Biol Chem. 2009;284(49):33763–72.

40. Gilio K, Munnix IC, Mangin P, et al. Non-redundant roles of phosphoinositide 3-kinase isoforms alpha and beta in glycoprotein VI-induced platelet signaling and thrombus formation. J Biol Chem. 2009;284(49):33750–62.

41. Knezevic I, Borg C, Le Breton GC. Identification of Gq as one of the G-proteins which copurify with human platelet thromboxane A2/prostaglandin H2 receptors. J Biol Chem. 1993;268(34):26011–7.

42. Djellas Y, Manganello JM, Antonakis K, Le Breton GC. Identification of Galpha13 as one of the G-proteins that couple to human platelet thromboxane A2 receptors. J Biol Chem. 1999;274(20):14325–30.

43. Offermanns S. Activation of platelet function through G protein-coupled receptors. Circ Res. 2006;99(12):1293–304.

44. Storey RF, Sanderson HM, White AE, May JA, Cameron KE, Heptinstall S. The central role of the P(2T) receptor in amplification of human platelet activation, aggregation, secretion and procoagulant activity. Br J Haematol. 2000;110(4):925–34.

45. Hechler B, Leon C, Vial C, et al. The P2Y1 receptor is necessary for adenosine 5′-diphosphate-induced platelet aggregation. Blood. 1998;92(1):152–9.

46. Hirsch E, Bosco O, Tropel P, et al. Resistance to thromboembolism in PI3Kgamma-deficient mice. FASEB J. 2001;15(11):2019–21.

47. Coughlin SR. Protease-activated receptors in hemostasis, thrombosis and vascular biology. J Thromb Haemost. 2005;3(8):1800–14.

48. Coughlin SR. Thrombin signalling and protease-activated receptors. Nature. 2000;407(6801):258–64.

49. Cohen S, Braiman A, Shubinsky G, Isakov N. Protein kinase C-theta in platelet activation. FEBS Lett. 2011;585(20):3208–15.

50. Li Z, Delaney MK, O'Brien KA, Du X. Signaling during platelet adhesion and activation. Arterioscler Thromb Vasc Biol. 2010;30(12):2341–9.

51. Offermanns S, Toombs CF, Hu YH, Simon MI. Defective platelet activation in G alpha(q)-deficient mice. Nature. 1997;389(6647):183–6.

52. Martin V, Guillermet-Guibert J, Chicanne G, et al. Deletion of the p110beta isoform of phosphoinositide 3-kinase in platelets reveals its central role in Akt activation and thrombus formation in vitro and in vivo. Blood. 2010;115(10):2008–13.

53. Kim S, Jin J, Kunapuli SP. Relative contribution of G-protein-coupled pathways to protease-activated receptor-mediated Akt phosphorylation in platelets. Blood. 2006;107(3):947–54.

8

54. Senis YA, Mazharian A, Mori J. Src family kinases: at the forefront of platelet activation. Blood. 2014;124(13):2013–24.
55. Hall KJ, Jones ML, Poole AW. Coincident regulation of PKCdelta in human platelets by phosphorylation of Tyr311 and Tyr565 and phospholipase C signalling. Biochem J. 2007;406(3):501–9.
56. Vanhaesebroeck B, Whitehead MA, Pineiro R. Molecules in medicine mini-review: isoforms of PI3K in biology and disease. J Mol Med (Berl). 2016;94(1):5–11.
57. Fougerat A, Gayral S, Malet N, Briand-Mesange F, Breton-Douillon M, Laffargue M. Phosphoinositide 3-kinases and their role in inflammation: potential clinical targets in atherosclerosis? Clin Sci (Lond). 2009;116(11):791–804.
58. Woulfe DS. Akt signaling in platelets and thrombosis. Expert Rev Hematol. 2010;3(1):81–91.
59. Iwanaga S, Miyata T, Tokunaga F, Muta T. Molecular mechanism of hemolymph clotting system in limulus. Thromb Res. 1992;68(1):1–32.
60. Takahashi H, Azumi K, Yokosawa H. Hemocyte aggregation in the solitary ascidian Halocynthia roretzi: plasma factors, magnesium ion, and Met-Lys-bradykinin induce the aggregation. Biol Bull. 1994;186(3):247–53.
61. Schneider W, Gattermann N. Megakaryocytes: origin of bleeding and thrombotic disorders. Eur J Clin Investig. 1994;24(Suppl 1):16–20.
62. McKenzie SE, Taylor SM, Malladi P, et al. The role of the human Fc receptor Fc gamma RIIA in the immune clearance of platelets: a transgenic mouse model. J Immunol. 1999;162(7):4311–8.
63. Andonegui G, Kerfoot SM, McNagny K, Ebbert KV, Patel KD, Kubes P. Platelets express functional toll-like receptor-4. Blood. 2005;106(7):2417–23.
64. Cognasse F, Hamzeh H, Chavarin P, Acquart S, Genin C, Garraud O. Evidence of toll-like receptor molecules on human platelets. Immunol Cell Biol. 2005;83(2):196–8.
65. Assinger A, Kral JB, Yaiw KC, et al. Human cytomegalovirus-platelet interaction triggers toll-like receptor 2-dependent proinflammatory and proangiogenic responses. Arterioscler Thromb Vasc Biol. 2014;34(4):801–9.
66. Panigrahi S, Ma Y, Hong L, et al. Engagement of platelet toll-like receptor 9 by novel endogenous ligands promotes platelet hyperreactivity and thrombosis. Circ Res. 2013;112(1):103–12.
67. Anabel AS, Eduardo PC, Pedro Antonio HC, et al. Human platelets express toll-like receptor 3 and respond to poly I:C. Hum Immunol. 2014;75(12):1244–51.
68. Koupenova M, Vitseva O, MacKay CR, et al. Platelet-TLR7 mediates host survival and platelet count during viral infection in the absence of platelet-dependent thrombosis. Blood. 2014;124(5):791–802.
69. Worth RG, Chien CD, Chien P, Reilly MP, McKenzie SE, Schreiber AD. Platelet FcgammaRIIA binds and internalizes IgG-containing complexes. Exp Hematol. 2006;34(11):1490–5.
70. Boilard E, Pare G, Rousseau M, et al. Influenza virus H1N1 activates platelets through FcgammaRIIA signaling and thrombin generation. Blood. 2014;123(18):2854–63.
71. Kraemer BF, Campbell RA, Schwertz H, et al. Novel anti-bacterial activities of beta-defensin 1 in human platelets: suppression of pathogen growth and signaling of neutrophil extracellular trap formation. PLoS Pathog. 2011;7(11):e1002355.
72. White JG. Platelets are covercytes, not phagocytes: uptake of bacteria involves channels of the open canalicular system. Platelets. 2005;16(2):121–31.
73. Dashty M, Akbarkhanzadeh V, Zeebregts CJ, et al. Characterization of coagulation factor synthesis in nine human primary cell types. Sci Rep. 2012;2:787.
74. McMichael M. New models of hemostasis. Top Companion Anim Med. 2012;27(2):40–5.
75. Dahlback B. Blood coagulation. Lancet. 2000;355(9215):1627–32.
76. Mackman N. The role of tissue factor and factor VIIa in hemostasis. Anesth Analg. 2009;108(5):1447–52.
77. Gale AJ. Continuing education course #2: current understanding of hemostasis. Toxicol Pathol. 2011;39(1):273–80.
78. Lane DA, Philippou H, Huntington JA. Directing thrombin. Blood. 2005;106(8):2605–12.
79. Schmaier AH. The contact activation and kallikrein/kinin systems: pathophysiologic and physiologic activities. J Thromb Haemost. 2016;14(1):28–39.
80. Renne T, Schmaier AH, Nickel KF, Blomback M, Maas C. In vivo roles of factor XII. Blood. 2012;120(22):4296–303.
81. Smith SA. The cell-based model of coagulation. J Vet Emerg Crit Care (San Antonio). 2009;19(1):3–10.
82. Morrissey JH. Plasma factor VIIa: measurement and potential clinical significance. Haemostasis. 1996;26(Suppl 1):66–71.

83. Tie JK, Carneiro JD, Jin DY, Martinhago CD, Vermeer C, Stafford DW. Characterization of vitamin K-dependent carboxylase mutations that cause bleeding and nonbleeding disorders. Blood. 2016;127(15):1847–55.

84. Mosesson MW. Fibrinogen and fibrin structure and functions. J Thromb Haemost. 2005;3(8):1894–904.

85. Foley JH, Kim PY, Mutch NJ, Gils A. Insights into thrombin activatable fibrinolysis inhibitor function and regulation. J Thromb Haemost. 2013;11(Suppl 1):306–15.

86. Swami A, Kaur V. von Willebrand disease: a concise review and update for the practicing physician. Clin Appl Thromb Hemost. 2017;23(8):900–10.

87. James PD, Notley C, Hegadorn C, et al. The mutational spectrum of type 1 von Willebrand disease: results from a Canadian cohort study. Blood. 2007;109(1):145–54.

88. Sanders YV, Fijnvandraat K, Boender J, et al. Bleeding spectrum in children with moderate or severe von Willebrand disease: relevance of pediatric-specific bleeding. Am J Hematol. 2015;90(12):1142–8.

89. Kessler CM, Knobl P. Acquired haemophilia: an overview for clinical practice. Eur J Haematol. 2015;95(Suppl 81):36–44.

90. Cote HC, Lord ST, Pratt KP. Gamma-chain dysfibrinogenemias: molecular structure-function relationships of naturally occurring mutations in the gamma chain of human fibrinogen. Blood. 1998;92(7):2195–212.

91. Walton BL, Getz TM, Bergmeier W, Lin FC, Uitte de Willige S, Wolberg AS. The fibrinogen gammaA/gamma' isoform does not promote acute arterial thrombosis in mice. J Thromb Haemost. 2014;12(5):680–9.

92. Wypasek E, Undas A. Protein C and protein S deficiency - practical diagnostic issues. Adv Clin Exp Med. 2013;22(4):459–67.

93. Castoldi E, Rosing J. APC resistance: biological basis and acquired influences. J Thromb Haemost. 2010;8(3):445–53.

94. Nishimura Y, Takagi Y. Strategy for cardiovascular surgery in patients with antithrombin III deficiency. Ann Thorac Cardiovasc Surg. 2018;24(4):187–92.

95. Cilia La Corte AL, Philippou H, Ariens RA. Role of fibrin structure in thrombosis and vascular disease. Adv Protein Chem Struct Biol. 2011;83:75–127.

96. Ariens RA. Fibrin(ogen) and thrombotic disease. J Thromb Haemost. 2013;11(Suppl 1):294–305.

97. Hoylaerts M, Rijken DC, Lijnen HR, Collen D. Kinetics of the activation of plasminogen by human tissue plasminogen activator. Role of fibrin. J Biol Chem. 1982;257(6):2912–9.

98. Chapin JC, Hajjar KA. Fibrinolysis and the control of blood coagulation. Blood Rev. 2015;29(1):17–24.

99. Cesarman-Maus G, Hajjar KA. Molecular mechanisms of fibrinolysis. Br J Haematol. 2005;129(3):307–21.

100. Hajjar KA, Hamel NM. Identification and characterization of human endothelial cell membrane binding sites for tissue plasminogen activator and urokinase. J Biol Chem. 1990;265(5):2908–16.

101. Hayward CPM. How I investigate for bleeding disorders. Int J Lab Hematol. 2018;40(Suppl 1):6–14.

102. Sobczak AIS, Pitt SJ, Stewart AJ. Glycosaminoglycan neutralization in coagulation control. Arterioscler Thromb Vasc Biol. 2018;38(6):1258–70.

103. Pearson JD. Endothelial cell function and thrombosis. Baillieres Best Pract Res Clin Haematol. 1999;12(3):329–41.

104. Vlodavsky I, Eldor A, Haimovitz-Friedman A, et al. Expression of heparanase by platelets and circulating cells of the immune system: possible involvement in diapedesis and extravasation. Invasion Metastasis. 1992;12(2):112–27.

105. Valentijn KM, van Driel LF, Mourik MJ, et al. Multigranular exocytosis of Weibel-Palade bodies in vascular endothelial cells. Blood. 2010;116(10):1807–16.

106. van Hinsbergh VW. Endothelium--role in regulation of coagulation and inflammation. Semin Immunopathol. 2012;34(1):93–106.

107. Gu JM, Katsuura Y, Ferrell GL, Grammas P, Esmon CT. Endotoxin and thrombin elevate rodent endothelial cell protein C receptor mRNA levels and increase receptor shedding in vivo. Blood. 2000;95(5):1687–93.

108. Wu HL, Lin CI, Huang YL, et al. Lysophosphatidic acid stimulates thrombomodulin lectin-like domain shedding in human endothelial cells. Biochem Biophys Res Commun. 2008;367(1):162–8.

109. Esmon CT. The protein C pathway. Chest. 2003;124(3 Suppl):26S–32S.

110. Osterud B, Bajaj MS, Bajaj SP. Sites of tissue factor pathway inhibitor (TFPI) and tissue factor expression under physiologic and pathologic conditions. On behalf of the subcommittee on Tissue Factor Pathway Inhibitor (TFPI) of the scientific and standardization committee of the ISTH. Thromb Haemost. 1995;73(5):873–5.

111. Jasuja R, Furie B, Furie BC. Endothelium-derived but not platelet-derived protein disulfide isomerase is required for thrombus formation in vivo. Blood. 2010;116(22):4665–74.

112. Bae JS, Yang L, Rezaie AR. Receptors of the protein C activation and activated protein C signaling pathways are colocalized in lipid rafts of endothelial cells. Proc Natl Acad Sci U S A. 2007;104(8):2867–72.

113. Delabranche X, Helms J, Meziani F. Immunohaemostasis: a new view on haemostasis during sepsis. Ann Intensive Care. 2017;7(1):117.

114. Lipinska-Gediga M. Coagulopathy in sepsis - a new look at an old problem. Anaesthesiol Intensive Ther. 2016;48(5):352–9.

115. Singer M, Deutschman CS, Seymour CW, et al. The third international consensus definitions for sepsis and septic shock (sepsis-3). JAMA. 2016;315(8):801–10.

116. Davis RP, Miller-Dorey S, Jenne CN. Platelets and coagulation in infection. Clin Transl Immunology. 2016;5(7):e89.

117. Opal SM, Kessler CM, Roemisch J, Knaub S. Antithrombin, heparin, and heparan sulfate. Crit Care Med. 2002;30(5 Suppl):S325–31.

118. Yamauchi T, Umeda F, Inoguchi T, Nawata H. Antithrombin III stimulates prostacyclin production by cultured aortic endothelial cells. Biochem Biophys Res Commun. 1989;163(3):1404–11.

119. Bajaj MS, Kuppuswamy MN, Saito H, Spitzer SG, Bajaj SP. Cultured normal human hepatocytes do not synthesize lipoprotein-associated coagulation inhibitor: evidence that endothelium is the principal site of its synthesis. Proc Natl Acad Sci U S A. 1990;87(22):8869–73.

120. Warn-Cramer BJ, Rao LV, Maki SL, Rapaport SI. Modifications of extrinsic pathway inhibitor (EPI) and factor Xa that affect their ability to interact and to inhibit factor VIIa/tissue factor: evidence for a two-step model of inhibition. Thromb Haemost. 1988;60(3):453–6.

121. Stalboerger PG, Panetta CJ, Simari RD, Caplice NM. Plasmin proteolysis of endothelial cell and vessel wall associated tissue factor pathway inhibitor. Thromb Haemost. 2001;86(3):923–8.

122. Esmon CT. The endothelial cell protein C receptor. Thromb Haemost. 2000;83(5):639–43.

# Biologically Active Lipids in Vascular Biology

*Clint Upchurch and Norbert Leitinger*

© Springer Nature Switzerland AG 2019
M. Geiger (ed.), *Fundamentals of Vascular Biology*, Learning Materials in Biosciences,
https://doi.org/10.1007/978-3-030-12270-6_9

**What You Will Learn in This Chapter**

Bioactive lipids have numerous and diverse roles in regulating vascular function. Throughout this chapter we will introduce diverse classes of enzymatically and nonenzymatically formed bioactive lipids. We will first discuss many of the known roles of lipids in the maintenance of vascular homeostasis. Next, we will describe the current understanding of how bioactive lipids, and especially oxidized lipids, contribute to redox regulation and metabolic function to drive a variety of vascular pathologies. Finally, we will provide a brief overview of historical and current approaches used for the isolation, separation, and determination of bioactive lipids that have allowed for the elucidation for their varied activities.

**❓ Questions**
  - How are bioactive lipids produced and recognized by cells?
  - What are the effects of bioactive lipids on cells of the vascular wall?
  - Can the structures and biological effects of these lipids be exploited for therapeutic purposes?

## 9.1  Introduction/Overview

9

Biologically active lipids can be roughly divided into two groups: (1) enzymatically formed and (2) nonenzymatically formed, bioactive lipids (◻ Fig. 9.1). Enzymatically formed, biologically active lipids include a diverse set of species that are involved in intracellular signaling as well as in cell-to-cell cross talk. Enzymatically formed bioactive lipids can be produced rapidly without involving transcription and translation, as is required for peptide and protein mediators, due to the high concentration of triglycerides, diacylglycerides, and phospholipids within cells. The rate-limiting step is typically the action of a phospholipase that liberates a "precursor lipid," mostly a fatty acid, from the glycerol backbone, which is then converted by one or more enzymes to a biologically active compound. Formation of bioactive lipids usually occurs within minutes after initial cellular stimulus. Conversely, the production of protein mediators requires several hours, unless they are released from preformed granules, such as the release of von Willebrand factor and P-selectin from Weibel-Palade bodies in endothelial cells. Key enzymatic pathways that produce such lipid mediators are the cyclooxygenase, lipoxygenase, and epoxygenase, as well as the sphingolipid/ceramide pathway, and the generation of platelet-activating factor (PAF). Furthermore, intracellular lipid mediators that are produced to either initiate or facilitate intracellular signaling pathways include phosphoinositides and diacylglycerol (DAG). Eicosanoids and sphingolipids, as well as PAF, act on highly specific G protein-coupled receptors that are expressed on cells of the vascular wall and selectively transmit signals depending on their G protein coupling. The individual cellular response to lipid ligands is further diversified by the fact that there are several receptor subtypes for each lipid mediator. For example, there are four subtypes of the prostaglandin E2 receptor (EP1-4) and five receptors for sphingosine-1-phosphate (S1P), each of which can elicit different signaling responses.

While enzymatically formed biologically active lipids have been implicated in a variety of physiological functions, including angiogenesis, vasculogenesis, control of vascular reactivity, endothelial barrier function, as well as maintenance of endothelial integrity, nonenzymatically formed, biologically active lipids such as oxidatively modified lipids are

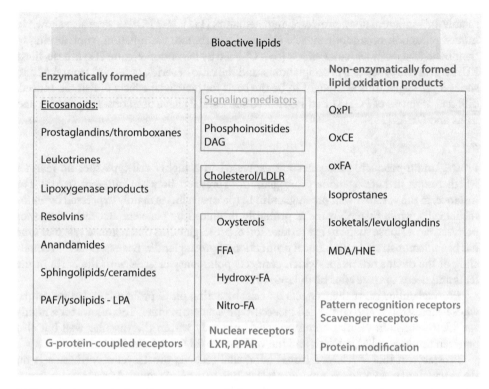

| Bioactive lipids | | |
|---|---|---|
| **Enzymatically formed** | | **Non-enzymatically formed lipid oxidation products** |
| Eicosanoids: | Signaling mediators | OxPl |
| Prostaglandins/thromboxanes | Phosphoinositides DAG | OxCE |
| Leukotrienes | | oxFA |
| Lipoxygenase products | Cholesterol/LDLR | Isoprostanes |
| Resolvins | Oxysterols | Isoketals/levuloglandins |
| Anandamides | FFA | MDA/HNE |
| Sphingolipids/ceramides | Hydroxy-FA | |
| PAF/lysolipids - LPA | Nitro-FA | Pattern recognition receptors Scavenger receptors |
| | Nuclear receptors | |
| G-protein-coupled receptors | LXR, PPAR | Protein modification |

☐ **Fig. 9.1** Enzymatically and nonenzymatically formed biologically active lipids

generally considered danger signals (danger-associated molecular patterns or DAMPs [94]) and have been implicated in numerous pathological settings including atherosclerosis and other chronic inflammatory conditions. Oxidized phospholipids and oxidized cholesterol esters are recognized by scavenger receptors or by pattern recognition receptors, such as SRB1 and CD36, or Toll-like receptors TLR4 or TLR2, respectively. Moreover, oxysterols and modified fatty acids, such as hydroxyl fatty acids or nitro fatty acids, can solicit nuclear signals by binding to nuclear hormone receptors, thereby regulating transcriptional programs involved in cellular metabolism.

## 9.2  Regulation of Vascular Physiology by Lipids

### 9.2.1  Eicosanoids

Eicosanoids are products of arachidonic acid derived by the action of cyclooxygenases 1 and 2, lipoxygenases, and P450 epoxygenases [97]. The rate-limiting step in the release of arachidonic acid from phospholipids is by the action of phospholipase A2, which by itself is controlled by calcium and protein kinases and phospholipase C. Arachidonic acid is then converted to prostaglandin H which is utilized by a variety of enzymes to produce the respective prostaglandin species [103].

Omega-3 polyunsaturated fatty acids (n-3 PUFA) eicosapentaenoic acid and docosahexaenoic acid are also released from phospholipids via PLA2 and are precursors to mainly anti-inflammatory, pro-resolving mediators [111]. N-3 PUFAs were also shown to induce relaxation of smooth muscle cells and to promote vasodilation, contributing to their beneficial cardiovascular effects [83]. Utilized by PG endoperoxide H synthase, these PUFAs give rise to series 3 prostaglandins, and the ratio of series 2 (derived from AA) and series 3 prostaglandins is thought to be the basis for beneficial effects of fish oil. Indeed, different affinities of PGHS1 and PGHS2 (COX 1 and 2) have been observed for AA and EPAs and DHAs [138].

### 9.2.1.1 Prostanoids

Prostaglandin-producing enzymes are expressed in a highly cell-type-specific manner which results in heterogenous prostaglandin expression between different tissues. For instance, in the vascular wall prostaglandin I2 (prostacyclin) is mainly expressed by endothelial cells, while thromboxane is primarily produced by platelets. Recently, a role for perivascular adipose tissue in the production of prostaglandins that regulate vascular tone has been demonstrated [104]. Local production of prostaglandin E is essential for remodeling of the ductus arteriosus, which connects pulmonary arteries with the aorta in the fetus and needs to close after birth [150, 151].

Prostaglandins exert their effects on cells by acting on G protein-coupled receptors, whose expression is also controlled in a cell-type-specific manner. This allows for a highly specific response in cellular communication not only within the vascular wall but also between immune cells, platelets, and the vasculature (◘ Fig. 9.2 and Table 9.1).

Prostacyclin that is released from endothelial cells via prostacyclin synthase acts on the vasculature in an autocrine manner by binding to the GαS-coupled prostacyclin receptor; however, recent evidence suggests that PGI2 acts via alternative pathways as well, including involvement of nuclear receptors such as PPARs [108]. Prostacyclin induces an anti-inflammatory response in ECs predominantly by increasing cyclic AMP levels, thereby providing the vasculature with an antithrombotic surface. Together with nitric oxide released by smooth muscle cells, the endothelial cell-produced prostacyclin contributes to the control of vascular homeostasis [51]. On the other hand, there is a significant contribution of other prostaglandins such as PGE2, PGF2, and thromboxane to the pro-inflammatory, pro-contractile, and procoagulant activation of endothelial cells. For instance, EP4, TP, and IP receptors expressed in the renal artery all contribute to regulation of renal blood flow and vascular resistance [40]. Furthermore, microsomal prostaglandin E synthase-1 (mPGES-1)-derived PGE2 was recently shown to mediate pathophysiological effects on the vasculature in hypertension [6].

Historically, the prostaglandin synthesis pathway has proven an excellent pharmacological target. The major benefits of nonsteroidal anti-inflammatory drugs (NSAIDs) are to prevent the formation of these pro-inflammatory lipid mediators via COX inhibition and, in the case of aspirin, also in the production of pro-resolving mediators [111]. Some of the detrimental effects that have been observed in patients treated with highly selective COX-2 inhibitors may be ascribed to shifting the balance toward production of pro-inflammatory eicosanoids [116, 124].

### 9.2.1.2 Leukotrienes

In the context of vascular biology, leukotrienes are mainly considered to be pro-inflammatory mediators. Leukotrienes are formed by immune cells and can be produced locally within atherosclerotic lesions [7]. They are thought to contribute to lesion forma-

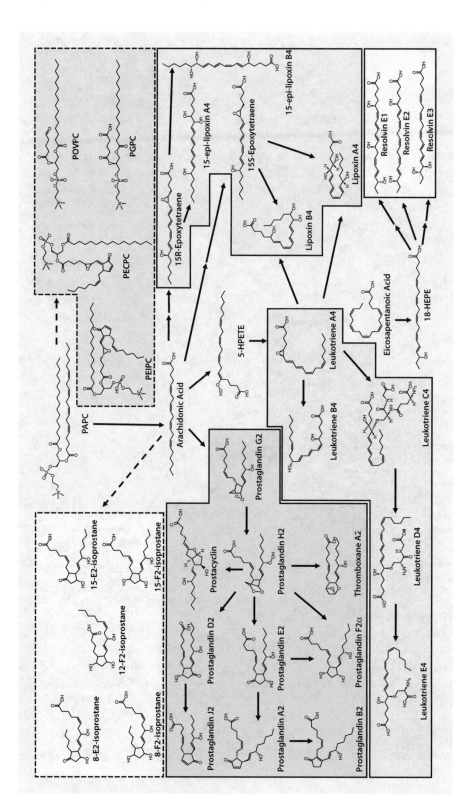

□ Fig. 9.2   Vasoactive lipid species enzymatically- or nonenzymatically-derived from eicosanoids

**Table 9.1** Prostanoids

| Vasoactive lipid | Precursor | Enzyme | Receptor | G protein coupling | Function |
|---|---|---|---|---|---|
| Prostaglandin A2 | Prostaglandin E2 | Dehydration | --- | --- | --- |
| Prostaglandin B2 | Prostaglandin A2 | Dehydration | TBXA2R | --- | Increased pulmonary blood pressure |
| Prostaglandin D2 | Prostaglandin H2 | PTGDS1 PTGDS2 H-PGDS | PTGDR1 | G$\alpha$s | Vasodilation, anti-inflammatory |
| | | | PTGDR2 | G$\alpha$i | Pro-inflammatory chemotaxis |
| | | | | | Vasoconstrictive |
| Prostaglandin E2 | Prostaglandin H2 | PGE synthase (3 isozymes) | PTGER1 | G$\alpha$q | Promotes inflammatory T-cell differentiation |
| | | | PTGER2 | G$\alpha$s | Anti-inflammatory |
| | | | PTGER3 | G$\alpha$i or G12 | Anti-inflammatory |
| | | | PTGER4 | G$\alpha$s | Pro-inflammatory |
| Prostaglandin F2$\alpha$ | Prostaglandin H2 Prostaglandin E2 | AKR1B1 | PTGFR (2 splice variants) | G$\alpha$q | Smooth muscle cell contraction |
| Prostaglandin G2 | Arachidonic acid | --- | --- | --- | --- |
| Prostaglandin H2 | Prostaglandin G2 | COX-1 COX-2 | --- | --- | --- |
| Prostaglandin I2 | Prostaglandin H2 | Prostacyclin synthase | PTGIR (3 splice variants) | G$\alpha$s | Vasodilation, inhibited thrombosis, anti-inflammatory |
| Prostaglandin J2 | Prostaglandin D2 | Dehydration reaction | PTGDR1 | G$\alpha$s | Vasodilation, anti-inflammatory |
| | | | PTGDR2 | G$\alpha$i | Pro-inflammatory chemotaxis, vasoconstrictive |
| | | | PPAR$\gamma$ | --- | --- |
| Thromboxane A2 | Prostaglandin H2 | Thromboxane A synthase | TBXA2R (2 splice variants) | G$\alpha$q G$\alpha$s G12/13 | Promotes thrombosis, increased heart rate, vasoconstriction, pro-inflammatory |

Mitchell and Kirkby [97]

9

smooth muscle cells. Pharmacologic inhibition of leukotriene receptors was shown to decrease atherosclerosis in animal models, and leukotriene receptor antagonists are being tested in clinical trials.

While there are three different series of leukotrienes identified in vivo, the 4-series leukotrienes, arachidonic acid metabolites, have been the focus of most studies (◻ Table 9.2). The 4-series leukotrienes are derived from 5-HPETE, a direct metabolite of arachidonic acid, which forms leukotriene A4. Leukotriene A4 can be further metabolized to leukotriene B4 and C4 by LTB4 hydrolase and LTC4 synthase, respectively. Leukotriene B4 has been shown to induce neutrophil chemotaxis and vascular adherence as well as vascular permeability [88, 105, 126]. Leukotriene C4 serves as precursor of leukotriene D4 and promotes an increase in microvascular blood flow as well as vascular smooth muscle vasoconstriction [4, 15]. Leukotriene D4 has been shown to mediate similar affects as leukotriene C4 on the vasculature [14, 15]. Leukotriene E4 is produced from leukotriene D4 by LTD4 dipeptidase and has been shown to increase vascular permeability while promoting smooth muscle cell constriction and pulmonary vasoconstriction [43, 144].

◻ **Table 9.2**  Leukotrienes

| Vasoactive lipid | Precursor | Enzyme | Receptor | G protein coupling | Function |
|---|---|---|---|---|---|
| Leukotriene A4 | 5-HPETE | LTA synthase | --- | --- | --- |
| Leukotriene B4 | Leukotriene A4 | LTB4 hydrolase | BLT1 BLT2 | $G\alpha i$ | Promotes neutrophil chemotaxis and vascular adherence [88, 105] Increases vascular permeability [126] |
| Leukotriene C4 | Leukotriene A4 | LTC4 synthase | CysLT1 CysLT2 | $G\alpha q/11$ | Increase microvascular cutaneous blood flow [15] Promotes vascular smooth muscle vasoconstriction [4] |
| Leukotriene D4 | Leukotriene C4 | γ-Glutamyl transpeptidase | CysLT1 CysLT2 | $G\alpha q/11$ | Increase microvascular cutaneous blood flow [14, 15] Promotes smooth muscle constriction [4] |
| Leukotriene E4 | Leukotriene D4 | LTD4 dipeptidase | GPR99 | $G\alpha q$ | Increased vascular permeability [144] Promotes smooth muscle constriction [144] Promotes pulmonary vasoconstriction [43] |

### 9.2.1.3 The Cytochrome P450 (CYP) Pathway

Cytochrome P450 (CYP) converts AA into 20-hydroxyeicosatetraenoic acid (20-HETE). 20-HETE is produced in vascular smooth muscle cells and is a potent vasoconstrictor. It acts on smooth muscle cells by reducing the open-state probability of calcium-activated potassium channels, and a G protein-coupled receptor for 20-HETE was recently discovered. 20-HETE has been shown to regulate blood flow and vasoconstriction in vivo [114] and, most importantly, to control physiological responses in the kidney [42, 62]. 20-HETE can produce oxidative stress and inflammation, contributing to endothelial dysfunction and hypertension [140]. Conversely, 20-HETE can also induce antihypertensive effects by inhibiting sodium reabsorption by the kidney [154].

Cytochrome P450 epoxygenases can convert arachidonic acid to epoxyeicosatetraenoic acids (EETs), also known as endothelial-derived hyperpolarization factors (EDHF). In addition to arachidonic acid, these epoxygenases can also utilize the omega-3-PUFAs, eicosapentaenoic and docosahexaenoic acids, to produce the corresponding epoxides [44]. EETs are potent anti-inflammatory [34, 35] vascular mediators that induce vascular relaxation mainly by acting on vascular smooth muscle cells where they activate large conductance calcium-activated potassium channels. Additionally, EETs have been shown to regulate angiogenesis, smooth muscle cell proliferation, platelet aggregation, and fibrinolysis. The mechanisms by which ETTs activate endothelial and smooth muscle cells are subject of investigation; several receptors for EETs have been characterized and new receptors are being identified [106]. Moreover, some EETs directly control the activity of T-type Ca channels [29], and a variety of receptor-independent mechanisms have been suggested and include direct modification of effector proteins as well as incorporation of EETs into phospholipid membranes [129, 133].

EETs are enzymatically deactivated by soluble epoxide hydrolase (EH), which converts EETs to dihydroxyeicosatrienoic acids (DHETs). Soluble epoxide hydrolase represents an attractive, novel drug target that has been implicated in regulation of a variety of vascular pathologies including hypertension, diabetic pain, and diabetic retinopathy [64]. Inhibition of this enzyme leads to an increase in the half-life of EETs, and currently EH inhibitors are in clinical trials [18, 52, 66, 139, 141].

## 9.2.2 Resolution of Vascular Inflammation

The resolution of inflammation has been considered a passive process for a long time; however, recent evidence suggests that the resolution of acute inflammation requires a highly coordinated process that is controlled by a variety of mediators released from inflammatory and vascular cell types. In particular the endothelium plays a significant role in the resolution of inflammation. Resolution is orchestrated by a set of recently discovered lipid mediators derived from PUFAs ( Table 9.3). Among these are the four classes of specialized pro-resolving lipid mediators (SPMs): lipoxins, resolvins, protectins, and the recently discovered maresins. These SPMs can be produced from arachidonic acid in various ways ( Fig. 9.2). Most interestingly, the production of some SPMs can be triggered by the action of aspirin. The first class of SPMs, lipoxins, have long been identified as anti-inflammatory lipid mediators [122]. Lipoxin A4 and B4, as well as 15-epi-lipoxin A4 and B4, are derived from arachidonic acid and have been shown to promote neutrophil chemotaxis and resolution of both macrophages and neutrophils.

◻ **Table 9.3**  Pro-resolving lipid mediators

| Vasoac-tive lipid | Precursor | Enzyme | Receptor | Function |
|---|---|---|---|---|
| Resolvin D1 | DHA | 15-Lipoxygenase | FPR2 GPR32 ResoDR1 DRV1 ALX FPR2 | Protects against endotoxin-induced impairment of barrier function |
| Resolvin D2 | DHA | 15-Lipoxygenase | GPR18 DRV2 | Inhibits thrombosis in deep dermal vascular network [21] Promote macrophage-dependent clot remodeling [39] |
| Resolvin D3 | 4S(5)-Epoxy-17S-hydroxy-DHA | 5-Lipoxygenase | --- | Promote macrophage-dependent clot remodeling [39] |
| Resolvin D4 | 4S(5)-Epoxy-17S-hydroxy-DHA | 5-Lipoxygenase | --- | No known vascular function |
| Resolvin D5 | | | GPR32 | Promote macrophage-dependent clot remodeling [39] |
| Resolvin D6 | | | --- | No known vascular function |
| Resolvin E1 | EPA | 5-Lipoxygenase | CMKLR1 RER1 BLT1 | Promote a protective phenotypic switch in vascular smooth muscle cells in the setting of atherosclerosis [59] Promotes neutrophil clearance [5] Inhibits platelet aggregation [37] |
| Resolvin E2 | EPA | 5-Lipoxygenase | CMKLR1 BLT1 | Dampened PAF response [102] Promotes neutrophil clearance [5] |
| Resolvin E3 | EPA | 12/15-Lipoxygenase | | Inhibits neutrophil chemotaxis and tissue infiltration [67] |
| Maresin 1 | DHA | 12-Lipoxygenase | | Enhances macrophage phagocytosis [123] Inhibits neutrophil infiltration [32, 123] |
| Maresin 2 | DHA | 12-Lipoxygenase | | Enhances macrophage phagocytosis [36] Inhibits neutrophil infiltration [36] |

(continued)

**◘ Table 9.3** (continued)

| Vasoactive lipid | Precursor | Enzyme | Receptor | Function |
|---|---|---|---|---|
| Lipoxin A4 | Arachidonic acid | 5/12-Lipoxygenase | ALX/FPR2 GPR32 | Promote neutrophil chemotaxis<br>Promote neutrophil and macrophage resolution at sites of inflammation |
| Lipoxin B4 | Arachidonic acid | 5/12-Lipoxygenase | ALX/FPR2 GPR32 | Promote neutrophil chemotaxis<br>Promote neutrophil and macrophage resolution at sites of inflammation |
| 15-Epi-lipoxin A4 | Arachidonic acid | 5-lipoxygenase | ALX/FPR2 GPR32 | Promote neutrophil chemotaxis<br>Promote neutrophil and macrophage resolution at sites of inflammation |
| 15-Epi-lipoxin B4 | Arachidonic acid | 5-Lipoxygenase | ALX/FPR2 GPR32 | Promote neutrophil chemotaxis<br>Promote neutrophil and macrophage resolution at sites of inflammation |
| Protectin D1 | DHA | 15-Lipoxygenase | | Inhibits leukocyte infiltration [61] |
| Protectin DX | DHA | 15-Lipoxygenase | | Inhibits platelet aggregation-mediated activation of neutrophils [85] |

Resolvins demonstrate a highly potent anti-inflammatory and pro-resolving effect in a variety of disease models [130]. Resolvins are categorized into classes based on their lipid progenitor. The D-class resolvins are derived from DHA, and to date, six unique species have been identified four of which have documented effects on the vasculature. Resolvin D1 has been shown to protect impairment of vascular barrier function by endotoxin challenge [153, 155]. Resolvin D2 has been shown to inhibit thrombosis and, with its counterparts D3 and D5, has been shown to induce macrophage-dependent clot remodeling [21, 39]. Three of the E-class resolvins, synthesized from EPA, also have identified effects on the vasculature. Resolvin E1 has been shown to promote an athero-protective smooth muscle cell phenotype while mediating neutrophils and inhibiting platelet clearance [5, 37, 59]. Similarly, resolvin E2 promotes neutrophil clearance but also inhibits the response to platelet-activating factor [5, 102]. Resolvin E3 has been shown to inhibit neutrophil chemotaxis and tissue infiltration [67]. Similarly, maresins 1 and 2 have been shown to enhance macrophage phagocytosis and inhibit neutrophil infiltration [32, 36, 123]. Furthermore, protectin D1 has been shown to inhibit leukocyte infiltration, while protectin DX has been shown to inhibit neutrophil activation [61, 85].

## 9.2.3 **Lysophospholipids: S1P and LPA**

In the context of vascular biology, the bioactive lysophospholipids sphingosine-1-phosphate (S1P) and lysophosphatidic acid (LPA) were shown to control a variety of cellular functions both in vascular physiology and in pathology [60] such as atherosclerosis and myocardial infarction [1]. A lipid phosphate phosphatase that metabolizes LPA has been implied as a risk factor in coronary artery disease [127], and strategies to influence receptor dependent signaling and LPA metabolism are being considered as therapeutic approaches.

On the other hand, S1P has mainly been shown to have beneficial activities on the vasculature, by promoting endothelial integrity and angiogenesis [57], in addition to controlling lymphocytes trafficking and inflammation. S1P produced by red blood cells and platelets is secreted from cells by specific transporters [136]. In the circulation S1P forms gradients across the endothelium, which seem to be essential for lymphocyte trafficking and maintenance of vascular homeostasis [147].

S1P is produced from ceramide via a series of enzymatic reactions [117] (◻ Fig. 9.3) and acts on one of five different subtypes of the S1P receptor, termed S1P1-5 (previously EDG receptors) [86]. Activation of S1P1 on endothelial cells results in control of angiogenesis, cell proliferation, and endothelial integrity. Furthermore, S1P is one of two lipid mediators (the other is oxidized phospholipids) known to enhance endothelial barrier function [71]. S1P3 was recently shown to be involved in the pathogenesis of atheroscle-

◻ **Fig. 9.3** Biochemical pathway for synthesis of sphingosine-1-phosphate

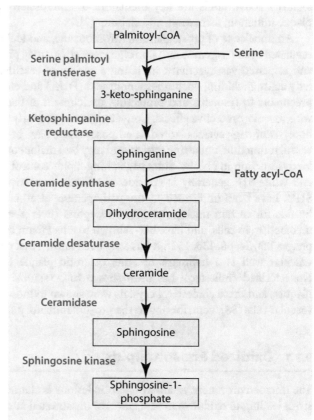

rosis [73]. Less is known about the intracellular functions of S1P, some of which may be exerted through direct binding to intracellular protein binding partners.

S1P receptors are promising therapeutic targets for vascular diseases. Fingolimod, a modulator of S1P receptors that was recently approved for the treatment of relapsing multiple sclerosis, has multiple effects. While its main activity includes the reduction off lymphocyte egress, its action on cardiac cells results in transient suppression of heart rate and mild reduction in blood pressure for some patients [27].

## 9.3  Vascular Damage/Atherosclerosis

Vascular damage is thought to be initiated by disruption of endothelial integrity, accompanied by inflammatory endothelial activation and loss of barrier function [75, 92]. In large arteries the development of atherosclerosis is characterized by the accumulation of lipids in the subendothelial space that results in the formation of an atherosclerotic lesion or plaque [50, 77, 81, 84, 92, 99, 132]. Plaque lipids are mainly derived from oxLDL that accumulates in macrophages after being taken up by scavenger receptors. These macrophages develop into so-called foam cells that are the hallmark of atherosclerotic lesions. As a consequence, atherosclerotic plaques contain many different species of biologically active lipids, and recent evidence suggests that the microenvironment within an atherosclerotic lesion is characterized by the relative abundance of these lipids. Moreover, biologically active lipids are key mediators of atherosclerotic pathology, contributing to plaque initiation, formation, and stability [2].

An imbalance of prostaglandin, thromboxane, and leukotriene formation due to dysregulated cyclooxygenase and lipoxygenase activity may promote endothelial dysfunction and impaired vasoreactivity, initiating a variety of vascular pathologies. Since omega-3 fatty acids, including docosahexaenoic acid (DHA) and eicosapentaenoic acid (EPA), are precursors to resolvins and protectins, enrichment of these PUFAs has been associated with vasculo-protective effects. Conversely, saturated fatty acids, which can be released by lipolysis of triglycerides stored in adipose tissue, have been ascribed pro-inflammatory, vascular function-impairing effects, and may be a major contributing factor driving vascular dysfunction in obesity. Similarly, sphingolipids control a variety of vascular functions, and while S1P generally is considered vasculo-protective, certain receptors, including S1P3, have been implicated in the pathogenesis of atherosclerosis [73]. Oxysterols are ligands for certain nuclear hormone receptors (liver X receptors), which control lipid metabolism in cells, and have been shown to affect foam cell formation in plaque macrophages [23, 24, 68, 135, 152]. It has long been known that cholesterol accumulation in the vascular wall is a hallmark of atherosclerotic plaque formation and destabilization. Nonesterified cholesterol has been shown to crystallize in the core of atherosclerotic plaques, and these cholesterol crystals were shown to induce inflammasome activation in vascular cells [38], contributing to the pro-inflammatory state of atherosclerotic plaques.

### 9.3.1  Oxidized Phospholipids

The microenvironment in atherosclerotic lesions is characterized by increased oxidative stress leading to oxidative modification of unsaturated fatty acids and cholesterol in both accumulating lipoproteins as well as in membranes of cells undergoing apoptosis and other forms of cell death [22, 118, 119]. Furthermore, oxidized lipids are shed from damaged membranes in the form of microvesicles and apoptotic blebs which further contribute to the oxidative environment of the atherosclerotic plaque [65, 112, 137]. Biologically active compounds such as 4-hydroxynonenal (HNE) can be derived from lipid oxidation [78]. These oxidation derived biologically active lipids were initially observed in rat livers [109], on erythrocyte membranes [110], and on oxidized lipoproteins [41, 69]. Direct free radical-induced oxidation of AA leads to the formation of isoprostanes, which have been shown to modulate inflammation in atherosclerotic plaques. Recent work has shown that serum levels of isoprostanes can be used as biomarkers for oxidative tissue damage and systemic oxidative stress [33, 96, 100]. Both the 8-series E2 and F2 isoprostanes induce vasoconstriction and anti-angiogenic effects; however, 8-series F2-isoprostanes have also been shown to disrupt the endothelial cell barrier [8, 56, 98]. Similarly, 12-series F2-isoprostanes have been shown to induce vasoconstriction in neural vasculature [63]. The 15-series F2- and E2-isoprostanes have been shown to exhibit both vasoconstriction and vasodilation. These opposing effects can be attributed to agonism of a diverse set of receptors by these lipid species [49, 95] (◘ Table 9.4).

Oxidized phospholipids have been identified as the biologically active components of minimally modified LDL [142] and are also present on the surface of apoptotic cells [30, 65]. On the one hand, some of these lipids, like PECPC and PEIPC, increase endothelial barrier properties [71], while on the other hand, other species, such as POVPC and PGPC, induce endothelial dysfunction by uncoupling NO synthase [146] and activating endothelial cells to bind monocytes, the initiating step in the development of atherosclerosis [11, 80]. Endothelial cells respond to these lipids by upregulation of a specific gene expression program that is largely dependent on Nrf2 [3, 115]. Previously, it has been shown that POVPC induces neutrophil-endothelial cell binding, while PGPC induces binding of both monocytes and neutrophils [80]. Furthermore, in the setting of atherosclerosis, PGPC and POVPC have been demonstrated to induce a pro-atherogenic switch in smooth muscle cells [31, 107]. Recently, it was shown that oxidized phospholipids also affect cellular metabolism in both endothelial cells [58, 76] and macrophages [121]. How these lipids affect cellular bioenergetics and metabolism is a subject of ongoing investigation. Oxidized phospholipids promote phenotypic switching in macrophages [120]. Mice that overexpress an antibody that captures oxidized phospholipids, E06, are protected against the development of atherosclerosis [113] and myocardial ischemia-reperfusion injury [149]. Furthermore, oxidized phospholipids have been shown to potently influence platelet activation [9, 16, 89, 90, 125] and coagulation [79, 87, 125].

Mechanisms by which cells recognize and respond to oxidized phospholipids include pattern recognition receptors (CD36, TLR2, CD14) as well as intracellular sensors such as the redox-sensitive Keap1/Nrf2 system and potentially nuclear hormone receptors such as peroxisome proliferator-activated receptors. OxPLs also have anti-inflammatory [91] and possibly pro-resolving properties [19, 20, 48]. Some of these properties can be ascribed to the formation of cyclopentenones that are potent activators of Nrf2, while others can be attributed to the induction of pro-resolving mediator synthesis, as is the case for LXA4 [72]. In humans, OxPLs are carried mainly on LP(a) lipoproteins [10], and their levels have been correlated with risk of cardiovascular [12, 26, 46, 74, 93] and calcific aortic valve disease [28, 70].

9

◻ **Table 9.4**    Biologically active oxidized lipids

| Vasoactive lipid | Precursor | Receptor | G protein coupling | Function |
|---|---|---|---|---|
| 8-Series F2-isoprostane | Arachidonic acid | TP receptor | | Vasoconstriction [98]<br>Anti-angiogenic [8]<br>Disrupts endothelial cell barrier [56] |
| | | PGE receptor | | |
| 12-Series F2-isoprostane | Arachidonic acid | | | Vasoconstriction of neural vasculature [63] |
| 15-Series F2-isoprostane | Arachidonic acid | TP receptor | | Vasoconstriction [49, 96] |
| | | EP2-4 | | Vasodilation [95] |
| 8-Series E2-isoprostane | Arachidonic acid | TP receptor | | Vasoconstriction [98]<br>Anti-angiogenic [8] |
| | | PGE receptor | | |
| 15-Series E2-isoprostane | Arachidonic acid | TP receptor | Gαq/11 | Vasoconstriction [49, 96]<br>Platelet aggregation [96] |
| | | EP2-4 | | Vasodilation [96] |
| Levuglandin E2 | Arachidonic acid | | | Disruption of the blood-brain barrier |
| PECPC | PAPC | | | Endothelial cell activator [131]<br>Increases vascular endothelium barrier function [13] |
| PEIPC | PAPC | EP2 | | Endothelial cell activator [131]<br>Increases vascular endothelium barrier function [13]<br>Promotes binding of monocytes to the endothelium [82, 143] |
| POVPC | PAPC | | | Selectively promotes monocyte-endothelial binding [80]<br>Impairs vascular endothelium barrier function [13]<br>Regulates vascular smooth cell phenotype potentiating atherosclerosis [31, 107] |
| PGPC | PAPC | | | Promotes monocyte- and neutrophil- endothelial cell binding [80]<br>Impairs vascular endothelium barrier function [13]<br>Regulates vascular smooth cell phenotype potentiating atherosclerosis [31, 107] |

## 9.4  Methods and Protocols for Lipid Analysis

The complexity and varying abundance of bioactive lipids necessitate further investigation into various roles lipids play in vascular physiology and pathology. Advances in analytical techniques have proven essential in determining the biological activity of lipids both in vitro and in vivo. In 1959, Bligh and Dyer introduced a rapid, convenient method of total lipid extraction from tissue [17], vastly improving upon the previously established method by Folch [45]. Colloquially known as the Bligh-Dyer method, this method allowed for rapid lipid extraction using minimal solvent to maximize recovery while minimizing lipid oxidation. In combination with thin-layer chromatography, the Bligh-Dyer method provided insight into the lipid composition of biological systems.

Thin-layer and column chromatography were used extensively to separate different classes of lipids [47]; however, these methods proved ineffective at identifying individual lipid species within a given class. At the same time, gas chromatography coupled with mass spectrometry (GC-MS) was commonly used to separate and identify volatile lipid species. GC-MS was limited to detection of volatile compounds and often required saponification of nonvolatile lipid species. Consequently, it was necessary to separate lipids by class, usually by high-performance liquid chromatography (HPLC), prior to GC-MS analysis. Furthermore, hard ionization, which fragments compounds, was used to identify analytes based on fragmentation patterns. These limitations restricted interrogation of the biological lipidome. To overcome these limitations, soft ionization approaches were developed. While many different soft ionization methods were originally developed, electrospray ionization (ESI) has become the ionization method of choice. For the first time, ESI provided a method to directly analyze the lipidome of biological samples without separation or derivatization. As a result, mass spectrometer-based methods for analyte separation were developed. One such method known as intrasource separation relied on modulating the mobile phase to produce preferential species ionization [53]. This was followed by precursor ion and neutral loss scanning which allowed for identification of lipid species based on a common fragment [25]. In combination, these methods led to the development of shotgun lipidomics, which integrates these approaches to identify lipid species based on class and acyl chain length [54, 55]. Integration of HPLC with ESI-MS (LC-MS) resulted in the enhanced sensitivity and provided a means to deconvolute isobaric species and is currently used as the standard approach of lipidome identification.

Currently, LC-MS lipidomic approaches can be grouped into two categories: targeted or untargeted [148]. Targeted LC-MS approaches are limited in scope and develop multiple reaction monitoring profiles (MRMs) for each lipid species of interest. Typically, these approaches use a triple-quadrupole mass spectrometer. An MRM is developed for each analyte in the panel by optimizing ionization for each analyte. This results in high sensitivity for each analyte. While this approach works well for low-abundance analytes, it is best suited for a small panel of analytes. Conversely, an untargeted approach allows for detection of all analytes in a sample. This is achieved by using a time-of-flight (TOF) detection method which determines the mass of a given analyte by the time required to travel a specified distance. Compared to triple-quadrupole instruments, TOF methodologies allow higher mass accuracy, which assists in deconvolution of isobaric peaks. Historically, this method has been limited in its ability to detect low-abundance species; however, recent advancements in detection methods have vastly improved these deficiencies. This has been further enhanced by the availability of comprehensive lipid databases such as LIPID

MAPS, which provide a compendium for identification of lipid species from mass spectra results. Despite these tremendous advances in lipid identification, previously described methods fail to retain the physical location of species within a tissue. Matrix-assisted laser desorption/ionization (MALDI) and desorption electrospray ionization (DESI) sources have provided insight into the biological location of lipid species within tissues and have proven integral for identifying in vivo lipid species location [101, 128, 134, 145].

---
**Take-Home Message**

Lipids are important messenger molecules in vascular biology. The relative abundance of different enzymatically and nonenzymatically formed lipid mediators determines vascular function and controls the initiation, progression, and resolution of inflammation. In pathological settings such as atherosclerosis, endothelial activation and phenotypic polarization of macrophages and smooth muscle cells are affected by bioactive plaque lipids. Furthermore, oxidatively modified lipids in the vascular milieu contribute largely to vascular dysfunction through effects on endothelial activation, leukocyte function, and cellular metabolism. Emerging mass spectrometry-based lipidomic approaches and the development of better in vivo models will allow to further investigate and more clearly define the multifaceted effects of these bioactive lipids on the vasculature.

---

**9**

## References

1. Abdel-Latif A, Heron PM, Morris AJ, Smyth SS. Lysophospholipids in coronary artery and chronic ischemic heart disease. Curr Opin Lipidol. 2015;26:432–7.
2. Adamson S, Leitinger N. Phenotypic modulation of macrophages in response to plaque lipids. Curr Opin Lipidol. 2011;22:335–42.
3. Afonyushkin T, Oskolkova OV, Philippova M, Resink TJ, Erne P, Binder BR, Bochkov VN. Oxidized phospholipids regulate expression of ATF4 and VEGF in endothelial cells via NRF2-dependent mechanism: novel point of convergence between electrophilic and unfolded protein stress pathways. Arterioscler Thromb Vasc Biol. 2010;30:1007–13.
4. Albert RK, Greenberg GM, Henderson W. Leukotriene C4 and D4 increase pulmonary vascular permeability in excised rabbit lungs. Chest. 1983;83:85S–6S.
5. Arita M, Ohira T, Sun YP, Elangovan S, Chiang N, Serhan CN. Resolvin E1 selectively interacts with leukotriene B4 receptor BLT1 and ChemR23 to regulate inflammation. J Immunol. 2007;178:3912–7.
6. Avendano MS, Garcia-Redondo AB, Zalba G, Gonzalez-Amor M, Aguado A, Martinez-Revelles S, Beltran LM, Camacho M, Cachofeiro V, Alonso MJ, et al. mPGES-1 (microsomal prostaglandin E synthase-1) mediates vascular dysfunction in hypertension through oxidative stress. Hypertension. 2018;72:492–502.
7. Back M, Weber C, Lutgens E. Regulation of atherosclerotic plaque inflammation. J Intern Med. 2015;278:462–82.
8. Benndorf RA, Schwedhelm E, Gnann A, Taheri R, Kom G, Didie M, Steenpass A, Ergun S, Boger RH. Isoprostanes inhibit vascular endothelial growth factor-induced endothelial cell migration, tube formation, and cardiac vessel sprouting in vitro, as well as angiogenesis in vivo via activation of the thromboxane A(2) receptor: a potential link between oxidative stress and impaired angiogenesis. Circ Res. 2008;103:1037–46.
9. Berger M, Wraith K, Woodward C, Aburima A, Raslan Z, Hindle MS, Moellmann J, Febbraio M, Naseem KM. Dyslipidemia-associated atherogenic oxidized lipids induce platelet hyperactivity through phospholipase Cγ2-dependent reactive oxygen species generation. Platelets. 2018:1–6.
10. Bergmark C, Dewan A, Orsoni A, Merki E, Miller ER, Shin MJ, Binder CJ, Horkko S, Krauss RM, Chapman MJ, et al. A novel function of lipoprotein [a] as a preferential carrier of oxidized phospholipids in human plasma. J Lipid Res. 2008;49:2230–9.

11. Berliner JA, Leitinger N, Tsimikas S. The role of oxidized phospholipids in atherosclerosis. J Lipid Res. 2009;50(Suppl):S207–12.

12. Bertoia ML, Pai JK, Lee JH, Taleb A, Joosten MM, Mittleman MA, Yang X, Witztum JL, Rimm EB, Tsimikas S, et al. Oxidation-specific biomarkers and risk of peripheral artery disease. J Am Coll Cardiol. 2013;61:2169–79.

13. Birukov KG, Bochkov VN, Birukova AA, Kawkitinarong K, Rios A, Leitner A, Verin AD, Bokoch GM, Leitinger N, Garcia JG. Epoxycyclopentenone-containing oxidized phospholipids restore endothelial barrier function via Cdc42 and Rac. Circ Res. 2004;95:892–901.

14. Bisgaard H, Kristensen JK. Effects of synthetic leukotriene D-4 on the local regulation of blood flow in human subcutaneous tissue. Prostaglandins. 1985;29:155–9.

15. Bisgaard H, Kristensen J, Sondergaard J. The effect of leukotriene C4 and D4 on cutaneous blood flow in humans. Prostaglandins. 1982;23:797–801.

16. Biswas S, Xin L, Panigrahi S, Zimman A, Wang H, Yakubenko VP, Byzova TV, Salomon RG, Podrez EA. Novel phosphatidylethanolamine derivatives accumulate in circulation in hyperlipidemic ApoE−/− mice and activate platelets via TLR2. Blood. 2016;127:2618–29.

17. Bligh EG, Dyer WJ. A rapid method of total lipid extraction and purification. Can J Biochem Physiol. 1959;37:911–7.

18. Blocher R, Wagner KM, Gopireddy RR, Harris TR, Wu H, Barnych B, Hwang SH, Xiang YK, Proschak E, Morisseau C, et al. Orally available soluble epoxide hydrolase/phosphodiesterase 4 dual inhibitor treats inflammatory pain. J Med Chem. 2018;61:3541–50.

19. Bochkov VN, Leitinger N. Anti-inflammatory properties of lipid oxidation products. J Mol Med. 2003;81:613–26.

20. Bochkov VN, Oskolkova OV, Birukov KG, Levonen AL, Binder CJ, Stockl J. Generation and biological activities of oxidized phospholipids. Antioxid Redox Signal. 2010;12:1009–59.

21. Bohr S, Patel SJ, Sarin D, Irimia D, Yarmush ML, Berthiaume F. Resolvin D2 prevents secondary thrombosis and necrosis in a mouse burn wound model. Wound Repair Regen. 2013;21:35–43.

22. Boullier A, Li Y, Quehenberger O, Palinski W, Tabas I, Witztum JL, Miller YI. Minimally oxidized LDL offsets the apoptotic effects of extensively oxidized LDL and free cholesterol in macrophages. Arterioscler Thromb Vasc Biol. 2006;26:1169–76.

23. Bradley MN, Hong C, Chen M, Joseph SB, Wilpitz DC, Wang X, Lusis AJ, Collins A, Hseuh WA, Collins JL, et al. Ligand activation of LXR beta reverses atherosclerosis and cellular cholesterol overload in mice lacking LXR alpha and apoE. J Clin Invest. 2007;117:2337–46.

24. Breevoort SR, Angdisen J, Schulman IG. Macrophage-independent regulation of reverse cholesterol transport by liver X receptors. Arterioscler Thromb Vasc Biol. 2014;34:1650–60.

25. Brugger B, Erben G, Sandhoff R, Wieland FT, Lehmann WD. Quantitative analysis of biological membrane lipids at the low picomole level by nano-electrospray ionization tandem mass spectrometry. Proc Natl Acad Sci U S A. 1997;94:2339–44.

26. Byun YS, Lee JH, Arsenault BJ, Yang X, Bao W, DeMicco D, Laskey R, Witztum JL, Tsimikas S, TNT Trial Investigators. Relationship of oxidized phospholipids on apolipoprotein B-100 to cardiovascular outcomes in patients treated with intensive versus moderate atorvastatin therapy: the TNT trial. J Am Coll Cardiol. 2015;65:1286–95.

27. Camm J, Hla T, Bakshi R, Brinkmann V. Cardiac and vascular effects of fingolimod: mechanistic basis and clinical implications. Am Heart J. 2014;168:632–44.

28. Capoulade R, Chan KL, Yeang C, Mathieu P, Bosse Y, Dumesnil JG, Tam JW, Teo KK, Mahmut A, Yang X, et al. Oxidized phospholipids, lipoprotein(a), and progression of calcific aortic valve stenosis. J Am Coll Cardiol. 2015;66:1236–46.

29. Cazade M, Bidaud I, Hansen PB, Lory P, Chemin J. 5,6-EET potently inhibits T-type calcium channels: implication in the regulation of the vascular tone. Pflugers Arch: Eur J Physiol. 2014;466:1759–68.

30. Chang MK, Binder CJ, Miller YI, Subbanagounder G, Silverman GJ, Berliner JA, Witztum JL. Apoptotic cells with oxidation-specific epitopes are immunogenic and proinflammatory. J Exp Med. 2004;200:1359–70.

31. Cherepanova OA, Pidkovka NA, Sarmento OF, Yoshida T, Gan Q, Adiguzel E, Bendeck MP, Berliner J, Leitinger N, Owens GK. Oxidized phospholipids induce type VIII collagen expression and vascular smooth muscle cell migration. Circ Res. 2009;104:609–18.

32. Dalli J, Zhu M, Vlasenko NA, Deng B, Haeggstrom JZ, Petasis NA, Serhan CN. The novel 13S,14S-epoxymaresin is converted by human macrophages to maresin 1 (MaR1), inhibits leukotriene A4 hydrolase (LTA4H), and shifts macrophage phenotype. FASEB J. 2013;27:2573–83.

33. Davies SS, Roberts LJ 2nd. F2-isoprostanes as an indicator and risk factor for coronary heart disease. Free Radic Biol Med. 2011;50:559–66.
34. Deng Y, Theken KN, Lee CR. Cytochrome P450 epoxygenases, soluble epoxide hydrolase, and the regulation of cardiovascular inflammation. J Mol Cell Cardiol. 2010;48:331–41.
35. Deng Y, Edin ML, Theken KN, Schuck RN, Flake GP, Kannon MA, DeGraff LM, Lih FB, Foley J, Bradbury JA, et al. Endothelial CYP epoxygenase overexpression and soluble epoxide hydrolase disruption attenuate acute vascular inflammatory responses in mice. FASEB J. 2011;25:703–13.
36. Deng B, Wang CW, Arnardottir HH, Li Y, Cheng CY, Dalli J, Serhan CN. Maresin biosynthesis and identification of maresin 2, a new anti-inflammatory and pro-resolving mediator from human macrophages. PLoS One. 2014;9:e102362.
37. Dona M, Fredman G, Schwab JM, Chiang N, Arita M, Goodarzi A, Cheng G, von Andrian UH, Serhan CN. Resolvin E1, an EPA-derived mediator in whole blood, selectively counterregulates leukocytes and platelets. Blood. 2008;112:848–55.
38. Duewell P, Kono H, Rayner KJ, Sirois CM, Vladimer G, Bauernfeind FG, Abela GS, Franchi L, Nunez G, Schnurr M, et al. NLRP3 inflammasomes are required for atherogenesis and activated by cholesterol crystals. Nature. 2010;464:1357–61.
39. Elajami TK, Colas RA, Dalli J, Chiang N, Serhan CN, Welty FK. Specialized proresolving lipid mediators in patients with coronary artery disease and their potential for clot remodeling. FASEB J. 2016;30:2792–801.
40. Eskildsen MP, Hansen PB, Stubbe J, Toft A, Walter S, Marcussen N, Rasmussen LM, Vanhoutte PM, Jensen BL. Prostaglandin I2 and prostaglandin E2 modulate human intrarenal artery contractility through prostaglandin E2-EP4, prostacyclin-IP, and thromboxane A2-TP receptors. Hypertension. 2014;64:551–6.
41. Esterbauer H, Jurgens G, Quehenberger O, Koller E. Autoxidation of human low density lipoprotein: loss of polyunsaturated fatty acids and vitamin E and generation of aldehydes. J Lipid Res. 1987;28:495–509.
42. Fan F, Roman RJ. Effect of cytochrome P450 metabolites of arachidonic acid in nephrology. J Am Soc Nephrol: JASN. 2017;28:2845–55.
43. Feddersen CO, Mathias M, Murphy RC, Reeves JT, Voelkel NF. Leukotriene E4 causes pulmonary vasoconstriction, not inhibited by meclofenamate. Prostaglandins. 1983;26:869–83.
44. Fleming I. The factor in EDHF: cytochrome P450 derived lipid mediators and vascular signaling. Vasc Pharmacol. 2016;86:31–40.
45. Folch J, Lees M, Sloane Stanley GH. A simple method for the isolation and purification of total lipides from animal tissues. J Biol Chem. 1957;226:497–509.
46. Fraley AE, Schwartz GG, Olsson AG, Kinlay S, Szarek M, Rifai N, Libby P, Ganz P, Witztum JL, Tsimikas S, et al. Relationship of oxidized phospholipids and biomarkers of oxidized low-density lipoprotein with cardiovascular risk factors, inflammatory biomarkers, and effect of statin therapy in patients with acute coronary syndromes: results from the MIRACL (Myocardial Ischemia Reduction With Aggressive Cholesterol Lowering) trial. J Am Coll Cardiol. 2009;53:2186–96.
47. Freeman CP, West D. Complete separation of lipid classes on a single thin-layer plate. J Lipid Res. 1966;7:324–7.
48. Friedli O, Freigang S. Cyclopentenone-containing oxidized phospholipids and their isoprostanes as pro-resolving mediators of inflammation. Biochim Biophys Acta. 2017;1862(4):382–92.
49. Fukunaga M, Takahashi K, Badr KF. Vascular smooth muscle actions and receptor interactions of 8-isoprostaglandin E2, an E2-isoprostane. Biochem Biophys Res Commun. 1993;195:507–15.
50. Glass CK, Witztum JL. Atherosclerosis. The road ahead. Cell. 2001;104:503–16.
51. Gryglewski RJ. Prostacyclin among prostanoids. Pharmacol Rep. 2008;60:3–11.
52. Hammock BD, Wagner K, Inceoglu B. The soluble epoxide hydrolase as a pharmaceutical target for pain management. Pain Manag. 2011;1:383–6.
53. Han X, Gross RW. Structural determination of picomole amounts of phospholipids via electrospray ionization tandem mass spectrometry. J Am Soc Mass Spectrom. 1995;6:1202–10.
54. Han X, Gross RW. Shotgun lipidomics: electrospray ionization mass spectrometric analysis and quantitation of cellular lipidomes directly from crude extracts of biological samples. Mass Spectrom Rev. 2005a;24:367–412.
55. Han X, Gross RW. Shotgun lipidomics: multidimensional MS analysis of cellular lipidomes. Expert Rev Proteomics. 2005b;2:253–64.
56. Hart CM, Karman RJ, Blackburn TL, Gupta MP, Garcia JG, Mohler ER 3rd. Role of 8-epi PGF2alpha, 8-isoprostane, in H2O2-induced derangements of pulmonary artery endothelial cell barrier function. Prostaglandins Leukot Essent Fatty Acids. 1998;58:9–16.

57. Hisano Y, Hla T. Bioactive lysolipids in cancer and angiogenesis. Pharmacol Ther. 2018;193:91–8.

58. Hitzel J, Lee E, Zhang Y, Bibli SI, Li X, Zukunft S, Pfluger B, Hu J, Schurmann C, Vasconez AE, et al. Oxidized phospholipids regulate amino acid metabolism through MTHFD2 to facilitate nucleotide release in endothelial cells. Nat Commun. 2018;9:2292.

59. Ho KJ, Spite M, Owens CD, Lancero H, Kroemer AH, Pande R, Creager MA, Serhan CN, Conte MS. Aspirin-triggered lipoxin and resolvin E1 modulate vascular smooth muscle phenotype and correlate with peripheral atherosclerosis. Am J Pathol. 2010;177:2116–23.

60. Holland WL, Summers SA. Sphingolipids, insulin resistance, and metabolic disease: new insights from in vivo manipulation of sphingolipid metabolism. Endocr Rev. 2008;29:381–402.

61. Hong S, Gronert K, Devchand PR, Moussignac RL, Serhan CN. Novel docosatrienes and 17S-resolvins generated from docosahexaenoic acid in murine brain, human blood, and glial cells. Autacoids in anti-inflammation. J Biol Chem. 2003;278:14677–87.

62. Hoopes SL, Garcia V, Edin ML, Schwartzman ML, Zeldin DC. Vascular actions of 20-HETE. Prostaglandins Other Lipid Mediat. 2015;120:9–16.

63. Hou X, Roberts LJ 2nd, Gobeil F Jr, Taber D, Kanai K, Abran D, Brault S, Checchin D, Sennlaub F, Lachapelle P, et al. Isomer-specific contractile effects of a series of synthetic f2-isoprostanes on retinal and cerebral microvasculature. Free Radic Biol Med. 2004;36:163–72.

64. Hu J, Dziumbla S, Lin J, Bibli SI, Zukunft S, de Mos J, Awwad K, Fromel T, Jungmann A, Devraj K, et al. Inhibition of soluble epoxide hydrolase prevents diabetic retinopathy. Nature. 2017;552:248–52.

65. Huber J, Vales A, Mitulovic G, Blumer M, Schmid R, Witztum JL, Binder BR, Leitinger N. Oxidized membrane vesicles and blebs from apoptotic cells contain biologically active oxidized phospholipids that induce monocyte-endothelial interactions. Arterioscler Thromb Vasc Biol. 2002;22:101–7.

66. Hwang SH, Wagner KM, Morisseau C, Liu JY, Dong H, Wecksler AT, Hammock BD. Synthesis and structure-activity relationship studies of urea-containing pyrazoles as dual inhibitors of cyclooxygenase-2 and soluble epoxide hydrolase. J Med Chem. 2011;54:3037–50.

67. Isobe Y, Arita M, Matsueda S, Iwamoto R, Fujihara T, Nakanishi H, Taguchi R, Masuda K, Sasaki K, Urabe D, et al. Identification and structure determination of novel anti-inflammatory mediator resolvin E3, 17,18-dihydroxyeicosapentaenoic acid. J Biol Chem. 2012;287:10525–34.

68. Joseph SB, McKilligin E, Pei L, Watson MA, Collins AR, Laffitte BA, Chen M, Noh G, Goodman J, Hagger GN, et al. Synthetic LXR ligand inhibits the development of atherosclerosis in mice. Proc Natl Acad Sci U S A. 2002;99:7604–9.

69. Jurgens G, Hoff HF, Chisolm GM 3rd, Esterbauer H. Modification of human serum low density lipoprotein by oxidation--characterization and pathophysiological implications. Chem Phys Lipids. 1987;45:315–36.

70. Kamstrup PR, Hung MY, Witztum JL, Tsimikas S, Nordestgaard BG. Oxidized phospholipids and risk of calcific aortic valve disease: the Copenhagen General Population Study. Arterioscler Thromb Vasc Biol. 2017;37:1570–8.

71. Karki P, Birukov KG. Lipid mediators in the regulation of endothelial barriers. Tissue Barriers. 2018;6:e1385573.

72. Ke Y, Zebda N, Oskolkova O, Afonyushkin T, Berdyshev E, Tian Y, Meng F, Sarich N, Bochkov VN, Wang JM, et al. Anti-inflammatory effects of OxPAPC involve endothelial cell-mediated generation of LXA4. Circ Res. 2017;121:244–57.

73. Keul P, Lucke S, von Wnuck Lipinski K, Bode C, Graler M, Heusch G, Levkau B. Sphingosine-1-phosphate receptor 3 promotes recruitment of monocyte/macrophages in inflammation and atherosclerosis. Circ Res. 2011;108:314–23.

74. Kiechl S, Willeit J, Mayr M, Viehweider B, Oberhollenzer M, Kronenberg F, Wiedermann CJ, Oberthaler S, Xu Q, Witztum JL, et al. Oxidized phospholipids, lipoprotein(a), lipoprotein-associated phospholipase A2 activity, and 10-year cardiovascular outcomes: prospective results from the Bruneck study. Arterioscler Thromb Vasc Biol. 2007;27:1788–95.

75. Kobayasi R, Akamine EH, Davel AP, Rodrigues MA, Carvalho CR, Rossoni LV. Oxidative stress and inflammatory mediators contribute to endothelial dysfunction in high-fat diet-induced obesity in mice. J Hypertens. 2010;28:2111–9.

76. Kuosmanen SM, Kansanen E, Kaikkonen MU, Sihvola V, Pulkkinen K, Jyrkkanen HK, Tuoresmaki P, Hartikainen J, Hippelainen M, Kokki H, et al. NRF2 regulates endothelial glycolysis and proliferation with miR-93 and mediates the effects of oxidized phospholipids on endothelial activation. Nucleic Acids Res. 2018;46:1124–38.

77. Kzhyshkowska J, Neyen C, Gordon S. Role of macrophage scavenger receptors in atherosclerosis. Immunobiology. 2012;217:492–502.

78. Lang J, Celotto C, Esterbauer H. Quantitative determination of the lipid peroxidation product 4-hydroxynonenal by high-performance liquid chromatography. Anal Biochem. 1985;150:369–78.

79. Lauder SN, Allen-Redpath K, Slatter DA, Aldrovandi M, O'Connor A, Farewell D, Percy CL, Molhoek JE, Rannikko S, Tyrrell VJ, et al. Networks of enzymatically oxidized membrane lipids support calcium-dependent coagulation factor binding to maintain hemostasis. Sci Signal. 2017;10(507):eaan2787.

80. Leitinger N, Tyner TR, Oslund L, Rizza C, Subbanagounder G, Lee H, Shih PT, Mackman N, Tigyi G, Territo MC, et al. Structurally similar oxidized phospholipids differentially regulate endothelial binding of monocytes and neutrophils. Proc Natl Acad Sci U S A. 1999;96:12010–5.

81. Ley K, Miller YI, Hedrick CC. Monocyte and macrophage dynamics during atherogenesis. Arterioscler Thromb Vasc Biol. 2011;31:1506–16.

82. Li R, Mouillesseaux KP, Montoya D, Cruz D, Gharavi N, Dun M, Koroniak L, Berliner JA. Identification of prostaglandin E2 receptor subtype 2 as a receptor activated by OxPAPC. Circ Res. 2006;98:642–50.

83. Limbu R, Cottrell GS, McNeish AJ. Characterisation of the vasodilation effects of DHA and EPA, n-3 PUFAs (fish oils), in rat aorta and mesenteric resistance arteries. PLoS One. 2018;13:e0192484.

84. Linton MF, Yancey PG, Davies SS, Jerome WGJ, Linton EF, Vickers KC. The role of lipids and lipoproteins in atherosclerosis. In: De Groot LJ, Chrousos G, Dungan K, Feingold KR, Grossman A, Hershman JM, Koch C, Korbonits M, McLachlan R, New M, et al., editors. Endotext. South Dartmouth: MDText.com, Inc; 2000.

85. Liu M, Boussetta T, Makni-Maalej K, Fay M, Driss F, El-Benna J, Lagarde M, Guichardant M. Protectin DX, a double lipoxygenase product of DHA, inhibits both ROS production in human neutrophils and cyclooxygenase activities. Lipids. 2014;49:49–57.

86. Mahajan-Thakur S, Bohm A, Jedlitschky G, Schror K, Rauch BH. Sphingosine-1-phosphate and its receptors: a mutual link between blood coagulation and inflammation. Mediat Inflamm. 2015;2015:831059.

87. Malleier JM, Oskolkova O, Bochkov V, Jerabek I, Sokolikova B, Perkmann T, Breuss J, Binder BR, Geiger M. Regulation of protein C inhibitor (PCI) activity by specific oxidized and negatively charged phospholipids. Blood. 2007;109:4769–76.

88. Malmsten CL, Palmblad J, Uden AM, Radmark O, Engstedt L, Samuelsson B. Leukotriene B4: a highly potent and stereospecific factor stimulating migration of polymorphonuclear leukocytes. Acta Physiol Scand. 1980;110:449–51.

89. Marathe GK, Davies SS, Harrison KA, Silva AR, Murphy RC, Castro-Faria-Neto H, Prescott SM, Zimmerman GA, McIntyre TM. Inflammatory platelet-activating factor-like phospholipids in oxidized low density lipoproteins are fragmented alkyl phosphatidylcholines. J Biol Chem. 1999;274:28395–404.

90. Marathe GK, Zimmerman GA, Prescott SM, McIntyre TM. Activation of vascular cells by PAF-like lipids in oxidized LDL. Vasc Pharmacol. 2002;38:193–200.

91. Mauerhofer C, Philippova M, Oskolkova OV, Bochkov VN. Hormetic and anti-inflammatory properties of oxidized phospholipids. Mol Asp Med. 2016;49:78–90.

92. Merched AJ, Ko K, Gotlinger KH, Serhan CN, Chan L. Atherosclerosis: evidence for impairment of resolution of vascular inflammation governed by specific lipid mediators. FASEB J. 2008;22:3595–606.

93. Merki E, Graham M, Taleb A, Leibundgut G, Yang X, Miller ER, Fu W, Mullick AE, Lee R, Willeit P, et al. Antisense oligonucleotide lowers plasma levels of apolipoprotein (a) and lipoprotein (a) in transgenic mice. J Am Coll Cardiol. 2011;57:1611–21.

94. Miller YI, Choi SH, Wiesner P, Fang L, Harkewicz R, Hartvigsen K, Boullier A, Gonen A, Diehl CJ, Que X, et al. Oxidation-specific epitopes are danger-associated molecular patterns recognized by pattern recognition receptors of innate immunity. Circ Res. 2011;108:235–48.

95. Milne GL, Yin H, Hardy KD, Davies SS, Roberts LJ 2nd. Isoprostane generation and function. Chem Rev. 2011;111:5973–96.

96. Milne GL, Dai Q, Roberts LJ 2nd. The isoprostanes--25 years later. Biochim Biophys Acta. 2015;1851:433–45.

97. Mitchell JA, Kirkby NS. Eicosanoids, prostacyclin and cyclooxygenase in the cardiovascular system. Br J Pharmacol. 2018.

98. Mobert J, Becker BF, Zahler S, Gerlach E. Hemodynamic effects of isoprostanes (8-iso-prostaglandin F2alpha and E2) in isolated guinea pig hearts. J Cardiovasc Pharmacol. 1997;29:789–94.

99. Moore KJ, Tabas I. Macrophages in the pathogenesis of atherosclerosis. Cell. 2011;145:341–55.

100. Morrow JD, Awad JA, Boss HJ, Blair IA, Roberts LJ 2nd. Non-cyclooxygenase-derived prostanoids (F2-isoprostanes) are formed in situ on phospholipids. Proc Natl Acad Sci U S A. 1992;89:10721–5.

101. Nordhoff E, Ingendoh A, Cramer R, Overberg A, Stahl B, Karas M, Hillenkamp F, Crain PF. Matrix-assisted laser desorption/ionization mass spectrometry of nucleic acids with wavelengths in the ultraviolet and infrared. Rapid Commun Mass Spectrom. 1992;6:771–6.

9

102. Oh SF, Dona M, Fredman G, Krishnamoorthy S, Irimia D, Serhan CN. Resolvin E2 formation and impact in inflammation resolution. J Immunol. 2012;188:4527–34.
103. Ozen G, Norel X. Prostanoids in the pathophysiology of human coronary artery. Prostaglandins Other Lipid Mediat. 2017;133:20–8.
104. Ozen G, Topal G, Gomez I, Ghorreshi A, Boukais K, Benyahia C, Kanyinda L, Longrois D, Teskin O, Uydes-Dogan BS, et al. Control of human vascular tone by prostanoids derived from perivascular adipose tissue. Prostaglandins Other Lipid Mediat. 2013;107:13–7.
105. Palmblad J, Malmsten CL, Uden AM, Radmark O, Engstedt L, Samuelsson B. Leukotriene B4 is a potent and stereospecific stimulator of neutrophil chemotaxis and adherence. Blood. 1981;58: 658–61.
106. Park SK, Herrnreiter A, Pfister SL, Gauthier KM, Falck BA, Falck JR, Campbell WB. GPR40 is a low-affinity epoxyeicosatrienoic acid receptor in vascular cells. J Biol Chem. 2018;293:10675–91.
107. Pidkovka NA, Cherepanova OA, Yoshida T, Alexander MR, Deaton RA, Thomas JA, Leitinger N, Owens GK. Oxidized phospholipids induce phenotypic switching of vascular smooth muscle cells in vivo and in vitro. Circ Res. 2007;101:792–801.
108. Pluchart H, Khouri C, Blaise S, Roustit M, Cracowski JL. Targeting the prostacyclin pathway: beyond pulmonary arterial hypertension. Trends Pharmacol Sci. 2017;38:512–23.
109. Poli G, Cecchini G, Biasi F, Chiarpotto E, Canuto RA, Biocca ME, Muzio G, Esterbauer H, Dianzani MU. Resistance to oxidative stress by hyperplastic and neoplastic rat liver tissue monitored in terms of production of unpolar and medium polar carbonyls. Biochim Biophys Acta. 1986;883:207–14.
110. Poli G, Biasi F, Chiarpotto E, Dianzani MU, De Luca A, Esterbauer H. Lipid peroxidation in human diseases: evidence of red cell oxidative stress after circulatory shock. Free Radic Biol Med. 1989;6: 167–70.
111. Poorani R, Bhatt AN, Dwarakanath BS, Das UN. COX-2, aspirin and metabolism of arachidonic, eicosapentaenoic and docosahexaenoic acids and their physiological and clinical significance. Eur J Pharmacol. 2016;785:116–32.
112. Purushothaman KR, Purushothaman M, Levy AP, Lento PA, Evrard S, Kovacic JC, Briley-Saebo KC, Tsimikas S, Witztum JL, Krishnan P, et al. Increased expression of oxidation-specific epitopes and apoptosis are associated with haptoglobin genotype: possible implications for plaque progression in human atherosclerosis. J Am Coll Cardiol. 2012;60:112–9.
113. Que X, Hung MY, Yeang C, Gonen A, Prohaska TA, Sun X, Diehl C, Maatta A, Gaddis DE, Bowden K, et al. Oxidized phospholipids are proinflammatory and proatherogenic in hypercholesterolaemic mice. Nature. 2018;558:301–6.
114. Roman RJ. P-450 metabolites of arachidonic acid in the control of cardiovascular function. Physiol Rev. 2002;82:131–85.
115. Romanoski CE, Che N, Yin F, Mai N, Pouldar D, Civelek M, Pan C, Lee S, Vakili L, Yang WP, et al. Network for activation of human endothelial cells by oxidized phospholipids: a critical role of heme oxygenase 1. Circ Res. 2011;109:e27–41.
116. Santilli F, Boccatonda A, Davi G, Cipollone F. The Coxib case: are EP receptors really guilty? Atherosclerosis. 2016;249:164–73.
117. Sasset L, Zhang Y, Dunn TM, Di Lorenzo A. Sphingolipid de novo biosynthesis: a rheostat of cardiovascular homeostasis. Trends Endocrinol Metab. 2016;27:807–19.
118. Scull CM, Tabas I. Mechanisms of ER stress-induced apoptosis in atherosclerosis. Arterioscler Thromb Vasc Biol. 2011;31:2792–7.
119. Seimon TA, Nadolski MJ, Liao X, Magallon J, Nguyen M, Feric NT, Koschinsky ML, Harkewicz R, Witztum JL, Tsimikas S, et al. Atherogenic lipids and lipoproteins trigger CD36-TLR2-dependent apoptosis in macrophages undergoing endoplasmic reticulum stress. Cell Metab. 2010;12:467–82.
120. Serbulea V, DeWeese D, Leitinger N. The effect of oxidized phospholipids on phenotypic polarization and function of macrophages. Free Radic Biol Med. 2017;111:156–68.
121. Serbulea V, Upchurch CM, Schappe MS, Voigt P, DeWeese DE, Desai BN, Meher AK, Leitinger N. Macrophage phenotype and bioenergetics are controlled by oxidized phospholipids identified in lean and obese adipose tissue. Proc Natl Acad Sci U S A. 2018;115:E6254–63.
122. Serhan CN. Lipoxin biosynthesis and its impact in inflammatory and vascular events. Biochim Biophys Acta. 1994;1212:1–25.
123. Serhan CN, Yang R, Martinod K, Kasuga K, Pillai PS, Porter TF, Oh SF, Spite M. Maresins: novel macrophage mediators with potent antiinflammatory and proresolving actions. J Exp Med. 2009; 206:15–23.
124. Sharma JN, Jawad NM. Adverse effects of COX-2 inhibitors. ScientificWorldJournal. 2005;5:629–45.

125. Slatter DA, Percy CL, Allen-Redpath K, Gajsiewicz JM, Brooks NJ, Clayton A, Tyrrell VJ, Rosas M, Lauder SN, Watson A, et al. Enzymatically oxidized phospholipids restore thrombin generation in coagulation factor deficiencies. JCI insight. 2018;3(6):e98459.
126. Smith MJ. Biological activities of leukotriene B4. Agents Actions. 1981;11:571–2.
127. Smyth SS, Mueller P, Yang F, Brandon JA, Morris AJ. Arguing the case for the autotaxin-lysophosphatidic acid-lipid phosphate phosphatase 3-signaling nexus in the development and complications of atherosclerosis. Arterioscler Thromb Vasc Biol. 2014;34:479–86.
128. Soltwisch J, Jaskolla TW, Hillenkamp F, Karas M, Dreisewerd K. Ion yields in UV-MALDI mass spectrometry as a function of excitation laser wavelength and optical and physico-chemical properties of classical and halogen-substituted MALDI matrixes. Anal Chem. 2012;84:6567–76.
129. Spector AA, Kim HY. Cytochrome P450 epoxygenase pathway of polyunsaturated fatty acid metabolism. Biochim Biophys Acta. 2015;1851:356–65.
130. Spite M, Serhan CN. Novel lipid mediators promote resolution of acute inflammation: impact of aspirin and statins. Circ Res. 2010;107:1170–84.
131. Subbanagounder G, Wong JW, Lee H, Faull KF, Miller E, Witztum JL, Berliner JA. Epoxyisoprostane and epoxycyclopentenone phospholipids regulate monocyte chemotactic protein-1 and interleukin-8 synthesis. Formation of these oxidized phospholipids in response to interleukin-1beta. J Biol Chem. 2002;277:7271–81.
132. Tabas I, Bornfeldt KE. Macrophage phenotype and function in different stages of atherosclerosis. Circ Res. 2016;118:653–67.
133. Tacconelli S, Patrignani P. Inside epoxyeicosatrienoic acids and cardiovascular disease. Front Pharmacol. 2014;5:239.
134. Takats Z, Wiseman JM, Gologan B, Cooks RG. Mass spectrometry sampling under ambient conditions with desorption electrospray ionization. Science. 2004;306:471–3.
135. Tontonoz P, Mangelsdorf DJ. Liver X receptor signaling pathways in cardiovascular disease. Mol Endocrinol. 2003;17:985–93.
136. Tukijan F, Chandrakanthan M, Nguyen LN. Mini-review: the signaling roles of S1P derived from red blood cells and platelets. Br J Pharmacol. 2018;175:3741.
137. van der Valk FM, Bekkering S, Kroon J, Yeang C, Van den Bossche J, van Buul JD, Ravandi A, Nederveen AJ, Verberne HJ, Scipione C, et al. Oxidized phospholipids on lipoprotein(a) elicit arterial wall inflammation and an inflammatory monocyte response in humans. Circulation. 2016;134:611–24.
138. Wada M, DeLong CJ, Hong YH, Rieke CJ, Song I, Sidhu RS, Yuan C, Warnock M, Schmaier AH, Yokoyama C, et al. Enzymes and receptors of prostaglandin pathways with arachidonic acid-derived versus eicosapentaenoic acid-derived substrates and products. J Biol Chem. 2007;282:22254–66.
139. Wagner KM, McReynolds CB, Schmidt WK, Hammock BD. Soluble epoxide hydrolase as a therapeutic target for pain, inflammatory and neurodegenerative diseases. Pharmacol Ther. 2017;180:62–76.
140. Waldman M, Peterson SJ, Arad M, Hochhauser E. The role of 20-HETE in cardiovascular diseases and its risk factors. Prostaglandins Other Lipid Mediat. 2016;125:108–17.
141. Waltenberger B, Garscha U, Temml V, Liers J, Werz O, Schuster D, Stuppner H. Discovery of potent soluble epoxide hydrolase (sEH) inhibitors by pharmacophore-based virtual screening. J Chem Inf Model. 2016;56:747–62.
142. Watson AD, Leitinger N, Navab M, Faull KF, Horkko S, Witztum JL, Palinski W, Schwenke D, Salomon RG, Sha W, et al. Structural identification by mass spectrometry of oxidized phospholipids in minimally oxidized low density lipoprotein that induce monocyte/endothelial interactions and evidence for their presence in vivo. J Biol Chem. 1997;272:13597–607.
143. Watson AD, Subbanagounder G, Welsbie DS, Faull KF, Navab M, Jung ME, Fogelman AM, Berliner JA. Structural identification of a novel pro-inflammatory epoxyisoprostane phospholipid in mildly oxidized low density lipoprotein. J Biol Chem. 1999;274:24787–98.
144. Welton AF, Crowley HJ, Miller DA, Yaremko B. Biological activities of a chemically synthesized form of leukotriene E4. Prostaglandins. 1981;21:287–96.
145. Wiseman JM, Puolitaival SM, Takats Z, Cooks RG, Caprioli RM. Mass spectrometric profiling of intact biological tissue by using desorption electrospray ionization. Angew Chem. 2005;44:7094–7.
146. Yan FX, Li HM, Li SX, He SH, Dai WP, Li Y, Wang TT, Shi MM, Yuan HX, Xu Z, et al. The oxidized phospholipid POVPC impairs endothelial function and vasodilation via uncoupling endothelial nitric oxide synthase. J Mol Cell Cardiol. 2017;112:40–8.
147. Yanagida K, Hla T. Vascular and immunobiology of the circulatory sphingosine 1-phosphate gradient. Annu Rev Physiol. 2017;79:67–91.

148. Yang K, Han X. Lipidomics: techniques, applications, and outcomes related to biomedical sciences. Trends Biochem Sci. 2016;41:954–69.
149. Yeang C, Hasanally D, Que X, Hung MY, Stamenkovic A, Chan D, Chaudhary R, Margulets V, Edel AL, Hoshijima M, et al. Reduction of myocardial ischemiareperfusion injury by inactivating oxidized phospholipids. Cardiovasc Res. 2018;115(1):179–89.
150. Yokoyama U. Prostaglandin E-mediated molecular mechanisms driving remodeling of the ductus arteriosus. Pediatr Int. 2015;57:820–7.
151. Yokoyama U, Minamisawa S, Ishikawa Y. The multiple roles of prostaglandin E2 in the regulation of the ductus arteriosus. In: Nakanishi T, Markwald RR, Baldwin HS, Keller BB, Srivastava D, Yamagishi H, editors. Etiology and morphogenesis of congenital heart disease: from gene function and cellular interaction to morphology. Tokyo: Springer; 2016. p. 253–8.
152. Zhang Y, Breevoort SR, Angdisen J, Fu M, Schmidt DR, Holmstrom SR, Kliewer SA, Mangelsdorf DJ, Schulman IG. Liver LXRalpha expression is crucial for whole body cholesterol homeostasis and reverse cholesterol transport in mice. J Clin Invest. 2012;122:1688–99.
153. Zhang X, Wang T, Gui P, Yao C, Sun W, Wang L, Wang H, Xie W, Yao S, Lin Y, et al. Resolvin D1 reverts lipopolysaccharide-induced TJ proteins disruption and the increase of cellular permeability by regulating IkappaBalpha signaling in human vascular endothelial cells. Oxidative Med Cell Longev. 2013;2013:185715.
154. Zhang C, Booz GW, Yu Q, He X, Wang S, Fan F. Conflicting roles of 20-HETE in hypertension and renal end organ damage. Eur J Pharmacol. 2018;833:190–200.
155. Zhao YL, Zhang L, Yang YY, Tang Y, Zhou JJ, Feng YY, Cui TL, Liu F, Fu P. Resolvin D1 protects lipopolysaccharide-induced acute kidney injury by down-regulating nuclear factor-kappa B signal and inhibiting apoptosis. Chin Med J. 2016;129:1100–7.

# Atherosclerosis

*Florian J. Mayer and Christoph J. Binder*

© Springer Nature Switzerland AG 2019
M. Geiger (ed.), *Fundamentals of Vascular Biology*, Learning Materials in Biosciences,
https://doi.org/10.1007/978-3-030-12270-6_10

**What Will You Learn in This Chapter?**

This chapter summarizes the current knowledge of the development of atherosclerotic disease. Well-established cardiovascular risk factors with a focus on lipids and lipoproteins are discussed, and you will learn how certain lifestyle factors can cause detrimental effects on your cardiovascular system. You will gain an understanding on how the endothelial barrier gets altered and how plaques develop until they are prone to rupture. You will also learn about selected immunological processes involved in atherogenesis with an emphasis on monocytes and T and B cells and how they interact with each other and arterial vascular properties. Finally, strategies for cardiovascular disease prevention are discussed.

Atherosclerosis is a specific type of arteriosclerosis, but the terms are often used interchangeably. The word comes from the Greek words "athero" (meaning gruel or paste) and "sclerosis" (hardness), and indeed, the disease is defined by the loss of arterial elasticity due to vessel thickening and stiffening. This chronic and complex disease probably already starts in childhood and progresses when people grow older. Atherosclerosis is caused by the deposition of lipids, immune cells and cell debris, calcium, and other substances in the inner lining (intima) of the vascular wall of large- and medium-sized arteries. The result of this buildup is called plaque and represents the underlying cause of coronary artery disease. If a plaque ruptures or if a blood clot is formed on the plaque's surface, the blood flow of the affected artery is disrupted, and the organs and tissues supplied by the blocked arteries stop receiving sufficient blood and oxygen to function properly. The drastic consequences are heart attacks and strokes. Due to the steady *increase* in life expectancy, atherosclerosis represents nowadays the leading cause of death worldwide [1]. In the last decades, a tremendous amount of research across all areas has focused on this disease, and many aspects of the processes leading to atherosclerosis have been unraveled.

## 10.1  Risk Factors for Atherosclerosis

The mechanisms that cause atherosclerosis involve more than one specific cause, and it is therefore considered a multifactorial disease. It has been clearly established that certain lifestyle habits, genetic traits, and other diseases raise the risk for the development of atherosclerosis. These conditions are generally described as risk factors. The more risk factors one has, the more likely it is to develop atherosclerosis. But, behavioral risk factors can be easily avoided in order to help delay or even prevent atherosclerotic disease.

### 10.1.1  Dyslipidemia

*The basics of lipidology are discussed in the book chapter "Lipids and Lipoproteins in Vascular Biology" by Norbert Leitinger.*

Cholesterol is the precursor for bile acids, steroid hormones, and vitamin D, but it seems that only very low cholesterol levels are needed to maintain these physiological functions [2]. A century of research links lipids with atherosclerosis [3, 4]. Dyslipidemia (i.e., deranged lipid parameters measured in serum) has been established as a causal factor for atherosclerosis-related diseases, such as coronary heart disease and peripheral vascular disease, and represents a target for several well-established and novel therapeutics in order to attenuate and reduce the patients' atherosclerotic burden.

Increased *total cholesterol* remains one of the major risk factors for the development of cardiovascular diseases [5]. Serum cholesterol levels are still, together with blood pressure, age, sex, and smoking status, a central component of various cardiovascular risk prediction models used in everyday clinical practice from the family physician to the cardiologist. But, while total cholesterol levels play an important role in general risk assessment, the individual lipoprotein particles (e.g., LDL, HDL) that transport cholesterol and the associated proteins (e.g., ApoB-100) play a more distinct role in the development and progression of atherosclerosis. Therefore, clinicians not merely look at total cholesterol levels alone but lipoprotein-associated cholesterol levels and associated variables implicated in cholesterol transport and metabolism.

The relationship of *triglyceride levels* in atherosclerosis is uncertain. While metabolic products of triglyceride metabolism have been shown toxic to endothelial cells, animal models have not fully established the fact that initiation or progression of atherosclerosis is induced simply by triglyceride-rich lipoproteins (i.e., VLDL and chylomicrons) [6]. Finally, interventional trial levels have not reached consistent evidence that reducing levels of triglycerides affects cardiovascular outcome [7].

### 10.1.1.1 Lipoproteins

Lipoproteins are heterogeneous with respect to size, density, composition, and physicochemical characteristics. They are pivotal for the transport of cholesterol in the blood. LDL in particular is the major vehicle that transports cholesterol all over the body in order to maintain the intracellular functions, the cell membrane, and the biosynthesis of steroid hormones.

*Low-density lipoprotein cholesterol (LDL)* has largely replaced total cholesterol as the primary lipid measurement for cardiovascular risk assessment [8]. A plethora of epidemiological studies suggest increased levels of LDL as an independent and robust risk factor for the development of atherosclerotic diseases [9]. Genetic studies of patients with familial hypercholesterolemia show that lifelong exposure to high levels of LDL cholesterol in the blood – often caused by mutations of the LDL receptor – results in markedly reduced life expectancy [10]. Mendelian randomization studies further provide evidence that LDL cholesterol is likely to be a causal factor for atherosclerotic plaque development and progression [11]. It is now well appreciated that the exposure to increased serum LDL cholesterol levels is a critical factor determining cardiovascular risk. Finally, therapeutic reduction of LDL cholesterol significantly reduces cardiovascular events and mortality. In fact, the lipid profile component most directly targeted for clinical intervention is LDL [12]. Unsurprisingly, much research on the pathophysiologic mechanisms of this disease has focused on this family of lipoprotein particles. LDL is catabolized by either of the two distinct pathways: a receptor-dependent pathway in the liver or a receptor-independent pathway in non-hepatic tissues. In the hepatic (LDL receptor dependent) pathway, LDLs interact with hepatocytes via high-affinity binding of the Apolipoprotein B-100 surface protein [13]. Importantly, even at low-plasma LDL levels, LDL receptors are saturated. Thus, when plasma LDL is increased, receptor-independent uptake is greater than the amount of LDL catabolized by the liver, and subsequently lipid deposition occurs – an essential step in the initiation, development, and progression of atherosclerosis. High levels of LDLs and by-products of oxidation reactions with LDL influence the response of the arterial wall, promote monocyte recruitment, modulate smooth muscle cell migration, reduce endothelial NO synthesis, and increase platelet activation, processes that all initiate atherosclerotic plaque formation [14–18].

*High-density lipoprotein (HDL)* is synthesized in the liver and intestine. HDL acquires apolipoproteins and lipids from hepatocytes and enterocytes as well as from the hydrolysis of triglyceride-rich lipoproteins, respectively. Cholesterol of HDL is then exchanged via the cholesteryl ester transfer protein (CETP) with VLDL or chylomicrons leading to a net transfer of cholesterol from HDL to triglyceride-rich lipoproteins followed by the hepatic removal of HDL ("reverse cholesterol transport") [19]. Strong inverse associations between HDL and coronary heart disease were first shown more than 50 years ago in the Framingham study, and the subsequent hypothesis that HDL is protective against atherosclerosis was further supported by various animal models [20, 21]. Macrophages have been shown to transfer cholesterol to HDL by the so-called cholesterol efflux mechanisms, which could explain the inhibitory effects on atherosclerotic lesions. These observational and experimental data formed the basis for the widely acknowledged perception of HDL as the "good cholesterol" and led to believe that HDL protects against atherosclerosis and that intervention to raise HDL reduces cardiovascular risk [21, 22]. Unfortunately, attempts to treat this residual risk, with, e.g., CETP inhibitors that dramatically raise HDL, have been shown ineffective or only met with modest success [23]. Today, HDL remains a useful biomarker for cardiovascular risk prediction, but, as yet, cannot be utilized for therapeutic interventions.

*Lp(a)* is commonly described as an LDL-like lipoprotein particle. Case-control and prospective epidemiological studies describe Lp(a) as a risk factor for myocardial infarction and stroke [24]. In addition, recent genetic studies found such a strong association between atherosclerosis and Lp(a) that it is now considered a causal factor in atherogenesis [25].

*Lipoprotein lipase (LPL)* is responsible for catalyzing lipolysis of triglycerides in lipoproteins. LPL of the vessel wall, which is mainly derived from macrophages, possess proatherogenic properties. LPL enhances the adhesion of monocytes to the endothelium, and lipolytic products of LPL such as free fatty acids act on endothelial cells to promote the entry of lipoproteins into arterial intima (see ▶ Sect. 10.2.1 "Endothelial Dysfunction") [26].

*Apolipoproteins* are the protein moiety of lipoprotein particles and are present mainly on the lipoprotein surface. Apolipoprotein B-100 (ApoB-100) is the main protein component of the (potentially) atherogenic lipoproteins LDL and Lp(a). ApoB-100 causes dysfunction of endothelium in the initial stage of atherogenesis by impairing endothelium-dependent vasodilation and proved to be a clinically useful early marker of atherosclerosis [27–29]. ApoE promotes clearance of triglyceride-rich lipoproteins by binding to LDLR to mediate lipolytic processing and endocytosis of triglyceride-rich lipoprotein remnant particles. A plethora of animal models have characterized anti-atherogenic functions of ApoE [30]. Apolipoprotein A-I (apoA-I) is the major protein associated with HDL. ApoA-I promotes reverse cholesterol transport from tissues to the liver for excretion [31].

### 10.1.1.2 Oxidized Lipids

The oxidation of lipids and lipoproteins is a complex process where both the lipids and the protein moiety undergo oxidative changes through enzymatic or nonenzymatic pathways and form a wide array of oxidized particles. Under conditions of oxidative stress, *oxidized low-density lipoproteins (OxLDL)* formation occurs in the extracellular space of the vessel wall, but LDL modifications can also develop due to the degradation of native LDLs by lysosomal enzymes within macrophages [32]. With further oxidization and protein modification of LDL, loss of recognition by the LDL receptor occurs. Subsequently, OxLDL binds with high affinity to several other receptors than the native LDL receptors including scavenger receptors CD36 and SRA-1 [33]. This results in the uptake of OxLDL by macrophages, cholesterol accumulation, and finally foam cell formation (see ▶ Sect. 10.3.1 "From Monocytes to Foam Cells").

Free radical-induced oxidation of membrane phospholipids further generates oxidized phospholipids (oxPLs). An abundance of complex oxPLs with diverse molecular characteristics and biological activities has been described. oxPLs were shown to promote adhesion of monocytes and neutrophils to endothelial cells by initiating chemokine expression as well as proliferation of smooth muscle cells [34]. In addition, they were found to promote the expression of tissue factor on endothelial cells, promote mitochondrial swelling and apoptosis, inhibit the binding HDL with hepatocytes and subsequently the delivery of cholesterol to the liver for excretion, and induce prothrombotic state via platelet activation [35]. Finally, targeted inhibition of oxPLs showed a significant reduction in the progression of atherosclerosis in mouse models [36].

## 10.1.2 Diabetes Mellitus

Diabetes mellitus, defined as a metabolic disease resulting from defects in insulin secretion, insulin actions, or both, is characterized by a chronic state of hyperglycemia. Diabetes mellitus is associated with increase morbidity and mortality, affects the patient's life quality, and is linked to some acute but mainly chronic complications, based on functional and structural damages to the blood vessels. The disease is unsurprisingly a major independent risk factor for the development of cardiovascular disease, and both type 1 and type 2 diabetes have been shown to accelerate the development of atherosclerosis.

Alterations of the vessel wall, due to endothelial and smooth muscle cell dysfunction, are the main characteristics of diabetic vasculopathy. Various mechanisms by which diabetes contributes to cardiovascular disease and atherosclerosis have been identified. Hyperglycemia increases the generation of superoxide anions, hydrogen peroxide, and hydroxyl radicals [37]. Subsequently lipid peroxidation and generation of oxidized free fatty acids occur, which contribute to the development of endothelial dysfunction and atherosclerosis [38]. Reduced insulin signaling also leads to endothelial dysfunction, by reducing endothelial nitric oxide synthase expression, nitric oxide production, and activation of enzymes that regulate the activity of nitric oxide [39].

During diabetic state, advanced glycation end products (AGEs) are generated, a diverse group of highly oxidant compounds formed by nonenzymatic reactions between the aldehydic group of reducing sugars with proteins, lipids, or nucleic acids. These harmful species promote vascular damage and acceleration of atherosclerotic plaque progression by altering functional properties of vessel wall extracellular matrix molecules and activation of cell receptor-dependent signaling [40]. Interaction between AGEs and its key receptor RAGE (receptor for advanced glycation end products), a transmembrane signaling receptor which is expressed in various cells relevant to atherosclerosis, alters cellular function and promotes gene expression and secretion of proinflammatory cytokines. The importance of the AGE-RAGE interaction and downstream pathways, leading to vessel wall injury and plaque development, has been verified in numerous animal models [41]. Varieties of oral antidiabetic drugs (e.g., metformin) are currently available and represent one of the biggest pharmaceutical domains. If oral antidiabetics do not suffice, because of either insufficient insulin synthesis or reduced insulin sensitivity, subcutaneous injection of insulin is required.

### 10.1.3  Behavioral Risk Factors

Even though various therapeutic options exist, one cannot emphasize enough that for most patients, the first strategy to treat atherosclerosis is to modify lifestyle habits.

#### 10.1.3.1  Smoking

Smoking is the most important preventable risk factor for the development of atherosclerosis.

Studies showed increased risk of developing atherosclerosis at all levels of cigarette smoking, and increased risks were found even for persons who smoked only one cigarette per day. The risks of myocardial infarction and death from stroke are lower among former smokers than among those who continue smoking [42, 43]. But, even though a lifelong residual risk still exists, after 10 years of smoking cessation, only a small difference can be found for the risk of cardiovascular events between former smokers and non-smokers [44–46]. Changing smoking habits remains one of the biggest challenges and one of the most demanding tasks for every clinician.

The pathophysiologic mechanisms by which tobacco smoke accelerates vascular disease are manifold and complex. Polycyclic aromatic hydrocarbons, oxidizing agents, particulate matter, and nicotine have been identified as potential contributing factors to atherogenesis. Via the release of catecholamines, nicotine increases heart rate, blood pressure, and platelet aggregability, adverse hemodynamic effects which are associated with progression of atherosclerosis and increased risk for atherothrombosis [47]. However, even though smoking has been investigated heavily, it is probably the least understood among all the risk factors for atherosclerosis. This is mainly due to the fact that cigarette smoke contains ≈4000 different chemicals and molecules of all sizes and variants. Apart from the abovementioned candidate compounds, the relevance of most other (toxic) compounds in cigarette smoke in the course of atherogenesis has not been studied or been poorly understood [48].

#### 10.1.3.2  Sedentary Lifestyle, Diet, and Obesity

Sedentary behaviors are defined as any waking time during which one is in seated, reclined, or lying posture, which produces low levels of energy [49]. This lifestyle has been identified as one of the leading risk factors for many chronic conditions and mortality during the last decade or so. Observational studies clearly show that lack of physical activity increases the risk of cardiovascular diseases such as type 2 diabetes, coronary heart disease, or stroke [50]. The increased risk of all-cause and CVD mortality is strongest for sitting time volumes greater than 6–8 hours per day, and a strong correlation between the development of diabetes mellitus type 2 and TV viewing time exists [51]. Physical activity of all sorts significantly reduces the risk of CVD including stroke and myocardial infarction. The increased cardiorespiratory fitness (due to physical activity) reduces blood pressure, improves endothelial function, and significantly lessens systemic inflammation [52]. On the other hand, sedentary behaviors lead to significantly reduced insulin sensitivity, increased serum triglycerides, impaired metabolic function, and attenuated endothelial function. Furthermore, a reduction in shear stress, as a result of physical inactivity, reduces nitric oxide bioavailability and leads to endothelial impairment, both highly proatherogenic factors [53]. Fatty acids and glucose are preferentially shuttled toward energy stores (i.e., adipose depots) if their use is reduced in the case of lower energy expenditure. But not only the reduced energy consumption due to physical inactivity but mainly the western diet led to an endemic disease which spread all over the world: obesity, defined as a body mass index of 30 kg/m² or greater. Obesity is associated with an increased risk of hypertension, diabetes mellitus type 2, metabolic syndrome, and dyslipidemia, all pivotal risk factors for the development of atherosclerosis [54]. The adipocyte not only functions as an energy reserve but also as an endocrine organ and might play a role in atherogenesis. Increased levels of leptin, an adipocyte-derived hormone that influences food intake and energy metabolism, might be a driving force for the development and progression of atherosclerosis. Leptin induces endothelial dysfunction and stimulates inflammatory reactions, oxidative stress, platelet aggregation and migration, as well as hypertrophy and proliferation of vascular smooth muscle cells [55]. Obesity has many other adverse effects for the cardiovascular system such as increased total blood volume and cardiac output, cardiac remodeling, impaired nitric oxide-dependent vasodilation, increased risk for atrial fibrillation, and increase in carbon monoxide in the blood [56, 57].

Diet plays a central role in the prevention of atherosclerosis. The relationship between eating vegetables and fruits and preventing stroke and myocardial infarction has been verified via an abundance of large-scale epidemiological studies [58, 59]. This has been partly attributed to the antioxidant properties of various molecules contained in fruit and vegetables. The consumption of seafood has also been contributed with a decreased risk for cardiovascular disease. N-3 polyunsaturated fatty acids (PUFAs) have been detected as the potential anti-atherogenic components in fish. PUFAs induce the synthesis of anti-inflammatory factors including protectin, an inhibitor of the complement cascade, and resolvins, which promote inflammatory resolution [60].

Since the early 1970s, it has been a widely accepted belief that dietary saturated fats and dietary cholesterol cause an increase in serum total cholesterol, as well as LDL, and thereby an increase in the risk of atherosclerosis. Based on these false assumptions, dietary fats were replaced with carbohydrates by food companies, which led to an increase of obesity, diabetes mellitus II, and coronary heart disease [61]. Even though lipids are a key player in atherosclerosis, dietary intake of fat (including cholesterol) is not the crucial factor in atherogenesis. Dietary cholesterol intake – unless excessive – has only a small effect on fasting plasma cholesterol levels and is not associated with coronary heart disease [62]. Lipids and lipoproteins are to a greater extent influenced by other factors than the dietary intake, such as body weight, metabolic state, genetic factors, and the individual inflammatory response. Carrying too much fat appears to be far more of a driver for atherosclerosis than eating it.

### 10.1.4 Arterial Hypertension

Significant numbers of patients with hypertension develop atherosclerosis, and the underlying mechanisms include endothelial dysfunction, oxidative stress, vascular remodeling, fibrosis, and alterations of immunoregulatory T cells [63]. Essential hypertension (which comprises 90% patients with arterial hypertension) is characterized by a defective endothelial nitric oxide pathway, impaired responsiveness to exogenous nitric oxide, and reduced generation of nitric oxide. The reduced nitric oxide bioavailability leads to loss of the protective properties of the endothelium and thus to the expression of a pro-inflammatory state, which subsequently turns into a pro-atherosclerotic environment. Notably, atherosclerosis is also a risk factor for hypertension, which promotes a vicious circle [64]. Increased body weight is likely the most important risk factor for arterial hypertension, and the correlation between body weight and arterial hypertension has been well described in the literature in large studies, including the Framingham cohort [65]. However, not all hypertensive patients are overweight, and other factors including genetic predisposition, age, and gender affect whether an individual develops arterial hypertension. Various therapeutic approaches are available for the treatment of hypertension, targeting different molecules involved such as the angiotensin-converting enzyme.

Questions Paragraph 1 (More than One Answer Can Be Correct)

1. **Which of the following are *modifiable* risk factors for the development of atherosclerosis?**
   A. Hypertension
   B. Hypertriglyceridemia
   C. Smoking
   D. Diabetes mellitus type I
   E. Physical inactivity

2. **Oxidation of lipids and lipoproteins leads to:**
   A. The development of advanced glycosylation end products (AGEs)
   B. The uptake of modified lipids and lipoproteins by macrophages via scavenger receptors
   C. Increased expression of inflammatory genes by endothelial cells
   D. The secretion of TGF-β by neutrophils
   E. Hypertriglyceridemia

## 10.2  Plaque Formation

### 10.2.1  Endothelial Dysfunction

Vascular endothelial cell dysfunction is the initiation step of atherosclerosis development. The initial hypothesis of endothelial dysfunction by Ross and Glomset, in the context of atherogenesis, postulated that the initiating event in the atherogenic process is a "response to injury," induced by noxious substances (e.g., aberrant cholesterol deposits) or hemodynamic forces (e.g., harmful shear stress of the blood flow) [66]. In the meantime various other factors involved in endothelial dysfunction have been elucidated, but the principle remains unchanged.

#### 10.2.1.1  Arterial Shear Stress and Endothelial Activation

Atherosclerosis is a systemic disease that affects all regions of the arterial tree. The most susceptible areas include branch points and inner curvatures of the coronary arteries, the carotid arteries, and the iliofemoral arteries. Even though these are the most common places where atherosclerotic plaques are found, because of its systemic nature, atherosclerosis can develop at any given location of the arterial tree. The predilection sites are characterized by changes in endothelial turnover, altered gene expression, and increased presence of subendothelial dendritic cells [67, 68]. The endothelial monolayer is directly exposed to blood flow and acts as a signal transduction interface for blood flow stimuli. Blood flow exerts biomechanical forces on the vasculature and affects the physiology and function of endothelial cells as well as subendothelial structures. Blood flow patterns can be categorized into laminar and disturbed flow. Laminar flow represents the unidirectional movement of blood within the vessel, whereas disturbed flow is characterized by areas of flow reversal and includes circulatory and turbulent blood flow. Under physiological conditions, the shear stress of laminar flow exhibits protective effects to the endothelium. When laminar flow changes into turbulent flow or circulatory flow, it produces shear stress harmful to the endothelium such as oscillating shear stress or low endothelial shear stress (◘ Fig. 10.1) [69, 70]. Under these constant disrupting conditions, the balanced endothelial regulation is altered and changes into a nonadaptive state. The endothelium is then activated and shows a proinflammatory transcription profile, resulting in expression of adhesion molecules, a higher permeability to plasma macromolecules, and a decrease in nitric oxide (NO), which finally results in endothelial dysfunction [69, 71, 72]. Endothelium-derived NO plays a key role in the physiological regulation of the cardiovascular system and is a powerful vasodilator. NO

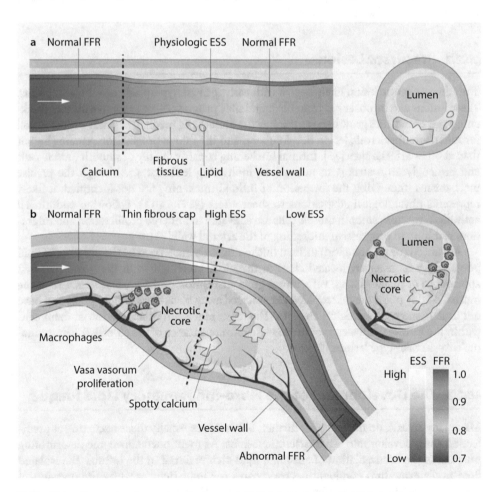

◘ **Fig. 10.1** Plaque development under low endothelial shear stress. **a** Stable lesion. **b** Rupture-prone vulnerable plaque The fractional flow reserve (FFR) is a physiological index of the severity of a stenosis in an (coronary) artery. It is determined by the pressure after a stenosis relative to the pressure before the stenosis. An FFR of a coronary artery lower than 0.8 is associated with myocardial ischemia [89]. ESS endothelial shear stress. (Modified after [269])

decreases the intracellular concentration via the stimulation of sGC to produce cGMP, causing relaxation of smooth muscle cells. A quiescent healthy endothelium constitutively
releases NO in response to laminar blood flow. Endothelial dysfunction is associated with a
significant reduction of NO availability. Decreased NO is caused by the uncoupling of eNOS
(nitric oxide synthase 3), reduced availability of its precursor l-arginine, enzyme dysfunction, and increased degradation. In patients with arterial hypertension, the nitric oxide-
mediated relaxation is severely blunted. Whether hypertension causes reduced NO synthesis
or the other way around is unclear, but evidence suggests that endothelial inflammatory
activation promotes hypertension development [73, 74]. Other important chronic conditions causally implicated in endothelial dysfunction are dyslipidemia and hyperglycemia
(see ▶ Sect. 10.1 "Risk Factors"). Taken together, the activation of endothelial cells is inher-
ently persistent since, at least in parts, it results from the arterial curvature and branching of
the arterial tree. Consequently, a persistent inflammatory response ensues – a prerequisite
for the formation of atherosclerotic plaques.

### 10.2.2 Precursor Lesions

The atherosclerotic lesion likely begins with early intimal hyperplasia near branch points.
These lesions can be observed from birth and may progress due to adaptive intimal thickening and sometimes grow to be as thick as the underlying media. They may provide a soil
for initial atherosclerotic lesion development, and the rate of progression remains higher
than at other arterial sites [75]. Intimal thickening consists mainly of smooth muscle cells
and proteoglycan matrix with little or no infiltrating leukocytes. Although the precise
mechanisms underlying the formation of intimal thickening are not identified, it likely
represents physiological adaptations to shear stress (◻ Fig. 10.1). Following endothelial
activation, vascular smooth muscle cells are activated to undergo proliferation and migration, resulting in progressive thickening of the arterial wall [76].

Advanced (or pathological) intimal thickening is characterized by the presence of lipid
pools which is generally located close to the medial wall and with the occasional first
appearance of macrophages, yet without gross disruption of the normal structure of the
intima. At this point the macrophages are located away from areas of lipid pools and have
not undergone morphological changes. The appearance and accumulation of foam cells
within the intima finally lead to the microscopic appearance of the so-called "xanthoma"
or "fatty-streak"-type lesions [77].

### 10.2.3 The Development of the Fibro-Inflammatory Lipid Plaque

Many fatty streaks do not progress further, but some, especially those occurring at predilection sites, develop into atherosclerotic lesions. As mentioned above, the determining
moment is the accumulation of acellular, lipid-rich material in the intima. The isolated
lipid pools grow into confluent necrotic cores (or lipid-rich cores) by the invasion of
macrophages. This process irreversibly interferes with the intimal structure and leads to
a disrupted extracellular matrix clogged with lipids and cell debris. When a necrotic core
is present, the lesion is commonly defined as a fibroatheroma. The early phase of the
development of the fibroatheroma is specifically characterized, besides the infiltration of
macrophages into the lipid pool, by the loss of matrix structures such as proteoglycans or

collagens. At later stages fibroatheromas are characterized by an accumulation of cellular debris, increased free cholesterol, and a depletion of extracellular matrix [78]. Another phonotypical feature of atherosclerosis is vascular calcification, a processes considered a consequence to vascular injury [79]. Vascular calcification is characterized by calcium deposition in the walls of the vasculature. Mechanisms modulating calcification are complex and involve smooth muscle cell apoptosis, osteochondrogenic differentiation, and matrix vesicle release [80]. In general, arterial calcification occurs in the intima of the vessel wall and in the aortic valve, and it is commonly found in the aorta, coronary, carotid, and renal arteries [81]. The calcification and consequential remodeling of the vasculature lead to arterial stiffness and fragile arterial plaques and contribute therefore significantly in the rapid worsening of atherosclerosis [82].

Lipid-rich foam cells contribute to the physical bulk of developing plaques. The expansion of plaque tissue mass leads to local hypoxic stress due to increases in both oxygen consumption and intercapillary distances. In response to tissue hypoxia, the vasa vasorum of the vessel wall are activated and undergo angiogenesis. While angiogenesis supports further plaque growth, the newly derived blood vessels are limited to the base and shoulder areas of a plaque, leaving an avascular core where large numbers of foam cells undergo apoptosis.

Most cells present within the atherosclerotic plaques, including endothelial cells, smooth muscle cells, lymphocytes, and macrophages, have been shown to undergo apoptosis [83]. During atherosclerosis apoptosis is induced by various factors such as oxidative stress, hypoxia, interferon-gamma, and cholesterol overload [84]. Macrophages represent the majority of dead cells in atherosclerotic lesions. The apoptosis of macrophages might have a potential beneficial effect in early lesion development since a decrease of these cells within the plaque would weaken the inflammatory response and lower the synthesis of matrix degrading enzymes. However, the loss of macrophages at a later stage decreases the ability to clear apoptotic cells and cellular debris [78]. Furthermore, under specific circumstances macrophages can secrete pro-inflammatory mediators during the ingestion of apoptotic bodies and might also contribute to the formation of secondary necrosis, an autolytic process of cell disintegration with the release of cellular components [85].

The very high numbers of apoptotic cells found in atherosclerotic lesions, suggest either increased cell death or insufficient mechanisms for the removal of apoptotic cells in the course of atherogenesis [86]. Accelerated and increased cell death may result from apoptotic triggers such as oxLDL particles and interferon gamma but also as a result of impaired efferocytosis. Evidence suggests that prophagocytic signals are reduced in atherosclerosis caused by inflammation, posttranslational modifications, and genetic variability. This further decreases the phagocytosis and clearance of cells within the atherosclerotic lesion. Uncleared apoptotic cells become secondary necrotic and release additional proinflammatory stimuli, thus promoting a vicious circle. The cell burden within the atherosclerotic plaque is therefore thought to be not the consequence of excess cell death but rather insufficient clearance of dead cells [87, 88]. The latter is a hallmark of an impaired resolution of inflammatory responses.

## 10.2.4  The Vulnerable Plaque

Atherosclerosis progresses slowly over decades, and the transition to an abrupt life-threatening event occurs infrequently. Thus, the majority of subjects with atherosclerosis develop asymptomatic disease, and the impact of slow progressive vessel narrowing and chronic ischemia is often compensated by the development of collateral vessels. The term "vulnerable plaque" (often synonymously referred to as "unstable plaque") describes an atherosclerotic lesion that is prone to rupture and may result in life-threatening events such as myocardial infarction or stroke. At advanced stages, atherosclerotic lesions are fragile protrusions filled with lipid droplets, leaky and unstable blood vessels, a large necrotic lipid core with abundant inflammatory cells and few smooth muscle cells, disintegrated extracellular matrix, and a thin fibrous cap on the arterial luminal side. In addition, intraplaque bleedings increase significantly the levels of free cholesterol and lead to rapid necrotic core expansion and plaque progression, further promoting its vulnerability. Such a rupture-prone plaque is often described as a thin-cap fibroatheroma (TCFA) [90]. Important clinical criteria defining vulnerable plaques include active inflammation, the presence of a thin cap with a large lipid core, endothelial denudation with superficial platelet aggregation, fissured plaques, the presence of high-grade coronary stenosis, arterial stiffness, and calcification [91, 92]. Notably, even though data support the importance of plaque composition in the risk for developing events, the only measurable plaque feature that actually predicts clinical outcome is the size of the plaque and the degree of the arterial stenosis [93]. Large-sized rupture-prone plaques might not be the only vulnerable plaques. Hemodynamically insignificant coronary plaques can rupture and produce cardiac events long before it produces relevant lumen narrowing with symptoms of angina pectoris [94].

Demographically, the etiology and pathophysiology of plaque formation might have shifted over the years. Changes in risk factor profiles, the reduced nicotine abuse, and especially the broad use of statin treatment could have led to relevant transformation of atherogenesis with significant clinical implications. Today, "typical" *patients* with the clinical *presentation* of *myocardial infarction or stroke not only consist of* middle-aged Caucasian males but a broad range of demographics including individuals with prevalent diabetes [95].

### 10.2.5 Plaque Rupture

Plaque destabilization is a biomechanical phenomenon depending on applied stresses, structural features, and biological processes that determine mechanical strength. The rupture of an atherosclerotic plaque is defined as a fibroatheroma with cap disruption in which a luminal thrombus communicates with the underlying necrotic core [96]. When the fibrous cap fails to resist hemodynamic stress, plaque rupture occurs, causing severe hemorrhage and thrombosis due to the collapse of intraplaque blood vessels. Plaque rupture is defined as a fibroatheroma with cap disruption, in which a luminal thrombus communicates with the underlying necrotic core [97]. Postmortem analyses revealed that disruption of the fibrous cap, leading to exposure of the thrombogenic lipid core to the bloodstream, is responsible for two-thirds of all coronary events [91]. In approximately one-third of all cases, the thrombus develops following intimal erosion [98]. In these patients thrombi occur over plaques with superficial endothelial erosion [99]. A defining aspect of plaque erosion is the absence of endothelial cells at the plaque-thrombus interface [100].

Depending on the location of the plaque rupture, blockage of arterial blood flow by the resultant thrombus may lead to myocardial infarction, stroke, or occlusion of a peripheral artery of the legs or arms. Thrombus formation on a ruptured or an eroded atherosclerotic plaque is the critical event that leads to atherothrombosis and vessel occlusion. But, not all cases of plaque disruption will lead to clinical events, given that thrombus growth processes are critical for the development of atherothrombosis and arterial embolism [101]. The thrombotic response to plaque rupture is regulated by several known (and likely a lot of yet unknown) factors. These include the thrombogenicity of plaque constituents (including lipids and tissue factor), local hemodynamics determined by the severity of the underlying stenosis, and shear induced platelet activation by systemic hemostatic activity [102]. Arterial thrombi are primarily composed of aggregated platelets due to their unique ability to adhere to the injured vessel wall as well as to other activated platelets under shear rate. During plaque rupture or plaque erosion, various subendothelial matrix proteins become exposed to blood, such as von Willebrand factor (vWF), fibrillar collagens, fibronectin, and laminin, which all promote platelet adhesion through specific receptors [103]. These adhesive interactions are responsible to the formation of stable aggregates and the promotion of thrombus growth. Tissue factor, mainly derived from the shedding of apoptotic tissue macrophages and activated leukocytes of the atherosclerotic plaque, is considered a critical factor for the formation and propagation of the thrombus [104].

After the plaque rupture, the organizing thrombus gets incorporated into the plaque, and re-endothelialization of the lesion surface occurs. Repeated cycles of plaque damage and thrombus formation promote progressive stenosis of the vessel lumen, leading to a marked reduction of blood flow and finally total occlusion of the artery [105, 106].

Questions Paragraph 2 (More than One Answer Can Be Correct)

**?** 3. Which of the following can cause endothelial dysfunction?
  A. Oscillating shear stress
  B. Low endothelial shear stress
  C. High endothelial shear stress
  D. Laminar flow
  E. Constant blood flow

**?** 4. Approximately one-third of all coronary events:
  A. Are caused by high endothelial shear stress
  B. Are not associated with atherosclerosis
  C. Are associated with insufficient LDL synthesis
  D. Are caused by intimal erosion of atherosclerotic plaques
  E. Involve platelet aggregation

## 10.3 Inflammation in Atherosclerosis

It has only been in the past two decades that the weight of evidence established atherosclerosis as an inflammatory disease rather than solely a disease associated with cholesterol accumulation and metabolism. Today, atherosclerosis is seen as a chronic inflammatory arterial disease driven by both innate and adaptive immune responses to modified lipoproteins and components of the dysfunctional vascular wall. Activation of the endothelium and the consequential leukocyte trafficking across the endothelial barrier supply the reacting tissue with effector cells. The primary leukocyte type recruited to the intima is the monocyte.

## 10.3.1 From Monocytes to Foam Cells

### 10.3.1.1 Monocyte Adhesion and Infiltration

Monocytes are hematopoietic cells and are usually defined by their cell surface expression of CSFR1 (CD115) and $\alpha M\beta 2$ integrin (CD11b). In humans, circulating monocytes of the peripheral blood can further be differentiated based on their cell surface expression in "classical" (CD14$^{++}$CD16$^-$) and "nonclassical" (CD14$^+$CD16$^{++}$) subtypes [107]. Additionally, a third population displaying an "intermediate" monocyte (CD14$^{++}$CD16$^+$) is also detectible in human blood and probably represents a transitional state between the two. In mice, monocytes are commonly distinguished based on their cell surface expression of the GPI-linked membrane protein Ly6C, in either Ly6C$^{high}$ (corresponding to the human CD14$^{++}$CD16$^-$ and CD14$^{++}$CD16$^+$) and Ly6C$^{low}$ (corresponding to the human CD14$^+$CD16$^{++}$) monocytes [108]. The classical monocytes are recruited to sites in case of infection where they differentiate to macrophages and dendritic cells and act as effector cells against a broad range of microorganisms [109]. Evidence further suggests that these cells are also capable of trafficking to lymph nodes and present antigen to T cells, functions similar to those of classical dendritic cells [110]. In contrast, nonclassical monocytes are often attributed a counterbalancing, anti-inflammatory role. These cells work as patrolling safeguards on the luminal surface of vascular endothelial cells during steady-state conditions [111], and up to one-third of all nonclassical monocytes have been suggested to be in contact with the blood vessel wall at all time [112]. Based on these functional characteristics, the classical and nonclassical monocytes are also referred to as "inflammatory monocytes" and "patrolling monocytes."

Monocytes play a pivotal role in the development and exacerbation of atherosclerosis. Large prospective clinical studies indicate that circulating monocyte levels are associated with cardiovascular outcome. As the atherosclerotic process worsens, the number of monocytes in the peripheral blood rises. CD 14$^{++}$16$^+$ monocytes were found increased in patients with coronary heart disease and myocardial infarction, probably due to an upregulation of proinflammatory mediators [113, 114]. Even in the general population without present cardiovascular disease, increased numbers of CD14$^{++}$ and CD16$^+$ monocytes were independently associated with cardiovascular events including myocardial infarction and stroke [115]. Besides a rise in monocyte numbers in the peripheral blood, an increase of monocyte activation leading to chemokine-dependent monocyte recruitment can be found in atherosclerosis [116].

Like all leukocytes, monocytes must transmigrate through the endothelium to reach their destination. Following chemotaxis, monocytes adhere and roll on endothelial cells through interaction with membrane-bound surface proteins. This so-called adhesion cascade is a sequence of adhesion and activation events that ends with the extravasation of monocytes (or in general leukocytes). The five major steps of the adhesion cascade include capture, rolling, slow rolling, firm adhesion, and transmigration. Within this process, each phase is conditional on the next. Every step appears to be necessary for effective leukocyte extravasation, since blocking any of the five can severely reduce leukocyte accumulation in the tissue. At any given moment, capture, rolling, slow rolling, firm adhesion, and transmigration all happen in parallel, involving different leukocytes in the same vessel.

The mechanisms, which enable circulating monocytes to detect sites of extravasation in inflamed or damaged tissues, were extensively investigated in the last decades or so. The upregulation of selectins on the surface of endothelial cells induced by inflammatory mediators provides a molecular basis for leukocytes to recognize endothelial cells in inflamed tissue and, at the same time, provides a mechanism for the docking and capture of these leukocytes. One of the most studied adhesion molecules in this context is P-selectin glycoprotein ligand-1 (PSGL-1), a transmembrane glycoprotein in leukocytes, which interacts not only with P-selectin but also with E-selectin and L-selectin. PSGL-1 is highly expressed on monocytes, especially the Ly6C$^{high}$ subset [117]. Once endothelial cells are activated, the Weibel-Palade bodies rapidly fuse with the plasma membrane and present P-selectin on the cell surface, initiating the immediate attachment and rapid rolling of leukocytes over endothelial cells [118]. Other important endothelial adhesive factors are the intercellular and vascular adhesion molecules (ICAM and VCAM) which belong to the cytokine-inducible Ig gene superfamilies' and bind leukocyte integrin. ICAM-1, ICAM-2, VCAM-1, and other CAM clustering events transduce multiple outside-in signals which all lead to activation of endothelial cells. These signaling cascades modulate numerous intracellular endothelial targets, including cytoskeletal remodeling machineries that facilitate junction opening and promote leukocyte crossing via endothelial cell junctions [119].

Chemokines and chemoattractants function as guidance cues for the transmigration step of monocytes by activation of integrins and thereby the cellular migration machinery. Leukocyte integrins must develop high affinity and avidity for their specific endothelial ligands in order to establish firm shear-resistant adhesions. The most important integrins for leukocyte extravasation are lymphocyte function-associated antigen 1 (LFA-1), which is expressed by all leukocytes, and macrophage antigen 1 (MAC1) as well as the β1 integrin very late antigen 4 (VLA4), which is highly expressed by inflammatory monocytes [120]. Successful leukocyte recruitment also depends on a transient loss of cell surface vascular endothelial (VE) cadherin [121]. VE cadherin plays a critical role in maintaining the integrity between neighboring cells, as well as in facilitating the barrier function of the endothelium to macromolecules and to extravasating leukocytes. When VLA-4 binds to endothelial VCAM-1, a signaling cascade is set in motion that causes VE-PTP (an endothelial membrane protein) to dissociate from VE - cadherin [122]. The dissociation of VE-PTP from VE cadherin is required for the opening of endothelial cell contacts during induction of leukocyte extravasation [123].

Paracellular leukocyte migration through endothelial cell junctions is the primary mode through which leukocytes finally breach through the endothelial cell barrier, but in vivo studies have illustrated that transcellular transmigration (i.e., through the body of the endothelium) of leukocytes also exists [124]. Via their integrins and adhesive ligands, leukocytes are also capable to crawl on the apical site of endothelial cells (even under high shear stress conditions) in search for potential exit cues [125].

Most of the abovementioned data about leukocyte or monocyte extravasation were observed in various in vitro and animal models under physiological or specific inflammatory conditions. Nevertheless, findings of atherosclerosis research indicate that the principles of leukocyte transmigration also apply for the development of endothelial dysfunction and atherosclerotic plaque formation. In vivo studies show that Ly6C$^{high}$ monocytes bind to activated endothelium and infiltrate atherosclerotic lesions better than Ly6C$^{low}$ monocytes [126]. Imaging of arterial vessels further showed that Ly6C$^{high}$ monocytes preferentially localize to lesion-prone sites such as the curvature of the aortic arch or arterial branch points [127]. Initially, monocytes accumulate at sites of endothelial dysfunction or atherosclerotic plaques with increased expression of various cell adhesion molecules such as ICAM-1, VCAM-1, or P-selectin. Mice with P-selectin deficiency or antibody-mediated inhibition of its function display reduced early plaque formation [128, 129], whereas the injection of recombinant murine P-selectin led to a shift toward a more unstable plaque phenotype [130]. In vitro systems (in the absence of flow) further highlight a prominent role of the endothelial cell adhesion molecule CD31, or platelet endothelial cell adhesion molecule-1 (PECAM-1), in the transmigration of monocytes [131]. Knockout of CD31 significantly reduces transmigration of monocytes into the vessel wall [132] and even mitigates atherosclerotic plaque formation in mice [133].

Other cellular components of the blood are also involved in the complex recruitment of leukocytes into atherosclerotic plaques. Platelets binding to the endothelium precede the appearance of adhesion leukocytes in plaques [134] and lead to surface expression of monocyte-attracting chemokines [135], enabling monocyte-platelet aggregation. These aggregates release further chemokines (e.g., CCL2, interleukin-1-beta), which promote monocyte recruitment to atherosclerotic lesions [136, 137]. Neutrophils might contribute to recruitment of classical monocytes by secreting cathelicidin, which induces integrin activation on monocytes, or cytoplasmic proteins such as calprotectin, which activates vascular endothelium and impairs endothelial integrity [138–142]. Neutrophil extracellular traps (NETs) – primarily described to catch and neutralize invading pathogens but recently spotlighted in sterile inflammation – may further promote lesional macrophage accumulation [143]. Finally, antibody-mediated depletion of neutrophils in mice lacking apolipoprotein E (Apoe$^{-/-}$) results in significantly reduced lesion burden in atherogenesis [144].

However, today, no direct evidence – such as in vivo imaging – exists, proving that the abovementioned mechanisms actually regulate leukocyte recruitment to human atherosclerotic lesions. Further investigations, including in vivo imaging studies, will be necessary to elucidate the path of leukocytes from blood to atherosclerotic plaques.

### 10.3.1.2 The Foam Cell Formation

The early stages of atherogenesis are characterized by the differentiation of recruited monocytes into macrophages, the orchestrating cells of the disease. Macrophages clear the excess modified lipoproteins that accumulate in the neointima, become engorged with lipids, and can therefore no longer emigrate from the plaque, contributing to the failure of inflammation resolution and to the maturation of a complicated atherosclerotic lesion.

Whereas the role of circulating monocyte subtypes in atherogenesis has only recently been investigated, the involvement of macrophages in atherosclerosis has first been described more than 30 years ago [145]. It then took another 10 years to recognize atherosclerotic lesional macrophages as (mostly) originally recruited monocytes, where they take up large amounts of cholesterol to generate the so-called foam cells filled with various cholesterol ester droplets [146]. Today we know of several origins and pathways by which macrophages develop and take place in the arterial branch. Resident vascular macrophages are likely recruited early in life, perhaps even before birth, and these resident macrophages are in parts independent from blood monocytes generated by medullary or extramedullary hematopoiesis [147, 148]. Their initial vascular seeding is likely dependent on endothelial cell activation, and they probably continuously self-renew under steady-state conditions throughout life [148]. How exactly residential arterial macrophages, of embryonic origin, interact with the vascular macrophage population of monocytic descent is yet unknown. Lesional macrophages might also stem from a phenotype switch from smooth muscle cells [149]. Notably, mature medial smooth muscle cells can undergo clonal expansion and transdifferentiate to macrophage-like cells during the development of atherosclerotic lesions [150]. The actual contribution in atherogenesis of cellular phenotype transition from smooth muscle cells to macrophage-like cells remains to be demonstrated.

The majority of macrophages within the atherosclerotic plaque derive from circulating monocytes [151]. Arteries recruit vascular macrophages mainly at sites predisposed to atherosclerosis, i.e., areas of basal endothelial activation that correspond to hemodynamic stress. In LDLR$^{-/-}$ mouse models of experimental atherosclerosis, the growth of athero-sclerotic plaque beyond the subendothelial macrophages can be observed within 2 weeks after introducing cholesterol in the diet [152]. It is at this time point that a robust influx of monocytes into plaques appears which subsequently triggers fatty streak formation and drives plaque progression (see ▸ Sect. 10.2 "Plaque Formation"). After recruitment, monocytes gain the ability to synthesize and secrete inflammatory mediators. They fur-ther begin to gain the properties of macrophages, with substantial additional changes that occur during inflammation-induced maturation. These include changes in metabolism and large increases in cell size and cytoplasmic complexity [153, 154]. After monocytes differentiate into macrophages, they become rapidly the most abundant cell type of the atherosclerotic plaque.

The phenotypes of lesional macrophages can likely change rapidly as the microenvi-ronment and intracellular signaling pathways change, for example, by increased exposure to lipoproteins or inflammatory stimuli (◻ Fig. 10.2). In the last decade, two major macrophage subtypes were unraveled [155]. The "proinflammatory" M1 macrophages are characterized by high production of NO and reactive oxygen intermediates and engage in

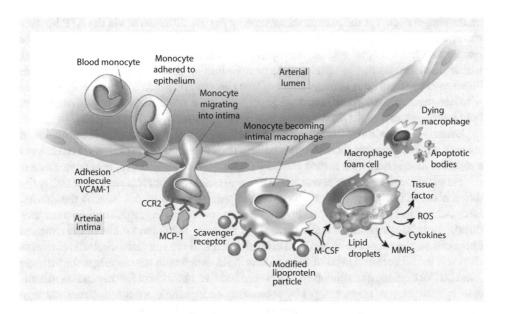

◻ **Fig. 10.2** Monocyte recruitment, foam cell formation, and macrophage apoptosis Monocytes migrate into the subendothelial layer of the intima where they differentiate into macrophages, accumulate lipids from retained (modified) lipoproteins, and turn into foam cells. In early lesions, the subsequent apoptosis of foams limits lesion cellularity and suppresses plaque progression. In advanced lesions, macrophages apoptosis promotes the development of the necrotic core. VCAM-1 vascular adhesion molecule 1, CCR-2 chemokine receptor type 2, M-CSF macrophage colony-stimulating factor, MMPs matrix metalloproteinases, ROS reactive oxygen species. (Modified after [270])

the defense against intracellular parasites and tumor development. The "anti-inflamma-tory" M2 (or alternative-activated) macrophages are mainly functional in tissue remodel-ing, angiogenesis, and tumor progression and are also involved in immunoregulation and allergic reactions. These two macrophage phenotypes have also been attributed a potential role in atherosclerosis [156]. However, it is unlikely that the phenotype of macrophages in atherosclerotic plaques can be classified into two distinct subsets. Lesional macrophages are a consequence of the continuous change of the (in)activation of specific intracellular signaling pathways as well as the microenvironmental change of the atherosclerotic plaque and can therefore best be regarded as a wide continuum of phenotypes and functions.

Although newly developed lesions mostly recruit monocytes from the blood, advanced atherosclerotic plaques rely primarily on locally proliferating macrophages, with a smaller contribution from additional monocyte recruitment [157]. This is probably also due to a change of mechanisms that orchestrate the influx and proliferation of monocyte-derived macrophages [126]. In early atherosclerosis, lesional proliferation is mediated by granulocyte-macrophage colony-stimulating factor (GM-CSF), whereas in established atherosclerosis, macrophage proliferation occurs independent of GM-CSF and is primarily regulated by other mediators such as type 1 macrophage scavenger receptor class A (Msr1) [157].

Macrophage proliferation in atherosclerotic lesions occurs mainly in response to exposure to oxidized low-density lipoprotein (oxLDL), oxidized phospholipids, and other lipid metabolites [158, 159]. Scavenger receptors, CD36 and SR class A (SR-A), are the principal receptors responsible for the binding and uptake of oxLDL in macrophages. Intracellular nonesterified free cholesterol traffics to the plasma membrane of the macro-phage and becomes available for efflux [160]. The final removal of cholesterol from the macrophages occurs at the plasma membrane by passive diffusion or via the ATP binding cassette (ABC) transporters [161]. The physiologic evolution of these processes may be related to a fundamental role of macrophages in efferocytosis (clearing apoptotic cells), which often results in the uptake of large amounts of cholesterol. The fragile balance between cholesterol ester storage, efflux of free cholesterol, and inflammation determines whether macrophages transform into foam cells. In a noninflammatory environment, lipid loading of macrophages may even prevent inflammatory activation [162].

Foam cell formation is the hallmark of atherogenesis. It all starts with the diffusion of atherogenic lipoproteins from the plasma into the subendothelial space and retention as a result of interaction with matrix proteoglycans. As discussed above, monocytes differ-entiate to macrophages after recruitment to the arterial wall, but it is unclear whether the differentiation process precedes lipoprotein ingestion, accompanies it, or is the driving force behind it. However, trapped lipoproteins in the intima aggregate and become oxi-dized. The subsequent uncontrolled scavenger receptor-mediated uptake of the oxidized lipoproteins, the excessive cholesterol esterification and the impaired cholesterol release all result in the accumulation of intracellular lipid droplets in macrophages and trigger foam cell formation. Accumulation of unesterified free cholesterol further causes inflam-matory activation of macrophages by promoting endoplasmic reticulum stress, calcium leakage into the cytosol, lysosome dysfunction, and inflammasome activation by choles-terol crystals [163–166]. OxLDL and oxidized phospholipids especially have the capacity to trigger NF-kB activation and chemokine expression following recognition by CD36-TLR4-TLR6 in macrophages [167]. Inflammation itself inhibits cholesterol efflux from macrophages and potentiates reverse cholesterol transport, suggesting that inflamma-tion and reduced cholesterol efflux participate in a vicious cycle [168]. Interestingly, macrophages are not the only cells which can transform into foams cells. Like intimal macrophages, intimal smooth muscle cells in atherosclerotic lesions accumulate excess amounts of cholesterol esters, and formation of lipid-laden smooth muscle cells has been repeatedly documented [169, 170]. Furthermore, resident intimal dendritic cells also ingest lipoproteins and differentiate into foam cells. Foam cells derived from dendritic cells probably even make up a large percentage of intimal foam cells and precede the appearance of macrophage-derived foam cells [171].

Lesional macrophages are also implicated in critical clinical events. Increase of macrophage apoptosis not only causes the enlargement but also the fragility of the lipid core. Deposition of cell debris drives proinflammatory responses and apoptotic signals for smooth muscle cells, endothelial cells, and leukocytes, transforming the lipid core into a necrotic core [172]. During this process inflammatory cytokines and reactive oxygen species (ROS) are produced, firing up further destabilizing processes within the advanced plaque. Finally, various enzymes secreted by foam cells such as matrix metalloproteinases (MMPs) degrade extracellular matrix which can finally result in plaque rupture and consequently atherothrombosis (see ▶ Sect. 10.2.3 "The Development of the Fibroinflammatory Lipid Plaque").

Nevertheless, even advanced atherosclerotic lesions can regress. The artificial increase of ApoA or the reduction of LDL leads to improved cholesterol efflux and decreased foam cell formation as well as macrophage content, at least in experimental mouse studies [173–175]. In those settings of plaque regression, the number of plaque macrophages not only declines, but the remaining cells transform in more reparative and less inflammatory phenotypes. The mostly reparative macrophages in regressing atherosclerotic lesions originate from the peripheral blood, indicating that even in the context of plaque regression, monocytes are continuously recruited to the arterial wall [176].

Targeting the macrophage count to limit inflammation or atherosclerotic plaque progression has proven beneficial in various atherosclerotic animal models [177, 178]. To date these preclinical data have not been successfully translated into the clinic, as prednisolone liposomes accumulated in human atherosclerotic plaque macrophages without showing any significant anti-inflammatory effects [179].

## 10.3.2 Dendritic Cells

*Dendritic cells (DCs)* play a critical role as specialized antigen-presenting cells and essential mediators in shaping immune reactivity and tolerance. It is therefore not surprising that DCs are frequently found in atherosclerotic lesions as well as in inflamed areas predisposed to atherosclerosis such as the aortic intima of healthy individuals [68, 180]. DCs are mainly located in the plaque shoulder, in rupture-prone lesions, or in marginal parts of the plaque core [180, 181]. Activated vascular dendritic cells promote the synthesis of tumor necrosis factor alpha and interferon gamma production, indicating that DCs are involved in T-cell activation. Indeed, disrupting the TGF-beta (an important regulator of T-cell function) signaling of DCs in Apoe$^{-/-}$ mice led to increased atherosclerotic lesion formation [182]. As mentioned above, dendritic cells can also absorb lipids and transform into foam cells [171]. DCs might even have a role in cholesterol metabolism, since the depletion of DCs leads to a significant elevation of cholesterol parameters [183]. Finally, clinical studies evaluating the count of peripheral DCs indicate that DCs are reduced in patients with (advanced) coronary artery disease, likely due to enhanced recruitment to atherosclerotic lesions [184, 185].

### 10.3.3 Lymphocytes in Atherosclerosis

It is now more than 30 years ago that the presence of a large number of CD3$^+$ T cells in atherosclerotic plaques was first described [186]. The majority of T cells in mouse and in human atherosclerotic plaques are CD4$^+$ T-helper cells (Th cells), but – even though to a much lesser amount – CD8$^+$ T cells can also be found in atherosclerotic plaques [187].

#### 10.3.3.1 T-Helper Cells

*Th1 cells* are the most abundant T-cell subtype in human atherosclerotic plaques (◘ Fig. 10.3) [188]. Th1 cells are generally involved in defense mechanism against intracellular pathogens but recently have also been causally implicated in several autoimmune and vascular inflammatory diseases [189]. The main mediators for Th1 differentiation are IFN-γ and interleukin-12. Upon activation, Th1 secrete various cytokines, such as TNF-α, IFN-γ, or IL-2. The most crucial cytokine produced by Th1 cells, IFN-γ, can exert diverse proatherogenic actions by activating macrophages and dendritic cells [190, 191]. IFN-γ further improves the efficiency of antigen presentation and promotes Th1 polarization. Reduction of IFN-γ in Apoe$^{-/-}$ mouse models leads to reduced lesion formation and enhanced plaque stability, whereas the administered of IFN-γ accelerates atherosclerotic plaque formation [192, 193]. The cellular effects of IFN-γ are mainly mediated through activation of the transcription factor STAT-1, and the inactivation of STAT-1 has repeatedly been shown to halt the atherogenic process [164, 194]. These studies, all performed in experimental mouse models though, indicate that atherosclerosis is a Th1-driven inflammatory disease [195]. However, IFN-γ is expressed not only by Th1 cells but also by natural killer T cells and natural killer cells as a part of the innate immune response.

*Th2 cells* play a crucial role in B-cell-mediated humoral responses, especially against extracellular pathogens. They secrete IL-4, IL-5, and IL-13, and Th2 differentiation is induced by dendritic cells via IL-6 and IL-13 secretion. The role of Th2 cells in atherosclerosis is highly controversial. Initial investigations attributed IL-4 a proatherogenic role [196, 197]. Yet, STAT-6 activation by IL-4 induces expression of the important Th2 differentiation transcription factor GATA-3, which in turn inhibits the production of IFN-γ [198]. Ergo, Th2 cells might counteract the proatherogenic Th1 effect. Indeed, further studies evaluating the effect of IL-4 on atherosclerosis suggest that the lack of Th2 responses does not promote atherosclerosis and may even be protective [199]. Another prototypical Th2 cytokine, IL-13, might has been shown to have anti-inflammatory and plaque-stabilizing properties [200]. Evidence further suggests that IL-5, secreted by Th2, has atheroprotective properties [201]. Finally, a recent study showed that genetic ablation of type 2 innate lymphoid cells, which under physiological conditions secrete IL-5 and IL-13, leads to accelerated atherosclerosis [195]. Taken together, Th2 cells are likely to have anti-atherogenic effects, but further research will be necessary to validate the role of Th2 cells in atherosclerosis.

*Th17 cells* play an important homeostatic role in the clearance of extracellular pathogens by recruiting neutrophils and by producing antimicrobial proteins and inflammatory factors [202]. TGF-β and the inflammatory cytokines IL-6, IL-21, IL-1β, and IL-23 play central roles in the generation and maintenance of Th17 cells. Th17 cells secrete apart from its main cytokine IL-17, also IL-21, GM-CSF, and IL-6. Apart from Th17 cells, IL-17 is also produced by other cell subtypes, including natural killer cells and natural killer T cells [203]. IL-17-positive cells were recently detected in aortic plaques of *Ldlr*$^{-/-}$ and *Apoe*$^{-/-}$ mice fed a western diet [204, 205]. Furthermore, cellular responses to oxLDL and

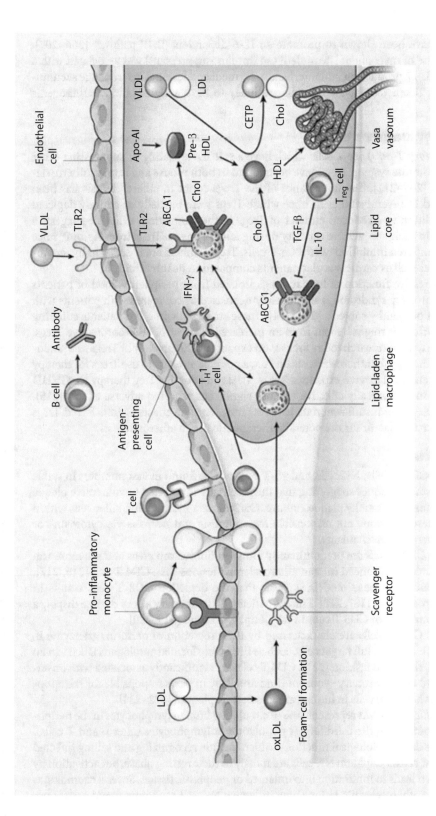

**Fig. 10.3** The role of T cells and B cells in the course of atherogenesis. During plaque development, not only proinflammatory monocytes but also T cells migrate into the atherosclerotic lesion. Among CD4+ T cells, T(h)1 cells are likely proatherogenic, and T(reg) cells are likely atheroprotective. After antigen-specific activation, T-helper 1 (T$_H$1) cells secrete interferon-γ, which activates smooth muscle cells and macrophages, and magnify and sustain the inflammatory response in the intima. Regulatory T (Treg) cells produce the anti-inflammatory cytokines IL-10 and TGF-β. The secretion of antibodies of certain B-cell subsets has atheroprotective properties. LDL low-density lipoprotein, VLDL very low-density lipoprotein, HDL high-de-nsity lipoprotein, oxLDL oxidized low-density lipoprotein, IFN-γ interferon gamma, TGF-β transforming growth factor-β, TLR Toll-like receptor, CETP cholesteryl ester transfer protein, Apo apolipoprotein AI, ABCA1, ABCG1 ATP-binding cassette transporters. (Modified after [271])

collagen V have been shown to promote an IL-6-dependent Th17 pathway [206, 207]. Finally, the use of mycophenolate mofetil (an immunosuppressant) was associated with a decrease of IL-17 production and interleukin-17-mediated lesional macrophage accumulation [208]. Taken together, Th17 cells are likely to have a significant proatherogenic characteristic.

### 10.3.3.2 Regulatory T Cells

*Tregs (regulatory T cells)* are T cells which have a role in regulating or suppressing other cells in the immune system. Tregs have been found in both mouse and human atherosclerotic lesions [209–211]. The importance of the Tregs count in atherosclerosis has been demonstrated in several studies in mice where Tregs were partially or entirely depleted [212]. In addition to a reduced number of Tregs in the circulation, evidence suggests a dysfunction in their suppressive capacity during atherogenesis. Tregs from $ApoE^{-po}$ mice displayed hampered inhibition of effector T cells. Treg numbers are reduced in hypercholesterolemia as well as cardiovascular patients compared to healthy controls.

The suppressive function of human Tregs isolated from peripheral blood of patients with acute coronary syndrome is significantly decreased as compared with patients with stable angina or healthy subjects [213, 214]. These studies suggest that patients suffering from atherosclerosis might benefit from an increased number of atheroprotective Tregs. In this context, most research today focuses on expanding the potency of Tregs as a potential immune therapy for atherosclerosis. In fact, a possible strategy to use Tregs for therapy is adoptive transfer of ex vivo expanded Tregs. Trials on adoptive Treg therapy for GVHD have proven to be clinically effective without significant reported adverse effects [215]. However, there is yet insufficient data to assess how long the effects of transferred Tregs last, likely a crucial factor for the potential therapeutic use in atherosclerosis.

### 10.3.3.3 Killer Cells

CD8 T cells, CD4 T cells, NK cells, and γδ-T cells can be found in vast numbers in stable as well as unstable plaques suggesting that these cells are implicated in vulnerable plaque development and potentially plaque rupture. Current data indicate that killer cells within the atherosclerotic plaque are responsible for apoptosis and necrosis via cytotoxin- or cytokine-dependent mechanisms.

*Cytotoxic CD8 T cells* are predominantly found in fibrous cap areas and are more frequently detected in advanced human atherosclerotic lesions than CD4 T cell [216, 217]. Recent evidence in mouse models suggests that the depletion CD8 T cells results in reduced atherosclerosis [187, 218]. Finally, patients with coronary artery disease display a significant increment of CD8 T cells in the peripheral blood [219, 220].

Cytotoxic CD4 T cells are characterized by their secretion of perforin, granzyme B, FasL, and TNF-alpha to kill target cells, such as EBV-transformed autologous B cells, in an MHC class II-restricted fashion [221]. CD4 T cells are significantly associated with unstable angina and acute coronary syndrome and are likely in parts responsible for the apoptosis of smooth muscle cells in human atherosclerotic plaques [222–224].

*Natural killer (NK) cells* represent 10% of circulating human lymphocytes in the peripheral blood, comprising the third largest population of lymphocytes (after B and T cells). NK cells have diverse biological functions, which include recognizing and killing infected and malignant cells. Peripheral NK cells are mostly in their resting phase, but activation by cytokines often leads to infiltration into inflamed or neoplastic tissues. Several chemokines present in the atherosclerotic lesions may influence NK cell recruitment and activation.

MCP-1 has been shown to be an important chemoattractant for NK cells, as well as Fractalkine or CX3CL1 (found in murine as well as human atherosclerotic lesions), which has been described to induce NK cell migration and activation via IFN-γ [225, 226].

Macrophages have been found in close proximity of NK cells within the atherosclerotic lesions, indicating that they are likely to interact during atherogenesis [227]. Further evidence that NK cells might play a role in (advanced) atherosclerotic disease is provided by patients with coronary heart disease displaying increased circulating numbers of NK cells [228].

### 10.3.3.4  B Cells

The involvement of *B cells* in atherosclerosis is complex. Three major pathways are currently known by which B cells could influence atherosclerosis: first through antibodies, second via the regulation of T-cell responses, and third by the production of chemokines and cytokines [229]. B cells can be found in in low numbers in atherosclerotic lesions and at adventitial sites close to lesions [230]. B-cell-deficient LDL$^{-/-}$ mice and splenectomized mice display increased atherosclerosis, but treatment with anti-CD20 antibody, which preferentially depletes B2 cells, as well as the specific genetic depletion of B2 cells, reduces atherosclerosis [231–234]. The proatherogenic characteristics of B2 cells are likely due to their significant implication in T-cell responses [235], but further research will be necessary to specify the role of B-cell subtypes in atherogenesis.

Even though intertwined with each other, the presence of antibodies within atherosclerotic plaques might be more meaningful than the B-cell count itself, since large numbers of total IgM and IgG can be detected in atherosclerotic lesions [236, 237]. IgM is likely to play an important role in the opsonization and noninflammatory removal of oxLDL particles and is protective against the development of atherosclerosis [238]. There are plenty of experimental and clinical studies which investigated various specific antibodies in patients with atherosclerosis, some of them having pro- and others anti-atherogenic properties. The probably best characterized antibodies in atherosclerosis are targeted against modified LDL particles. Anti-oxLDL IgM (and IgG) antibodies were found to have protective effects in atherosclerosis, and vaccines which induce production of anti-oxLDL antibodies may have protective effects against cardiovascular disease [239, 240].

Taken together, due to the diverse functions and wide-ranging influence of B cells on various immune responses, the distinct functional pathways by which B cells regulate atherosclerosis development remain highly debated and require further investigations.

## 10.3.4  Inflammatory Cytokines in Atherosclerosis

Since inflammation has been recognized as the hallmark of atherosclerosis, an abundance of clinical and experimental research has followed in order to elucidate inflammatory processes and molecules in the course of atherogenesis. ☐ Table 10.1 shows pro- and anti-inflammatory mediators with particular attention given to cytokines with demonstrated effects on atherosclerosis. The biological effects of proinflammatory cytokines that may account for their proatherogenic properties are diverse. It might start with the alteration of the endothelial barrier, followed by their role in the recruitment and activation of immune cells to the lesion, and could end with the destabilization of the atherosclerotic plaques by promoting cell apoptosis and matrix degradation. The most abundant pro-atherosclerotic cytokine in human atherosclerotic plaques are Th1 interleukins [188].

□ **Table 10.1** List of selected potentially pro- and anti-atherogenic cytokines. Even though for some cytokines the evidence is overwhelming of their proatherogenic (e.g., IL-1β) or anti-atherogenic (e.g., TGF-β) properties, most cytokines in the context of atherosclerosis have only been investigated in animal models

| Cytokine | Atherosclerotic animal models | Pro–/anti-atherogenic properties |
|---|---|---|
| IL-1β | Administration of IL-1β leads to increased plaque formation. IL-1β$^{-/-}$ mice display a reduced atherosclerotic burden [241]. IL-1β inhibition in humans showed reduced cardiovascular events [242] | Proatherogenic |
| IL-2 | IL-2 injection increases atherosclerotic lesions and anti-IL-2 antibody reduces lesion size [243] | Proatherogenic |
| Il-5 | IL-5 has atheroprotective properties, and genetic ablation of IL-5-secreting cells leads to accelerated atherosclerosis [201] | Likely anti-atherogenic |
| Il-6 | IL-6 injection increases plaque formation [244]. But, lesion formation also augmented in IL-6$^{-/-}$mice [245, 246]. | Uncertain |
| Il-8 | Il-8 receptor deficiency reduces the plaque progression [247]. Increased in fibrous plaques and foam cells [248, 249] | Likely proatherogenic |
| Il-10 | IL-10 blocks MAPK signaling and NF-κB activation in monocytes, macrophage, endothelial cells, and smooth muscle cells which subsequently results in the down-regulation of the synthesis of pro-inflammatory cytokines, such as IL-1, IL-6, and TNF-α [250, 251] | Anti-atherogenic |
| IL-12 | IL-12 administration accelerates atherosclerotic lesions; IL-12$^{-/-}$ mice have reduced lesion formation [252] | Proatherogenic |
| IL-13 | IL-13 has anti-inflammatory and plaque-stabilizing properties, and genetic ablation of IL-13-secreting cells leads to accelerated atherosclerosis [200] | Likely anti-atherogenic |
| IL-15 | IL-15 detected in plaques; IL-15 inhibition leads to reduction of atherosclerotic lesion size [253, 254] | Likely proatherogenic |
| IL-18 | IL-18 detectable in plaques; several animal models provide evidence that IL-18 has pro-atherosclerotic effects [255] | Proatherogenic |
| IL-33 | IL-33 reduces macrophage foam cell formation [256, 257] | Likely anti-atherogenic |
| TGF-β | TGF-β stimulates the chemotaxis of repair cells, restrains fibrosis through antiproliferative and apoptotic effects on fibrotic cells, and halts arterial calcification [258] | Anti-atherogenic |
| IFN-γ | IFN-γ exerts diverse proatherogenic actions by activating macrophages and dendritic cells [190, 191]. IFN-γ inhibition leads to reduced lesion formation and administration of IFN-γ accelerates atherosclerotic plaque formation [192, 193]. | Proatherogenic |

**10**

Questions Paragraph 3

**5.** The majority of macrophages of the early atherosclerotic lesion are derived from:
A. T cells
B. Smooth muscle cells
C. Circulating monocytes
D. Neutrophils
E. Apoptotic cells

**6.** Integrins are pivotal for leukocyte extravasation. They include:
A. LFA-1
B. Interferon gamma
C. MAC-1
D. VE cadherin
E. ICAM-1

**7.** Tregs have:
A. Mostly proatherogenic properties
B. Mostly anti-atherogenic properties
C. No (proven) impact on atherosclerosis
D. Mostly proinflammatory properties
E. Mostly anti-inflammatory properties

## 10.4  Prevention of Atherosclerosis

Atherosclerosis is still an incurable progressive disease. The change of unhealthy lifestyle habits and strict adherence to medication have been shown to drastically reduce cardiovascular mortality and morbidity, but no actual cure that entirely reverses atherosclerotic disease has yet been found [259]. It is therefore of highest relevance that prevention strategies for the development of atherosclerosis should be implemented not only in the medical field but also on a societal level. One of the most important tasks is the eradication of risk factors, which often stem from unhealthy lifestyle habits (see ▶ Sect. 10.1 "Risk factors"). A survey showed that only 5% of US adults meet the health standards for the three likely most crucial healthy behaviors for the prevention of cardiovascular disease: physical activity, vegetable and fruit consumption, and non-smoking. Considering the fact that atherosclerosis affects more and more children and young adults, and since unhealthy life habits often remain unchanged, it is likely that the atherosclerotic disease burden will not regress in the near future [260].

### 10.4.1  Lipid-Lowering Medications

Statins act by competitively inhibiting HMG-CoA reductase, the first and key rate-limiting enzyme of the endogenous cholesterol synthesis, by mimicking the natural substrate molecule HMG-CoA. In the presence of statins, HMG-CoA is not efficiently processed to produce the next molecule of the metabolic cycle (mevalonate), thereby blocking the pathway of cholesterol biosynthesis, resulting in lower intracellular cholesterol levels and an upregulation of LDL receptors on the cell surface. Thereby the clearance of LDL particles by the liver is promoted, and lower total cholesterol and LDL cholesterol levels in the blood are achieved [261]. Statins are the cornerstone of efforts to reduce the risk for cardiovascular events in patients with atherosclerosis. Large trials have undoubtedly demonstrated that LDL reduction with statins effectively reduces morbidity and mortality, irrespective of the cardiovascular risk [262].

The cholesterol absorption transport inhibitor, ezetimibe, which lowers LDL only moderately but is well tolerated and has few side effects, further reduces CVD when added to statin therapy compared to statin alone [263]. In addition, in the case of statin intolerance or severe side effects (e.g., rhabdomyolysis), ezetimibe represents a useful alternative as a lipid-lowering drug. Besides ezetimibe, older drugs which have been used for decades such as niacin, bile acid sequestrants, and fibrates are available for lowering of LDL. However, the lipid-lowering effects of these drugs are often marginal, and they do not provide any beneficial effect when added to statins [264]. Recently, a novel very effective LDL lowering drug class has been introduced to the market: PCSK-9 inhibitors. PCSK9 (proprotein convertase subtilisin/kexin type) is a crucial protein in LDL cholesterol metabolism by its pivotal role in the degradation of the LDL receptor. Blocking the activity of PCSK9 with monoclonal antibodies reduces the degradation of LDL receptors and increases the hepatic clearance of LDL cholesterol. PCSK9 inhibition is a fruitful new therapeutic option for the treatment of dyslipidemia and associated cardiovascular diseases [265].

### 10.4.2  Novel Therapeutic Strategies for Atherosclerosis

Inhibiting inflammatory responses might grant us novel pharmaceutical options for atherosclerosis (see ▶ Sect. 10.3 "Inflammation and Atherosclerosis"). In this context, new therapeutic approaches testing the idea that targeting inflammation can reduce the risk of adverse cardiovascular outcome in patients with chronic atherosclerotic diseases have been initiated. A recent study showed that inhibition of interleukin-1β (a potent inflammatory cytokine) reduces major cardiovascular events in patients with advanced coronary artery disease [266]. Another still ongoing study currently tests whether low-dose methotrexate, a folic acid inhibitor used as an anti-inflammatory agent in various rheumatic diseases, can reduce rates of cardiovascular events among stable post-myocardial infarction patients with type 2 diabetes mellitus or metabolic syndrome, comorbidities associated with an enhanced proinflammatory response [267]. Colchicine, an anti-inflammatory agent that is most commonly used to treat acute gout attacks, might also be a potential drug candidate for the inhibition of the inflammatory response in atherosclerosis [268]. The future will show whether and to which extent the modulation of inflammatory responses in patients with chronic atherosclerosis will be used in every day clinical practice.

Question Paragraph 4

❓ 8.  Reduction of serum LDL cholesterol levels is one of the most important clinical tasks for the prevention of heart attacks and stroke. Which of the following treatment options work by direct inhibition of the HMG-CoA reductase (one answer)?

    A. Statins
    B. PCSK-9 inhibitors
    C. Ezetimibe
    D. Niacin
    E. Fibrates

---

Take-Home Messages

- Atherosclerosis is a systemic disease that affects all regions of the arterial tree.
- Atherosclerosis is the underlying cause for myocardial infarction, stroke, and peripheral vascular disease.
- Atherosclerotic cardiovascular disease is the leading cause of death worldwide.
- Atherosclerosis is considered a chronic inflammatory condition driven by innate and adaptive immune responses to modified lipoproteins and components of the dysfunctional vascular wall.
- Multiple mechanisms promote atherosclerosis, and it is therefore considered a multifactorial disease.
- The most important cardiovascular risk factors include dyslipidemia, diabetes mellitus, hypertension, smoking, and obesity.
- Endothelial cell dysfunction is the initiation step of atherosclerosis development.
- Vulnerable plaques (or "unstable plaques") are atherosclerotic lesions prone to rupture.
- Plaque rupture leads to thrombus formation which can results in blockage of arterial blood flow.
- Prevention strategies include change of harmful lifestyle habits, pharmacological lipid lowering, and management of diabetes and hypertension.

## Answers

✓ 1.  A, B, C, E

✓ 2.  B, C

✓ 3.  A, B

✓ 4.  D

✓ 5.  C

✓ 6.  A,C

✓ 7.  B, E

✓ 8.  A

## References

1. Collaborators GBDCoD. Global, regional, and national age-sex specific mortality for 264 causes of death, 1980-2016: a systematic analysis for the Global Burden of Disease Study 2016. Lancet. 2017;390(10100):1151–210.
2. Olsson AG, Angelin B, Assmann G, Binder CJ, Bjorkhem I, Cedazo-Minguez A, et al. Can LDL cholesterol be too low? Possible risks of extremely low levels. J Intern Med. 2017;281(6):534–53.
3. Bailey CH. Observations on cholesterol-fed guinea pigs. Proc Soc Exper Biol. 1915;13:60–2.

4.  Steinberg D. Thematic review series: the pathogenesis of atherosclerosis. An interpretive history of the cholesterol controversy: part II: the early evidence linking hypercholesterolemia to coronary disease in humans. J Lipid Res. 2005;46(2):179–90.
5.  World Health Organization, Global Health Observatory. Raised cholesterol. Situation and trends. 2015.
6.  Goldberg IJ, Eckel RH, McPherson R. Triglycerides and heart disease: still a hypothesis? Arterioscler Thromb Vasc Biol. 2011;31(8):1716–25.
7.  Talayero BG, Sacks FM. The role of triglycerides in atherosclerosis. Curr Cardiol Rep. 2011;13(6):544–52.
8.  Ivanova EA, Myasoedova VA, Melnichenko AA, Grechko AV, Orekhov AN. Small dense low-density lipoprotein as biomarker for atherosclerotic diseases. Oxidative Med Cell Longev. 2017;2017:1273042.
9.  Morris PB, Ballantyne CM, Birtcher KK, Dunn SP, Urbina EM. Review of clinical practice guidelines for the management of LDL-related risk. J Am Coll Cardiol. 2014;64(2):196–206.
10. de Ferranti SD. Familial hypercholesterolemia in children and adolescents: a clinical perspective. J Clin Lipidol. 2015;9(5 Suppl):S11–9.
11. Jansen H, Samani NJ, Schunkert H. Mendelian randomization studies in coronary artery disease. Eur Heart J. 2014;35(29):1917–24.
12. Ponikowski P, Voors AA, Anker SD, Bueno H, Cleland JG, Coats AJ, et al. 2016 ESC Guidelines for the diagnosis and treatment of acute and chronic heart failure: the task force for the diagnosis and treatment of acute and chronic heart failure of the European Society of Cardiology (ESC). Developed with the special contribution of the Heart Failure Association (HFA) of the ESC. Eur J Heart Fail. 2016;18(8):891–975.
13. Li Y, Cam J, Bu G. Low-density lipoprotein receptor family: endocytosis and signal transduction. Mol Neurobiol. 2001;23(1):53–67.
14. Magwenzi S, Woodward C, Wraith KS, Aburima A, Raslan Z, Jones H, et al. Oxidized LDL activates blood platelets through CD36/NOX2-mediated inhibition of the cGMP/protein kinase G signaling cascade. Blood. 2015;125(17):2693–703.
15. Steffen Y, Vuillaume G, Stolle K, Roewer K, Lietz M, Schueller J, et al. Cigarette smoke and LDL cooperate in reducing nitric oxide bioavailability in endothelial cells via effects on both eNOS and NADPH oxidase. Nitric Oxide. 2012;27(3):176–84.
16. Badrnya S, Schrottmaier WC, Kral JB, Yaiw KC, Volf I, Schabbauer G, et al. Platelets mediate oxidized low-density lipoprotein-induced monocyte extravasation and foam cell formation. Arterioscler Thromb Vasc Biol. 2014;34(3):571–80.
17. Bujo H, Saito Y. Modulation of smooth muscle cell migration by members of the low-density lipoprotein receptor family. Arterioscler Thromb Vasc Biol. 2006;26(6):1246–52.
18. Riwanto M, Landmesser U. High density lipoproteins and endothelial functions: mechanistic insights and alterations in cardiovascular disease. J Lipid Res. 2013;54(12):3227–43.
19. Fisher EA, Feig JE, Hewing B, Hazen SL, Smith JD. High-density lipoprotein function, dysfunction, and reverse cholesterol transport. Arterioscler Thromb Vasc Biol. 2012;32(12):2813–20.
20. Kannel WB, Dawber TR, Friedman GD, Glennon WE, McNamara PM. Risk factors in coronary heart disease. An evaluation of several serum lipids as predictors of coronary heart disease; the Framingham Study. Ann Intern Med. 1964;61:888–99.
21. Hoekstra M, Van Eck M. Mouse models of disturbed HDL metabolism. Handb Exp Pharmacol. 2015;224:301–36.
22. Badimon JJ, Badimon L, Fuster V. Regression of atherosclerotic lesions by high density lipoprotein plasma fraction in the cholesterol-fed rabbit. J Clin Invest. 1990;85(4):1234–41.
23. Kingwell BA, Chapman MJ, Kontush A, Miller NE. HDL-targeted therapies: progress, failures and future. Nat Rev Drug Discov. 2014;13(6):445–64.
24. Farzianpour F, Rahimi Foroushani A, Shahidi Sadeghi N, Ansari Nosrati S. Relationship between patient's rights charter' and patients' satisfaction in gynecological hospitals. BMC Health Serv Res. 2016;16:476.
25. Kassner U, Schlabs T, Rosada A, Steinhagen-Thiessen E. Lipoprotein(a)--An independent causal risk factor for cardiovascular disease and current therapeutic options. Atheroscler Suppl. 2015;18:263–7.
26. Li Y, He PP, Zhang DW, Zheng XL, Cayabyab FS, Yin WD, et al. Lipoprotein lipase: from gene to atherosclerosis. Atherosclerosis. 2014;237(2):597–608.
27. Kraml P, Syrovatka P, Stipek S, Fialova L, Koprivova H, Potockova J, et al. Hyperlipoproteinemia impairs endothelium-dependent vasodilation. Physiol Res. 2004;53(5):471–80.
28. Lind L. Vasodilation in resistance arteries is related to the apolipoprotein B/A1 ratio in the elderly: the Prospective Investigation of the Vasculature in Uppsala Seniors (PIVUS) study. Atherosclerosis. 2007;190(2):378–84.

10

29. Schmidt C, Bergstrom G. Apolipoprotein B and apolipoprotein A-I in vascular risk prediction - a review. Curr Pharm Des. 2014;20(40):6289–98.

30. Vasquez EC, Peotta VA, Gava AL, Pereira TM, Meyrelles SS. Cardiac and vascular phenotypes in the apolipoprotein E-deficient mouse. J Biomed Sci. 2012;19:22.

31. Mei X, Atkinson D. Lipid-free Apolipoprotein A-I structure: insights into HDL formation and atherosclerosis development. Arch Med Res. 2015;46(5):351–60.

32. Wen Y, Leake DS. Low density lipoprotein undergoes oxidation within lysosomes in cells. Circ Res. 2007;100(9):1337–43.

33. Goyal T, Mitra S, Khaidakov M, Wang X, Singla S, Ding Z, et al. Current concepts of the role of oxidized LDL receptors in atherosclerosis. Curr Atheroscler Rep. 2012;14:150–9.

34. Heery JM, Kozak M, Stafforini DM, Jones DA, Zimmerman GA, McIntyre TM, et al. Oxidatively modified LDL contains phospholipids with platelet-activating factor-like activity and stimulates the growth of smooth muscle cells. J Clin Invest. 1995;96(5):2322–30.

35. Salomon RG. Structural identification and cardiovascular activities of oxidized phospholipids. Circ Res. 2012;111(7):930–46.

36. Que X, Hung MY, Yeang C, Gonen A, Prohaska TA, Sun X, et al. Oxidized phospholipids are proinflammatory and proatherogenic in hypercholesterolaemic mice. Nature. 2018;558(7709):301–6.

37. Fatehi-Hassanabad Z, Chan CB, Furman BL. Reactive oxygen species and endothelial function in diabetes. Eur J Pharmacol. 2010;636(1–3):8–17.

38. Suzuki K, Nakagawa K, Miyazawa T. Augmentation of blood lipid glycation and lipid oxidation in diabetic patients. Clin Chem Lab Med. 2014;52(1):47–52.

39. Kuboki K, Jiang ZY, Takahara N, Ha SW, Igarashi M, Yamauchi T, et al. Regulation of endothelial constitutive nitric oxide synthase gene expression in endothelial cells and in vivo: a specific vascular action of insulin. Circulation. 2000;101(6):676–81.

40. Fukami K, Yamagishi S, Okuda S. Role of AGEs-RAGE system in cardiovascular disease. Curr Pharm Des. 2014;20(14):2395–402.

41. Barlovic DP, Soro-Paavonen A, Jandeleit-Dahm KA. RAGE biology, atherosclerosis and diabetes. Clin Sci (Lond). 2011;121(2):43–55.

42. Aberg A, Bergstrand R, Johansson S, Ulvenstam G, Vedin A, Wedel H, et al. Cessation of smoking after myocardial infarction. Effects on mortality after 10 years. Br Heart J. 1983;49(5):416–22.

43. Gellert C, Schottker B, Muller H, Holleczek B, Brenner H. Impact of smoking and quitting on cardiovascular outcomes and risk advancement periods among older adults. Eur J Epidemiol. 2013;28(8): 649–58.

44. Kawachi I, Colditz GA, Stampfer MJ, Willett WC, Manson JE, Rosner B, et al. Smoking cessation in relation to total mortality rates in women. A prospective cohort study. Ann Intern Med. 1993;119(10): 992–1000.

45. Jacobs DR Jr, Adachi H, Mulder I, Kromhout D, Menotti A, Nissinen A, et al. Cigarette smoking and mortality risk: twenty-five-year follow-up of the Seven Countries Study. Arch Intern Med. 1999;159(7):733–40.

46. Mons U, Muezzinler A, Gellert C, Schottker B, Abnet CC, Bobak M, et al. Impact of smoking and smoking cessation on cardiovascular events and mortality among older adults: meta-analysis of individual participant data from prospective cohort studies of the CHANCES consortium. BMJ. 2015;350: h1551.

47. Messner B, Bernhard D. Smoking and cardiovascular disease: mechanisms of endothelial dysfunction and early atherogenesis. Arterioscler Thromb Vasc Biol. 2014;34(3):509–15.

48. Hackshaw A, Morris JK, Boniface S, Tang JL, Milenkovic D. Low cigarette consumption and risk of coronary heart disease and stroke: meta-analysis of 141 cohort studies in 55 study reports. BMJ. 2018;360:j5855.

49. Ainsworth BE, Haskell WL, Whitt MC, Irwin ML, Swartz AM, Strath SJ, et al. Compendium of physical activities: an update of activity codes and MET intensities. Med Sci Sports Exerc. 2000;32(9 Suppl):S498–504.

50. Endorsed by The Obesity Society, Young DR, Hivert MF, Alhassan S, Camhi SM, Ferguson JF, et al. Sedentary behavior and cardiovascular morbidity and mortality: a science advisory from the American Heart Association. Circulation. 2016;134(13):e262–79.

51. Patterson R, McNamara E, Tainio M, de Sa TH, Smith AD, Sharp SJ, et al. Sedentary behaviour and risk of all-cause, cardiovascular and cancer mortality, and incident type 2 diabetes: a systematic review and dose response meta-analysis. Eur J Epidemiol. 2018;33(9):811–29.

52. Lavie CJ, Arena R, Swift DL, Johannsen NM, Sui X, Lee DC, et al. Exercise and the cardiovascular system: clinical science and cardiovascular outcomes. Circ Res. 2015;117(2):207–19.
53. Thosar SS, Johnson BD, Johnston JD, Wallace JP. Sitting and endothelial dysfunction: the role of shear stress. Med Sci Monit. 2012;18(12):RA173–80.
54. Todd Miller M, Lavie CJ, White CJ. Impact of obesity on the pathogenesis and prognosis of coronary heart disease. J Cardiometab Syndr. 2008;3(3):162–7.
55. Beltowski J. Leptin and atherosclerosis. Atherosclerosis. 2006;189(1):47–60.
56. Mathew B, Francis L, Kayalar A, Cone J. Obesity: effects on cardiovascular disease and its diagnosis. J Am Board Fam Med. 2008;21(6):562–8.
57. Alpert MA, Omran J, Bostick BP. Effects of obesity on cardiovascular hemodynamics, cardiac morphology, and ventricular function. Curr Obes Rep. 2016;5(4):424–34.
58. He FJ, Nowson CA, MacGregor GA. Fruit and vegetable consumption and stroke: meta-analysis of cohort studies. Lancet. 2006;367(9507):320–6.
59. Joshipura KJ, Hu FB, Manson JE, Stampfer MJ, Rimm EB, Speizer FE, et al. The effect of fruit and vegetable intake on risk for coronary heart disease. Ann Intern Med. 2001;134(12):1106–14.
60. Calder PC. The role of marine omega-3 (n-3) fatty acids in inflammatory processes, atherosclerosis and plaque stability. Mol Nutr Food Res. 2012;56(7):1073–80.
61. Johnson RK, Appel LJ, Brands M, Howard BV, Lefevre M, Lustig RH, et al. Dietary sugars intake and cardiovascular health: a scientific statement from the American Heart Association. Circulation. 2009;120(11):1011–20.
62. McNamara DJ. Dietary cholesterol, heart disease risk and cognitive dissonance. Proc Nutr Soc. 2014;73(2):161–6.
63. Montecucco F, Pende A, Quercioli A, Mach F. Inflammation in the pathophysiology of essential hypertension. J Nephrol. 2011;24(1):23–34.
64. Pan WH, Bai CH, Chen JR, Chiu HC. Associations between carotid atherosclerosis and high factor VIII activity, dyslipidemia, and hypertension. Stroke. 1997;28(1):88–94.
65. Moore LL, Visioni AJ, Qureshi MM, Bradlee ML, Ellison RC, D'Agostino R. Weight loss in overweight adults and the long-term risk of hypertension: the Framingham study. Arch Intern Med. 2005;165(11):1298–303.
66. Ross R, Glomset JA. The pathogenesis of atherosclerosis (first of two parts). N Engl J Med. 1976;295(7):369–77.
67. Wentzel JJ, Chatzizisis YS, Gijsen FJ, Giannoglou GD, Feldman CL, Stone PH. Endothelial shear stress in the evolution of coronary atherosclerotic plaque and vascular remodelling: current understanding and remaining questions. Cardiovasc Res. 2012;96(2):234–43.
68. Millonig G, Niederegger H, Rabl W, Hochleitner BW, Hoefer D, Romani N, et al. Network of vascular-associated dendritic cells in intima of healthy young individuals. Arterioscler Thromb Vasc Biol. 2001;21(4):503–8.
69. Zhou J, Li YS, Chien S. Shear stress-initiated signaling and its regulation of endothelial function. Arterioscler Thromb Vasc Biol. 2014;34(10):2191–8.
70. Maurovich-Horvat P, Ferencik M, Voros S, Merkely B, Hoffmann U. Comprehensive plaque assessment by coronary CT angiography. Nat Rev Cardiol. 2014;11(7):390–402.
71. Chiu JJ, Usami S, Chien S. Vascular endothelial responses to altered shear stress: pathologic implications for atherosclerosis. Ann Med. 2009;41(1):19–28.
72. Zaragoza C, Marquez S, Saura M. Endothelial mechanosensors of shear stress as regulators of atherogenesis. Curr Opin Lipidol. 2012;23(5):446–52.
73. Tousoulis D, Kampoli AM, Tentolouris C, Papageorgiou N, Stefanadis C. The role of nitric oxide on endothelial function. Curr Vasc Pharmacol. 2012;10(1):4–18.
74. Li H, Horke S, Forstermann U. Vascular oxidative stress, nitric oxide and atherosclerosis. Atherosclerosis. 2014;237(1):208–19.
75. Kaprio J, Norio R, Pesonen E, Sarna S. Intimal thickening of the coronary arteries in infants in relation to family history of coronary artery disease. Circulation. 1993;87(6):1960–8.
76. Bentzon JF, Otsuka F, Virmani R, Falk E. Mechanisms of plaque formation and rupture. Circ Res. 2014;114(12):1852–66.
77. Stary HC, Chandler AB, Dinsmore RE, Fuster V, Glagov S, Insull W Jr, et al. A definition of advanced types of atherosclerotic lesions and a histological classification of atherosclerosis. A report from the Committee on Vascular Lesions of the Council on Arteriosclerosis, American Heart Association. Arterioscler Thromb Vasc Biol. 1995;15(9):1512–31.

10

78. Otsuka F, Kramer MC, Woudstra P, Yahagi K, Ladich E, Finn AV, et al. Natural progression of atherosclerosis from pathologic intimal thickening to late fibroatheroma in human coronary arteries: a pathology study. Atherosclerosis. 2015;241(2):772–82.

79. Shao JS, Cheng SL, Pingsterhaus JM, Charlton-Kachigian N, Loewy AP, Towler DA. Msx2 promotes cardiovascular calcification by activating paracrine Wnt signals. J Clin Invest. 2005;115(5):1210–20.

80. Otsuka F, Sakakura K, Yahagi K, Joner M, Virmani R. Has our understanding of calcification in human coronary atherosclerosis progressed? Arterioscler Thromb Vasc Biol. 2014;34(4):724–36.

81. Qasim AN, Rafeek H, Rasania SP, Churchill TW, Yang W, Ferrari VA, et al. Cardiovascular risk factors and mitral annular calcification in type 2 diabetes. Atherosclerosis. 2013;226(2):419–24.

82. Chen NX, Moe SM. Vascular calcification: pathophysiology and risk factors. Curr Hypertens Rep. 2012;14(3):228–37.

83. Stoneman VE, Bennett MR. Role of apoptosis in atherosclerosis and its therapeutic implications. Clin Sci (Lond). 2004;107(4):343–54.

84. Mallat Z, Tedgui A. Apoptosis in the vasculature: mechanisms and functional importance. Br J Pharmacol. 2000;130(5):947–62.

85. Fadok VA, Bratton DL, Konowal A, Freed PW, Westcott JY, Henson PM. Macrophages that have ingested apoptotic cells in vitro inhibit proinflammatory cytokine production through autocrine/paracrine mechanisms involving TGF-beta, PGE2, and PAF. J Clin Invest. 1998;101(4):890–8.

86. Kasikara C. The role of non-resolving inflammation in atherosclerosis. J Clin Invest. 2018;128(7): 2713–23.

87. Linton MF, Babaev VR, Huang J, Linton EF, Tao H, Yancey PG. Macrophage apoptosis and efferocytosis in the pathogenesis of atherosclerosis. Circ J. 2016;80(11):2259–68.

88. Viola J, Soehnlein O. Atherosclerosis - A matter of unresolved inflammation. Semin Immunol. 2015;27(3):184–93.

89. Pijls NH, De Bruyne B, Peels K, Van Der Voort PH, Bonnier HJ, Bartunek JKJJ, et al. Measurement of fractional flow reserve to assess the functional severity of coronary-artery stenoses. N Engl J Med. 1996;334(26):1703–8.

90. Pedrigi RM, de Silva R, Bovens SM, Mehta VV, Petretto E, Krams R. Thin-cap fibroatheroma rupture is associated with a fine interplay of shear and wall stress. Arterioscler Thromb Vasc Biol. 2014;34(10):2224–31.

91. Naghavi M, Libby P, Falk E, Casscells SW, Litovsky S, Rumberger J, et al. From vulnerable plaque to vulnerable patient: a call for new definitions and risk assessment strategies: Part II. Circulation. 2003;108(15):1772–8.

92. Naghavi M, Libby P, Falk E, Casscells SW, Litovsky S, Rumberger J, et al. From vulnerable plaque to vulnerable patient: a call for new definitions and risk assessment strategies: Part I. Circulation. 2003;108(14):1664–72.

93. Goncalves I, den Ruijter H, Nahrendorf M, Pasterkamp G. Detecting the vulnerable plaque in patients. J Intern Med. 2015;278(5):520–30.

94. Stefanadis C, Antoniou CK, Tsiachris D, Pietri P. Coronary atherosclerotic vulnerable plaque: current perspectives. J Am Heart Assoc. 2017;6(3):e005543.

95. Libby P, Pasterkamp G. Requiem for the 'vulnerable plaque'. Eur Heart J. 2015;36(43):2984–7.

96. Finn AV, Nakano M, Narula J, Kolodgie FD, Virmani R. Concept of vulnerable/unstable plaque. Arterioscler Thromb Vasc Biol. 2010;30(7):1282–92.

97. Hansson GK, Libby P, Tabas I. Inflammation and plaque vulnerability. J Intern Med. 2015;278(5): 483–93.

98. Jia H, Abtahian F, Aguirre AD, Lee S, Chia S, Lowe H, et al. In vivo diagnosis of plaque erosion and calcified nodule in patients with acute coronary syndrome by intravascular optical coherence tomography. J Am Coll Cardiol. 2013;62(19):1748–58.

99. Kanwar SS, Stone GW, Singh M, Virmani R, Olin J, Akasaka T, et al. Acute coronary syndromes without coronary plaque rupture. Nat Rev Cardiol. 2016;13(5):257–65.

100. Bentzon JF, Falk E. Plaque erosion: new insights from the road less travelled. Circ Res. 2017;121(1): 8–10.

101. Badimon L, Vilahur G. Thrombosis formation on atherosclerotic lesions and plaque rupture. J Intern Med. 2014;276(6):618–32.

102. Cosemans JM, Angelillo-Scherrer A, Mattheij NJ, Heemskerk JW. The effects of arterial flow on platelet activation, thrombus growth, and stabilization. Cardiovasc Res. 2013;99(2):342–52.

103. Lievens D, von Hundelshausen P. Platelets in atherosclerosis. Thromb Haemost. 2011;106(5):827–38.

104. Winckers K, ten Cate H, Hackeng TM. The role of tissue factor pathway inhibitor in atherosclerosis and arterial thrombosis. Blood Rev. 2013;27(3):119–32.
105. Ma X, Hibbert B, McNulty M, Hu T, Zhao X, Ramirez FD, et al. Heat shock protein 27 attenuates neo-intima formation and accelerates reendothelialization after arterial injury and stent implantation: importance of vascular endothelial growth factor up-regulation. FASEB J. 2014;28(2):594–602.
106. Versari D, Lerman LO, Lerman A. The importance of reendothelialization after arterial injury. Curr Pharm Des. 2007;13(17):1811–24.
107. Fernandez Pujol B, Lucibello FC, Gehling UM, Lindemann K, Weidner N, Zuzarte ML, et al. Endothelial-like cells derived from human CD14 positive monocytes. Differentiation. 2000;65(5):287–300.
108. Mitchell AJ, Roediger B, Weninger W. Monocyte homeostasis and the plasticity of inflammatory monocytes. Cell Immunol. 2014;291(1–2):22–31.
109. Shi C, Pamer EG. Monocyte recruitment during infection and inflammation. Nat Rev Immunol. 2011;11(11):762–74.
110. Jakubzick C, Gautier EL, Gibbings SL, Sojka DK, Schlitzer A, Johnson TE, et al. Minimal differentiation of classical monocytes as they survey steady-state tissues and transport antigen to lymph nodes. Immunity. 2013;39(3):599–610.
111. Auffray C, Fogg D, Garfa M, Elain G, Join-Lambert O, Kayal S, et al. Monitoring of blood vessels and tissues by a population of monocytes with patrolling behavior. Science. 2007;317(5838):666–70.
112. Carlin LM, Stamatiades EG, Auffray C, Hanna RN, Glover L, Vizcay-Barrena G, et al. Nr4a1-dependent Ly6C(low) monocytes monitor endothelial cells and orchestrate their disposal. Cell. 2013;153(2):362–75.
113. Heine GH, Ulrich C, Seibert E, Seiler S, Marell J, Reichart B, et al. CD14(++)CD16+ monocytes but not total monocyte numbers predict cardiovascular events in dialysis patients. Kidney Int. 2008;73(5):622–9.
114. Tapp LD, Shantsila E, Wrigley BJ, Pamukcu B, Lip GY. The CD14++CD16+ monocyte subset and mono-cyte-platelet interactions in patients with ST-elevation myocardial infarction. J Thromb Haemost. 2012;10(7):1231–41.
115. Berg KE, Ljungcrantz I, Andersson L, Bryngelsson C, Hedblad B, Fredrikson GN, et al. Elevated CD14++CD16- monocytes predict cardiovascular events. Circ Cardiovasc Genet. 2012;5(1):122–31.
116. Banai S, Finkelstein A, Almagor Y, Assali A, Hasin Y, Rosenschein U, et al. Targeted anti-inflammatory systemic therapy for restenosis: the Biorest Liposomal Alendronate with Stenting sTudy (BLAST)-a double blind, randomized clinical trial. Am Heart J. 2013;165(2):234–40.. e1
117. McEver RP, Cummings RD. Role of PSGL-1 binding to selectins in leukocyte recruitment. J Clin Invest. 1997;100(11 Suppl):S97–103.
118. Zarbock A, Ley K, McEver RP, Hidalgo A. Leukocyte ligands for endothelial selectins: specialized gly-coconjugates that mediate rolling and signaling under flow. Blood. 2011;118(26):6743–51.
119. Aricescu AR, Jones EY. Immunoglobulin superfamily cell adhesion molecules: zippers and signals. Curr Opin Cell Biol. 2007;19(5):543–50.
120. Herter J, Zarbock A. Integrin regulation during leukocyte recruitment. J Immunol. 2013;190(9):4451–7.
121. Weber C, Fraemohs L, Dejana E. The role of junctional adhesion molecules in vascular inflammation. Nat Rev Immunol. 2007;7(6):467–77.
122. Vockel M, Vestweber D. How T cells trigger the dissociation of the endothelial receptor phosphatase VE-PTP from VE-cadherin. Blood. 2013;122(14):2512–22.
123. Broermann A, Winderlich M, Block H, Frye M, Rossaint J, Zarbock A, et al. Dissociation of VE-PTP from VE-cadherin is required for leukocyte extravasation and for VEGF-induced vascular permeability in vivo. J Exp Med. 2011;208(12):2393–401.
124. Mamdouh Z, Mikhailov A, Muller WA. Transcellular migration of leukocytes is mediated by the endo-thelial lateral border recycling compartment. J Exp Med. 2009;206(12):2795–808.
125. Shulman Z, Shinder V, Klein E, Grabovsky V, Yeger O, Geron E, et al. Lymphocyte crawling and tran-sendothelial migration require chemokine triggering of high-affinity LFA-1 integrin. Immunity. 2009;30(3):384–96.
126. Swirski FK, Libby P, Aikawa E, Alcaide P, Luscinskas FW, Weissleder R, et al. Ly-6Chi monocytes dominate hypercholesterolemia-associated monocytosis and give rise to macrophages in atheromata. J Clin Invest. 2007;117(1):195–205.
127. Swirski FK, Pittet MJ, Kircher MF, Aikawa E, Jaffer FA, Libby P, et al. Monocyte accumulation in mouse atherogenesis is progressive and proportional to extent of disease. Proc Natl Acad Sci U S A. 2006;103(27):10340–5.

10

128. Manka D, Collins RG, Ley K, Beaudet AL, Sarembock IJ. Absence of p-selectin, but not intercellular adhesion molecule-1, attenuates neointimal growth after arterial injury in apolipoprotein e-deficient mice. Circulation. 2001;103(7):1000–5.

129. Phillips JW, Barringhaus KG, Sanders JM, Hesselbacher SE, Czarnik AC, Manka D, et al. Single injection of P-selectin or P-selectin glycoprotein ligand-1 monoclonal antibody blocks neointima formation after arterial injury in apolipoprotein E-deficient mice. Circulation. 2003;107(17): 2244–9.

130. Woollard KJ, Lumsden NG, Andrews KL, Aprico A, Harris E, Irvine JC, et al. Raised soluble P-selectin moderately accelerates atherosclerotic plaque progression. PLoS One. 2014;9(5):e97422.

131. Muller WA. The role of PECAM-1 (CD31) in leukocyte emigration: studies in vitro and in vivo. J Leukoc Biol. 1995;57(4):523–8.

132. Schenkel AR, Chew TW, Muller WA. Platelet endothelial cell adhesion molecule deficiency or block-ade significantly reduces leukocyte emigration in a majority of mouse strains. J Immunol. 2004;173(10):6403–8.

133. Harry BL, Sanders JM, Feaver RE, Lansey M, Deem TL, Zarbock A, et al. Endothelial cell PECAM-1 pro-motes atherosclerotic lesions in areas of disturbed flow in ApoE-deficient mice. Arterioscler Thromb Vasc Biol. 2008;28(11):2003–8.

134. Massberg S, Brand K, Gruner S, Page S, Muller E, Muller I, et al. A critical role of platelet adhesion in the initiation of atherosclerotic lesion formation. J Exp Med. 2002;196(7):887–96.

135. van Gils JM, Zwaginga JJ, Hordijk PL. Molecular and functional interactions among monocytes, platelets, and endothelial cells and their relevance for cardiovascular diseases. J Leukoc Biol. 2009;85(2):195–204.

136. von Hundelshausen P, Schmitt MM. Platelets and their chemokines in atherosclerosis-clinical appli-cations. Front Physiol. 2014;5:294.

137. Seizer P, Gawaz M, May AE. Platelet-monocyte interactions--a dangerous liaison linking thrombosis, inflammation and atherosclerosis. Curr Med Chem. 2008;15(20):1976–80.

138. Doring Y, Drechsler M, Wantha S, Kemmerich K, Lievens D, Vijayan S, et al. Lack of neutrophil-derived CRAMP reduces atherosclerosis in mice. Circ Res. 2012;110(8):1052–6.

139. Wantha S, Alard JE, Megens RT, van der Does AM, Doring Y, Drechsler M, et al. Neutrophil-derived cathelicidin promotes adhesion of classical monocytes. Circ Res. 2013;112(5):792–801.

140. Pedersen L, Nybo M, Poulsen MK, Henriksen JE, Dahl J, Rasmussen LM. Plasma calprotectin and its association with cardiovascular disease manifestations, obesity and the metabolic syndrome in type 2 diabetes mellitus patients. BMC Cardiovasc Disord. 2014;14:196.

141. Eue I, Langer C, Eckardstein A, Sorg C. Myeloid related protein (MRP) 14 expressing monocytes infil-trate atherosclerotic lesions of ApoE null mice. Atherosclerosis. 2000;151(2):593–7.

142. Croce K, Gao H, Wang Y, Mooroka T, Sakuma M, Shi C, et al. Myeloid-related protein-8/14 is critical for the biological response to vascular injury. Circulation. 2009;120(5):427–36.

143. Doring Y, Manthey HD, Drechsler M, Lievens D, Megens RT, Soehnlein O, et al. Auto-antigenic pro-tein-DNA complexes stimulate plasmacytoid dendritic cells to promote atherosclerosis. Circulation. 2012;125(13):1673–83.

144. Drechsler M, Megens RT, van Zandvoort M, Weber C, Soehnlein O. Hyperlipidemia-triggered neutro-philia promotes early atherosclerosis. Circulation. 2010;122(18):1837–45.

145. Ross R. The pathogenesis of atherosclerosis--an update. N Engl J Med. 1986;314(8):488–500.

146. Ross R. Atherosclerosis--an inflammatory disease. N Engl J Med. 1999;340(2):115–26.

147. Yona S, Kim KW, Wolf Y, Mildner A, Varol D, Breker M, et al. Fate mapping reveals origins and dynamics of monocytes and tissue macrophages under homeostasis. Immunity. 2013;38(1):79–91.

148. Ensan S, Li A, Besla R, Degousee N, Cosme J, Roufaiel M, et al. Self-renewing resident arterial macro-phages arise from embryonic CX3CR1(+) precursors and circulating monocytes immediately after birth. Nat Immunol. 2016;17(2):159–68.

149. Shankman LS, Gomez D, Cherepanova OA, Salmon M, Alencar GF, Haskins RM, et al. KLF4-dependent phenotypic modulation of smooth muscle cells has a key role in atherosclerotic plaque pathogene-sis. Nat Med. 2015;21(6):628–37.

150. Feil S, Fehrenbacher B, Lukowski R, Essmann F, Schulze-Osthoff K, Schaller M, et al. Transdifferentia-tion of vascular smooth muscle cells to macrophage-like cells during atherogenesis. Circ Res. 2014;115(7):662–7.

151. Tacke F, Alvarez D, Kaplan TJ, Jakubzick C, Spanbroek R, Llodra J, et al. Monocyte subsets differentially employ CCR2, CCR5, and CX3CR1 to accumulate within atherosclerotic plaques. J Clin Invest. 2007;117(1):185–94.

152. Liu P, Yu YR, Spencer JA, Johnson AE, Vallanat CT, Fong AM, et al. CX3CR1 deficiency impairs dendritic cell accumulation in arterial intima and reduces atherosclerotic burden. Arterioscler Thromb Vasc Biol. 2008;28(2):243–50.

153. Gautier EL, Ivanov S, Lesnik P, Randolph GJ. Local apoptosis mediates clearance of macrophages from resolving inflammation in mice. Blood. 2013;122(15):2714–22.

154. Kratofil RM, Kubes P, Deniset JF. Monocyte conversion during inflammation and injury. Arterioscler Thromb Vasc Biol. 2017;37(1):35–42.

155. Mills CD. Anatomy of a discovery: m1 and m2 macrophages. Front Immunol. 2015;6:212.

156. Colin S, Chinetti-Gbaguidi G, Staels B. Macrophage phenotypes in atherosclerosis. Immunol Rev. 2014;262(1):153–66.

157. Robbins CS, Hilgendorf I, Weber GF, Theurl I, Iwamoto Y, Figueiredo JL, et al. Local proliferation dominates lesional macrophage accumulation in atherosclerosis. Nat Med. 2013;19(9):1166–72.

158. Hamilton JA, Myers D, Jessup W, Cochrane F, Byrne R, Whitty G, et al. Oxidized LDL can induce macrophage survival, DNA synthesis, and enhanced proliferative response to CSF-1 and GM-CSF. Arterioscler Thromb Vasc Biol. 1999;19(1):98–105.

159. Que X, Hung MY, Yeang C, Gonen A, Prohaska TA, Sun X, et al. Oxidized phospholipids are proinflammatory and proatherogenic in hypercholesterolaemic mice. Nature. 2018;561(7724):E43.

160. Brown MS, Ho YK, Goldstein JL. The cholesteryl ester cycle in macrophage foam cells. Continual hydrolysis and re-esterification of cytoplasmic cholesteryl esters. J Biol Chem. 1980;255(19):9344–52.

161. Westerterp M, Murphy AJ, Wang M, Pagler TA, Vengrenyuk Y, Kappus MS, et al. Deficiency of ATP-binding cassette transporters A1 and G1 in macrophages increases inflammation and accelerates atherosclerosis in mice. Circ Res. 2013;112(11):1456–65.

162. Spann NJ, Garmire LX, McDonald JG, Myers DS, Milne SB, Shibata N, et al. Regulated accumulation of desmosterol integrates macrophage lipid metabolism and inflammatory responses. Cell. 2012;151(1):138–52.

163. Li Y, Schwabe RF, DeVries-Seimon T, Yao PM, Gerbod-Giannone MC, Tall AR, et al. Free cholesterol-loaded macrophages are an abundant source of tumor necrosis factor-alpha and interleukin-6: model of NF-kappaB- and map kinase-dependent inflammation in advanced atherosclerosis. J Biol Chem. 2005;280(23):21763–72.

164. Lim WS, Timmins JM, Seimon TA, Sadler A, Kolodgie FD, Virmani R, et al. Signal transducer and activator of transcription-1 is critical for apoptosis in macrophages subjected to endoplasmic reticulum stress in vitro and in advanced atherosclerotic lesions in vivo. Circulation. 2008;117(7):940–51.

165. Duewell P, Kono H, Rayner KJ, Sirois CM, Vladimer G, Bauernfeind FG, et al. NLRP3 inflammasomes are required for atherogenesis and activated by cholesterol crystals. Nature. 2010;464(7293):1357–61.

166. Emanuel R, Sergin I, Bhattacharya S, Turner J, Epelman S, Settembre C, et al. Induction of lysosomal biogenesis in atherosclerotic macrophages can rescue lipid-induced lysosomal dysfunction and downstream sequelae. Arterioscler Thromb Vasc Biol. 2014;34(9):1942–52.

167. Stewart CR, Stuart LM, Wilkinson K, van Gils JM, Deng J, Halle A, et al. CD36 ligands promote sterile inflammation through assembly of a Toll-like receptor 4 and 6 heterodimer. Nat Immunol. 2010;11(2):155–61.

168. McGillicuddy FC, de la Llera MM, Hinkle CC, Joshi MR, Chiquoine EH, Billheimer JT, et al. Inflammation impairs reverse cholesterol transport in vivo. Circulation. 2009;119(8):1135–45.

169. Frontini MJ, O'Neil C, Sawyez C, Chan BM, Huff MW, Pickering JG. Lipid incorporation inhibits Src-dependent assembly of fibronectin and type I collagen by vascular smooth muscle cells. Circ Res. 2009;104(7):832–41.

170. Choi HY, Rahmani M, Wong BW, Allahverdian S, McManus BM, Pickering JG, et al. ATP-binding cassette transporter A1 expression and apolipoprotein A-I binding are impaired in intima-type arterial smooth muscle cells. Circulation. 2009;119(25):3223–31.

171. Paulson KE, Zhu SN, Chen M, Nurmohamed S, Jongstra-Bilen J, Cybulsky MI. Resident intimal dendritic cells accumulate lipid and contribute to the initiation of atherosclerosis. Circ Res. 2010;106(2):383–90.

172. Tabas I. Consequences and therapeutic implications of macrophage apoptosis in atherosclerosis: the importance of lesion stage and phagocytic efficiency. Arterioscler Thromb Vasc Biol. 2005;25(11):2255–64.

10

173. Shah PK, Yano J, Reyes O, Chyu KY, Kaul S, Bisgaier CL, et al. High-dose recombinant apolipoprotein A-I(milano) mobilizes tissue cholesterol and rapidly reduces plaque lipid and macrophage content in apolipoprotein e-deficient mice. Potential implications for acute plaque stabilization. Circulation. 2001;103(25):3047–50.

174. Tian F, Wang L, Arias A, Yang M, Sharifi BG, Shah PK. Comparative antiatherogenic effects of intravenous AAV8- and AAV2-mediated ApoA-IMilano gene transfer in hypercholesterolemic mice. J Cardiovasc Pharmacol Ther. 2015;20(1):66–75.

175. Wang L, Tian F, Arias A, Yang M, Sharifi BG, Shah PK. Comparative effects of diet-induced lipid lowering versus lipid lowering along with Apo A-I milano gene Therapy on regression of atherosclerosis. J Cardiovasc Pharmacol Ther. 2016;21(3):320–8.

176. Rahman K, Vengrenyuk Y, Ramsey SA, Vila NR, Girgis NM, Liu J, et al. Inflammatory Ly6Chi monocytes and their conversion to M2 macrophages drive atherosclerosis regression. J Clin Invest. 2017;127(8):2904–15.

177. Lobatto ME, Fayad ZA, Silvera S, Vucic E, Calcagno C, Mani V, et al. Multimodal clinical imaging to longitudinally assess a nanomedical anti-inflammatory treatment in experimental atherosclerosis. Mol Pharm. 2010;7(6):2020–9.

178. Duivenvoorden R, Tang J, Cormode DP, Mieszawska AJ, Izquierdo-Garcia D, Ozcan C, et al. A statin-loaded reconstituted high-density lipoprotein nanoparticle inhibits atherosclerotic plaque inflammation. Nat Commun. 2014;5:3065.

179. van der Valk FM, van Wijk DF, Lobatto ME, Verberne HJ, Storm G, Willems MC, et al. Prednisolone-containing liposomes accumulate in human atherosclerotic macrophages upon intravenous administration. Nanomedicine. 2015;11(5):1039–46.

180. Yilmaz A, Lochno M, Traeg F, Cicha I, Reiss C, Stumpf C, et al. Emergence of dendritic cells in rupture-prone regions of vulnerable carotid plaques. Atherosclerosis. 2004;176(1):101–10.

181. Erbel C, Sato K, Meyer FB, Kopecky SL, Frye RL, Goronzy JJ, et al. Functional profile of activated dendritic cells in unstable atherosclerotic plaque. Basic Res Cardiol. 2007;102(2):123–32.

182. Oh SA, Li MO. TGF-beta: guardian of T cell function. J Immunol. 2013;191(8):3973–9.

183. Gautier EL, Huby T, Saint-Charles F, Ouzilleau B, Pirault J, Deswaerte V, et al. Conventional dendritic cells at the crossroads between immunity and cholesterol homeostasis in atherosclerosis. Circulation. 2009;119(17):2367–75.

184. Yilmaz A, Weber J, Cicha I, Stumpf C, Klein M, Raithel D, et al. Decrease in circulating myeloid dendritic cell precursors in coronary artery disease. J Am Coll Cardiol. 2006;48(1):70–80.

185. Yilmaz A, Ratka J, Rohm I, Pistulli R, Goebel B, Asadi Y, et al. Decrease in circulating plasmacytoid dendritic cells during short-term systemic normobaric hypoxia. Eur J Clin Investig. 2016;46(2):115–22.

186. Jonasson L, Holm J, Skalli O, Bondjers G, Hansson GK. Regional accumulations of T cells, macrophages, and smooth muscle cells in the human atherosclerotic plaque. Arteriosclerosis. 1986;6(2):131–8.

187. Kyaw T, Winship A, Tay C, Kanellakis P, Hosseini H, Cao A, et al. Cytotoxic and proinflammatory CD8+ T lymphocytes promote development of vulnerable atherosclerotic plaques in apoE-deficient mice. Circulation. 2013;127(9):1028–39.

188. Frostegard J, Ulfgren AK, Nyberg P, Hedin U, Swedenborg J, Andersson U, et al. Cytokine expression in advanced human atherosclerotic plaques: dominance of pro-inflammatory (Th1) and macrophage-stimulating cytokines. Atherosclerosis. 1999;145(1):33–43.

189. Lintermans LL, Stegeman CA, Heeringa P, Abdulahad WH. T cells in vascular inflammatory diseases. Front Immunol. 2014;5:504.

190. Schoenborn JR, Wilson CB. Regulation of interferon-gamma during innate and adaptive immune responses. Adv Immunol. 2007;96:41–101.

191. Bradley LM, Dalton DK, Croft M. A direct role for IFN-gamma in regulation of Th1 cell development. J Immunol. 1996;157(4):1350–8.

192. Whitman SC, Ravisankar P, Elam H, Daugherty A. Exogenous interferon-gamma enhances atherosclerosis in apolipoprotein E−/− mice. Am J Pathol. 2000;157(6):1819–24.

193. Gupta S, Pablo AM, Jiang X, Wang N, Tall AR, Schindler C. IFN-gamma potentiates atherosclerosis in ApoE knock-out mice. J Clin Invest. 1997;99(11):2752–61.

194. Agrawal S, Febbraio M, Podrez E, Cathcart MK, Stark GR, Chisolm GM. Signal transducer and activator of transcription 1 is required for optimal foam cell formation and atherosclerotic lesion development. Circulation. 2007;115(23):2939–47.

195. Buono C, Binder CJ, Stavrakis G, Witztum JL, Glimcher LH, Lichtman AH. T-bet deficiency reduces atherosclerosis and alters plaque antigen-specific immune responses. Proc Natl Acad Sci U S A. 2005;102(5):1596–601.
196. Davenport P, Tipping PG. The role of interleukin-4 and interleukin-12 in the progression of atherosclerosis in apolipoprotein E-deficient mice. Am J Pathol. 2003;163(3):1117–25.
197. King VL, Szilvassy SJ, Daugherty A. Interleukin-4 deficiency decreases atherosclerotic lesion formation in a site-specific manner in female LDL receptor–/– mice. Arterioscler Thromb Vasc Biol. 2002;22(3):456–61.
198. Maier E, Duschl A, Horejs-Hoeck J. STAT6-dependent and -independent mechanisms in Th2 polarization. Eur J Immunol. 2012;42(11):2827–33.
199. King VL, Cassis LA, Daugherty A. Interleukin-4 does not influence development of hypercholesterolemia or angiotensin II-induced atherosclerotic lesions in mice. Am J Pathol. 2007;171(6): 2040–7.
200. Cardilo-Reis L, Gruber S, Schreier SM, Drechsler M, Papac-Milicevic N, Weber C, et al. Interleukin-13 protects from atherosclerosis and modulates plaque composition by skewing the macrophage phenotype. EMBO Mol Med. 2012;4(10):1072–86.
201. Binder CJ, Hartvigsen K, Chang MK, Miller M, Broide D, Palinski W, et al. IL-5 links adaptive and natural immunity specific for epitopes of oxidized LDL and protects from atherosclerosis. J Clin Invest. 2004;114(3):427–37.
202. Miossec P, Korn T, Kuchroo VK. Interleukin-17 and type 17 helper T cells. N Engl J Med. 2009;361(9): 888–98.
203. Takatori H, Kanno Y, Watford WT, Tato CM, Weiss G, Ivanov II, et al. Lymphoid tissue inducer-like cells are an innate source of IL-17 and IL-22. J Exp Med. 2009;206(1):35–41.
204. Smith E, Prasad KM, Butcher M, Dobrian A, Kolls JK, Ley K, et al. Blockade of interleukin-17A results in reduced atherosclerosis in apolipoprotein E-deficient mice. Circulation. 2010;121(15):1746–55.
205. Ma T, Gao Q, Zhu F, Guo C, Wang Q, Gao F, et al. Th17 cells and IL-17 are involved in the disruption of vulnerable plaques triggered by short-term combination stimulation in apolipoprotein E-knockout mice. Cell Mol Immunol. 2013;10(4):338–48.
206. Lim H, Kim YU, Sun H, Lee JH, Reynolds JM, Hanabuchi S, et al. Proatherogenic conditions promote autoimmune T helper 17 cell responses in vivo. Immunity. 2014;40(1):153–65.
207. Dart ML, Jankowska-Gan E, Huang G, Roenneburg DA, Keller MR, Torrealba JR, et al. Interleukin-17-dependent autoimmunity to collagen type V in atherosclerosis. Circ Res. 2010;107(9):1106–16.
208. von Vietinghoff S, Koltsova EK, Mestas J, Diehl CJ, Witztum JL, Ley K. Mycophenolate mofetil decreases atherosclerotic lesion size by depression of aortic T-lymphocyte and interleukin-17-mediated macrophage accumulation. J Am Coll Cardiol. 2011;57(21):2194–204.
209. Heller EA, Liu E, Tager AM, Yuan Q, Lin AY, Ahluwalia N, et al. Chemokine CXCL10 promotes atherogenesis by modulating the local balance of effector and regulatory T cells. Circulation. 2006;113(19):2301–12.
210. Maganto-Garcia E, Tarrio ML, Grabie N, Bu DX, Lichtman AH. Dynamic changes in regulatory T cells are linked to levels of diet-induced hypercholesterolemia. Circulation. 2011;124(2):185–95.
211. Dietel B, Cicha I, Voskens CJ, Verhoeven E, Achenbach S, Garlichs CD. Decreased numbers of regulatory T cells are associated with human atherosclerotic lesion vulnerability and inversely correlate with infiltrated mature dendritic cells. Atherosclerosis. 2013;230(1):92–9.
212. Ait-Oufella H, Salomon BL, Potteaux S, Robertson AK, Gourdy P, Zoll J, et al. Natural regulatory T cells control the development of atherosclerosis in mice. Nat Med. 2006;12(2):178–80.
213. Mor A, Planer D, Luboshits G, Afek A, Metzger S, Chajek-Shaul T, et al. Role of naturally occurring CD4+ CD25+ regulatory T cells in experimental atherosclerosis. Arterioscler Thromb Vasc Biol. 2007;27(4):893–900.
214. Mor A, Luboshits G, Planer D, Keren G, George J. Altered status of CD4(+)CD25(+) regulatory T cells in patients with acute coronary syndromes. Eur Heart J. 2006;27(21):2530–7.
215. Di Ianni M, Falzetti F, Carotti A, Terenzi A, Castellino F, Bonifacio E, et al. Tregs prevent GVHD and promote immune reconstitution in HLA-haploidentical transplantation. Blood. 2011;117(14): 3921–8.
216. Paul VS, Paul CM, Kuruvilla S. Quantification of various inflammatory cells in advanced atherosclerotic plaques. J Clin Diagn Res. 2016;10(5):EC35–8.
217. Rossmann A, Henderson B, Heidecker B, Seiler R, Fraedrich G, Singh M, et al. T-cells from advanced atherosclerotic lesions recognize hHSP60 and have a restricted T-cell receptor repertoire. Exp Gerontol. 2008;43(3):229–37.

10

218. Cochain C, Koch M, Chaudhari SM, Busch M, Pelisek J, Boon L, et al. CD8+ T cells regulate monopoiesis and circulating Ly6C-high monocyte levels in atherosclerosis in mice. Circ Res. 2015;117(3): 244–53.

219. Bergstrom I, Backteman K, Lundberg A, Ernerudh J, Jonasson L. Persistent accumulation of interferon-gamma-producing CD8+CD56+ T cells in blood from patients with coronary artery disease. Atherosclerosis. 2012;224(2):515–20.

220. Kolbus D, Ljungcrantz I, Andersson L, Hedblad B, Fredrikson GN, Bjorkbacka H, et al. Association between CD8+ T-cell subsets and cardiovascular disease. J Intern Med. 2013;274(1):41–51.

221. Marshall NB, Swain SL. Cytotoxic CD4 T cells in antiviral immunity. J Biomed Biotechnol. 2011;2011:954602.

222. Liuzzo G, Goronzy JJ, Yang H, Kopecky SL, Holmes DR, Frye RL, et al. Monoclonal T-cell proliferation and plaque instability in acute coronary syndromes. Circulation. 2000;101(25):2883–8.

223. Dumitriu IE, Araguas ET, Baboonian C, Kaski JC. CD4+ CD28 null T cells in coronary artery disease: when helpers become killers. Cardiovasc Res. 2009;81(1):11–9.

224. Sato K, Niessner A, Kopecky SL, Frye RL, Goronzy JJ, Weyand CM. TRAIL-expressing T cells induce apoptosis of vascular smooth muscle cells in the atherosclerotic plaque. J Exp Med. 2006;203(1): 239–50.

225. Allavena P, Bianchi G, Zhou D, van Damme J, Jilek P, Sozzani S, et al. Induction of natural killer cell migration by monocyte chemotactic protein-1, −2 and −3. Eur J Immunol. 1994;24(12):3233–6.

226. Umehara H, Bloom ET, Okazaki T, Nagano Y, Yoshie O, Imai T. Fractalkine in vascular biology: from basic research to clinical disease. Arterioscler Thromb Vasc Biol. 2004;24(1):34–40.

227. Selathurai A, Deswaerte V, Kanellakis P, Tipping P, Toh BH, Bobik A, et al. Natural killer (NK) cells augment atherosclerosis by cytotoxic-dependent mechanisms. Cardiovasc Res. 2014;102(1):128–37.

228. Backteman K, Andersson C, Dahlin LG, Ernerudh J, Jonasson L. Lymphocyte subpopulations in lymph nodes and peripheral blood: a comparison between patients with stable angina and acute coronary syndrome. PLoS One. 2012;7(3):e32691.

229. Sage AP, Mallat Z. Multiple potential roles for B cells in atherosclerosis. Ann Med. 2014;46(5):297–303.

230. Hamze M, Desmetz C, Berthe ML, Roger P, Boulle N, Brancherau P, et al. Characterization of resident B cells of vascular walls in human atherosclerotic patients. J Immunol. 2013;191(6):3006–16.

231. Ait-Oufella H, Herbin O, Bouaziz JD, Binder CJ, Uyttenhove C, Laurans L, et al. B cell depletion reduces the development of atherosclerosis in mice. J Exp Med. 2010;207(8):1579–87.

232. Sage AP, Tsiantoulas D, Baker L, Harrison J, Masters L, Murphy D, et al. BAFF receptor deficiency reduces the development of atherosclerosis in mice--brief report. Arterioscler Thromb Vasc Biol. 2012;32(7):1573–6.

233. Major AS, Fazio S, Linton MF. B-lymphocyte deficiency increases atherosclerosis in LDL receptor-null mice. Arterioscler Thromb Vasc Biol. 2002;22(11):1892–8.

234. Caligiuri G, Nicoletti A, Poirier B, Hansson GK. Protective immunity against atherosclerosis carried by B cells of hypercholesterolemic mice. J Clin Invest. 2002;109(6):745–53.

235. Lund FE, Randall TD. Effector and regulatory B cells: modulators of CD4+ T cell immunity. Nat Rev Immunol. 2010;10(4):236–47.

236. Lewis MJ, Malik TH, Ehrenstein MR, Boyle JJ, Botto M, Haskard DO. Immunoglobulin M is required for protection against atherosclerosis in low-density lipoprotein receptor-deficient mice. Circulation. 2009;120(5):417–26.

237. Yla-Herttuala S, Palinski W, Butler SW, Picard S, Steinberg D, Witztum JL. Rabbit and human atherosclerotic lesions contain IgG that recognizes epitopes of oxidized LDL. Arterioscler Thromb. 1994;14(1):32–40.

238. Kyaw T, Tay C, Krishnamurthi S, Kanellakis P, Agrotis A, Tipping P, et al. B1a B lymphocytes are atheroprotective by secreting natural IgM that increases IgM deposits and reduces necrotic cores in atherosclerotic lesions. Circ Res. 2011;109(8):830–40.

239. Suthers B, Hansbro P, Thambar S, McEvoy M, Peel R, Attia J. Pneumococcal vaccination may induce anti-oxidized low-density lipoprotein antibodies that have potentially protective effects against cardiovascular disease. Vaccine. 2012;30(27):3983–5.

240. Tsiantoulas D, Perkmann T, Afonyushkin T, Mangold A, Prohaska TA, Papac-Milicevic N, et al. Circulating microparticles carry oxidation-specific epitopes and are recognized by natural IgM antibodies. J Lipid Res. 2015;56(2):440–8.

241. Kirii H, Niwa T, Yamada Y, Wada H, Saito K, Iwakura Y, et al. Lack of interleukin-1beta decreases the severity of atherosclerosis in ApoE-deficient mice. Arterioscler Thromb Vasc Biol. 2003;23(4):656–60.

242. Ridker PM. Canakinumab for residual inflammatory risk. Eur Heart J. 2017;38(48):3545–8.

243. Upadhya S, Mooteri S, Peckham N, Pai RG. Atherogenic effect of interleukin-2 and antiatherogenic effect of interleukin-2 antibody in apo-E-deficient mice. Angiology. 2004;55(3):289–94.
244. Huber SA, Sakkinen P, Conze D, Hardin N, Tracy R. Interleukin-6 exacerbates early atherosclerosis in mice. Arterioscler Thromb Vasc Biol. 1999;19(10):2364–7.
245. Schieffer B, Selle T, Hilfiker A, Hilfiker-Kleiner D, Grote K, Tietge UJ, et al. Impact of interleukin-6 on plaque development and morphology in experimental atherosclerosis. Circulation. 2004;110(22):3493–500.
246. Madan M, Bishayi B, Hoge M, Amar S. Atheroprotective role of interleukin-6 in diet- and/or pathogen-associated atherosclerosis using an ApoE heterozygote murine model. Atherosclerosis. 2008;197(2):504–14.
247. Boisvert WA, Curtiss LK, Terkeltaub RA. Interleukin-8 and its receptor CXCR2 in atherosclerosis. Immunol Res. 2000;21(2–3):129–37.
248. Rus HG, Vlaicu R, Niculescu F. Interleukin-6 and interleukin-8 protein and gene expression in human arterial atherosclerotic wall. Atherosclerosis. 1996;127(2):263–71.
249. Liu Y, Hulten LM, Wiklund O. Macrophages isolated from human atherosclerotic plaques produce IL-8, and oxysterols may have a regulatory function for IL-8 production. Arterioscler Thromb Vasc Biol. 1997;17(2):317–23.
250. Rajasingh J, Bord E, Luedemann C, Asai J, Hamada H, Thorne T, et al. IL-10-induced TNF-alpha mRNA destabilization is mediated via IL-10 suppression of p38 MAP kinase activation and inhibition of HuR expression. FASEB J. 2006;20(12):2112–4.
251. Heiskanen M, Kahonen M, Hurme M, Lehtimaki T, Mononen N, Juonala M, et al. Polymorphism in the IL10 promoter region and early markers of atherosclerosis: the Cardiovascular Risk in Young Finns Study. Atherosclerosis. 2010;208(1):190–6.
252. Lee TS, Yen HC, Pan CC, Chau LY. The role of interleukin 12 in the development of atherosclerosis in ApoE-deficient mice. Arterioscler Thromb Vasc Biol. 1999;19(3):734–42.
253. Wuttge DM, Eriksson P, Sirsjo A, Hansson GK, Stemme S. Expression of interleukin-15 in mouse and human atherosclerotic lesions. Am J Pathol. 2001;159(2):417–23.
254. van Es T, van Puijvelde GH, Michon IN, van Wanrooij EJ, de Vos P, Peterse N, et al. IL-15 aggravates atherosclerotic lesion development in LDL receptor deficient mice. Vaccine. 2011;29(5):976–83.
255. Blankenberg S, Tiret L, Bickel C, Peetz D, Cambien F, Meyer J, et al. Interleukin-18 is a strong predictor of cardiovascular death in stable and unstable angina. Circulation. 2002;106(1):24–30.
256. Miller AM, Xu D, Asquith DL, Denby L, Li Y, Sattar N, et al. IL-33 reduces the development of atherosclerosis. J Exp Med. 2008;205(2):339–46.
257. McLaren JE, Michael DR, Salter RC, Ashlin TG, Calder CJ, Miller AM, et al. IL-33 reduces macrophage foam cell formation. J Immunol. 2010;185(2):1222–9.
258. Toma I, McCaffrey TA. Transforming growth factor-beta and atherosclerosis: interwoven atherogenic and atheroprotective aspects. Cell Tissue Res. 2012;347(1):155–75.
259. Piepoli MF, Corra U, Adamopoulos S, Benzer W, Bjarnason-Wehrens B, Cupples M, et al. Secondary prevention in the clinical management of patients with cardiovascular diseases. Core components, standards and outcome measures for referral and delivery: a policy statement from the cardiac rehabilitation section of the European Association for Cardiovascular Prevention & Rehabilitation. Endorsed by the Committee for Practice Guidelines of the European Society of Cardiology. Eur J Prev Cardiol. 2014;21(6):664–81.
260. Magnussen CG, Niinikoski H, Juonala M, Kivimaki M, Ronnemaa T, Viikari JS, et al. When and how to start prevention of atherosclerosis? Lessons from the Cardiovascular Risk in the Young Finns Study and the Special Turku Coronary Risk Factor Intervention Project. Pediatr Nephrol. 2012;27(9):1441–52.
261. Stancu C, Sima A. Statins: mechanism of action and effects. J Cell Mol Med. 2001;5(4):378–87.
262. Ebrahim S, Taylor FC, Brindle P. Statins for the primary prevention of cardiovascular disease. BMJ. 2014;348:g280.
263. Cannon CP, Blazing MA, Giugliano RP, McCagg A, White JA, Theroux P, et al. Ezetimibe added to statin therapy after acute coronary syndromes. N Engl J Med. 2015;372(25):2387–97.
264. Group AS, Ginsberg HN, Elam MB, Lovato LC, Crouse JR 3rd, Leiter LA, et al. Effects of combination lipid therapy in type 2 diabetes mellitus. N Engl J Med. 2010;362(17):1563–74.
265. Ajufo E, Rader DJ. Recent advances in the pharmacological management of hypercholesterolaemia. Lancet Diabetes Endocrinol. 2016;4(5):436–46.

266. Ridker PM, Everett BM, Thuren T, MacFadyen JG, Chang WH, Ballantyne C, et al. Antiinflammatory therapy with Canakinumab for atherosclerotic disease. N Engl J Med. 2017;377(12):1119–31.

267. Sparks JA, Barbhaiya M, Karlson EW, Ritter SY, Raychaudhuri S, Corrigan CC, et al. Investigating methotrexate toxicity within a randomized double-blinded, placebo-controlled trial: rationale and design of the Cardiovascular Inflammation Reduction Trial-Adverse Events (CIRT-AE) Study. Semin Arthritis Rheum. 2017;47(1):133–42.

268. Leung YY, Yao Hui LL, Kraus VB. Colchicine – Update on mechanisms of action and therapeutic uses. Semin Arthritis Rheum. 2015;45(3):341–50.

269. Maurovich-Horvat P, et al. Comprehensive plaque assessment by coronary CT angiography. Nat Rev Cardiol. 2014;11(7):390–402.

270. Libby P. Inflammation in atherosclerosis. Nature. 2002;420(6917):868–74.

271. Libby P, et al. Progress and challenges in translating the biology of atherosclerosis. Nature. 2011;473(7347):317–25.

# Venous Thromboembolism

*Thomas Gary*

© Springer Nature Switzerland AG 2019
M. Geiger (ed.), *Fundamentals of Vascular Biology*, Learning Materials in Biosciences,
https://doi.org/10.1007/978-3-030-12270-6_11

**What You Will Learn in This Chapter**
In this chapter you will learn about the pathogenesis, the diagnosis, and the treatment of venous thrombosis and pulmonary embolism. Furthermore you will learn about special clinical situations (i.e., cancer, elderly patients, and pregnancy), which are associated with a high thrombotic risk. Current therapeutic concepts for these groups of patients will also be discussed.

## 11.1  Introduction

Venous thromboembolism (VTE) is composed of deep vein thrombosis (DVT) – a clot formation in the deep veins mainly of the lower limbs – and, as a possible complication of this, pulmonary embolism (PE), a migration of clot material with the blood flow to the arterial pulmonary circulation.

The incidence of VTE is estimated to be 1 per 1000 people annually, with DVT accounting for approximately two-thirds of these events [1]. PE occurs in up to one-third of cases and is the primary contributor to mortality. The treatment of VTE includes acute therapy as well as long-term prophylaxis. This therapy usually consists in the inhibition of the plasmatic coagulation system (anticoagulation).

## 11.2  Pathogenesis

Virchow's triad, first described in 1856, implicates three contributing factors for thrombus formation: venous stasis, vascular injury, and hypercoagulability [2]. Venous stasis is the most prominent of the three factors, but stasis alone appears to be insufficient to cause thrombus formation. However, the concurrent presence of venous stasis and vascular injury or hypercoagulability greatly increases the risk for clot formation. The clinical conditions most closely associated with DVT are fundamentally related to the elements of Virchow's triad. These conditions include surgery, trauma, malignancy, prolonged immobility, pregnancy, advancing age, and a history of VTE.

## 11.3  Diagnosis

The clinical presentation of DVT varies with the extent and location of the thrombus formation. The cardinal signs and symptoms of DVT include asymmetrical swelling, warmth, or pain in an extremity. A number of scoring systems have been developed to estimate the pretest probability of DVT. The most widely used scoring system is the Wells criteria [3]. Similar to the Wells scoring criteria, the D-dimer assay has a high sensitivity and relatively low specificity for the diagnosis of a DVT [4]. The test measures D-dimer, a prominent fibrin degradation product, which is generated by the fibrinolytic response to thrombus formation in the body. Elevation of D-dimer is not unique to venous thrombosis, as it can be increased in various pathologic conditions including malignancy, inflammatory conditions, pregnancy, and liver disease. Given its high sensitivity, the D-dimer assay can help to rule out DVT especially in low-risk patients.

11

### 11.3.1   Deep Vein Thrombosis

The diagnosis of DVT has changed little in the last years. The gold standard decades ago was the phlebography. This examination has several disadvantages. Due to the invasiveness of the procedure and the need for a contrast agent, phlebography was already replaced years ago by the compression ultrasound (CUS) [5]. With CUS the deep and superficial veins of the leg can be examined easily from the groin to the ankle without contrast agent and radiation within minutes. The veins are examined in cross-sectional view. If there is no thrombosis in the veins, the vessel can be compressed easily with low pressure. In the case of DVT, this is not possible, and in the cross-sectional view, a thrombus formation can be detected. By means of the echogenicity of the thrombus material, the examiner can also judge about the age of the DVT.

### 11.3.2   Pulmonary Embolism

The diagnosis of a PE is usually done with a computed tomography pulmonary angiography (CTPA) [6]. CTPA has high sensitivity and specificity and is minimally invasive and fast with scan duration of less than 1 second nowadays. CTPA can also reveal other reasons for chest pain and shortness of breath such as musculoskeletal injuries, pericardial abnormalities, pneumonia, and various vascular pathologies. Most recognized and of frequent concern of using CT is the theoretical risk of cancer as a result of ionizing radiation. However, advances in protocols and technique can nowadays minimize the amount of ionizing radiation. CTPA is performed with intravenous contrast agent, which is associated with contrast-induced nephropathy (CIN) and may not be suitable for patients with renal impairment.

Lung scintigraphy (LS) could be used as one further diagnostic procedure for PE and as one alternative for CTPA [7]. LS refers to the use of radioisotopes for ventilation, perfusion, or both. Ventilation (V) and perfusion (Q) scans are used in the evaluation of PE. Although CTPA is the current gold standard, there are some clinical situations in which VQ scan is preferred, particularly renal failure and allergies to CT contrast agents. Ventilation agents include aerosolized technetium-99 m (Tc-99 m)-labeled agents [diethylenetriaminepentaacetic acid (DTPA), sulfur colloid, and ultrafine carbon suspensions] and radioactive noble gases [krypton-81 m and xenon-133 (Xe-133)].

### 11.4   Treatment

In the acute situation in which a medication with immediate onset of action is needed, the anticoagulation is carried out with a low-molecular-weight heparin (LMWH), the pentasaccharide fondaparinux, or the direct-acting oral anticoagulants rivaroxaban (Xarelto®) or apixaban (Eliquis®) [8].

Due to the different study designs for two further direct-acting oral anticoagulants (dabigatran, Pradaxa®, and edoxaban, Lixiana®), a 5-day heparin therapy is required prior to the treatment start (the so-called heparin lead-in; see also ◻ Table 11.1).

Long-term secondary prophylaxis is preferably carried out with a non-vitamin K oral anticoagulant (NOAC; see ◻ Table 11.1) or, in individual cases, with a vitamin K antagonist (VKA) (e.g., severe renal impairment). In the case of a cancer-associated VTE

⬛ **Table 11.1** NOAC medication for the immediate treatment of VTE events

| | Dabigatran[a] | Rivaroxaban[b] | Apixaban[b] | Edoxaban[b] |
|---|---|---|---|---|
| Renally cleared | 80% | 35% | 27% | 50% |
| Standard dose | 150 mg 1-0-1 | 15 mg 1-0-1 for 3 weeks; afterward 20 mg 0-1-0, intake with food mandatory! Optional after 6 months reduction to 10 mg 0-1-0 (see text) | 10 mg 1-0-1 for 1 week; afterward 5 mg 1-0-1 for 6 months; afterward 2.5 mg 1-0-1 | 60 mg 0-1-0; 30 mg in case of CrCl 30–50 ml/min or body-weight <60 kg |
| Reduced dose | 110 mg 1-0-1 | 15 mg 0-1-0 | See above | See above |
| Not in case of… | CrCl <30 ml/min | CrCl <15 ml/min | CrCl <15 ml/min | CrCl <15 ml/min |
| Heparin lead-in (at least 5 days) | Yes | No | No | Yes |

*CrCl* creatinine clearance (parameter for kidney function)
[a]Direct thrombin inhibitor
[b]Direct factor Xa inhibitor

**11**

(CAT), the treatment with LMWH has been superior to other forms of therapy for years now. However, the Hokusai VTE Cancer Trial, published in 2018, was able to show that the oral substance edoxaban was non-inferior to the subcutaneous-applied LMWH dalteparin in the treatment of CAT patients [9]. Therefore, since 2018 edoxaban can be used as an alternative to LMWH in the therapy of CAT. This therapy is also tested and seems to be safe in patients with low platelet count down to 50,000 G/L.

### 11.4.1  Fibrinolysis

The latest European Society of Cardiology (ESC) VTE guidelines published in 2014 has divided patients with acute PE into a high-risk group, an intermediate-risk group, and a low-risk group, depending on the mortality [10].

**High-risk group:** this group is characterized by a high mortality. The following changes are apparent:
- Clinical signs: shock or hypotension with or without signs of right ventricular dysfunction or myocardial damage
- Treatment: reperfusion of the pulmonary circulation with thrombolysis or embolectomy

Intermediate-risk group:  no shock, no hypotension, and signs of right ventricular dysfunction.

Intermediate high-risk group:  both right ventricular dysfunction in the imaging procedure and also laboratory findings (TropT or NT-proBNP); treatment: anticoagulation; if the patient gets hemodynamically unstable, thrombolytic therapy may also be considered.

Intermediate low-risk group:  right ventricular dysfunction either by imaging or laboratory findings (TropT or NT-proBNP); treatment: anticoagulation.

Low-risk group:  no shock, no hypotension, no signs of right ventricular dysfunction, and no signs of myocardial damage; treatment: anticoagulation.

- Fibrinolytic Therapy

Fibrinolytic therapy is restricted to patients with a high mortality, usually the patients in the ESC high-risk group. Mostly recombinant tissue plasminogen activator (rtPA) is used for the systemic fibrinolytic treatment. The dosage of rtPA should be 100 mg over 2 hours. As the patients are in a massively reduced condition, rtPA can also be given with 0.6 mg/kg bodyweight over 15 minutes (maximum dosage 50 mg).

Definition hypotension:  systolic RR below 90 mmHg or fall of RR > 40 mmHg over 15 minutes not caused by arrhythmia, sepsis, or hypovolemia.

Definition marker for right ventricular dysfunction:  typical signs in echocardiography or CT and BNP or proBNP increase

Marker for myocardial damage:  elevated troponin T or troponin I

## 11.5  Duration of Anticoagulant Treatment

### 11.5.1  Provoked VTE

A provoked VTE is a VTE caused by a specific risk factor (e.g., immobilization, surgery, stroke, hospitalization). The initial anticoagulant treatment should be given for 3–6 months [11]. After this time period, the anticoagulant treatment can be stopped when the underlying factor (e.g., immobilization) does not exist anymore. In some cases (postthrombotic syndrome, etc.), further anticoagulant treatment can be evaluated.

### 11.5.2  Spontaneous (Idiopathic) VTE

Anticoagulation is given for 3–6 months, followed by a risk-benefit assessment of the anticoagulant medication. If the patient is in a stable condition and there are no contraindications for further anticoagulation, anticoagulation should be continued. The risk of VTE recurrence can be determined with different calculation models (among others the Vienna prediction model) [12]. In case of a high risk of bleeding, a dose reduction can be performed for two NOACs (rivaroxaban and apixaban) after 6 months (see ◻ Table 11.1).

In general the decision to continue the anticoagulant medication is made individually, depending on the comorbidities, the bleeding risk, and the general condition of the patient.

## 11.6  Special Situations

### 11.6.1  Cancer

VTE in cancer patients is a common complication and significantly increases the morbidity and mortality of the patients [13]. VTE is the second leading cause of death in cancer patients, after the cancer itself.

Though the association between cancer and thrombosis has been apparent for over 150 years, the mechanisms of CAT are multifactorial and incompletely understood. Cancer type, cancer stage, tumor-derived factors, and genetics affect CAT risk [13]. The presence of metastasis increases the CAT risk. Patients with the highest 1-year incidence rate of VTE are those with cancers of the brain, lung, uterus, bladder, pancreas, stomach, and kidney. For these cancer types, CAT risk increases 4–13-fold in patients with metastases as compared with those with localized disease.

In cancer patients, both the thrombotic risk and the bleeding risk are elevated. In these patients, various factors including thrombocytopenia, renal insufficiency, chemotherapy-associated mucositis, gastrointestinal lesions, concomitant medication, and nausea and vomiting must be considered.

So far various guidelines are published on the treatment of CAT. The ESMO (European Society for Medical Oncology) guidelines recommend a 6 months' treatment with LMWH (low-molecular-weight heparin) [14]. The ESC (European Society of Cardiology) guidelines recommend anticoagulation with LMWH for the first 3–6 months [10]. Thereafter, the therapy should be continued with LMWH or VKA depending on the activity of the malignant disease. The most recently updated guidelines of the ACCP (American College of Chest Physicians) recommend a CAT treatment with LMWH over VKA or NOAC treatment [15].

However, data from newer NOAC studies were not included in these guidelines. The Hokusai VTE Cancer Trial was able to show that edoxaban is non-inferior for the treatment with LMWH in CAT patients [9]. However, the rate of severe bleeding in edoxaban-treated patients was higher mainly reflecting non-life-threatening bleeding complications. Furthermore, these bleeding complications were mainly seen in patients with gastrointestinal cancer.

After CAT anticoagulant medication should be given for at least 3–6 months. After this time the anticoagulant treatment should be continued in case of active cancer or ongoing anticancer medication. If the cancer is resumed, the anticoagulant medication can be stopped in most cases.

### 11.6.2  Elderly Patients

The incidence of VTE increases with age. There are several reasons for this phenomenon: the blood flow is reduced in frail patients by the low physical activity, the vessel wall can be damaged in case of surgical procedures, and the coagulation tendency of the blood can be increased due to an aging process of the coagulation system.

In the treatment of VTE, immediate initiation of anticoagulation to prevent progression of the thrombotic event is of utmost importance. Nowadays all societies agree that due to the very good efficacy and a significant reduction of especially intracranial bleeding complications, NOAC medication is the treatment of choice even in elderly patients.

The exact assessment of the risk of bleeding in VTE patients has been difficult for years, as appropriate scoring models were lacking. The HAS-BLED score, which can be used to estimate the risk of bleeding in atrial fibrillation (AF) patients, was investigated recently in VTE patients as well [16]. Furthermore, especially in older patients, antiplatelet medication is given frequently which increases the risk of bleeding complications. Antiplatelet medication in addition to anticoagulant medication should only be given in case of an arterial intervention in the last 12 months.

As an additional factor, the dose reduction recommendations of the substances should also be taken into account (see ◨ Table 11.1). Since dabigatran is eliminated 80% renally, the substance should not be given in patients with a creatinine clearance below 30 ml/min. The factor X inhibitors can be given down to a creatinine clearance of 15 ml/min.

An additional reduction in bleeding risk can be achieved by reducing the dose of rivaroxaban and apixaban after 6 months (see ◨ Table 11.1) in patients after spontaneous VTE or recurrent VTE. In the case of apixaban, the treatment should generally be reduced to 2.5 mg twice daily after 6 months [17]. In case of rivaroxaban, after 6 months, the dosage can be continued with 20 mg in case of a high thrombotic risk, or it can also be reduced to a prophylactic dose of 10 mg in case of a high bleeding risk [18].

### 11.6.3 Pregnancy

Due to the changes in the coagulation system, pregnancy is a highly thrombophilic situation. VTE is a frequent cause of maternal death in western countries. Compared with the (nonpregnant) general population, the incidence of VTE in a pregnant woman before birth is about 5-fold higher and increases to 15-fold after the delivery. The VTE risk of a pregnant woman is about 0.2%, and the risk of a fatal pulmonary embolism is about 0.002% [19].

The VTE risk is approximately the same from the beginning of pregnancy throughout all trimesters. After the delivery the VTE risk increases and is highest in the first 6 weeks postpartum. This increase is primarily explained by peripartal vascular damage and the release of tissue factor. In the course of cesarean section, the additional tissue damage increases the risk of thrombosis again by a factor of 3–5.

There are many reasons for the increased VTE risk during pregnancy; hypercoagulability is certainly the most significant. These changes lead to a physiological protection against bleeding complications during childbirth.

Other reasons for the increased risk of VTE are the hormonal changes in pregnancy and the compressive effect of the gravid uterus on the pelvic vessels and the inferior caval vein. The latter causes a decreased venous runoff with stasis of the blood flow in the venous system.

The physiological changes in the coagulation system start with the conception and last up to 6–8 weeks postpartum. These changes include an increased concentration of coagulation factors, decrease in protein S, increase in plasminogen activator inhibitor (PAI) type 1, increase in placental PAI-2 produced in the last trimester of pregnancy, elevation of prothrombin fragment 1 + 2, and increase of thrombin-antithrombin complexes.

The most important additional VTE risk factor is a positive VTE history. Up to 25% of thromboembolic events occurring during pregnancy are recurrent thrombosis. An underlying hereditary thrombophilia is another major risk factor, as well as an autoimmune disease, possibly associated with an antiphospholipid antibody syndrome.

In case of a heterozygous factor V Leiden mutation, the risk of VTE in pregnancy increases 8-fold, while a homozygous mutation increases the risk 34-fold.

Of practical importance are the following defects: antithrombin deficiency, combination defects (heterozygous factor V and prothrombin mutation), hetero- or homozygous defects (factor V or prothrombin mutation), and an antiphospholipid antibody syndrome. A reduced protein S is physiological during pregnancy. Hereditary protein S and protein C deficiency are extremely rare. However, if present, they represent a potent thrombophilic risk factor, which is further enhanced during pregnancy.

Thrombophilia screening is warranted in women with habitual abortions to exclude antiphospholipid antibody syndrome. In these women, treatment with low-molecular-weight heparin (LMWH) may be considered in combination with acetylsalicylic acid to avoid recurrent abortion.

The compression ultrasound is also the method of choice for the diagnosis of a DVT during pregnancy. In case of a suspected PE, diagnosis is much more difficult. Due to the radiation exposure, CTPA and LS are difficult, especially in early pregnancy. In case of negative compression ultrasound and suspected PE, LS and CTPA are possible even in pregnant women, especially in hemodynamically unstable condition [20].

In case of anticoagulant treatment during pregnancy, LMWH is the treatment of choice. These substances do not pass the placenta, and the efficacy and safety of LMWH for VTE prophylaxis and for the treatment of VTE events in pregnant women have been well studied. NOAC should not be used on pregnant or nursing women.

**11**

Take-Home Message

Correct diagnosis and immediate start of anticoagulant treatment are of utmost importance in VTE patients. Nowadays direct-acting oral anticoagulants are the treatment of choice for most of these patients. Further research is needed especially in elderly patients to optimize the exact duration of anticoagulant treatment to avoid bleeding complications

## References

1. Beckman MG, Hooper WC, Critchley SE, et al. Venous thromboembolism: a public health concern. Am J Prev Med. 2010;38:S495–501.
2. Kumar DR, Hanlin E, Glurich I, et al. Virchow's contribution to the understanding of thrombosis and cellular biology. Clin Med Res. 2010;8:168–72.
3. Wells PS, Anderson DR, Bormanis J, et al. Value of assessment of pretest probability of deep-vein thrombosis in clinical management. Lancet. 1997;350:1795–8.
4. Wells PS, Anderson DR, Rodger M, et al. Evaluation of D-dimer in the diagnosis of suspected deep-vein thrombosis. N Engl J Med. 2003;349:1227–35.
5. Kearon C, Julian JA, Newman TE, et al. Noninvasive diagnosis of deep venous thrombosis. McMaster diagnostic imaging practice guidelines initiative. Ann Intern Med. 1998;128:663–77.
6. Goldhaber SZ, Bounameaux H. Pulmonary embolism and deep vein thrombosis. Lancet. 2012; 379:1835–46.

7. Parker JA, Coleman RE, Grady E, et al. SNM practice guideline for lung scintigraphy 4.0. J Nucl Med Technol. 2012;40:57–65.
8. Streiff MB, Agnelli G, Connors JM, et al. Guidance for the treatment of deep vein thrombosis and pulmonary embolism. J Thromb Thrombolysis. 2016;41:32–67.
9. Raskob GE, van Es N, Verhamme P, et al. Edoxaban for the treatment of cancer-associated venous thromboembolism. N Engl J Med. 2018;378(7):615–24.
10. Konstantinides SV, Torbicki A, Agnelli G, et al. 2014 ESC guidelines on the diagnosis and management of acute pulmonary embolism. Eur Heart J. 2014;35(43):3033–69.
11. Konstantinides S, Torbicki A. Management of venous thrombo-embolism: an update. Eur Heart J. 2014;35:2855–63.
12. Marcucci M, Iorio A, Douketis JD, et al. Risk of recurrence after a first unprovoked venous thromboembolism: external validation of the Vienna Prediction Model with pooled individual patient data. J Thromb Haemost. 2015;13(5):775–81.
13. Khorana AA. Venous thromboembolism and prognosis in cancer. Thromb Res. 2010;125:490–3.
14. Mandalà M, Falanga A, Roila F, et al. Management of venous thromboembolism (VTE) in cancer patients: ESMO clinical practice guidelines. Ann Oncol. 2011;22(Suppl 6):vi85–92.
15. Kearon C, Akl EA, Ornelas J, et al. Antithrombotic therapy for VTE disease CHEST guideline and expert panel report. Chest. 2016;149(2):315–52.
16. Rief P, Raggam RB, Hafner F, et al. Calculation of HAS-BLED score is useful for early identification of venous thromboembolism patients at high risk for major bleeding events: a prospective outpatients cohort study. Semin Thromb Hemost. 2018;44(4):348–52.
17. Agnelli G, Buller HR, Cohen A, et al. Apixaban for extended treatment of venous thromboembolism. N Engl J Med. 2013;368:699–708.
18. Weitz I, Lensing AWA, Prins MH, et al. Rivaroxaban or aspirin for extended treatment of venous thromboembolism. N Engl J Med. 2017;376:1211–22.
19. Liu S, Rouleau J, Joseph KS, et al. Epidemiology of pregnancy associated venous thromboembolism: a population-based study in Canada. J Obstet Gynaecol Can. 2009;31:611–20.
20. Wan T, Skeitha L, Karovitch A, et al. Guidance for the diagnosis of pulmonary embolism during pregnancy: consensus and controversies. Thromb Res. 2017;157:23–8.

# Genetics of Vascular Diseases

*Christine Mannhalter*

© Springer Nature Switzerland AG 2019
M. Geiger (ed.), *Fundamentals of Vascular Biology*, Learning Materials in Biosciences,
https://doi.org/10.1007/978-3-030-12270-6_12

**What You Will Learn in This Chapter**

In this book ▶ Chap. 1 will present an overview over genetic variants with proven or at least convincingly suggested causal relation to vascular diseases. I will discuss polymorphisms in genes with potential association with secondary risk factors and address functional consequences of genetic variations. The polymorphisms that are causally involved in the pathogenesis of vascular dysfunction typically impair proteins relevant for the blood coagulation system, hypertension, obesity, or diabetes. I will address selected genetic factors responsible for thrombotic diseases, hypertension, diabetes, and obesity. I will also briefly cover the basics of genome-wide association studies (GWAS) and next-generation sequencing (NGS). After reading this article, the reader shall understand the importance of the potential heritability of the vascular risk which is a key issue in preventive medicine and is of particular importance for the development of good strategies to preserve vascular health.

## 12.1 Introduction

Arterial and venous thrombotic disorders with the clinical manifestations myocardial infarction (MI), stroke, peripheral arterial disease, and venous thromboembolism (VTE) are major causes of morbidity and mortality. Age, sex, systolic blood pressure, hypertension, total and high-density lipoprotein, cholesterol levels, smoking, diabetes, prohormone B-type natriuretic peptide levels, chronic kidney disease, leukocyte count, C-reactive protein levels, homocysteine levels, uric acid levels, coronary artery calcium [CAC] scores, carotid intima-media thickness, existing peripheral arterial disease, and pulse wave velocity are all important risk factors for arterial thrombosis [1]. However, these established risk factors do not fully capture the overall risk of the diseases. Today we know that the highly complex pathological phenotypes of vascular diseases are for a large part determined by genetic predisposition. There is agreement that occlusive coronary thrombus formation is highly important for myocardial infarction (MI) and stroke. Today, a large body of data is available associating single nucleotide polymorphisms (SNPs) in candidate genes with ischemic stroke or coronary artery disease. However, the results for individual SNPs and genes are controversial, and for many recently discovered genetic factors, the proof of their functional role is still missing.

## 12.2 Genetic Variations in the Genome: Single Nucleotide Polymorphisms/Variants (SNPs, SNVs)

Understanding the relationship between genotype and phenotype is a major biological challenge. There is great hope that genetic, genomic, and epigenetic discoveries will enhance this understanding, improve the diagnostic capability, and identify new treatment options [2]. The modern unit of genetic variations is the single nucleotide polymorphism (SNP). SNPs are frequent variations of single base pairs in the genome occurring with >1% prevalence in a given population. Approximately three million SNPs are found in every individual, indicating a frequency of 1 SNP per 1000 nucleotides. SNPs occur frequently in cytosine-rich regions. They are responsible for about 90% of the genetic heterogeneity. SNPs can be located in the coding region and lead to changes of amino acids (non-synonymous SNPs), but mostly they do not alter the sequence for an amino acid (synonymous SNPs). Non-synonymous SNPs can have functional (gain of function or loss of function) effects, but frequently these are relatively small. SNPs may also influence the regulation of gene transcription (regulatory SNPs), splicing, or RNA processing and via these mechanisms affect the protein concentration. The variants are typically used as markers of a genomic region, and the large majority of them have only a minimal impact on biological systems. Most of the known SNPs reside outside of coding regions and/or in genomic regions lacking annotated genes. Often, SNPs are considered as successful mutations because they prevailed in a population. The scientific importance of SNPs is based on their high prevalence and their simple detectability. Although SNPs have become popular for genetic mapping, discovery and application of SNPs in humans are not easy.

## 12.3 Technological Platforms for the Detection of Genetic Variants

### 12.3.1 Genome-Wide Association (GWA)

GWA studies (GWAS) are used to detect associations between genetic variants and traits in samples from different populations. They are a powerful tool for the investigation of the genetic mechanisms of human diseases. GWAS became possible after the chip-based microarray technology was developed which allowed testing for one million or more SNPs simultaneously. Two platforms, one from Illumina (San Diego, CA) and one from Affymetrix (Santa Clara, CA), are most frequently used. While the Affymetrix platform has short DNA sequences bound to a chip that recognizes a specific SNP allele, Illumina uses a bead-based technology. Each platform selects different SNPs, and this can be important when specific human populations are being studied. SNPs with better overall genomic coverage are preferable when analysing African populations, because African genomes had more time to recombine and therefore more SNPs are needed to capture variations across the African genome.

Up to now, about 10,000 strong associations have been reported between genetic variants and complex traits. From GWAS results one can conclude that for complex traits, many loci contribute to the genetic variation. Thus, for most traits and diseases, polymorphisms in many genes have been associated with a phenotype, and often this is population dependent [3, 4]. Therefore, on average, the contribution of a single variant to a disease is small.

Importantly, the technology for analysing genomic variations is changing rapidly. Chip-based genotyping platforms will soon be replaced by sequencing the entire genome by NGS.

### 12.3.2 Next-Generation Sequencing (NGS)

Sequencing allows the determination of the precise order of nucleotides – adenine, guanine, cytosine, and thymine – within a DNA molecule. The development of rapid DNA sequencing methods has greatly enhanced biological and medical research and discovery. NGS analyses have revolutionized our understanding of biological processes. The rapid speed and the moderate price of modern DNA sequencing technologies have been instrumental for the sequencing of complete genomes of numerous species, including humans, animal species, plants, or microorganisms. With NGS one can sequence the entire genome in an automated process after fragmentation of the genome into small pieces [5, 6]. In many basic science or clinical studies, substantive insight has been gained by comparing the DNA sequences of genes in different groups of subjects. Today, there are many software tools for the computational analysis of NGS data, each with its own algorithm.

It has been shown that a small proportion of the variants detected by NGS cannot be reproduced by Sanger sequencing [7]. Therefore, before any firm conclusions are drawn from an NGS study, potential variants identified by NGS should be confirmed by another method. Furthermore, certain regions of the human genome are not accurately readable by current NGS technologies. These include regions of high homology (e.g. segmental duplications) for which reads cannot be uniquely mapped, regions where the reference genome contains errors or has not been annotated yet, regions with multiple reference haplotypes, and tandem repeats that extend beyond the sequenced read length.

### 12.3.3  Importance of the Study Design

NGS lead to the identification of numerous variants, for some of which the clinical significance is difficult to ascertain based on our current knowledge. One study found out that even among laboratories experienced in genetic testing, there was low concordance in designating genetic variants as pathogenic. In an unselected population, the putatively pathogenic genetic variants were not associated with an abnormal phenotype in all participants. These findings raised the question what implication the notification of patients about incidental genetic findings may have [8, 9].

Today, scientists and clinicians agree that the success of any well-conducted case/control study depends on the definition of phenotype criteria. Standardized phenotype rules are particularly critical for multicentre studies to prevent introduction of a site-based effect into the study. Even when established phenotype criteria are used, there may be variability among clinicians how the criteria are applied to assign case and control status. High inter-rater agreement means that phenotype rules are consistently implemented across multiple sites, which is highly important. The definition of the criteria for cases and controls is particularly important for multicentre studies of relatively rare diseases. To better understand the genetic basis of a relatively rare form of a vascular disease, one has to collect a large number of well-characterized samples of cases and controls. Then, one can perform genome-wide association analyses to assess the association and impact of common and low-frequency genetic variants on, e.g. cerebral venous thrombosis (CVT) risk by using a case-control study design. Replication of the results is important to confirm putative findings [10].

### 12.3.4  Statistics

Statistical tests are generally considered significant, and the null hypothesis is rejected if the p-value falls below a predefined alpha value, which is nearly always set to 0.05. In case of GWAS, hundreds of thousands to millions of tests are done, each with a false positivity probability. The likelihood to find false positives over the entire GWAS analysis is therefore high, and correction for multiple testing is essential. One of the simplest approaches to correct for multiple testing is the Bonferroni correction. For a typical GWAS using 500,000 SNPs, statistical significance of a SNP association would be set at $10^{-7}$.

GWA studies represent an enormous opportunity to examine interactions among genetic variants throughout the genome; however, multilocus analysis presents numerous computational, statistical, and logistical challenges [8].

## 12.4 The Haemostatic System, Gene Polymorphisms, and Atherothrombotic Diseases

Today, we know that hereditary and acquired factors influence the risk of thrombosis, and a number of environmental factors add to this [11]. The majority of DNA sequence changes that influence haemostatic and thrombotic phenotypes are either SNPs resulting in missense, nonsense, or splice site mutations, small micro-insertions, and micro-deletions which disrupt the reading frame. Several SNPs are located in genes/proteins that had not been associated with haemostasis, and for some of them, the function is not well known. However, in association studies with GWAS or NGS, certain SNPs were found more frequently in patients with atherothrombotic diseases than in healthy controls. A list of selected genes and the coded proteins which have been reported in several studies is included in ◻ Table 12.1. For many of these variants, sequence data, phenotypic annotations, and original literature citations are available from the Human Gene Mutation Database (HGMD®). Since the 1990s it is clear that besides mutations and polymorphisms in coding regions of genes, genetic variations in regulatory regions, particularly in the promoter regions, have an important impact on blood coagulation because of their effect on the concentration of the proteins [12–14]. For a number of coagulation factors (platelet components and plasmatic proteins), reproducible data exists regarding the contribution of SNPs to the variability of their plasma concentration and possible effects on the clinical phenotype (◻ Tables 12.2, 12.3, 12.4, and 12.5).

### 12.4.1 How Is a Thrombus Formed?

The rupture of the endothelial lining leads to the exposure of subendothelial matrix proteins, which can interact with blood platelets and plasmatic proteins. Thus, platelets play an important role, and several polymorphisms have been reported to influence thrombus formation (◻ Table 12.2) [15–17]. Besides platelets, von Willebrand Factor (VWF) is needed for the initiation of coagulation which mediates the binding of platelets to the subendothelium in primary haemostasis. The adhering platelets become activated and expose negative charges. These allow the interaction of haemostatic proteins with the activated platelets. Calcium is required for the interaction with the negative charges [18].

The coagulation system comprises several enzymes, synthesized as proenzymes, and cofactors which interact with each other in a waterfall cascade [11] (◻ Fig. 12.1). The proenzymes are converted to active enzymes eventually leading to thrombin formation. Coagulation can be progressing via two intertwined pathways, the intrinsic system comprising factors XII, prekallikrein, high-molecular-weight kininogen, and factors XI, IX, and VIII at negatively charged surfaces and the extrinsic pathway via activation of factor VII in the presence of tissue factor. Both pathways converge at a common stage involving factors X, V, and II. As a result thrombin is formed, which cleaves fibrinogen (factor I) and forms fibrin. The crosslinking of fibrin by activated factor XIII increases clot stability and its resistance to fibrinolysis [20–23].

**□ Table 12.1**   Genes, corresponding proteins, functional effects

**Association with atrial fibrillation, body mass index**

| | |
|---|---|
| PITX2 | Paired-like homeodomain transcription factor 2 |
| ZFHX3 | Zinc finger homeobox 3 |

**Association with coronary artery disease, ischemic stroke, blood pressure, myocardial remodeling atherosclerotic stroke**

| | |
|---|---|
| ABO | Blood group, alpha 1–3-N-acetylgalactosaminyltransferase and alpha 1–3- galactosyltransferase |
| HDAC9 | Histone deacetylase 9 |
| ALDH2 | Aldehyde dehydrogenase 2 |

**Association with smooth muscle cell development, ischemic stroke in Chinese Han population**

| | |
|---|---|
| FOXF2 | Forkhead box F2 |

**Association with coagulation, pulmonary hypertension, early-onset ischemic stroke**

| | |
|---|---|
| HABP2 | Hyaluronan-binding protein 2 |

**Association with carotid plaque formation, ischemic stroke**

| | |
|---|---|
| MMP12 | Matrix metallopeptidase 12 |

**Association with neuro-inflammation, blood pressure regulation, oligodendrocyte differentiation**

| | |
|---|---|
| TSPAN2 | Tetraspanin 2 |

**Association with venous thromboembolism, in vivo regulation of the A disintegrin and metalloproteinase 10, transfusion-related acute lung injury**

| | |
|---|---|
| TSPAN15 | Tetraspanin15 |
| SCL44A2 | Solute carrier family 44, member 2 |
| KNG1 | Kininogen 1, high-molecular-weight kininogen |
| PROC | Protein C |
| PROCR | Protein C receptor |
| SERPINC1 | Serpin family C member 1, antithrombin |
| STXBP5 | Syntaxin-binding protein 5 |

**Association with myocardial infarction**

| | |
|---|---|
| SERPINE1 | Plasminogen activator inhibitor-1 |
| PLAT | Tissue plasminogen activator |
| THBD | Thrombomodulin |
| PROCR | Endothelial protein C receptor |
| FGA | Fibrinogen alpha-chain |
| FGB | Fibrinogen beta-chain |

**12**

**◘ Table 12.1** (continued)

| Association with VWF levels | |
|---|---|
| STXBP5 | Syntaxin-binding protein 5 |
| STX2 | Syntaxin 2 |
| TC2N | Tandem C2 domains |
| CLEC4M | C-type lectin domain family 4 member M |
| SNARE | Small NF90 (ILF3)-associated RNA E |
| **Association with FVIII levels** | |
| SCARA5 | Scavenger receptor class A member 5 |
| STAB2 | Stabilin 2 |
| **Associated with decreased risk of coronary heart disease** | |
| PCSK9 | Proprotein convertase subtilisin/kexin type 9 |
| NPC1L1 | NPC1-like intracellular cholesterol transporter 1 |
| APOC3 | Apolipoprotein C3 |
| APOA5 | Apolipoprotein A5 |
| **Association with blood pressure** | |
| LSP1 | Lymphocyte-specific protein 1 |
| TNNT3 | Troponin type 3 |
| MTHFR | Methylene tetrahydrofolate reductase |
| NPPB | Natriuretic peptide B |
| AGT | Angiotensinogen |
| ATP2B1 | ATPase plasma membrane Ca2+ transporting 1 |
| STK39 | Serine/threonine kinase 39 |
| CYP17A1 | Cytochrome P450 family 17 subfamily A member 1 |
| NPPA | Natriuretic peptide A |
| NPPB | Natriuretic peptide B |
| CSK | C-terminal Src kinase |
| ZNF652 | Zinc finger protein 652 |
| UMOD | Uromodulin |
| CACNB2 | Calcium voltage-gated channel auxiliary subunit beta 2 |
| PLEKHA7 | Pleckstrin homology domain containing A7 |
| SH2B3 | SH2B adaptor protein 3 |

(continued)

**Table 12.1** (continued)

| | |
|---|---|
| *TBX3, TBX5* | T-box 3, T-box 5 |
| *ULK4* | unc-51 like kinase 4 |
| *ULK3* | unc-51 like kinase 3 |
| *CYP1A2* | Cytochrome P450 family 1 subfamily A member 2 |
| *NT5C2* | 5′-nucleotidase, cytosolic II |
| *PLCD3* | Phospholipase C delta 3 |
| *ATXN2* | Ataxin 2 |
| *CACNB2* | Calcium voltage-gated channel auxiliary subunit beta 2 |
| *HFE* | Homeostatic iron regulator |
| *WNK4* | WNK lysine deficient protein kinase 4 |
| *BDKRB2* | Bradykinin receptor B2 |
| **Association with systolic blood pressure** | |
| *ACE* | Angiotensin-converting enzyme |
| *ADD1* | Alpha adducin-1 |
| *ADRB2* | Adrenergic beta-2 receptor |
| *CYP11B2* | Cytochrome P450 family 11 subfamily B member 2 |
| *GNB3* | Guanine nucleotide-binding protein 3 |
| *NOS3* | Nitric oxide synthase 3 |
| **Association with body mass index, obesity risk, diabetes type 2** | |
| *FTO* | FTO, alpha-ketoglutarate-dependent dioxygenase, Fat Mass and Obesity Associated |

**Table 12.2** Selected polymorphisms in platelet glycoproteins [20]

| Quantitative trait loci | Structure/function | Unknown |
|---|---|---|
| Integrin $\alpha_2$ 807 C>T | Integrin $\beta_3$ Leu33Pro | $\alpha_2$ AR long/short |
| Integrin $\alpha_2$–52 T/C | Fc RIIA | $P_2Y_{12}$ |
| Integrin $\alpha_2$–92 T/C | GPIbα Ko | Serotonin receptor T102C; 44-base pair insertion/deletion |
| GPVI | GPIbα VNTR | c-mpl 550 C/A |
| GPIb α–5T/C | $\alpha_2$ AR Asn251Lys GNB3 Integrin $\alpha_2$ Thr799Met | PAR1 IVS-14A/T |

**□ Table 12.3**  Selected polymorphisms in genes of the haemostatic system

| Gene | Polymorphism | rs number | Functional consequences | References |
|------|-------------|-----------|------------------------|------------|
| Fibrino-gen | −455G > A<br>Thr331Ala | rs1800790<br>rs6050 | Increased fibrinogen plasma levels<br>Modestly lower plasma levels | [53, 79, 80] |
| FXIII | Val35Leu | rs5985 | Increased FXIII activity, decreased clot stability, resistance to fibrinolysis | [81–83] |
| PAI-1 | −675(4G/5G) | rs34857375 | Lower PAI-1 levels | [30, 84, 85] |
| FXII | −4C > T | rs1801020 | Association with lower FXII plasma levels | [19, 86] |
| FVII | Arg413Gln<br>−401G > T<br>−402G > A | rs6046<br>rs510335<br>rs510317 | Associated with lower FVII activity<br>Associated with lower FVII activity 89<br>Associated with higher FVII activity and antigen | [87] |
| Pro-thrombin | 20210G > A | rs1799963 | Associated with Increased prothrombin plasma level | [88, 89] |
| FVIII | Asp1260Glu | rs1800291 | Associated with FVIII activity | [71, 90, 91] |
| Protein C | −1654C/A<br>−1641 G/T<br>1386 T > C, | rs2069901 | Association of CG haplotype with Protein C levels | [92, 93] |
| GPIa (ITGA2) | Glu534Lys | rs1801106 | Influences GPIa activity, affects expression density | [94–96] |
| P-selectin | Thr715Pro | rs6136 | Associated with soluble P-selectin levels | [97–100] |

**□ Table 12.4**  Selected polymorphisms in the vWF gene and their functional consequences

| Gene | SNP | Variants | Functional consequence | Reference |
|------|-----|----------|------------------------|-----------|
| vWF | | Thr789Ala | CAD in patients with diabetes mellitus I | [67] |
| VWF<br>vWF | | A(−1185)G<br>G(−1051)A | MI <75 years | [68] |
| vWF | | Sma I polymorphism in intron 2 | stroke, MI | [69] |

**□ Table 12.5    Polymorphisms associated with VTE [46]**

| Locus | SNP | OR | Phenotype |
|-------|-----|-----|-----------|
| *ABO* | [O, A2] vs [A1, B], rs2519093 | 1.50 | ↑ VWF, ↑ VIII |
| *F2* | rs1799963 | 2.50 | ↑ F2 |
| *F5* | rs6025 | 3.00 | Resistance to APC |
| *F11* | rs2036914, rs2289252 | 1.35 | ↑ F11 |
| *FGG* | rs2066865 | 1.47 | ↓ FGG |
| *GP6* | rs1613662 | 1.15 | ↑ platelet activity |
| *KNG1* | rs710446 | 1.20 | ↓ aPTT |
| *PROCR* | rs867186 | 1.22 | ↑ PC |
| *SLC44A2* | rs2288904 | 1.19 | unknown |
| *STXBP5* | rs1039084 | 1.11 | ↑ VWF |
| *TSPAN15* | rs78707713 | 1.28 | unknown |
| *VWF* | rs1063856 | 1.15 | ↑ VWF |

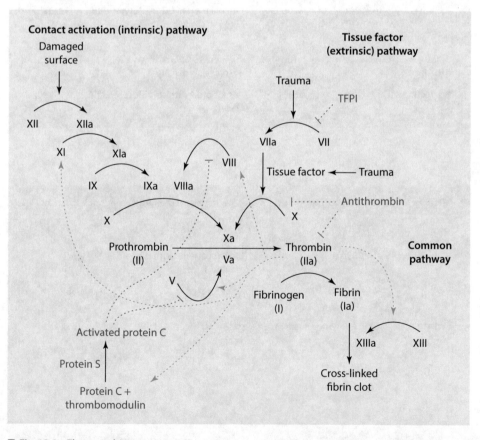

**□ Fig. 12.1**   The coagulation system [11]

## 12.4.2  **Atherothrombosis, Vascular Diseases, and Genetics**

Vascular diseases represent highly complex pathological phenotypes that are influenced by genetic susceptibility as well as by modifiable (e.g. smoking) and non-modifiable (e.g. age, gender) risk factors. Much of the burden of vascular diseases can be explained by "classic" risk factors (e.g. smoking, blood pressure). However, the contribution of genetic risk factors to vascular diseases is increasingly recognized and understood. The identification of underlying mechanisms facilitates the detection of novel therapeutic targets and contributes to an optimal design of prevention strategies. Genome-wide analysis (genotyping several hundred thousands or millions of genetic variants distributed across the genome) allows an estimation of the heritability of diseases even in the absence of familial information [24].

Only a small proportion of strokes or CAD are attributable to monogenic conditions, the vast majority is multifactorial, with multiple genetic and environmental risk factors of small effect size each [25]. Understanding genetic factors and heritability of the vascular risk is a key issue in preventive medicine and is of particular importance for the development of cost-efficient strategies to preserve vascular health.

Evidence is accumulating from genetic studies and clinical trials, which associations between genetic factors and vascular diseases are causal. However, noncausal risk factors may still have their value, particularly, when added to cardiovascular risk scores [26, 27].

GWAS and NGS generated a large body of information on the genomics of cardiovascular disease [28]. Many articles reported GWAS results presenting new genetic determinants of cardiovascular disease, including coronary artery disease. This expanded our knowledge on the molecular players in cardiovascular pathophysiology considerably. As pointed out before, it is important to understand the fundamentals of these new technologies, e.g. the relationship between GWAS and standard epidemiologic study design, the concepts of DNA sequence variation and linkage disequilibrium, statistical considerations in studies involving many independent variables and large sample sizes, as well as the biologic and clinical significance of GWAS-based discoveries.

Unfortunately, published results for individual polymorphisms and genes are often controversial due to the inequality of studied cohorts, the influence of the study design, the presence of modulating factors in the population, the genetic background, etc. Therefore, in spite of successful research, our understanding of the disease pathogenesis is still incomplete. When standardized study design will become self-evident, there is great hope that in the near future, genetic, genomic, and epigenetic discoveries will broaden the diagnostic capability and revolutionize treatment modalities.

## 12.4.3  **Cerebrovascular Disease**

Recently, heritability of stroke has been shown based on data from large genome-wide association studies, and the current estimates of heritability may even increase if phenotyping becomes more accurate [29–31]. However, heritability, morbidity, and mortality are heterogeneous and depend in part on stroke subtypes. Genetic polymorphisms and environmental risk factors may interact with each other, and different combinations may affect the risk differently [32]. Given the small effect size of genetic risk variants for stroke, large numbers of individuals have to be studied to reach sufficient statistical power. The identification of functionally relevant variants requires large collaborative efforts, which became possible through international consortia [33–36]. While initial studies identified risk loci for specific stroke subtypes, recent studies revealed loci associated with stroke in general. This shows that GWAS and large international consortia have been instrumental in finding genetic risk factors for stroke.

Some new risk factors have been reported to be associated with atrial fibrillation (e.g. *PITX2* and *ZFHX3*). *PITX2*, for instance, was previously known to play a crucial role in embryogenesis, ontogenesis, growth, and development via the *Wnt/beta-catenin* and *POU1F1* pathways. In addition, deleterious mutations in this gene had been reported to cause serious heart defects in humans. Now, *PITX2* was found associated with atrial fibrillation. *ZFHX3*, a known tumour suppressor gene, was also shown to be involved in the pathogenesis of atrial fibrillation. For coronary artery disease, *ABO*, *chr9p21*, *HDAC9*, and *ALDH2*, for blood pressure regulation *ALDH2 and HDAC9*, for smooth muscle cell development *FOXF2*, for coagulation *HABP2*, for carotid plaque formation *MMP12*, and for neuro-inflammation *TSPAN2* were identified [25]. However, another analysis of genetic variants in stroke patients of 6 cohorts gave different results. No single locus of significance (approximately $p < 10^7$) could be identified [37]. In a recent meta-analysis which evaluated 136 biomarkers for their relevance in stroke, 3 biomarkers (C-reactive protein, P-selectin, and homocysteine) could differentiate between ischemic stroke and healthy control subjects. High levels of admission glucose and high fibrinogen levels were strong predictors of poor prognosis after ischemic stroke and symptomatic intracerebral haemorrhage post-thrombolysis. D-dimer concentrations predicted in-hospital death. The study concluded that only few biomarkers have meaningful clinical value. Furthermore, several of the identified genetic variants that contributed to vascular problems in mice could not be confirmed in humans [38, 39]. Nevertheless, some genetic markers have been re-evaluated and confirmed and seem to be important in certain populations [40].

### 12.4.4 Venous Thromboembolism (VTE)

Venous thrombosis (VT) is a multifactorial disease with a clear genetic component that was first suspected nearly 60 years ago. VTE comprises deep vein thrombosis, pulmonary embolism, or both. Today we know that a number of putative conditions interact and finally contribute to increase the individual risk and ultimately lead to venous occlusive disorders. Thrombophilia is commonly defined as hypercoagulable state attributable to inherited or acquired disorders of blood coagulation or fibrinolysis. VTE has a strong genetic basis. The genetically determined thrombophilic conditions include deficiencies of natural anticoagulants such as antithrombin, protein C, and protein S, increased functional activity of clotting factors (especially factor VIII), as well as prothrombotic polymorphisms in gene encoding factor V (i.e. *factor V Leiden*) and prothrombin. Until 1993 we knew only antithrombin, protein C, and protein S deficiency as risk factors for VTE. These deficiencies were caused by mutations in the three genes, but they explained VTE just in a small number of patients (less than 10%). In 1993, Dahlbäck et al. reported a family with poor anticoagulant response due to resistance to activated protein C [41]. Shortly thereafter it was shown that the resistance was caused by a point mutation in the *F5* gene at nt1691, leading to an Arg > Gln amino acid exchange at amino acid position 506 (*FV 506R > Q, FV Leiden, rs6025*) (◻ Fig. 12.2) [42]. Coagulation factor V (FV) acts as cofactor of FXa and plays an important role in the regulation of the coagulation process. To develop its procoagulatory activity, FV has to be activated either by thrombin or FXa

**◻ Fig. 12.2** Factor V Leiden mutation

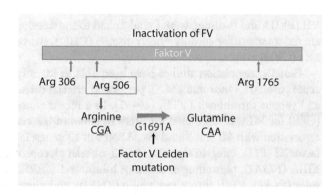

Inactivation of FV

Faktor V

↑ ↑          ↑
Arg 306   Arg 506        Arg 1765

↓
Arginine    ➡️    Glutamine
CGA    G1691A    CAA
↑
Factor V Leiden
mutation

by limited proteolysis of several peptide bonds. For maintenance of the haemostatic balance, inactivation of FVa has to occur by proteolytic cleavage at Arg506, Arg306, and Arg679 by activated protein C (APC). In the presence of the *FV Leiden* mutation, the haemostatic process is no longer in balance. *Factor V Leiden* is frequent in the Caucasian population (~5%) and represents one of the most important risk factors for inherited venous thrombosis [11, 43]. Large meta-analyses investigated the role of the *FV 506R > Q* mutation in myocardial infarction and reported an odds ratio of ~1.3 for carriers of the *FV 506Q* allele [44]. These results suggest that screening for the *FV 506R > Q* polymorphism in unselected patients at risk for MI is not indicated. However, according to the literature, it still cannot be excluded that the *FV 506Q* mutation has a role in selected patient groups. Despite the identification of various inherited risk factors associated with VTE, a clear cause for the disease is still missing in almost 50% of patients. So far, GWAS has not been able to identify genetic variants with implications for clinical care, and unexplained heritability remains. In the past years, the development of several NGS platforms has been offering the possibility of generating fast, inexpensive, and accurate genomic information. However, the use of NGS for VTE studies has been limited [45].

Recently, new SNPs have been identified in genome-wide association studies, such as those in *TSPAN15* (rs78707713) and *SCL44A2* (rs2288904) genes. These two unexpected loci have been previously shown to associate with transfusion related acute lung injury. Both variants did not associate with known haemostatic plasma markers and had not been associated with other cardiovascular diseases nor were they related to quantitative biomarkers. Evidently, meta-analyses of GWAS data can uncover unexpected actors of VTE aetiology and pave the way for novel mechanistic concepts of VTE pathophysiology. Deepened knowledge of all potential risk factors and the clear understanding of their role in the pathophysiology of venous thrombosis are essential to help achieve a faster and more efficient diagnosis as well as a more effective prophylaxis of patients at higher risk. By now, genetic variants in 17 genes are well established; however, additional candidate genes that deserve further validation are likely to exist [46–48].

## 12.4.5 **Coronary Artery Disease**

The relationship of haemostasis and thrombosis with atherothrombotic vascular disease has been extensively studied. Elevated levels of haemostatic factors, such as fibrinogen, plasminogen activator inhibitor (PAI-1), VWF, tissue plasminogen activator (tPA), factor

VII (FVII), and d-dimer, were found linked to the development of atherothrombosis and are risk markers for coronary heart disease (CHD), stroke, and other cardiovascular disease (CVD) events.

Genetic association studies have been able to identify gene variants associated with acute coronary syndrome (ACS), peripheral arterial disease (PAD), ischemic stroke (IS), and venous thrombosis (VTE) [49–51]. In a recent study, a weighted genetic risk score (GRS) for MI was calculated for SNPs in genes of the coagulation system. Evidence for association with MI was found for 35 SNPs in 12 genes: factor V (*F5*), protein S (*PROS1*), factor XI (*F11*), integrin alpha 2 (*ITGA2*, platelet glycoprotein Ia), factor XII (*F12*), factor XIIIA (*F13A1*), plasminogen activator inhibitor-1 (*SERPINE1*), tissue plasminogen activator (*PLAT*), *VWF*, thrombomodulin (*THBD*), endothelial protein C receptor (*PROCR*), and factor IX (*F9*). The GRS differed significantly between cases and controls, and subjects in the highest quintile had a 2.69-fold increased risk for MI compared with those in the lowest quintile [52]. It is well known that fibrinogen levels vary interindividually and high levels have been reported to play a role in CVD. It has been shown that genetic factors contribute to the variability of fibrinogen plasma levels. The two most frequently studied polymorphisms are the −455G > A (rs1800790) polymorphism, located in the promoter region of the fibrinogen beta-chain (*FGB*), and the *312Thr > Ala* (rs6050) polymorphism in the fibrinogen alpha-chain (*FGA*). −455G > A is in complete linkage disequilibrium with the −148C > T polymorphism in the IL-6 responsive element and leads to increased plasma fibrinogen levels, while 312Thr > Ala is associated with modestly lower plasma fibrinogen levels (◘ Table 12.3). The *FGA rs6050* single nucleotide polymorphism (SNP) indicates a decreased risk, whereas the *FGB* -455G > A SNP seems to increase the risk of stroke. The risk of myocardial infarction does not seem to be altered by either of those SNPs [53]. Interestingly, an association between post-stroke mortality and the 312Thr > Ala polymorphism in patients with atrial fibrillation had been observed by Carter et al. [54]. Thus, although the general risk for MI or stroke does not seem to be significantly affected by fibrinogen polymorphisms, they may play a role in subgroups of patients. The importance of fibrinogen polymorphisms in thrombotic disease has still to be proven, but meta-analyses are under way to obtain convincing data [55].

The *fibrinolytic system* is essential for the dissolution of blood clots and the maintenance of a patent vascular system (◘ Fig. 12.3) [11, 56]. Abnormalities in the fibrinolytic system have been implicated in the pathogenesis of myocardial infarction and stroke [57]. Two plasminogen activators exist in blood: tissue-type plasminogen activator (tPA) and urokinase. Activation of plasminogen by tissue-type plasminogen activator (tPA) is enhanced in the presence of fibrin or at the endothelial cell surface. Inhibition of fibrinolysis occurs either at the level of plasminogen activation by a plasminogen activator inhibitor (PAI-1) or at the level of plasmin by alpha-2 antiplasmin [11] (◘ Fig. 12.3). PAI-1 is the key regulator of the fibrinolytic system and has the greatest inhibitory effect. It plays a crucial role in various physiological processes including fibrinolysis, tissue repair, blood coagulation, thrombolysis, ovulation, embryogenesis, angiogenesis, and cell adhesion and migration. Polymorphisms in the *PAI-1* gene affect PAI-1 levels and have been associated with myocardial infarction and stroke [58]. In the promoter region of the *PAI-1* gene, a common single guanine nucleotide insertion/deletion polymorphism (4G/5G), situated 675 bp from the transcription start site, has been identified (◘ Table 12.3). The 5G allele contains a repressor site and binds both, an enhancer and a repressor, while the 4G site binds only an enhancer. This leads to a lower transcription of *PAI-1* in carriers of the 5G genotype. A recently published meta-analysis comprising 41 studies including 12,461

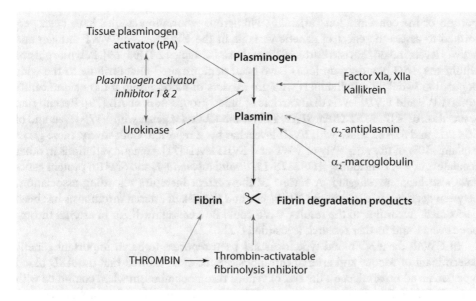

Plasminogen activator inhibitor 1 & 2

Tissue plasminogen activator (tPA)

Plasminogen

Factor XIa, XIIa Kallikrein

Urokinase

Plasmin

$\alpha_2$-antiplasmin

$\alpha_2$-macroglobulin

Fibrin   Fibrin degradation products

THROMBIN ⟶ Thrombin-activatable fibrinolysis inhibitor

☐ **Fig. 12.3**   The fibrinolytic system [11]

cases and 14,993 controls identified the *PAI-1 4G* allele as weak risk factor for MI in Caucasian, Asian, and African populations. The MI risk for the *4G/4G* genotype was increased compared to *5G/5G* genotype and the *5G* allele, with an odds ratio at 1.16 [59]. Elevation in PAI-1 and reduction in tPA levels were observed in stroke patients. In some studies including ours, the *4G/4G* genotype conferred a reduced risk of stroke, possibly due to a more stable clot and less likelihood of embolic events [60]. Interestingly, a nested case-control study in two independent Swedish cohorts reported an association of the *4G* allele with an increased risk of future ischemic stroke [61].

The *PAI-1 4G/5G* polymorphism has also been associated with hypertension. In 815 unrelated Spanish individuals, the *4G/4G PAI-1* genotype conferred an elevated relative risk of developing arterial hypertension, regardless of PAI-1 levels and other hypertension-related factors [62]. Recently, it was reported that central arterial blood pressure was higher in women carrying the *PAI-1 4G/4G* genotype compared to females with the *5G/5G* genotype [63]. Thus, even though there are many studies suggesting a contribution of PAI-1 to the risk of vascular events, the exact mechanisms and the affected disease entities are still unclear.

As pointed out before, *VWF and factor VIII* are highly important for an efficient initiation and activation of the haemostatic system. FVIII is a large glycoprotein which is synthesized in the liver. It is elevated in acute phase conditions like stress, inflammation, etc. The *F8* gene is located on the X-chromosome, and mutations within the gene cause haemophilia A. In plasma, FVIII is bound to VWF, a large plasma glycoprotein synthesized in endothelial cells and megakaryocytes. VWF is particularly important under high shear stress, where platelet adhesion and primary haemostasis are fully dependent on VWF [64]. Elevated plasma levels of FVIII and VWF have been associated with an increased risk for MI and stroke [65]. Until now, the causes for elevated FVIII remain unclear, but there is no doubt that high FVIII levels are not merely a consequence of inflammation but are partly genetically determined. Plasma VWF levels are also to a large extent genetically determined, and sequence variations within the *VWF* gene explain a portion of the concentration variability. Numerous association studies have been performed to assess the effect of genetic variants in the *VWF* gene on VWF antigen and activity levels and on the risk of arterial thrombosis (◼ Table 12.4) [66–69]. Polymorphisms within the *ABO* blood-group locus have been demonstrated to contribute to the wide population variability. Individuals who are carriers of blood group 0 have significantly lower VWF and F VIII levels than carriers of blood groups A, B, or AB [70]. Recently, an association of SNPs in *STXBP5*, *STX2*, *TC2N*, and *CLEC4M* genes with VWF levels and of *SCARA5* and *STAB2* genes with FVIII levels has been reported. Collectively, these genes explain ~10% of the variability of VWF and FVIII levels [71]. Genetic variations in other regulators of VWF including *ADAMTS-13*, *thrombospondin-1*, and *SNARE* protein genes have also been investigated. A review of the current literature regarding associations between genetic variations in the *VWF* gene and the risk of arterial thrombosis has been performed. According to the results, VWF could be a causal mediator of arterial thrombotic events, and further research is justified [72].

In GWAS the *ABO* blood type locus has been reported to be an important genetic determinant of venous and arterial thrombosis. Sode et al. showed that the *ABO* blood type had an additive effect on the risk of venous thromboembolism when combined with *factor V Leiden* R506Q and *prothrombin G20210A* mutations; blood type was the most important risk factor for venous thromboembolism in the general population [73].

While the contribution of the *FV* 506R > Q mutation to the risk of venous thrombosis has been shown in many studies, its role in patients with cardiovascular or cerebrovascular thrombosis has not yet been proven. As reported by Lalouschek et al., *FV Leiden* only confers a significantly increased risk for stroke in female smokers (OR 8.8; P = 0.004). No interaction between the mutation, smoking, and risk of stroke was observed in men [74].

As pointed out before, the genetic predisposition to deep vein thrombosis (DVT) is only partially explained by known genetic risk variants. High-throughput genotyping and sequencing technologies allow the identification of several new genetic variants. It seems that the *ADAMTS13* and *VWF* genes also contribute to DVT. An investigation of 186 haemostatic/pro-inflammatory genes by NGS identified rare (frequency of <1%) and low-frequency (<5%) non-synonymous variants in *ADAMTS13* as risk factors for DVT. The mutations were associated with lower plasma levels of ADAMTS-13 activity [75]. As mentioned above, 17 other genes were confirmed to contain genetic variations associated with VT risk: *ABO*, *F2*, *F5*, *F9*, *F11*, *FGG*, *GP6*, *KNG1*, *PROC*, *PROCR*, *PROS1*, *SERPINC1*, *SLC44A2*, *STXBP5*, *THBD*, *TSPAN15*, and *VWF*. However, the common polymorphisms in these genes account only for a modest part (~5%) of the VT heritability (◼ Table 12.5) [46].

Already in 2007, Knowles et al. tested 21 SNPs in coagulation factor and platelet glycoprotein genes and evaluated their association with MI in the ADVANCE study population [76]. They did not find significant associations of the chosen polymorphisms with acute MI. However, they did not test, e.g. the *4G/5G* polymorphism in *PAI-1* (rs1799889), which had been associated with MI. In a recent meta-analysis, the highest quantile of the blood PAI-1 level was associated with higher CHD risk compared to the lowest quantile. A Mendelian randomization analyses suggested a causal effect of increased PAI-1 level on CHD risk possibly mediated by a causal effect of PAI-1 on elevation of blood glucose and high-density lipoprotein cholesterol [77]. Recent advances in the field of multifactorial genetics, in particular GWASs and their meta-analyses, now provide the statistical power to identify and replicate many previously discovered genetic variants [78].

I could not discuss all known genetic risk factors for vascular diseases. However, I wanted to list some SNPs in genes of the haemostatic system for which the functional

**12**

effects are known (▫ Table 12.3). The application of NGS to vascular diseases is still limited by considerable computational burden associated with the analysis of the large numbers of samples. The identification of rare loss-of-function variants in genes such as *PCSK9*, *NPC1L1*, *APOC3*, and *APOA5*, which cause a decreased risk of CHD and no adverse side effects, illustrates the necessity of translating genetic findings into mechanistic information.

## 12.5 Hypertension and Vascular Dysfunction

It has long been known that elevated blood pressure (BP) is an important risk factor for stroke and ischemic heart disease. Pathophysiology and molecular mechanisms involved in hypertension regulation are not very well known. Recently, NGS technology has identified hundreds of gene loci associated with cardiovascular pathologies, including blood pressure (BP) regulation. Expectations grew that new pathways and genetic mechanisms underlying BP regulation would be identified by these new technologies. Fewer than half the people affected with hypertension are aware of their condition, many others are aware but either not treated or inadequately treated, although it has been shown that successful treatment of hypertension reduces the global burden of disease and mortality [101]. Hypertension represents a serious health burden with nine million people dying as a consequence of hypertension-related complications. Elevated blood pressure is a complex trait caused by multifactorial genetic inheritance together with environmental factors [102]. The heritability of blood pressure (BP) is estimated to be 30–50%, and several genes with large effects have been identified in familial forms of hypertension [103]. To identify the common relevant biomarkers, studies of patients with hypertension have to comprise accurate standardized blood pressure (BP) measurement, assessment of the patients' predicted risk of atherosclerotic CVD, detection of secondary causes of hypertension, as well as the presence of comorbidities (e.g. kidney disease). A number of GWA studies were performed with the aim to identify genetic variants affecting BP levels. This approach led to the identification of multiple genetic variants which explain only 2–3% of the genetic variance of hypertension. Presumably, some variants are too rare to be detected by GWAS. NGS facilitated the discovery of additional causative variants. Undoubtedly, there is a need for replication studies within larger consortia to better understand the functional effects of the new genes and to learn how they might be used in patient care [104]. Today, it is well accepted that systemic arterial hypertension is a modifiable risk factor for all-cause morbidity and mortality of cardiovascular disease (CVD). In spite of some positive results by NGS, research on the genetics of hypertension has been disappointing. Genotyping tens of thousands of individuals and meta-analysing dozens of cross-sectional, population-based studies on systolic and diastolic blood pressure (BP) showed that hypertension is influenced by environmental and lifestyle factors as well as by many genetic loci, each of which has only a small effect on blood pressure regulation. Even though high-throughput genomic technologies have identified many SNPs involved in BP regulation, only some of them have been replicated and validated. Forty-seven distinct genetic variants were found to be robustly associated with BP, but collectively they explain only a few percent of the heritability for BP phenotypes. Polymorphisms in the following genes could be replicated: *LSP1*, *TNNT3*, *MTHFR*, *NPPB*, *AGT*, and *ATP2B1*. It is known that the Val/Val *(677TT) MTHFR* genotype leads to higher plasma homocysteine levels. According to recent data, *MTHFR 677TT* appears to be independently associated with hypertension. If prospective trials can confirm that the *MTHFR* mutation conveys a predisposition to hypertension which can be corrected by low-dose riboflavin, the findings could have important implications for the management of hypertension [105]. A meta-analysis of 11 studies reported loci in 130 genes of which some were repeatedly identified: *STK39*, *CYP17A1*, *MTHFR-NPPA*, *MTHFR-NPPB*, *ATP2B1*, *CSK*, *ZNF652*, *UMOD*, *CACNB2*, *PLEKHA7*, *SH2B3*, *TBX3-TBX5*, *ULK4*, *ULK3*, *CYP1A2*, *NT5C2*, *PLCD3*, *ATXN2*, and *HFE*. Unfortunately, only two genes, *WNK4* and *BDKRB2*, overlapped between genetic and epigenetic studies [106]. For these two genes, the functional effect on blood pressure could be proven. *WNK4* could be shown to be involved in the regulation of the renal NaCl cotransporter, the major salt transport pathway. For the bradykinin B2 receptor *BDKRB2*, several studies have reported that polymorphisms in this gene are associated with transcription of the receptor and an association with the risk of hypertension was substantiated by meta-analyses.

It was interesting to find an association between a common SNP in the *PAI-1* gene and hypertension [107]. Impaired fibrinolytic function, characterized by increased PAI-1 levels and decreased tissue plasminogen activator (tPA) activity, has been detected in patients with hypertension. Data from the literature indicate that antihypertensive drugs vary in their influence on fibrinolysis. Angiotensin-converting enzyme inhibitors (ACE-I) have been shown to improve the fibrinolytic balance by reducing plasma PAI-1 levels, while calcium channel blockers (CCB) have been reported to increase tPA activity. The positive effect of ACE-I on the fibrinolytic system has been related to: (1) inhibition of angiotensin II, which stimulates PAI-1 expression; (2) inhibition of degradation of bradykinin, a potent stimulus for tPA production; and (3) improvement of insulin sensitivity.

In a Japanese association study, selected candidate gene variants were evaluated for genetic associations with systolic BP (SBP)/diastolic BP (DBP) in 19,426 individuals. Associations with *ACE*, *ADD1*, *ADRB2*, *AGT*, *CYP11B2*, *GNB3*, and *NOS3* were tested. BP trait associations at two loci (*AGT rs699 and CYP11B2 rs1799998*) were confirmed. The most significant association was found for *CYP11B2 rs1799998*. This study provides evidence for two variants in genes with clinical and physiological relevance that are likely to account in part for BP variance in the general population and are worth following up via a target gene approach [108].

## 12.6 Obesity and Type 2 Diabetes

Obesity is a complex and multifactorial disease that occurs as a result of the interaction between environmental factors and genetic components. Genetic susceptibility appears to determine the individual risk to develop obesity by more than 40% [109]. The search for genetic variations, which, combined with environmental factors, give rise to pathologically relevant BMI levels, was initially hypothesis driven. Candidate gene studies revealed a small number of relevant genes and causative gene variants, which were robustly associated with obesity and provided the basis for rare, monogenic forms of the disease. The causal genes for monogenic obesity could be confirmed using animal models and linkage studies. But these approaches were not helpful for polygenic obesity. Also for obesity research, GWA approaches and NGS have brought a breakthrough with the discovery of genetic variants and tens of new susceptibility loci. The use of exome genotyping arrays and deep sequencing of candidate loci helped to identify rare variants that may be important [110]. In a previous review, it was reported that candidate genes and GWAS have led to the discovery of 58 loci contributing to polygenic obesity. These loci explain only a small fraction of the heritability for obesity, and many genes remain to be discovered. Some genes are involved in the regulation of food intake. Furthermore, genes predisposing to obesity are participating in the control of energy balance. Interestingly, it has been shown that there is a partial overlap between monogenic and polygenic forms of obesity [111]. Several studies reported that the Fat Mass and Obesity Associated (*FTO*) gene is the major contributor to polygenic obesity. The predisposing *FTO* variant was associated with increased total and fat dietary intake as well as diminished satiety and/or increased feeling of hunger in children and adults [112, 113]. Nevertheless, heterozygous loss-of-function mutations in *FTO* are found in both lean and obese subjects. This indicates that mutations in susceptibility genes, although causing defined cellular dysfunctions, can give rise to multiple phenotypes (in combination with secondary risk factors). This overlap in genetic basis of disease can create combined phenotypes representing a malign constellation of multiple "cardiometabolic" risk factors specifically regarding the "metabolic syndrome" [114]. Indeed, the *FTO* gene was also associated with type 2 diabetes (T2DM). It codes for an approx. 60kD 2-oxoglutarate-dependent nucleic acid demethylase. Despite the known strong impact of *FTO* polymorphisms on obesity, the underlying molecular mechanism is still not well understood. Recent attempts to unravel the functional coupling of *FTO* to cellular signalling pathways revealed a potential linkage to neuronal plasticity [115]. So far, this neuronal mechanism is rather speculative and needs confirmation by further functional studies. *FTO* has been suggested as common risk determinant for both obesity and T2DM, and linkage of FTO expression to insulin resistance has been reported [116, 117]. Another interesting aspect is the impact of *FTO* risk alleles on physical activity. Physical activity may specifically attenuate the obesity risk in A-allele carriers of the *FTO* rs9939609 genotype [118].

---

**Take-Home Message**

Although our knowledge is still far from being complete, it is clear that genetic research is necessary and important for the better understanding of the pathomechanisms of vascular diseases and the identification of potential therapeutic targets. Data from candidate gene studies, GWAS, and sequencing experiments have to be carefully analysed to identify the potential candidates for therapeutic approaches. Many results showed that it is now absolutely necessary to perform functional studies to translate the genetic findings into clinical practice.

---

# References

1. Libby P, Ridker P, Maseri A. Inflammation and atherosclerosis. Circulation. 2002;105:1135–43.
2. Mannhalter C. New developments in molecular biological diagnostic. Haemostaseologie. 2017;37:138–51.
3. Bush WS, Moore JH. Genome-wide association studies. PLoS Comput Biol. 2012;8:e1002822.
4. Visscher PM, Wray NR, Zhang Q, Sklar P, McCarthy MI, Brown MA, Yang J. 10 years of GWAS discovery: biology, function, and translation. Am J Hum Genet. 2017;101:5–22.
5. Church GM. Genomes for all. Sci Am. 2006;294:46–54.
6. Pareek CS, Smoczynski R. Tretyn A Sequencing technologies and genome sequencing. J Appl Genet. 2011;52:413–35.
7. van den Akker J, Mishne G, Zimmer AD, Zhou AY. A machine learning model to determine the accuracy of variant calls in capture-based next generation sequencing. BMC Genomics. 2018;19(1):263.
8. Moore JH, Ritchie MD. The challenges of whole-genome approaches to common diseases. JAMA. 2004;291:1642–3.

9. Van Driest SL, Wells QS, Stallings S, Bush WS, Gordon A, Nickerson DA, Kim JH, Crosslin DR, Jarvik GP, Carrell DS, Ralston JD, Larson EB, Bielinski SJ, Olson JE, Ye Z, Kullo IJ, Abul-Husn NS, Scott SA, Bottinger E, Almoguera B, Connolly J, Chiavacci R, Hakonarson H, Rasmussen-Torvik LJ, Pan V, Persell SD, Smith M, Chisholm RL, Kitchner TE, He MM, Brilliant MH, Wallace JR, Doheny KF, Shoemaker MB, Li R, Manolio TA, Callis TE, Macaya D, Williams MS, Carey D, Kapplinger JD, Ackerman MJ, Ritchie MD, Denny JC, Roden DM. Association of arrhythmiar-related genetic variants with phenotypes documented in electronic medical records. JAMA. 2016;315:47–57.

10. Cotlarciuc I, Marjot T, Khan MS, Hiltunen S, Haapaniemi E, Metso TM, Putaala J, Zuurbier SM, Brouwer MC, Passamonti SM, Bucciarelli P, Pappalardo E, Patel T, Costa P, Colombi M, Canhão P, Tkach A, Santacroce R, Margaglione M, Favuzzi G, Grandone E, Colaizzo D, Spengos K, Arauz A, Hodge A, Ditta R, Debette S, Zedde M, Pare G, Ferro JM, Thijs V, Pezzini A, Majersik JJ, Martinelli I, Coutinho JM, Tatlisumak T, Sharma P, ISGC (International Stroke Genetics Consortium) and BEAST investigators. Towards the genetic basis of cerebral venous thrombosis -the BEAST Consortium: a study protocol. BMJ Open. 2016;6:e012351.

11. Mannhalter C. Biomarkers for arterial and venous thrombotic disorders. Haemostaseologie. 2014;34:115–30.

12. van't Hooft FM, Silveira A, Tornvall P, Iliadou A, Ehrenborg E, Eriksson P, Hamsten A. Two common functional polymorphisms in the promoter region of the coagulation factor VII gene determining plasma factor VII activity and mass concentration. Blood. 1999;93(10):3432–41.

13. Arnaud E, Barbalat V, Nicaud V, Cambien F, Evans A, Morrison C, Arveiler D, Luc G, Ruidavets JB, Emmerich J, Fiessinger JN, Aiach M. Polymorphisms in the 5′ regulatory region of the tissue factor gene and the risk of myocardial infarction and venous thromboembolism: the ECTIM and PATHROS studies. Etude Cas-Témoins de l'Infarctus du Myocarde. Paris Thrombosis case-control Study. Arterioscler Thromb Vasc Biol. 2000;20:892–8.

14. Blake GJ, Schmitz C, Lindpaintner K, Ridker PM. Mutation in the promoter region of the beta-fibrinogen gene and the risk of future myocardial infarction, stroke and venous thrombosis. Eur Heart J. 2001;22:2262–6.

15. Heemskerk JW, Mattheij NJ, Cosemans JM. Platelet-based coagulation: different populations, different functions. J Thromb Haemost. 2013;11:2–16.

16. Travers RJ, Smith SA, Morrissey JH. Polyphosphate, platelets, and coagulation. Int J Lab Hematol. 2015;37(Suppl 1):31–5.

17. Vijayan KV, Bray PF. Molecular mechanisms of prothrombotic risk due to genetic variations in platelet genes: enhanced outside-in signaling through the Pro33 variant of integrin beta3. Exp Biol Med. 2006;231:505–13.

18. Winter WE, Flax SD, Harris NS. Coagulation testing in the core laboratory. Lab Med. 2017;48:295–313.

19. Schousboe I. Pharmacological regulation of factor XII activation may be a new target to control pathological coagulation. Biochem Pharmacol. 2008;75:1007–13.

20. Periayah MH, Halim AS, Mat Saad AZ. Mechanism action of platelets and crucial blood coagulation pathways in hemostasis. Int J Hematol Oncol Stem Cell Res. 2017;11:319–27.. Review.

21. Müller F, Mutch NJ, Schenk WA, Smith SA, Esterl L, Spronk HM, Schmidbauer S, Gahl WA, Morrissey JH, Renné T. Platelet polyphosphates are proinflammatory and procoagulant mediators in vivo. Cell. 2009;139:1143–56.

22. Müller F, Renné T. Platelet polyphosphates: the nexus of primary and secondary hemostasis. Scand J Clin Lab Invest. 2011;71:82–6.

23. Puy C, Tucker EI, Wong ZC, Gailani D, Smith SA, Choi SH, Morrissey JH, Gruber A, McCarty OJ. Factor XII promotes blood coagulation independent of factor XI in the presence of long-chain polyphosphates. J Thromb Haemost. 2013;11:1341–52.

24. Yang J, Benyamin B, McEvoy BP, et al. Common SNPs explain a large proportion of the heritability for human height. Nat Genet. 2010;42:565–9.

25. Chauhan G, Debette S. Genetic risk factors for ischemic and hemorrhagic stroke. Curr Cardiol Rep. 2016;18:124.

26. Keavney B, Danesh J, Parish S, Palmer A, Clark S, Youngman L, Delépine M, Lathrop M, Peto R, Collins R. Fibrinogen and coronary heart disease: test of causality by Mendelian randomization. Int J Epidemiol. 2006;35:935–43.

27. Lacey B, Herrington WG, Preiss D, Lewington S, Armitage J. The role of emerging risk factors in cardiovascular outcomes. Curr Atheroscler Rep. 2017;19:28.

28. Dubé JB, Hegele RA. Genetics 100 for cardiologists: basics of genome-wide association studies. Can J Cardiol. 2013;29:10–7.

12

29. Liao D, Myers R, Hunt S, et al. Familial history of stroke and stroke risk. The Family Heart Study. Stroke. 1997;28:1908–12.
30. Jood K, Ladenvall C, Rosengren A, Blomstrand C, Jern C. Family history in ischemic stroke before 70 years of age: the Sahlgrenska Academy Study on Ischemic Stroke. Stroke. 2005;36:1383–7.
31. Devan WJ, Falcone GJ, Anderson CD, et al. Heritability estimates identify a substantial genetic contribution to risk and outcome of intracerebral hemorrhage. Stroke. 2013;44:1578–83.
32. Zondervan KT, Cardon LR. Designing candidate gene and genome-wide case–control association studies. Nat Protoc. 2007;2:2492–501.
33. Wang X, Cheng S, Brophy VH, Erlich HA, Mannhalter C, Berger K, Lalouschek W, Browner WS, Shi Y, Ringelstein EB, Kessler C, Luedemann J, Lindpaintner K, Liu L, Ridker PM, Zee RY. Cook NR; RMS Stroke SNP Consortium. A meta-analysis of candidate gene polymorphisms and ischemic stroke in 6 study populations: association of lymphotoxin-alpha in nonhypertensive patients. Stroke. 2009;40:683–95.
34. Psaty BM, O'Donnell CJ, Gudnason V, Lunetta KL, Folsom AR, Rotter JI, Uitterlinden AG, Harris TB, Witteman JC, Boerwinkle E, CHARGE Consortium. Cohorts for Heart and Aging Research in Genomic Epidemiology (CHARGE) Consortium: design of prospective meta-analyses of genome-wide association studies from 5 cohorts. Circ Cardiovasc Genet. 2009;2:273–80.
35. Eicher JD, Chami N, Kacprowski T, Nomura A, Chen MH, Yanek LR, Tajuddin SM, Schick UM, Slater AJ, Pankratz N, Polfus L, Schurmann C, Giri A, Brody JA, Lange LA, Manichaikul A, Hill WD, Pazoki R, Elliot P, Evangelou E, Tzoulaki I, Gao H, Vergnaud AC, Mathias RA, Becker DM, Becker LC, Burt A, Crosslin DR, Lyytikäinen LP, Nikus K, Hernesniemi J, Kähönen M, Raitoharju E, Mononen N, Raitakari OT, Lehtimäki T, Cushman M, Zakai NA, Nickerson DA, Raffield LM, Quarells R, Willer CJ, Peloso GM, Abecasis GR, Liu DJ, Global Lipids Genetics Consortium, Deloukas P, Samani NJ, Schunkert H, Erdmann J, CARDIoGRAM Exome Consortium, Myocardial Infarction Genetics Consortium, Fornage M, Richard M, Tardif JC, Rioux JD, Dube MP, de Denus S, Lu Y, Bottinger EP, Loos RJ, Smith AV, Harris TB, Launer LJ, Gudnason V, Velez Edwards DR, Torstenson ES, Liu Y, Tracy RP, Rotter JI, Rich SS, Highland HM, Boerwinkle E, Li J, Lange E, Wilson JG, Mihailov E, Mägi R, Hirschhorn J, Metspalu A, Esko T, Vacchi-Suzzi C, Nalls MA, Zonderman AB, Evans MK, Engström G, Orho-Melander M, Melander O, O'Donoghue ML, Waterworth DM, Wallentin L, White HD, Floyd JS, Bartz TM, Rice KM, Psaty BM, Starr JM, Liewald DC, Hayward C, Deary IJ, Greinacher A, Völker U, Thiele T, Völzke H, van Rooij FJ, Uitterlinden AG, Franco OH, Dehghan A, Edwards TL, Ganesh SK, Kathiresan S, Faraday N, Auer PL, Reiner AP, Lettre G, Johnson AD. Platelet-related variants identified by exomechip meta-analysis in 157,293 individuals. Am J Hum Genet. 2016;99:40–55.
36. Weymann A, Sabashnikov A, Ali-Hasan-Al-Saegh S, Popov AF, Jalil Mirhosseini S, Baker WL, Lotfaliani M, Liu T, Dehghan H, Yavuz S, de Oliveira Sá MP, Jang JS, Zeriouh M, Meng L, D'Ascenzo F, Deshmukh AJ, Biondi-Zoccai G, Dohmen PM, Calkins H, Cardiac Surgery And Cardiology-Group Imcsc-Group IM20. Predictive role of coagulation, fibrinolytic, and endothelial markers in patients with atrial fibrillation, stroke, and thromboembolism: a meta-analysis, meta-regression, and systematic review. Med Sci Monit Basic Res. 2017;23:97–140.
37. Lanktree MB, Dichgans M, Hegele RA. Advances in genomic analysis of stroke: what have we learned and where are we headed? Stroke. 2010;41:825–32.
38. Hasan N, McColgan P, Bentley P, Edwards RJ, Sharma P. Towards the identification of blood biomarkers for acute stroke in humans: a comprehensive systematic review. Br J Clin Pharmacol. 2012;74:230–40.
39. Pasterkamp G, van der Laan SW, Haitjema S, Foroughi Asl H, Siemelink MA, Bezemer T, van Setten J, Dichgans M, Malik R, Worrall BB, Schunkert H, Samani NJ, de Kleijn DP, Markus HS, Hoefer IE, Michoel T, de Jager SC, Björkegren JL, den Ruijter HM, Asselbergs FW. Human validation of genes associated with a murine atherosclerotic phenotype. Arterioscler Thromb Vasc Biol. 2016;36:1240–6.
40. Chen M, Mao BY, Wang D, Cheng X, Xu CX. Association between rs1801133 polymorphism and risk of adult ischemic stroke: meta-analysis based on case-control studies. Thromb Res. 2016;137:17–25.
41. Dahlbäck B, Carlsson M, Svensson PJ. Familial thrombophilia due to a previously un-recognized mechanism characterized by poor anticoagulant response to activated protein C: prediction of a cofactor to activated protein C. Proc Natl Acad Sci U S A. 1993;90:1004–8.
42. Bertina RM, Koeleman BP, Koster T, Rosendaal FR, Dirven RJ, de Ronde H, van der Velden PA, Reitsma PH. Mutation in blood coagulation factor V associated with resistance to activated protein C. Nature. 1994;369(6475):64–7.
43. Endler G, Mannhalter C. Polymorphisms in coagulation factor genes and their impact on arterial and venous thrombosis. Clin Chim Acta. 2003;330:31–55.. Review
44. Juul K, Tybjaerg-Hansen A, Steffensen R, Kofoed S, Jensen G, Nordestgaard BG. Factor V Leiden: the Copenhagen City Heart Study and 2 meta-analyses. Blood. 2002;100:3–10.

45. Cunha MLR, Meijers JCM, Middeldorp S. Introduction to the analysis of next generation sequencing data and its application to venous thromboembolism. Thromb Haemost. 2015;114:920–32.
46. Morange PE, Suchon P, Trégouët DA. Genetics of venous thrombosis: update in 2015. Thromb Haemost. 2015;114:910–9.
47. Crous-Bou M, Harrington LB, Kabrhel C. Environmental and genetic risk factors associated with venous thromboembolism. Semin Thromb Hemost. 2016;42:808–20.
48. Trégouët DA, Morange PE. What is currently known about the genetics of venous thromboembolism at the dawn of next generation sequencing technologies. Br J Haematol. 2018;180:335–45.
49. Kathiresan S, Yang Q, Larson MG, Camargo AL, Tofler GH, Hirschhorn JN, Gabriel SB, O'Donnell CJ. Common genetic variation in five thrombosis genes and relations to plasma hemostatic protein level and cardiovascular disease risk. Arterioscler Thromb Vasc Biol. 2006;26:1405–12.
50. Yang Q, Kathiresan S, Lin JP, Tofler GH, O'Donnell CJ. Genome-wide association and linkage analyses of hemostatic factors and hematological phenotypes in the Framingham Heart Study. BMC Med Genet. 2007;8(Suppl 1):S12.
51. Lovely RS, Yang Q, Massaro JM, Wang J, D'Agostino RB Sr, O'Donnell CJ, Shannon J, Farrell DH. Assessment of genetic determinants of the association of γ' fibrinogen in relation to cardiovascular disease. Arterioscler Thromb Vasc Biol. 2011;31:2345–52.
52. Guella I, Duga S, Ardissino D, Merlini PA, Peyvandi F, Mannucci PM, Asselta R. Common variants in the haemostatic gene pathway contribute to risk of early-onset myo-cardial infarction in the Italian population. Thromb Haemost. 2011;106:655–64.
53. Siegerink B, Rosendaal FR, Algra A. Genetic variation in fibrinogen; its relationship to fibrinogen levels and the risk of myocardial infarction and ischemic stroke. J Thromb Haemost. 2009;7:385–90.
54. Carter AM, Catto AJ, Grant PJ. Association of the alpha-fibrinogen Thr312Ala poly-morphism with poststroke mortality in subjects with atrial fibrillation. Circulation. 1999;99:2423–6.
55. de Vries PS, Chasman DI, Sabater-Lleal M, Chen MH, Huffman JE, Steri M, Tang W, Teumer A, Marioni RE, Grossmann V, Hottenga JJ, Trompet S, Müller-Nurasyid M, Zhao JH, Brody JA, Kleber ME, Guo X, Wang JJ, Auer PL, Attia JR, Yanek LR, Ahluwalia TS, Lahti J, Venturini C, Tanaka T, Bielak LF, Joshi PK, Rocanin-Arjo A, Kolcic I, Navarro P, Rose LM, Oldmeadow C, Riess H, Mazur J, Basu S, Goel A, Yang Q, Ghanbari M, Willemsen G, Rumley A, Fiorillo E, de Craen AJ, Grotevendt A, Scott R, Taylor KD, Delgado GE, Yao J, Kifley A, Kooperberg C, Qayyum R, Lopez LM, Berentzen TL, Räikkönen K, Mangino M, Bandinelli S, Peyser PA, Wild S, Trégouët DA, Wright AF, Marten J, Zemunik T, Morrison AC, Sennblad B, Tofler G, de Maat MP, de Geus EJ, Lowe GD, Zoledziewska M, Sattar N, Binder H, Völker U, Waldenberger M, Khaw KT, Mcknight B, Huang J, Jenny NS, Holliday EG, Qi L, Mcevoy MG, Becker DM, Starr JM, Sarin AP, Hysi PG, Hernandez DG, Jhun MA, Campbell H, Hamsten A, Rivadeneira F, Mcardle WL, Slagboom PE, Zeller T, Koenig W, Psaty BM, Haritunians T, Liu J, Palotie A, Uitterlinden AG, Stott DJ, Hofman A, Franco OH, Polasek O, Rudan I, Morange PE, Wilson JF, Kardia SL, Ferrucci L, Spector TD, Eriksson JG, Hansen T, Deary IJ, Becker LC, Scott RJ, Mitchell P, März W, Wareham NJ, Peters A, Greinacher A, Wild PS, Jukema JW, Boomsma DI, Hayward C, Cucca F, Tracy R, Watkins H, Reiner AP, Folsom AR, Ridker PM, O'Donnell CJ, Smith NL, Strachan DP, Dehghan A. A meta-analysis of 120 246 individuals identifies 18 new loci for fibrinogen concentration. Hum Mol Genet. 2016;25:358–70.
56. Preissner KT. Physiology of blood coagulation and fibrinolysis: biochemistry. [Article in German]. Haemostaseologie. 2008;28:259–71.
57. Babu MS, Prabha TS, Kaul S, Al-Hazzani A, Shafi G, Roy S, Balakrishna N, Jyothy A, Munshi A. Association of genetic variants of fibrinolytic system with stroke and stroke subtypes. Gene. 2012;495:76–80.
58. Jood K, Ladenvall P, Tjärnlund-Wolf A, Ladenvall C, Andersson M, Nilsson S, Blomstrand C, Jern C. Fibrinolytic gene polymorphism and ischemic stroke. Stroke. 2005;36:2077–81.
59. Gong LL, Peng JH, Han FF, Zhu J, Fang LH, Wang YH, Du GH, Wang HY, Liu LH. Association of tissue plasminogen activator and plasminogen activator inhibitor polymorphism with myocardial infarction: a meta-analysis. Thromb Res. 2005;130:e43–51.
60. Endler G, Lalouschek W, Exner M, Mitterbauer G, Häring D, Mannhalter C. The 4G/4G genotype at nucleotide position −675 in the promotor region of the plasminogen activator inhibitor 1 (PAI-1) gene is less frequent in young patients with minor stroke than in controls. Br J Haematol. 2000;110:469–71.
61. Wiklund PG, Nilsson L, Ardnor SN, Eriksson P, Johansson L, Stegmayr B, Hamsten A, Holmberg D, Asplund K. Plasminogen activator inhibitor-1 4G/5G polymorphism and risk of stroke: replicated findings in two nested case-control studies based on independent cohorts. Stroke. 2005;36(8):1661–5.

12

62. Martínez-Calatrava MJ, González-Sánchez JL, Zabena C, Martínez-Larrad MT, Luque-Otero M, Serrano-Ríos M. Is the plasminogen activator inhibitor-1 gene a candidate gene predisposing to hypertension? Results from a population-based study in Spain. J Hypertens. 2007;25:773–7.

63. Björck HM, Eriksson P, Alehagen U, De Basso R, Ljungberg LU, Persson K, Dahlström U, Länne T. Gender-specific association of the plasminogen activator inhibitor-1 4G/5G polymorphism with central arterial blood pressure. Am J Hypertens. 2011;24:802–8.

64. Savage B, Saldivar E, Ruggeri Z. Initiation of platelet adhesion by arrest onto fibrinogen or translocation on von Willebrand factor. Cell. 1996;84:289–97.

65. Saito I, Folsom AR, Brancati FL, Duncan BB, Chambless LE, McGovern PG. Nontraditional risk factors for coronary heart disease incidence among persons with diabetes: the Atherosclerosis Risk in Communities (ARIC) Study. Ann Intern Med. 2000;133:81–91.

66. Kamphuisen PW, Houwing-Duistermaat JJ, van Houwelingen HC, Eikenboom JC, Bertina RM, Rosendaal FR. Familial clustering of factor VIII and von Willebrand factor levels. Thromb Haemost. 1998;79:323–7.

67. Lacquemant C, Gaucher C, Delorme C, Chatellier G, Gallois Y, Rodier M, Passa P, Balkau B, Mazurier C, Marre M, Froguel P. Association between high von willebrand factor levels and the Thr789Ala vWF gene polymorphism but not with nephropathy in type I diabetes. The GENEDIAB Study Group and the DESIR Study Group. Kidney Int. 2000;57:1437–43.

68. Di Bitondo R, Cameron CL, Daly ME, Croft SA, Steeds RP, Channer KS, Samani NJ, Lillicrap D, Winship PR. The −1185 A/G and −1051 G/A dimorphisms in the von Willebrand factor gene promoter and risk of myocardial infarction. Br J Haematol. 2001;115:701–6.

69. Dai K, Gao W, Ruan C. The Sma I polymorphism in the von Willebrand factor gene associated with acute ischemic stroke. Thromb Res. 2001;104:389–95.

70. Ay C, Thom K, Abu-Hamdeh F, Horvath B, Quehenberger P, Male C, Mannhalter C, Pabinger I. Determinants of factor VIII plasma levels in carriers of haemophilia A and in control women. Haemophilia. 2010;16:111–7.

71. Antoni G, Oudot-Mellakh T, Dimitromanolakis A, Germain M, Cohen W, Wells P, Lathrop M, Gagnon F, Morange PE, Tregouet DA. Combined analysis of three genome-wide association studies on vWF and FVIII plasma levels. BMC Med Genet. 2011;12:102.

72. Van Schie MC, van Loon JE, de Maat MP, Leebeek FW. Genetic determinants of von Willebrand factor levels and activity in relation to the risk of cardiovascular disease: a review. J Thromb Haemost. 2011;9:899–908.

73. Sode BF, Allin KH, Dahl M, Gyntelberg F, Nordestgaard BG. Risk of venous thromboembolism and myocardial infarction associated with factor V Leiden and prothrombin mutations and blood type. CMAJ. 2013;185:E229–37.

74. Lalouschek W, Schillinger M, Hsieh K, Endler G, Tentschert S, Lang W, Cheng S, Mannhalter C. Matched case-control study on factor V Leiden and the prothrombin G20210A mutation in patients with ischemic stroke/transient ischemic attack up to the age of 60 years. Stroke. 2005;36:1405–9.

75. Lotta LA, Tuana G, Yu J, Martinelli I, Wang M, Yu F, Passamonti SM, Pappalardo E, Valsecchi C, Scherer SE, Hale W 4th, Muzny DM, Randi G, Rosendaal FR, Gibbs RA, Peyvandi F. Next-generation sequencing study finds an excess of rare, coding single-nucleotide variants of ADAMTS13 in patients with deep vein thrombosis. J Thromb Haemost. 2013;11:1228–39.

76. Knowles JW, Wang H, Itakura H, Southwick A, Myers RM, Iribarren C, Fortmann SP, Go AS, Quertermous T, Hlatky MA. Association of polymorphisms in platelet and hemo-stasis system genes with acute myocardial infarction. Am Heart J. 2007;154:1052–8.

77. Song C, Burgess S, Eicher JD, O'Donnell CJ, Johnson AD. Causal effect of plasminogen activator inhibitor type 1 on coronary heart disease. J Am Heart Assoc. 2017;6:pii: e004918.

78. Orho-Melander M. Genetics of coronary heart disease: towards causal mechanisms, novel drug targets and more personalized prevention. J Intern Med. 2015;278:433–46.

79. Humphries SE. Genetic regulation of fibrinogen. Eur Heart J. 1995;16(Suppl A):16–9.

80. de Maat MP. Effects of diet, drugs, and genes on plasma fibrinogen levels. Ann N Y Acad Sci. 2001;936:509–21.

81. Slowik A, Dziedzic T, Pera J, Figlewicz DA, Szczudlik A. Coagulation factor XIII Val34Leu polymorphism in patients with small vessel disease or primary intracerebral hemorrhage. Cerebrovasc Dis. 2005;19:165–70.

82. Ariens RA, Philippou H, Nagaswami C, Weisel JW, Lane DA, Grant PJ. The factor XIII V34L polymorphism accelerates thrombin activation of factor XIII and affects cross-linked fibrin structure. Blood. 2000;96:988–95.

83. Corral J, Gonzalez-Conejero R, Iniesta JA, Rivera J, Martinez C, Vicente V. The FXIII Val34Leu polymorphism in venous and arterial thromboembolism. Haematologica. 2000;85:293–7.

84. Dawson S, Hamsten A, Wiman B, Henney A, Humphries S. Genetic variation at the plasminogen activator inhibitor-1 locus is associated with altered levels of plasma plasminogen activator inhibitor-1 activity. Arterioscler Thromb. 1991;11:183–90.

85. Erikson P, Kallin B, Van'T Hooft FM, Bovenholm P, Hamsten A. Allele-specific increase in basal transcription of plasminogen activator inhibitor 1 gene is associated with myocardial infarction. Proc Natl Acad Sci U S A. 1995;92:1851–5.

86. Kanaji T, Okamura T, Osaki K, Kuroiwa M, Shimoda K, Hamasaki N, Niho Y. A common genetic polymorphism (46C to T substitution) in the 5′-untranslated region of the coagulation factor XII gene is associated with low translation efficiency and decrease in plasma factor XII level. Blood. 1998;91:2010–4.

87. Iacoviello L, Di Castelnuovo A, De Knijff P, D'Orazio A, Amore C, Arboretti R, Kluft C, Benedetta Donati M. Polymorphisms in the coagulation factor VII gene and the risk of myocardial infarction. N Engl J Med. 1998;338:79–85.

88. Poort SR, Rosendaal FR, Reitsma PH, Bertina RM. A common genetic variation in the 3′-untranslated region of the prothrombin gene is associated with elevated plasma prothrombin levels and an increase in venous thrombosis. Blood. 1996;88:3698–703.

89. Ceelie H, Spaargaren-van Riel CC, Bertina RM, Vos HL. G20210A is a functional mutation in the prothrombin gene; effect on protein levels and 3′-end formation. J Thromb Haemost. 2004;2(1):119–27.

90. Viel KR, Machiah DK, Warren DM, Khachidze M, Buil A, Fernstrom K, Souto JC, Peralta JM, Smith T, Blangero J, Porter S, Warren ST, Fontcuberta J, Soria JM, Flanders WD, Almasy L. Howard TEA sequence variation scan of the coagulation factor VIII (FVIII) structural gene and associations with plasma FVIII activity levels. Blood. 2007;109:3713–24.

91. Huffman JE, de Vries PS, Morrison AC, Sabater-Lleal M, Kacprowski T, Auer PL, Brody JA, Chasman DI, Chen MH, Guo X, Lin LA, Marioni RE, Müller-Nurasyid M, Yanek LR, Pankratz N, Grove ML, de Maat MP, Cushman M, Wiggins KL, Qi L, Sennblad B, Harris SE, Polasek O, Riess H, Rivadeneira F, Rose LM, Goel A, Taylor KD, Teumer A, Uitterlinden AG, Vaidya D, Yao J, Tang W, Levy D, Waldenberger M, Becker DM, Folsom AR, Giulianini F, Greinacher A, Hofman A, Huang CC, Kooperberg C, Silveira A, Starr JM, Strauch K, Strawbridge RJ, Wright AF, McKnight B, Franco OH, Zakai N, Mathias RA, Psaty BM, Ridker PM, Tofler GH, Völker U, Watkins H, Fornage M, Hamsten A, Deary IJ, Boerwinkle E, Koenig W, Rotter JI, Hayward C, Dehghan A, Reiner AP, O'Donnell CJ, Smith NL. Rare and low-frequency variants and their association with plasma levels of fibrinogen, FVII, FVIII, and vWF. Blood. 2015;126:e19–29.

92. Aiach M, Nicaud V, Alhenc-Gelas M, Gandrille S, Arnaud E, Amiral J, Guize L, Fiessinger JN, Emmerich J. Complex association of protein C gene promoter polymorphism with circulating protein C levels and thrombotic risk. Arterioscler Thromb Vasc Biol. 1999;19:1573–6.

93. Scopes D, Berg LP, Krawczak M, Kakkar VV, Cooper DN. Polymorphic variation in the human protein C (PROC) gene promoter can influence transcriptional efficiency in vitro. Blood Coagul Fibrinolysis. 1995;6:317–21.

94. Nurden AT. Polymorphisms of human platelet membrane glycoproteins: structure and clinical significance. Thromb Haemost. 1995;74:345–51.

95. Kunicki TJ. The influence of platelet collagen receptor polymorphisms in hemostasis and thrombotic disease. Arterioscler Thromb Vasc Biol. 2002;22:14–20.

96. Lewandowski K, Swierczyńska A, Kwaśnikowski P, Elikowski W, Rzeźniczak M. The prevalence of C807T mutation of glycoprotein Ia gene among young male survivors of myocardial infarction: a relation with coronary angiography results. Kardiol Pol. 2005;63:107–13.

97. Herrmann SM, Ricard S, Nicaud V, Mallet C, Evans A, Ruidavets JB, Arveiler D, Luc G, Cambien F. The P-selectin gene is highly polymorphic: reduced frequency of the Pro715 allele carriers in patients with myocardial infarction. Hum Mol Genet. 1998;7:1277–84.

98. Volcik KA, Ballantyne CM, Coresh J, Folsom AR, Wu KK, Boerwinkle E. P-selectin Thr715Pro polymorphism predicts P-selectin levels but not risk of incident coronary heart disease or ischemic stroke in a cohort of 14595 participants: the Atherosclerosis Risk in Communities Study. Atherosclerosis. 2006;186(1):74–9.

99. Gremmel T, Kopp CW, Steiner S, Seidinger D, Ay C, Koppensteiner R, Mannhalter C, Panzer S. The P-selectin gene Pro715 allele and low levels of soluble P-selectin are associated with reduced P2Y12 adenosine diphosphate receptor reactivity in clopidogrel-treated patients. Atherosclerosis. 2011;217:135–8.

12

100. Subramanian H, Gambaryan S, Panzer S, Gremmel T, Walter U, Mannhalter C. The Thr715Pro variant impairs terminal glycosylation of P-selectin. Thromb Haemost. 2012;108:963–72.
101. Oparil S, Acelajado MC, Bakris GL, Berlowitz DR, Cífková R, Dominiczak AF, Grassi G, Jordan J, Poulter NR, Rodgers A, Whelton PK. Hypertension. Nat Rev Dis Primers. 2018;4:18014.
102. Costa A, Franco OL. Impact and influence of "omics" technology on hypertension studies. Int J Cardiol. 2017;228:1022–34.
103. Vehaskari VM. Heritable forms of hypertension. Pediatr Nephrol. 2009;24:1929–37.
104. Russo A, Di Gaetano C, Cugliari G, Matullo G. Advances in the genetics of hypertension: the effect of rare variants. Int J Mol Sci. 2018;19:pii: E688.
105. Johnson T, Gaunt TR, Newhouse SJ, et al. Blood pressure loci identified with a gene-centric array. Am J Hum Genet. 2011;89:688–700.
106. Natekar A, Olds RL, Lau MW, Min K, Imoto K, Slavin TP. Elevated blood pressure: our family's fault? The genetics of essential hypertension. World J Cardiol. 2014;6:327–37.
107. Fogari R, Zoppi A. Antihypertensive drugs and fibrinolytic function. Am J Hypertens. 2006;19:1293–9.
108. Takeuchi F, Yamamoto K, Katsuya T, Sugiyama T, Nabika T, Ohnaka K, Yamaguchi S, Taka-yanagi R, Ogihara T, Kato N. Reevaluation of the association of seven candidate genes with blood pressure and hypertension: a replication study and meta-analysis with a larger sample size. Hypertens Res. 2012;35:825–31.
109. Maes HH, Neale MC, Eaves LJ. Genetic and environmental factors in relative body weight and human obesity. Behav Genet. 1997;27:325–51.
110. Apalasamy YD, Mohamed Z. Obesity and genomics: role of technology in unraveling the complex genetic architecture of obesity. Hum Genet. 2015;134:361–74.
111. Choquet H, Meyre D. Genetics of obesity: what have we learned? Curr Genomics. 2011;12:169–79.
112. Speakman JR, Rance KA, Johnstone AM. Polymorphisms of the FTO gene are associated with variation in energy intake, but not energy expenditure. Obesity (Silver Spring). 2008;16:1961–5.
113. Wardle J, Carnell S, Haworth CM, Farooqi IS, O'Rahilly S, Plomin R. Obesity associated genetic variation in FTO is associated with diminished satiety. J Clin Endocrinol Metab. 2008;93:3640–3.
114. Mottillo S, Filion KB, Genest J, Joseph L, Pilote L, Poirier P, Rinfret S, Schiffrin EL, Eisenberg MJ. The metabolic syndrome and cardiovascular risk a systematic review and me-ta-analysis. J Am Coll Cardiol. 2010;56:1113–32.
115. Rask-Andersen M, Almén MS, Olausen HR, Olszewski PK, Eriksson J, Chavan RA, Levine AS, Fredriksson R, Schiöth HB. Functional coupling analysis suggests link between the obesity gene FTO and the BDNF-NTRK2 signaling pathway. BMC Neurosci. 2011;12:117.
116. Perry JR, Frayling TM. New gene variants alter type 2 diabetes risk predominantly through reduced beta-cell function. Curr Opin Clin Nutr Metab Care. 2008;11:371–7.
117. Frayling TM, Timpson NJ, Weedon MN, Zeggini E, Freathy RM, Lindgren CM, Perry JR, Elliott KS, Lango H, Rayner NW, Shields B, Harries IW, Barrett JC, Ellard S, Groves CJ, Knight B, Patch AM, Ness AR, Ebrahim S, Lawlor DA, Ring SM, Ben-Shlomo Y, Jarvelin MR, Sovio U, Bennett AJ, Melzer D, Ferrucci L, Loos RJ, Barroso I, Wareham NJ, Karpe F, Owen KR, Cardon LR, Walker M, Hitman GA, Palmer CN, Doney AS, Morris AD, Smith GD, Hattersley AT, McCarthy MI. A common variant in the FTO gene is associated with body mass index and predisposes to childhood and adult obesity. Science. 2007;316(5826):889–94.
118. Andreasen CH, Stender-Petersen KL, Mogensen MS, Torekov SS, Wegner L, Andersen G, Nielsen AL, Albrechtsen A, Borch-Johnsen K, Rasmussen SS, Clausen JO, Sandbaek A, Lauritzen T, Hansen L, Jørgensen T, Pedersen O, Hansen T. Low physical activity accentuates the effect of the FTO rs9939609 polymorphism on body fat accumulation. Diabetes. 2008;57:95–101.

# Animal Models in Cardiovascular Biology

*Helga Bergmeister, Ouafa Hamza, Attila Kiss, Felix Nagel, Patrick M. Pilz, Roberto Plasenzotti, and Bruno K. Podesser*

© Springer Nature Switzerland AG 2019
M. Geiger (ed.), *Fundamentals of Vascular Biology*, Learning Materials in Biosciences,
https://doi.org/10.1007/978-3-030-12270-6_13

## What You Will Learn from This Chapter

Animal models have significantly contributed to our understanding of vascular biology and cardiac function over the last century. In fact, pioneering experimental models such as the isolated heart preparation according to Oskar Langendorff have paved the grounds for modern physiology of the heart [1]. Similarly, with the help of experiments on the myograph, nitric oxide (NO) was discovered and introduced into modern vascular biology [2]. According to a survey of living Nobel laureates, 97% responded that animal experiments have been vital to the discovery and development of many advances in physiology and medicine, and 92% felt strongly that animal models are still crucial to the investigations and developments of many medical treatments [3].

This chapter will focus on a general description of what an animal model is and what are the legal requirements to perform animal experiments. In short, a history of animal used in biomedical research will be presented. Finally, we will present the most commonly used small and large animal models in cardiovascular biology. It is important to stress that the chapter will direct its attention on the use of vertebrate animals only.

## 13.1 What Is an Animal Model?

Already in 1620 Francis Bacon proposed a process of scientific discovery based on a collection of observations, followed by a systematic evaluation of these observations in an effort to demonstrate their truthfulness. Thereby he set the tenets for modern hypothesis-driven research [4]. It can be argued that hypothesis testing is an inefficient mechanism for discovery – however, it generally produces meaningful and most important reproducible results and therefore is a cornerstone of modern science.

There are many types of different models in biomedical research, e.g., in vitro assays, computer simulation, mathematical models, and animal models. In general, every model serves as a surrogate and is not necessarily identical to the subject being modeled. We assume in this chapter that the human biological system is the subject being modeled. In general, animals may model one process to another (analogous processes) or reflect the counterpart of a genetic sequence (homologous processes). While the first does not causally link genotype to phenotype, the second does. The most widely used homologous model is the genetically manipulated mouse. Another useful concept in modeling concerns one-to-one modeling versus many-to-many modeling. In the first the model is reflecting a similar phenotype to that which is being modeled. Examples of one-to-one modeling include many infectious and monogenetic diseases. In contrast, many-to-many modeling results from a process in an organism at different level, e.g., subcellular, cell, tissue, organ, or system. Examples of many-to-many modeling include complex diseases such as cancer, obesity, cardiovascular disease, etc. Most of these are polygenetic with environmental influences. Therefore, the many-to-many model is more commonly used and requires the use of multiple model systems including computational modeling and in vitro and in vivo modeling followed by population-based studies. The high-throughput techniques such as next-generation sequencing and omics technologies will further facilitate this process.

On the following pages, animal models will be classified as spontaneous or induced. Spontaneous models may be normal animals with phenotypic similarities to those of humans or abnormal members of a species that arise through spontaneous mutations. In contrast, animals submitted to surgical, genetic, chemical, X-ray, or other manipulations resulting in an alteration to their normal physiology are induced models.

## 13.2  What Are the Legal Requirements to Perform an Animal Experiment?

Our current concept of animal experiments is based on the fundamental work of W.M.S. Russel and R.L. Burch from 1959 with the inspiring title *The Principles of Humane Experimental Technique* [5]. In this small book, the authors define what is today the integral part of all national and international directives or laws dedicated to animal experiments: the *3R principles replace, reduce, refine* [6]. Each European country had to follow this directive and implemented national laws [7]. They describe in detail what type of animal experiments is allowed and which is not. They also clarify the requirements for persons who want to perform animal experiments and who want to breed or house animals. The prerequisite for starting experiments is the submission of a respective research proposal to the local or institutional animal welfare commission. This proposal should be hypothesis-driven, describing the state-of-the-art status of the respective scientific problem, and give detailed information about experimental procedure and severity of burden to the animals. After permission the experiments can be started at a certified institution. Recently, the *3Rs* have been expanded by a *4th R* representing *respect*. This indicates that our ethical position is further developing toward empathy for our mammalian cousins [8], devoting their lives for ours.

## 13.3  Short History of Animal Experiments [9]

Domestication goes back to the Neolithic revolution and includes buffalo, cattle, sheep, and dog. The first scientific approaches were made in ancient Greece. In a climate of scepticism, all natural phenomena were questioned and not explained by mysticism and demonology any more. Lighthouses such as Plato, Aristotle, and Hippocrates paved the way for modern philosophy and medicine. In 304 BC the anatomist Erasistratus observed the relation between food intake and weight gain in birds. Galen in the second-century AD showed that arteries contain blood. He was among those physicians, who carried out careful anatomical dissections of animals and humans. The information gained was remarkably precise and described the anatomical conditions of the, e.g., eye and the optic nerve and its connection to the brain. For centuries these observations have been unparalleled.

The Greeks and Romans used this information also in medical schools and trained skilled doctors and surgeons (who were not considered doctors at that time) for civil and military purposes. With the decline of the Roman Empire, also this knowledge disappeared, was kept alive only in the Islamic world, and found back to Europe via Salerno and its medical school in the tenth century. However, dissections of human corps were forbidden until the thirteenth century. Instead animals were dissected, and it is therefore no wonder that Leonardo da Vinci's drawings of the heart show not a human but the heart of an ox.

With the age of enlightenment, medical science made a tremendous step forward. In part because parallel to medicine, new technical developments such as the invention of the microscope by Leeuwenhoek went hand in hand. Using his sophisticated instruments, he was able to confirm what Malpighi had described earlier, the circulation of red blood cells in the rabbits' ear. Also W. Harvey's description of the circulation of blood in 1628 was based on experimental work in animals. By the middle of the eighteenth century, Priestly had discovered that the life-promoting constituent of air was oxygen. Hales made the first recordings of blood pressure in a horse in 1733. Crawford was the first to measure the metabolic heat of an animal using water calorimetry in 1788. In 1815 Laennec had developed the stethoscope using animals.

The nineteenth-century developments in medicine can be summarized as a common fight against infections. The French scientist Louis Pasteur was the first to realize that microscopic particles, he called them vibrions, were the cause of a fatal disease in silkworms. By eliminating these vibrions, he cured the worms. Consequently, he started to look for other causative agents to proof his germ theory of disease. By isolating these agents and processing them with high temperature, he was able to show that reinjection of these attenuated organism would protect against the disease. Pasteur referred to this process as vaccination in homage to the English surgeon Edward Jenner, who discovered that injection of matter from cowpox lesions into humans protects against smallpox. In Germany, Robert Koch developed in vitro culture of bacteria, thereby reducing the number of animals for research. His postulates that a specific agent is responsible for a specific disease are still valid and led Koch to the discovery of the *Mycobacterium* tuberculosis. As a consequence, he developed tuberculin to identify infected animals and people.

For our current understanding of biomedical developments, two other important discoveries are at least as important. In 1859, the English naturalist Ch. Darwin published *On the Origin of Species*, in which he hypothesized that all life evolves by selection of traits that give one species an advantage over the other. This hypothesis stands behind the current understanding that fish and reptiles are common ancestors of birds and mammals or in the words of E. Haeckel from 1866 that the ontogenesis displays the phylogenesis. In the same year, G. Mendel demonstrated that specific traits are inherited in a predictable way. This was the birthday of modern genetics.

At the beginning of the twentieth century, the genetic code was still not known. Researchers such as A. Kossel discovered the nucleic acids in salmon sperm and human leukocytes. Later P. Levine discovered the nucleotides before J. Watson and F. Crick were able to describe the double helix structure of DNA.

## 13.4 Surgically Induced Cardiac Small Animal Models

The understanding of various human pathologies and their treatment has developed over the past decades. One of the most powerful reasons for that progress was the upcoming use of animal models copying human syndromes [10]. In cardiovascular research, animal models of acute or chronic myocardial infarction, cardiac hypertrophy, and heart failure are very common, but also futuristic interventions like heterotrophic heart transplantations can be performed [10]. Therefore, different small rodents or mammalians are commonly used, like mice, rats, guinea pigs, hamsters, and rabbits [10]. From anatomical and physiological view, each species has its pros and cons, but this way of research is the most comparable one to human settings. About 90% of genes and many organs like the heart, lungs, brain, or liver are shared between humans and mammalians [11]. Via genetic modification it is even possible to mimic human conditions even closer. Depending on the scientific issue of interest and research, it is necessary to replicate human diseases in detail. For that reason, there are multiple surgical procedures and techniques to achieve comparable results.

For detailed descriptions, see also the reviews by Patten et al. in 2009, Zaragoza et al. in 2011, Ramzy et al. in 2005, and Camacho et al. in 2016 [12–15].

### 13.4.1  Myocardial Infarction: Acute MI

(Acute) MI is still the no. 1 contributor to mortality and morbidity worldwide [16]. Although therapies and drugs are improving, further research needs to be done to find treatments against subsequent events like restenosis, remodeling of the myocardium, heart failure, or even death [17]. Surgical models are commonly used and are not limited to a special species. Mice, hamster, guinea pigs, rats, and even rabbits are applied. The largest number of interventions is performed in rats but also mice; even neonates are popular [10, 15].

By inducing an ischemic injury to the myocardium using surgical occlusion of a coronary vessel, with or without following reperfusion, it is possible to mimic acute MI [18, 19]. Therefore, the chest and pericardium are opened. Then the left anterior descending artery [20] is identified and either ligated permanently (developing chronic MI, leading to HF) or temporarily (mimic an acute infarction with subsequent reperfusion) using a tourniquet. This procedure allows to evaluate inflammation, fibrosis, degradation of ECM, apoptosis, as well as necrosis and remodeling after MI [15]. Pfeffer et al. was one of the first who established this model to show the correlation of infarction size, LV chamber dimensions, and LV function [19]. A pitfall or limitation of this model is the variability of the coronary vessels in mice [15, 21]. While an ischemia without reperfusion is a good opportunity to learn more about pathological mechanisms of anaerobic conditions and their treatment, a setting with subsequent reperfusion presents a chance to get a better understanding of reoxygenation and reperfusion injury.

### 13.4.2  Heart Failure (HF) and Chronic MI

Worldwide, the number of patients presenting HF is about 26 millions, but due to the improvement of treatments and drugs, incidence will be stable, while prevalence is increasing due to higher age and comorbidities of the patients [22]. Therefore, new strategies and therapeutic approaches need to be evaluated. This model is similar to the acute MI model. Rats and mice are the most common used species. As explained above a chronic myocardial infarction is achieved by a permanent ligation of the coronary artery without subsequent reperfusion. This leads to a distinct fibrosis and scar tissue, inflammation, and loss of contractility, resulting in structural, hemodynamic, and molecular changes of the ventricle. A stable phase of scar formation is reached after about 2–3 weeks [23].

### 13.4.3  Cardiac Hypertrophy and Hypertension

In 2015, 1.13 billion people were suffering from hypertension worldwide. Hypertension often leads to other cardiovascular diseases such as cardiac hypertrophy and finally to heart failure. Animal models are commonly performed in mice and rats. There are various surgical techniques to induce pressure overload, respectively arterial hypertension and to achieve cardiac hypertrophy [12]. Rockman et al. introduced the so-called transverse aortic constriction (TAV) in mice, and Litwin et al. first described a procedure called ascending aortic banding (AAB) in rats [24, 25]. Both of these procedures were performed to obtain a constriction of the aorta, in different sections, mimicking an aortic valve stenosis (LV outflow narrowing) or an increase of afterload, resulting in hypertrophy of the LV and arterial hypertension. A ligation is tied around the aorta and a standardized 27G cannula and fixed with knots. The cannula is removed quickly and so a uniform stenosis is achieved in all animals.

In the so-called debanding – procedure describing the removal of the constricting ligature after a defined time – the reverse remodeling can be modeled [26].

## 13.4.4  Heterotopic Heart Transplantation (HTX)

HTX is a standard surgical procedure of terminal HF. According to the Global Observatory on Donation and Transplantation (GODT), there were about 7000 heart transplantations in 2015 worldwide – number increasing [27]. Although this treatment improves constantly and shows improving survival rates, many underlying mechanisms of immune rejection response or tolerance, cardiac allograft injury, and post-HTX infections are not fully understood yet [28]. Therefore, mice models are mostly common to mimic the clinical situation. First techniques were published by Drs. Corry *and* Russell in 1973 [29, 30]. Due to better understanding of patho-mechanisms and new opportunities as knockout mice, the interventions have been modified and improved as well [31]. The heart is transplanted heterotopically either into the abdomen of the mouse or rat. The donor heart is anastomosed with the abdominal aorta and the inferior vena cava [31]. This is technically extremely challenging. Alternatively, the heterotopic transplantation can also be performed to the cervical vessels. In this case, the aorta and the main pulmonary artery of the donor were connected to the external jugular vein and common carotid artery of the recipient [32].

## 13.4.5  Conditioning Models

Despite improvements in medical therapy, acute myocardial infarction (MI) remains a major health problem worldwide with an incidence in Austria of 300 myocardial infarctions per 100.000 inhabitants per year [33]. Rapid restoration of coronary blood flow by means of either primary percutaneous intervention or thrombolysis is standard treatment for patients with acute ST-elevation myocardial infarction. However, reperfusion of the jeopardized myocardium results in a cascade of harmful events, referred to as myocardial ischemia reperfusion injury, which largely contribute for the development of congestive heart failure [34]. Thus, there is an emerging need to identify clinically feasible and cost-effective therapeutic approaches to limit myocardial IRI and improve cardiac function in patients with MI.

### 13.4.5.1  Local and Remote Ischemic Conditioning

In 1986 Murry et al. [35] first demonstrated that multiple brief episodes of IR (usually 5 min) applied on the left anterior descending coronary artery in dogs prior to the sustained more prolonged coronary artery occlusion protected the heart against IRI. This phenomenon is termed ischemic preconditioning (IPC). Later, the conditioning concept moved toward to postconditioning (IPOSTC), which refers to the ability of a series of brief occlusions of either the coronary arterial circulation after a severe ischemic insult to protect against ischemic reperfusion injury of the myocardium [36]. In the following years, IPOSTC was further enhanced by developing the idea that the conditioning stimulus could also be applied to a distant organ or tissue, which is called remote ischemic conditioning (RIC) [37]. Remote ischemic conditioning, defined as nonlethal IR insult in distal organs or site (forearm, hind limb), protects the heart against the subsequent acute myocardial IR injury as well as reduced infarct size. The efficacy of RIC on reduction myocardial infarct size and increase myocardial salvage in humans was demonstrated in a pioneering study [38]. The mechanisms underlying ischemic conditioning have not been completely elucidated. Considerable progress has been made toward the identification of potential triggers, intracellular signaling pathways, and end effectors involved in ischemic conditioning. There are numbers of studies suggesting the potential role for endogenous paracrine mediators (adenosine, acetylcholine, bradykinin, endothelin, opioids, and reactive oxygen species), released during the brief period of IR and acting on receptors, as triggers of the cardioprotective effects of ischemic conditioning [39].

In clinical scenario, effect of remote ischemic preconditioning on clinical outcomes in *patients undergoing coronary artery bypass graft surgery* (ERICCA) phase III clinical trial including 1612 patients failed to prove evidence of the cardioprotective efficacy of RIC [40]. In another recent randomized controlled trial including 516 patients confirmed the cardioprotective effect of RIC in patients with *ST-elevation MI*, supporting the clinically usefulness of RIC [41]. Therefore, RIC should be considered a potential therapeutic strategy for limiting MI size and adverse left ventricular remodeling in patients with MI. However further studies are warranted to prove its cardiac and vascular protective efficacy.

## 13.5  Diet-Induced Cardiac Small Animal Models

### 13.5.1  The Dahl Salt-Sensitive and Salt-Resistant Rat

The Dahl salt-sensitive (DS) rat is an established experimental model of hypertrophy [42]. Since the DSl rat has a mutation of the alpha1-Na-K-ATPase, sodium excretion from the myocytes is impaired. DS rats develop hypertension with low renin and aldosterone levels and compensated LVH. The Dahl salt-resistant (DR) rat develops neither hypertension nor hypertrophy and serves as age- and sex-matched control group. To induce hypertension, both strains have to be fed with high-salt diet (7.8% NaCl) and water ad libitum for 4 weeks [43]. If rats are fed for more than 12 weeks, progression to heart failure can be observed.

## 13.6  Genetically Induced Cardiac Small Animal Models

In contrast to the high prevalence of atherosclerosis among humans, none of the common wild-type laboratory rodent strains spontaneously develop atherosclerotic plaques even when they are maintained on diets similar to those consumed by humans in western society (high-fat western diet, 21% fat and 0.15% cholesterol); the normal chow of a mouse is 4.5% fat and 0.022% cholesterol. During the last decades, technologic advancement in the genetic modifications in mice generated multiple models, which exhibit pathologic conditions comparable to human cardiovascular disease, like the genesis of atherosclerotic lesions of mice lacking APO lipoprotein E or lacking low-density lipoprotein. In these genetically modified animals, the gene of interest related to a specific physiological or pathological question is overexpressed, silenced, or deleted [44]. Consequently, different rodent strains are established as models for complex cardiovascular research. Despite the generation of these new models, little is known about the physiologic cardiovascular condition in mice or potential differences that may exist between strains of mice.

For example, the normal blood pressure (BP) and heart rate (HR) fluctuate throughout the day. Mice are nocturnal active animals with a diurnal rhythm of BP and HR peak values occurring during the middle of light and dark period. The reference level of heartbeats per minute is set between 350 and 450 bpm during sleep and rises up to 750–800 bpm after a placement in a different cage with the corresponding systolic pressure from 102 to 140 mmHg [45, 46].

The average plasma cholesterol level of wild-type mice on a regular mouse chow diet is approximately 80 mg/dl, and most of this cholesterol is carried out by high-density lipoproteins, in contrast to humans, where LDL removes the plasma cholesterol.

### 13.6.1  The ApoE Knockout Mouse

The first models of atherosclerosis with a defective gene coding for apolipoprotein E were developed by Nobuyo Maeda and Jan Breslow in 1992 [47–50]. ApoE plays a central role in the lipoprotein metabolism and is required for the efficient receptor-mediated plasma clearance of chylomicron remnants and VLDL remnant particles by the liver [51, 52]. The lack of ApoE in the mutant mice results in an increased accumulation of cholesterol-enriched remnant particles and an elevation of plasma cholesterol levels to about 400 mg/dl even on a regular mouse chow. Most of the plasma cholesterol is found in the atherogenic lipoprotein fractions, namely, the very low-density lipoprotein (VLDL), intermediate-density lipoprotein (IDL), and low-density lipoprotein (LDL) fractions; this profile can easily be aggraded by the use of a "western diet" [53]. The apoE-deficient mice spontaneously develop aortic arteriosclerotic plaques similar to those seen in humans [54, 55].

Despite the models for atherosclerosis like the ApoE mouse, there are models with a targeted deletion of muscle protein to mimic cardiomyopathic diseases.

### 13.6.2  Muscle LIM Protein Knockout Mouse

One model is the muscle LIM protein (MLP) knockout mouse. Total deletion of the muscle LIM protein in the MLP −/− mice led to the development of fatigue after postnatal day 5 in 50–70% of the young mice. All these mice die within 20–30 hours from the onset of these symptoms [56]. The adult phenotype shows severe defects in cardiac function and structure. This cardiac phenotype in the adult MLP-deficient mice reproduces the clinical features of cardiomyopathy and heart failure in humans [12, 57, 58].

### 13.6.3  Muscular Dystrophy Mice, Rats, and Rabbits

The muscular dystrophies are rare diseases with progressive muscle weakness and cycles of muscle necrosis [59]. Among the various dystrophy types, Duchenne muscular dystrophy (DMD) is a progressive, fatal, X-linked monogenic muscle disorder that results in progressive muscle wasting and fibro-fatty replacement of muscle. Boys with DMD (approximately 1:3500) usually show motor difficulties by early age of life (>6). Thereafter, muscle weakness progresses and leaves patients wheelchair-bound by their teens. Death usually occurs before DMD patients reach 40 years of age. The gene defect for DMD was mapped to an X chromosome gene that encodes for the intracellular protein dystrophin. Of importance, besides skeletal muscle degeneration, DMD is also often associated with severe cardiovascular complications including cardiomyopathy development, cardiac arrhythmias, and vascular dysfunctions [60]. Of importance, the incidence of cardiomyopathy in DMD patients surviving into their third decade is nearly 100%, which significantly contributes to mortality and morbidity. A number of various approaches are being taken for the development of targeted therapies for DMD disorders, with the aim of preventing disease progression or reversing some of the disease-associated pathology. However, compounds can be screened in cell and tissue cultures; the use of relevant animal models is key to understanding the potential efficacy of different DMD therapeutic approaches.

Different small animal models have been extensively used to dissect disease mechanisms of cardiac and skeletal muscle disorders in DMD and to test therapeutic strategies. The most widely used animal model of DMD is the mdx mouse which has a nonsense mutation in exon 23 of the Dmd gene [61]. In addition, several other strains with different Dmd mutations, including a targeted deletion of exon 526, have been developed [61]. Furthermore, mdx mice have been crossed with utrophin knockout (KO) mice to produce a double KO, which has a more severe dystrophic phenotype (within 8–10 weeks of age) than the mdx mouse [62].

More recently, a novel rat model of DMD with a progression of cardiac and skeletal muscle dysfunction has been demonstrated using TALENs targeting exon 23 [63]. Intriguingly, a recent study published a description of a rabbit model for DMD using a CRISPR-Cas9 to target exon 51 of the Dmd gene to ablate dystrophin expression in a New Zealand rabbit [63].

## 13.7 Surgically, Diet-, or Genetically Induced Cardiac Large Animal Models

Large animal models provide the researcher with a more similar anatomical and physiological testing platform compared to rodents. Heart rate, adrenergic receptor ratios, oxygen consumption, and response to loss of regulatory proteins in mice were found to be different to humans as well as the contractile protein expression [64, 65].

### 13.7.1 Myocardial Infarction and Ischemic Cardiomyopathy

#### 13.7.1.1 Coronary Artery Ligation

Coronary artery ligation models can be useful to study the resulting myocardial infarction as well as heart failure mechanisms secondary to ischemic cardiomyopathy. Ischemic/reperfusion lesions can be applied using a suture around the coronary artery that is fixed with a tourniquet and released after a defined protocol. The first ligation model was established in dogs [66] and was used to study different drug therapies in ischemic dilated cardiomyopathy. The downside of the dog ligation models is the abundance of collateral coronary network resulting in a small and non-reproducible infarct size on top of the high mortality rate of more than 50 due to arrhythmias during the acute phase.

A dilated ischemic cardiomyopathy was established in dogs by repetitive microembolization over 10 weeks [67] and was used to explore different medical regiments for heart failure as well as the neurohormonal activation in this setting.

Although canine models were the first to be used for preclinical testing, their limitations because of the collateral network encouraged exploration of other species like pigs and sheep.

Besides open chest animal models (surgical models), catheter-based models started to slowly be used. These models are more convenient because they are less invasive and allow reoperation of the animals to test devices without facing adherence complications as seen in open chest models.

Total occlusion of the coronary arteries was achieved by either coil deployment [68], autothrombus injection, microsphere embolization [69], or ethanol injection [70]. Ischemic/reperfusion injuries were established by inflating a balloon for 30–90 minutes then retrieving it [71].

### 13.7.1.2  Hydraulic Occluder and Ameroid Constrictor

They are used in larger animals allowing for complete or partial occlusion of coronary arteries. These models allow to study myocardial infarction as well as hibernating myocardium. They are used as well for heart failure studies given the ischemic cardiomyopathy that they result in [20, 72].

### 13.7.1.3  Atherosclerosis

Reproduction of atherosclerosis in large animals was tempted in pigs. Although these models were more suitable to study coronary lesions rather than ischemic myocardial dysfunction because of the long time needed to develop a plaque big enough to produce a hemodynamically significant stenosis, pigs and nonhuman primates also share similar characteristics of lipoprotein metabolism including cholesterol distributions and enzymatic activities. Furthermore they are more comparable to humans as both pigs and nonhuman primates are omnivorous and diurnal. Elderly farm pigs and nonhuman primates even develop spontaneous atherosclerosis [73, 74], but for ethical issue, nonhuman primates are restricted.

## 13.7.2  Muscular Dystrophy in Large Animals

The pig and dog are an attractive option for translational studies, as they have a very similar size and anatomy to humans and could thus be a valuable test of potential problems when scaling up a therapy first tested in mice before translating it for use in humans. There are numerous studies describing genetically modified pigs (delete exon 52 of the porcine *Dmd* gene) and dogs (mutations of Dmd gene were identified in golden retriever) to better understand the disease mechanism of DMD [63]. However, large animal such as pig with DMD dies prematurely, often in the first week of life, and survival appears to largely depend on the level of utrophin expression.

### 13.7.3   Valvular Heart Disease

#### 13.7.3.1   Mitral Regurgitation

Although spontaneous myxomatous mitral regurgitation can be used for mechanistic insights like in dogs [75], different provoked mitral regurgitation models were established. Chordae tendineae rupture was used either surgically or percutaneously to produce a primary mitral regurgitation [76–78]. Other models used another approach which consisted of creating a hole in the mitral leaflet [79]. Ischemic mitral regurgitation was also achieved by occluding marginal [80] branches in an ovine model. A combination of both mechanisms has been established in an ovine model, where chordal cutting and ameroid constrictor was placed on the marginal branches resulting in a severe heart failure model both ischemic and volume overload reproducing [81].

#### 13.7.3.2   Aortic Stenosis

Besides high-fat diet model, other aortic stenosis models were attempted. Under cardiopulmonary bypass aortic valve annulus and leaflets were injected with cyanoacrylate to create an acute aortic stenosis in minipigs [82]. A supravalvular stenosis mimicked by a banding surgery was established [83] in order to study the effect of an ejection obstacle on the left ventricle as a pressure overload model. Although this model does not represent an aortic stenosis model *stricto sensu*, it allows to better understand the consequence of volume overload and identify the timing for surgery where the myocardium damage is still reversible.

## 13.8   Animal Models for Vascular Graft Testing and Vein Graft Disease

Coronary artery disease (CAD) is one of the main causes of morbidity and mortality and is predominantly induced by atherosclerotic changes of the supplying blood vessels of the human heart [84]. Significant advances have been made in the last decades to immediately restore the patency of obstructed coronary vessels utilizing catheter-based interventional techniques. However, surgical revascularization using coronary artery bypass grafting (CABG) is still mandatory when interventional therapy is not successful in the long term or when multi-coronary vessels are diseased.

Autologous tissue is the preferred material for small diameter vessel reconstructions because of low thrombogenicity and favorable biomechanical properties. In the clinics, vascular substitutes derived from saphenous vein, internal thoracic artery, and radial artery are currently applied to bypass diseased vascular segments of patients [85]. Grafts derived from radial artery and internal thoracic artery show sufficient outcomes for coronary artery bypass procedures with high patency rates of more than 90% after 10 years of implantation. They are superior to vein grafts with an average patency of 50%. Vein grafts, which are the most common bypass materials used, are often susceptible to thrombosis, atherosclerosis, and intimal hyperplasia due to tissue injury during harvesting and after graft arterialization. The limited life expectancy of these grafts implies that patients must be interventionally or surgically retreated. To improve the performance of venous replacement materials, efforts have been made to understand the pathophysiological processes which lead to graft failure in the acute, subacute, and late postoperative period [86]. Preclinical models are therefore essential in mimicking the vasculature of human patients to provide new insights into vein graft disease and to evaluate new therapeutic strategies which increase long-term patency [87]. In this regard, it has been shown that perioperative antiplatelet therapy is successful to reduce thrombogenic events and that the process of atherosclerosis can be successfully diminished by intensive lipid-lowering therapy. Different therapeutic tools have been further evaluated in animals to prevent the development of intimal hyperplasia in vein grafts, which is caused by the loss of the endothelial layer during harvesting and by the remodeling process of the venous wall to resist high arterial pressures.

However, the autologous approach is often failing in coronary patients because of insufficient vessel quality due to comorbidities or previous vessel harvesting. Clinically available synthetic substitutes reveal inadequate in vivo performance because of innate thrombogenicity and intimal hyperplasia development. Therefore, important efforts have been made in designing new prosthetic vascular substitutes which should be equivalent to autologous vessels. Tissue-engineered vascular grafts show potential to overcome the limitations of synthetic artificial vascular grafts by matching the biological activity and biomechanical properties of native vessels [88]. However, these conduits have not been approved for routine use, and animals serving as nonhuman models are necessary to assess the performance of the newly designed coronary bypass conduits [89].

Successful in vivo performance of bypass grafts in humans is often limited by three main factors: the inability to endothelialize spontaneously prosthetic surfaces, the tendency to hypercoagulability, and the development of neointimal hyperplasia. Animal models, which shall mimic very closely the human patient, should reflect these limitations. Further considerations, which are closely associated with the respective conduit, are the localization of the implant and the length and diameter of the host vessel at the anastomotic site. To depict the human situation, the blood flow at the implantation site should be carefully considered for the evaluation of small-caliber coronary bypass grafts because most of the implants in humans will be implanted in with low blood flow regions susceptible to thrombotic events. Host vessels and coronary conduits should be matched closely regarding their wall thickness and diameter. Trans-anastomotic endothelialization in animals is highly accelerated (up to more than ten times) when compared to humans. Therefore, the length of the implant should be chosen appropriately to assess endothelialization for the clinical application. Paired-site implantation with suitable controls should be used for the evaluation of graft performance to exclude effects due to the variability between recipients. Especially for the evaluation of tissue-engineered conduits including human tissue, the immunogenicity of the host has to be carefully considered. Further considerations concern the costs of the preclinical trial, the availability of the animal model, and the compliance of the animals to different investigation procedures. Last but not least ethical considerations regarding the relevant animal species are very important.

### 13.8.1  Small Rodents

Rats and mice are often used to simulate humane disease pathophysiology and screen potential medications and candidate materials. Genetically modified strains and immune-incompetent animals are available, and the costs of experiments are low even when large numbers of animals are used. However, small rodent models have limitations regarding their size, life expectancy, and comparative cardiovascular physiology and hemostasis mechanisms. Mouse models can be advantageous for the study of pathophysiologic mechanisms because there are many genetically modified strains with targeted mutations in cardiovascular disease available. They develop atherosclerotic lesions, which are very similar to humans and can be generated within a short period of time. Immune-deficient strains (scid, scid beige, nude mice) are further favorable for the evaluation of human tissue containing designed tissue-engineered conduits. The small vessel size of mice makes the insertion of conduits challenging. Rats are much more preferable than mice because of their greater body weight and larger vessel size. They are physiologically and genetically more similar to humans than mice. Bypass grafts have been implanted in both species into the aorta, carotid, or femoral artery.

### 13.8.2    Rabbits

The New Zealand White rabbit is often used in biomedical research because of reasonable costs for purchase and maintenance. Models with heritable disease (hypercholesteremic Watanabe rabbit) are available as in- and outbred strains. In healthy animals, hypercholesteremic status can be induced by feeding high cholesterol diets. However, the number of genetically modified animals is low. The size of blood vessels of the rabbit is much more appropriate for graft insertion than in small rodents, and comparative physiology regarding hemostasis and endothelialization is more similar to humans than in rats and mice. Grafts have been applied into the carotid artery, the femoral artery, and the aorta.

### 13.8.3    Large Animal Models

Large animal models like sheep, pigs, dogs, and nonhuman primates resemble closer human vasculature anatomy and physiology and thus enable the implantation of appropriately sized vascular grafts. Sheep and nonhuman primates show greater similarity to humans regarding their thrombogenicity mechanisms and endothelialization behavior. Pigs and nonhuman primates have similarities with humans regarding their lipoprotein metabolism and cholesterol levels. Large animal models with heritable disease or experimentally induced pathologies are available to investigate implants under conditions which are similar to human cardiovascular disease.

Dogs have been the most used animal model in cardiovascular research in the past. Canines represent a more stringent model for vascular graft investigations due to their lack of spontaneous endothelialization. They differ from humans in terms of thrombogenicity. Assessment of platelet aggregability is recommended previous beginning of the study to involve animals with similar thrombocyte aggregation. Dogs have further a potent fibrinolytic system. Canines provide a range of implantation sites with the possibility of paired prostheses evaluation and have appropriate vessels for large (aorta 8 mm) and small diameter applications. Because of ethical considerations due to the companion status of dogs, the use of dogs is decreasing.

Sheep and goat are appropriate models for cardiovascular research because anatomic and hemodynamic conditions approximate the human situation. Sheep have the closest similarity with humans regarding the activity of their coagulation and fibrinolysis system. Sheep has been often used as models for surgical interventions because arterial and venous vessels of the long neck region and the femoral region have appropriate diameter and length. Endothelialization and neointimal formation is very similar to the human patient. Humanized, genetically modified, animals are currently not available.

Pigs share many similarities with regard to physiology and anatomy with humans and are therefore one of the most used animal species in translational. Most of the strains used are crossbred farm pigs which reveal high growth rates and are not easy to handle. Miniature and micro pigs have been developed which stay relatively small and are much more compliant. Because of their convenient bodyweight, senescent models can be used for long-term observations. The size of the heart and arteries in mature pigs is very similar to humans. Humans and pigs reveal a similar blood supply by the coronary arteries. Arteries of pigs are fragile and show a great tendency to muscular spasms, and the intima is highly susceptible to denudation. Like in humans, pigs have similar response to vascular injury and develop intimal hyperplasia in the same extent as humans. Elderly pigs can develop spontaneous atherosclerosis. Diet-accelerated atherosclerosis has been used to establish models of coronary atherosclerosis.

## 13.9 Abdominal Aortic Aneurysm Models in Small Animals

Abdominal aortic aneurysms (AAA) are a chronic and potentially life-threatening disease in case of rupture. Currently, endovascular aortic repair and open surgerical repair have been only established to treat AAA. In order to develop a causal medical treatment, a detailed mechanistic description of AAA formation requiring accurate animal models is needed. The following list describes basic AAA models in small animals. Combined with each other or with other chemicals and knockout models, AAA progression and spontaneous dissection formation can be further enhanced (a more detailed list and less frequently used models can be found in the cited reviews) [90–92].

The elastase model: Pressure-perfused intraluminal application of porcine pancreatic elastase leads to elastin degradation as well as adventitial macrophage infiltration and AAA formation. After laparotomy, dissection of the infrarenal aorta, and clamping of this segment, the aorta is perfused with elastase via a microcatheter. After the infusion, the aortotomy has to be sutured. Thus, to avoid aortotomy, AAA induction by adventitial application of elastase has been developed as well.

The calcium chloride model: Adventitial application of calcium chloride administered with a $CaCl_2$-soaked gauze patch (e.g., $CaCl_2$ at 0.5 M for 15 minutes) after laparotomy leads to inflammatory cell infiltration, elastin degradation, and calcification of the aortic wall as well as AAA formation. Not needing an aortotomy, the model is relatively simple to establish.

Angiotensin II in Apo E−/− mice on high-fat diet: Continuous angiotensin II perfusion via an osmotic pump in Apo E −/− on high-fat diet induces spontaneous aortic aneurysm as well as aortic dissection formation of the whole aorta. The aneurysms show more common pathophysiological features to human AAA than the other models. However, death rate during angiotensin II perfusion and incomplete AAA induction rate remain the disadvantages of the model.

Aortic diameter measurement: In the elastase and calcium chloride model, measuring the aortic diameter during AAA induction and at organ harvesting with the surgical microscope is a simple method to calculate individual growth rates. Moreover, ultrasound is used to determine growth rate, as well as an estimation of aortic stiffness and wall stress. Alternatively, computed tomography and magnetic resonance imaging can be performed for aortic size measurements.

## 13.10    Spontaneous Animal Models

Models for cardiovascular disease which are developed via breeding selection of spontaneously developed pathological phenotypes from laboratory animals are rare. Many of
these pathological phenotypes seen in the spontaneous models are related to the field of
the metabolic diseases with cardiovascular symptoms.

### 13.10.1    The Nude Mouse

Described for the first time in 1966 by Flanagan, the nude mouse occurred as a spontaneous mutation, in which there is developmental failure of the thymus, resulting in mice
devoid of circulating functional t-cells [93].

### 13.10.2    The SCID Mouse

This mouse was described for the first time by Bosma et al. in 1983. Due to a mutation of
the *rag1* gene encoding for recombinase activity, this mouse failed to rearrange immunoglobulin or T cell receptor genes, and this developed a combined T and B cell immunodeficiency [94]. The use of these mice certainly helped to describe the cellular basis of
immunity.

### 13.10.3    Spontaneous Hypertensive Rats

There is one rat strain existing which develops a spontaneous hypertension. The spontaneously hypertensive rat (SHR) was discovered in 1959 by Okamoto and Aoki [95]. Since
then the strain has been maintained by selective sibling mating.

   Hypertension is present in 100% of the animals.

### 13.10.4    BB Wistar Rat

In 1997, researchers from the Bio Breeding Laboratories found animals with hyperglycemia and ketoacidosis in an outbred stock of Wistar rats. Till now the BB Wistar rat is the
oldest, best known, and most extensively studied rat strain in type 1 diabetes [96, 97]. The
effects of the diabetes in the BB rat are limited to mild effects in the myocardium with
progressive loss of myofilaments [98], contractile dysfunction in the perfused heart [99],
and impaired angiogenic sprouting of the aorta [100].

### 13.10.5    NOD Mouse

A mouse model for type 1 diabetes is the nonobese diabetic mouse (NOD mouse).
Developed in 1974 at the Shionogi Research Laboratories in Osaka, this phenotype was
found during a selective breeding searching for a new model for diabetes [101, 102].
Despite the T cell-mediated insulitis in young mice and the development of a type 1

diabetes, several risk factors for cardiovascular disease are elevated in this mouse strain. Bred with a major histocompatibility complex class II ß chain (MHC) – ko strain – the NOD mouse can be a model for autoimmune cardiomyopathy [15, 103]. But there is one publication existing which argues against the use of the NOD mouse as a model for cardiovascular changes in human diabetic autonomic neuropathy [104].

## 13.10.6  Watanabe Heritable Hyperlipidemic (WHLL) Rabbit

Rabbit myocardium shares more similarities with human myocardium than small rodent myocardium, but differences remain. For example, rabbits might not serve as the best animal model for studying effects of exercise on the cardiovascular system mainly because its heart rate reserve is much less than in humans [105].

The Watanabe heritable hyperlipidemic (WHLL) rabbit is a model for hypercholesterolemia and atherosclerosis [106–108].

A rabbit spontaneous MI model could be obtained from the selective breeding descendants of coronary atherosclerosis – prone WHHL-MI rabbits. Because the coronary lumen area stenosis is enhanced in these rabbits, the incidence of spontaneous WHHL-MI was increased in proportion to the serial nearly occluded coronary lesions. However, observations indicate that the mechanisms for MI in WHHL rabbits are different from those in humans [109].

> **Take-Home Message**
>
> Preclinical evaluation has a high translational value for the transfer of new therapeutics from bench to bedside. The selection of appropriate animal models is essential because the validity of such preclinical models depends on how close these models resemble the human situation. Because a single animal model will not meet all criteria, proper study design and knowledge of comparative anatomy and physiology of animal models are important to significantly increase the impact of preclinical studies and compensate the fact that these models do not perfectly replicate the clinical situation. Among the criteria that might be important when selecting your experimental model are the following: (1) Is the animal model appropriate for its intended use? (2) Can the model be developed and maintained at reasonable costs? (3) Is the model interesting for not only one limited kind of research? (4) Is the model reproducible and reliable? and (5) Is the model reasonably available and accessible? [9].

## References

1. Langendorff O. Untersuchungen am überlebenden Säugethierherzen. In: Pflügers Archive - European Journal of Physiology: Springer; 1895. p. 291–332.
2. Furchgott RF. Studies on relaxation of rabbit aorta by sodium nitrite: the basis for the proposal that the acid-activatable inhibitory factor from retractor penis is inorganic nitrite and the endothelium-derived relaxing factor is nitric oxide Vasodilatation: vascular smooth muscle, peptides, autonomic nerves and endotheliumed. New York: Raven Press; 1988. p. 401–14.
3. SIMR. Seriously ill for medical research: centenary survey of Nobel laureates in physiology or medicine. Dunstable: SIMR; 1998.

4. Bacon F. Novum organum/The four idols: aphorisms concerning the interpretation of nature and the kingdom of man. Chicago, IL: Encyclopedia Britannica; 1990.
5. Russell WM, Burch RL, Hume CW. The principles of humane experimental technique. London: Methuen; 1959.
6. Union, E., DIRECTIVE 2010/63/EU OF THE EUROPEAN PARLIAMENT AND OF THE COUNCIL of 22 September 2010 on the protection of animals used for scientific purposes. Offical Journal of the European Union, 2010.
7. Bundesgesetz über Versuche an lebenden Tieren (Tierversuchsgesetz 2012 – TVG 2012). 2012; Available from: https://www.jusline.at/gesetz/tvg_2012.
8. Max-Planck-Gesellschaft, White paper on animal research in the max Planck Society, Max-Planck-Gesellschaft, Editor. 2016.
9. Maurer Kea. Animal models in biomedical research in laboratory animal medicine, 1505–107. 3rd ed: Elsevier; 2015. https://doi.org/10.1016/B978-0-12-409527-4.00044-4.
10. Hasenfuss G. Animal models of human cardiovascular disease, heart failure and hypertrophy. Cardiovasc Res. 1998;39(1):60–76.
11. Mullins LJ, Mullins JJ. Insights from the rat genome sequence. Genome Biol. 2004;5(5):221.
12. Patten RD, Hall-Porter MR. Small animal models of heart failure: development of novel therapies, past and present. Circ Heart Fail. 2009;2(2):138–44.
13. Zaragoza C, et al. Animal models of cardiovascular diseases. J Biomed Biotechnol. 2011;2011:497841.
14. Ramzy D, et al. Cardiac allograft vasculopathy: a review. Can J Surg. 2005;48(4):319–27.
15. Camacho P, et al. Small mammalian animal models of heart disease. Am J Cardiovasc Dis. 2016;6(3): 70–80.
16. WHO. Global Health Estimates 2015: deaths by cause, age, sex, by country and by region, 2000–2015. W.H. Organization, Editor. 2016.
17. Bulluck H, Yellon DM, Hausenloy DJ. Reducing myocardial infarct size: challenges and future opportunities. Heart. 2016;102(5):341–8.
18. Chimenti S, et al. Myocardial infarction: animal models. Methods Mol Med. 2004;98:217–26.
19. Pfeffer MA, et al. Myocardial infarct size and ventricular function in rats. Circ Res. 1979;44(4):503–12.
20. Ishikawa K, et al. Development of a preclinical model of ischemic cardiomyopathy in swine. Am J Physiol Heart Circ Physiol. 2011;301(2):H530–7.
21. Michael LH, et al. Myocardial ischemia and reperfusion: a murine model. Am J Phys. 1995;269(6 Pt 2):H2147–54.
22. Savarese G, Lund LH. Global public health burden of heart failure. Card Fail Rev. 2017;3(1):7–11.
23. Yang F, et al. Myocardial infarction and cardiac remodelling in mice. Exp Physiol. 2002;87(5):547–55.
24. Litwin SE, et al. Serial echocardiographic-Doppler assessment of left ventricular geometry and function in rats with pressure-overload hypertrophy. Chronic angiotensin-converting enzyme inhibition attenuates the transition to heart failure. Circulation. 1995;91(10):2642–54.
25. Rockman HA, et al. ANG II receptor blockade prevents ventricular hypertrophy and ANF gene expression with pressure overload in mice. Am J Phys. 1994;266(6 Pt 2):H2468–75.
26. Merino D, et al. Experimental modelling of cardiac pressure overload hypertrophy: modified technique for precise, reproducible, safe and easy aortic arch banding-debanding in mice. Sci Rep. 2018;8(1):3167.
27. Transplantation, G.O.o.D.a. Number of transplanted organs worldwide in 2015. 2016; Available from: http://www.transplant-observatory.org/contador1/.
28. Bedi DS, et al. Animal models of chronic allograft injury: contributions and limitations to understanding the mechanism of long-term graft dysfunction. Transplantation. 2010;90(9):935–44.
29. Corry RJ, Winn HJ, Russell PS. Heart transplantation in congenic strains of mice. Transplant Proc. 1973;5(1):733–5.
30. Corry RJ, Winn HJ, Russell PS. Primarily vascularized allografts of hearts in mice. The role of H-2D, H-2K, and non-H-2 antigens in rejection. Transplantation. 1973;16(4):343–50.
31. Liu F, Kang SM. Heterotopic heart transplantation in mice. J Vis Exp. 2007;(6):238.
32. Pilat N, et al. Blockade of adhesion molecule lymphocyte function-associated antigen-1 improves long-term heart allograft survival in mixed chimeras. J Heart Lung Transplant. 2018;37(9):1119–30.
33. Basra SS, et al. Acute coronary syndromes: unstable angina and non-ST elevation myocardial infarction. Heart Fail Clin. 2016;12(1):31–48.
34. Hausenloy DJ, Yellon DM. Myocardial ischemia-reperfusion injury: a neglected therapeutic target. J Clin Invest. 2013;123(1):92–100.

13

35. Murry CE, Jennings RB, Reimer KA. Preconditioning with ischemia: a delay of lethal cell injury in ischemic myocardium. Circulation. 1986;74(5):1124–36.

36. Zhao ZQ, et al. Inhibition of myocardial injury by ischemic postconditioning during reperfusion: comparison with ischemic preconditioning. Am J Physiol Heart Circ Physiol. 2003;285(2):H579–88.

37. Przyklenk K, Whittaker P. Remote ischemic preconditioning: current knowledge, unresolved questions, and future priorities. J Cardiovasc Pharmacol Ther. 2011;16(3–4):255–9.

38. Botker HE, et al. Remote ischaemic conditioning before hospital admission, as a complement to angioplasty, and effect on myocardial salvage in patients with acute myocardial infarction: a randomised trial. Lancet. 2010;375(9716):727–34.

39. Schmidt MR, Redington A, Botker HE. Remote conditioning the heart overview: translatability and mechanism. Br J Pharmacol. 2015;172(8):1947–60.

40. Hausenloy DJ, et al. Remote ischemic preconditioning and outcomes of cardiac surgery. N Engl J Med. 2015;373(15):1408–17.

41. Gaspar A, et al. Randomized controlled trial of remote ischaemic conditioning in ST-elevation myocardial infarction as adjuvant to primary angioplasty (RIC-STEMI). Basic Res Cardiol. 2018;113(3):14.

42. Nagata K, et al. Early changes in excitation-contraction coupling: transition from compensated hypertrophy to failure in Dahl salt-sensitive rat myocytes. Cardiovasc Res. 1998;37(2):467–77.

43. Podesser BK, et al. Unveiling gender differences in demand ischemia: a study in a rat model of genetic hypertension. Eur J Cardiothorac Surg. 2007;31(2):298–304.

44. Gordon JW, Ruddle FH. Integration and stable germ line transmission of genes injected into mouse pronuclei. Science. 1981;214(4526):1244.

45. Doevendans PA, et al. Cardiovascular phenotyping in mice. Cardiovasc Res. 1998;39(1):34–49.

46. The mouse in biomedical research. 2nd ed. Normative biology, husbandry and models. Vol. 3. Burlington: American College of Laboratory Animal Medicine Series; 2007.

47. Piedrahita JA, et al. Generation of mice carrying a mutant apolipoprotein E gene inactivated by gene targeting in embryonic stem cells. Proc Natl Acad Sci U S A. 1992;89(10):4471–5.

48. Zhang SH, et al. Spontaneous hypercholesterolemia and arterial lesions in mice lacking apolipoprotein E. Science. 1992;258(5081):468.

49. Maeda N. History of discovery: development of apolipoprotein E-deficient mice. Arterioscler Thromb Vasc Biol. 2011;31(9):1957–62.

50. Getz GS, Reardon CA. ApoE knockout and knockin mice: the history of their contribution to the understanding of atherogenesis. J Lipid Res. 2016;57(5):758–66.

51. Mahley RW. Apolipoprotein E: cholesterol transport protein with expanding role in cell biology. Science. 1988;240(4852):622.

52. Getz GS, Reardon CA. Apoprotein E as a lipid transport and signaling protein in the blood, liver, and artery wall. J Lipid Res. 2009;50(Suppl):S156–61.

53. Vasquez EC, et al. Cardiac and vascular phenotypes in the apolipoprotein E-deficient mouse. J Biomed Sci. 2012;19(1):22.

54. Ross R. The pathogenesis of atherosclerosis: a perspective for the 1990s. Nature. 1993;362:801.

55. Stoll G, Bendszus M. Inflammation and atherosclerosis: novel insights into plaque formation and destabilization. Stroke. 2006;37(7):1923–32.

56. Lorenzen-Schmidt I, et al. Young MLP deficient mice show diastolic dysfunction before the onset of dilated cardiomyopathy. J Mol Cell Cardiol. 2005;39(2):241–50.

57. Arber S, et al. MLP-deficient mice exhibit a disruption of cardiac cytoarchitectural organization, dilated cardiomyopathy, and heart failure. Cell. 1997;88(3):393–403.

58. Ross J Jr. Dilated cardiomyopathy concepts derived from gene deficient and transgenic animal models. Circ J. 2002;66(3):219–24.

59. Emery AE. Population frequencies of inherited neuromuscular diseases--a world survey. Neuromuscul Disord. 1991;1(1):19–29.

60. Hermans MC, et al. Hereditary muscular dystrophies and the heart. Neuromuscul Disord. 2010;20(8):479–92.

61. Im WB, et al. Differential expression of dystrophin isoforms in strains of mdx mice with different mutations. Hum Mol Genet. 1996;5(8):1149–53.

62. Perkins KJ, Davies KE. The role of utrophin in the potential therapy of Duchenne muscular dystrophy. Neuromuscul Disord. 2002;12(Suppl 1):S78–89.

63. Wells DJ. Tracking progress: an update on animal models for Duchenne muscular dystrophy. Dis Model Mech. 2018;11(6):dmm035774.

64. Ginis I, et al. Differences between human and mouse embryonic stem cells. Dev Biol. 2004;269(2): 360–80.
65. Haghighi K, et al. Human phospholamban null results in lethal dilated cardiomyopathy revealing a critical difference between mouse and human. J Clin Invest. 2003;111(6):869–76.
66. Hood WB Jr, McCarthy B, Lown B. Myocardial infarction following coronary ligation in dogs. Hemodynamic effects of isoproterenol and acetylstrophanthidin. Circ Res. 1967;21(2):191–9.
67. Sabbah HN, et al. A canine model of chronic heart failure produced by multiple sequential coronary microembolizations. Am J Phys. 1991;260(4 Pt 2):H1379–84.
68. Biondi-Zoccai G, et al. A novel closed-chest porcine model of chronic ischemic heart failure suitable for experimental research in cardiovascular disease. Biomed Res Int. 2013;2013:410631.
69. Dariolli R, et al. Development of a closed-artery catheter-based myocardial infarction in pigs using sponge and lidocaine hydrochloride infusion to prevent irreversible ventricular fibrillation. Physiol Rep. 2014;2(8):e12121.
70. Shi W, et al. A swine model of percutaneous intracoronary ethanol induced acute myocardial infarction and ischemic mitral regurgitation. J Cardiovasc Transl Res. 2017;10(4):391–400.
71. Suzuki Y, et al. In vivo porcine model of reperfused myocardial infarction: in situ double staining to measure precise infarct area/area at risk. Catheter Cardiovasc Interv. 2008;71(1):100–7.
72. Teramoto N, et al. Experimental pig model of old myocardial infarction with long survival leading to chronic left ventricular dysfunction and remodeling as evaluated by PET. J Nucl Med. 2011;52(5): 761–8.
73. Clarkson TB, et al. Nonhuman primate models of atherosclerosis: potential for the study of diabetes mellitus and hyperinsulinemia. Metabolism. 1985;34(12 Suppl 1):51–9.
74. Jensen TW, et al. A cloned pig model for examining atherosclerosis induced by high fat, high cholesterol diets. Anim Biotechnol. 2010;21(3):179–87.
75. Pedersen HD, Haggstrom J. Mitral valve prolapse in the dog: a model of mitral valve prolapse in man. Cardiovasc Res. 2000;47(2):234–43.
76. Choi H, et al. Quantification of mitral regurgitation using proximal isovelocity surface area method in dogs. J Vet Sci. 2004;5(2):163–71.
77. Spinale FG, et al. Structural basis for changes in left ventricular function and geometry because of chronic mitral regurgitation and after correction of volume overload. J Thorac Cardiovasc Surg. 1993;106(6):1147–57.
78. Tsutsui H, et al. Effects of chronic beta-adrenergic blockade on the left ventricular and cardiocyte abnormalities of chronic canine mitral regurgitation. J Clin Invest. 1994;93(6):2639–48.
79. Leroux AA, et al. Animal models of mitral regurgitation induced by mitral valve chordae tendineae rupture. J Heart Valve Dis. 2012;21(4):416–23.
80. Llaneras MR, et al. Large animal model of ischemic mitral regurgitation. Ann Thorac Surg. 1994;57(2):432–9.
81. Cui YC, et al. A pig model of ischemic mitral regurgitation induced by mitral chordae tendinae rupture and implantation of an ameroid constrictor. PLoS One. 2014;9(12):e111689.
82. Anderson CA, et al. An acute animal model of aortic stenosis: initial attempts at leaflet modification. J Heart Valve Dis. 2012;21(2):172–4.
83. Wisenbaugh T, et al. Contractile function, myosin ATPase activity and isozymes in the hypertrophied pig left ventricle after a chronic progressive pressure overload. Circ Res. 1983;53(3):332–41.
84. Benjamin EJ, et al. Heart disease and stroke statistics-2018 update: a report from the American Heart Association. Circulation. 2018;137(12):e67–e492.
85. Gaudino M, et al. Mechanisms, consequences, and prevention of coronary graft failure. Circulation. 2017;136(18):1749–64.
86. Parang P, Arora R. Coronary vein graft disease: pathogenesis and prevention. Can J Cardiol. 2009;25(2):e57–62.
87. Schachner T, Laufer G, Bonatti J. In vivo (animal) models of vein graft disease. Eur J Cardiothorac Surg. 2006;30(3):451–63.
88. Naito Y, et al. Vascular tissue engineering: towards the next generation vascular grafts. Adv Drug Deliv Rev. 2011;63(4–5):312–23.
89. Byrom MJ, et al. Animal models for the assessment of novel vascular conduits. J Vasc Surg. 2010;52(1):176–95.
90. Lysgaard Poulsen J, Stubbe J, Lindholt JS. Animal models used to explore abdominal aortic aneurysms: a systematic review. Eur J Vasc Endovasc Surg. 2016;52(4):487–99.

13

91. Patelis N, et al. Animal models in the research of abdominal aortic aneurysms development. Physiol Res. 2017;66(6):899–915.
92. Senemaud J, et al. Translational relevance and recent advances of animal models of abdominal aortic aneurysm. Arterioscler Thromb Vasc Biol. 2017;37(3):401–10.
93. Flanagan SP. Nude', a new hairless gene with pleiotropic effects in the mouse. Genet Res. 1966;8(3):295–309.
94. Bosma GC, Custer RP, Bosma MJ. A severe combined immunodeficiency mutation in the mouse. Nature. 1983;301(5900):527–30.
95. Okamoto Aoki K. Development of a strain of spontanously hypertensive rats. Jpn Circ J. 1963;27:11.
96. Mordes JP, et al. Rat models of type 1 diabetes: genetics, environment, and autoimmunity. ILAR J. 2004;45(3):278–91.
97. Chappel CI, Chappel WR. The discovery and development of the BB rat colony: an animal model of spontaneous diabetes mellitus. Metabolism. 1983;32(7, Supplement 1):8–10.
98. Hsiao Y-C, et al. Ultrastructural alterations in cardiac muscle of diabetic BB Wistar rats. Virchows Archiv A. 1987;411(1):45–52.
99. Broderick TL, Hutchison AK. Cardiac dysfunction in the euglycemic diabetic-prone BB Wor rat. Metabolism. 2004;53(11):1391–4.
100. Onuta G, et al. Angiogenic sprouting from the aortic vascular wall is impaired in the BB rat model of autoimmune diabetes. Microvasc Res. 2008;75(3):420–5.
101. Hanafusa T, Miyagawa J-i, Nakajima H, Tomita K, Kuwajima M, Matsuzawa Y, Tarui S. The NOD mouse. Diabetes Res Clin Pract. 1994;24:4.
102. Solomon M, Sarvetnick N. The pathogenesis of diabetes in the NOD mouse. Adv Immunol. 2004; Academic Press. 84:239–64.
103. Neuman JC, Nieman KM, Schalinske KL. Characterization of methyl group metabolism in the non-obese diabetic (NOD) mouse. FASEB J. 2010;24(1_supplement):915.11.
104. Gross V, et al. Cardiovascular autonomic regulation in Non-Obese Diabetic (NOD) mice. Auton Neurosci. 2008;138(1):108–13.
105. Milani-Nejad N, Janssen PML. Small and large animal models in cardiac contraction research: advantages and disadvantages. Pharmacol Ther. 2014;141(3):235–49.
106. Watanabe Y. Serial inbreeding of rabbits with hereditary hyperlipidemia (WHHL-Rabbit). Atherosclerosis. 1980;36:7.
107. The tale of the Watanabe rabbit. Lab Anim. 2012;41:277.
108. Shiomi M, Ito T. The Watanabe heritable hyperlipidemic (WHHL) rabbit, its characteristics and history of development: a tribute to the late Dr. Yoshio Watanabe. Atherosclerosis. 2009;207(1):1–7.
109. Shiomi M, et al. Development of an animal model for spontaneous myocardial infarction (WHHLMI rabbit). Arterioscler Thromb Vasc Biol. 2003;23(7):1239–44.

# Endothelial Cell Isolation and Manipulation

*Christine Brostjan*

© Springer Nature Switzerland AG 2019
M. Geiger (ed.), *Fundamentals of Vascular Biology*, Learning Materials in Biosciences,
https://doi.org/10.1007/978-3-030-12270-6_14

**What You Will Learn in This Chapter**

In vitro characterization of primary vascular cells requires their isolation and purification from tissue. Considering the diversity in composition and function between the distinct types of aortic, arterial, capillary, venous, and lymphatic vessels, it is essential to retrieve vascular cells from the vessel type of investigation and conduct experiments at early passages to largely retain their differentiation. Hence, a variety of protocols for *endothelial cell isolation* from distinct species, tissues, and vessel types have been developed which will be listed in the following chapter, including two detailed standard protocols for the *isolation of microvascular and lymphatic endothelial cells* from human skin and for the *isolation of human umbilical vein endothelial cells*. Furthermore, mechanistic studies frequently require the overexpression or silencing of a single gene which is achieved by cell transfection or transduction methods. Adaptation of protocols for primary *endothelial cells* has been required as they are not readily amenable to gene transfer. In the following, several chemical, physical *transfection*, and viral *transduction* techniques will be introduced which were successfully applied to vascular cells in vitro and in vivo. The latter paved the route to vascular gene therapy, which is an area of particular interest that will be discussed in more detail. Again, a standard protocol for in vitro *endothelial* manipulation by *electroporation* is provided.

## 14.1   Isolation of Endothelial Cells

Optimization of protocols for the isolation of endothelial cells (ECs) started as early as in the 1960s [1, 2] and was repeatedly reviewed over the last decades [3–6], with a particular emphasis on human [7, 8] and rodent cells [9] but also dog-derived cells [10].

While outgrowth of ECs can be applied to larger vessels [11], the initial isolation step usually relies on mechanical and/or enzymatic digestion of tissue. Endothelial cells then need to be separated from the remaining tissue components and other tissue-resident cell types. To this end, unique features and surface markers of ECs were employed to then physically isolate the cells of interest by either immunopanning [12], immunomagnetic beads [13–18], or fluorescence-activated cell sorting [19–21]. While the latter yields high purity of early passage cells, it requires the access to a sorting device. Hence, immunolabeling of ECs and separation by magnetic beads has proved to be a widely available and applied method of cell isolation.

Early attempts of EC tagging were relying on the uptake of fluorescently labeled acetylated low-density lipoprotein [22] or the selective endothelial binding of lectins like *Ulex europaeus* I [14, 17]. However, the majority of EC isolation protocols has applied antibodies against endothelial surface markers such as CD31 [16, 18, 23–25], CD105 [26], VE-cadherin [27], ICAM-2 [28], VCAM-1 [29], or E-selectin after pro-inflammatory EC activation [30]. To further separate lymphatic from blood vessel endothelial cells in mixtures of tissue microvessels, specific antigens on lymphatic endothelial cells were used such as podoplanin [31], LYVE-1 [20], and vascular endothelial growth factor receptor 3 (VEGFR-3) [21].

The isolation of endothelium from macrovessels [32] is distinct from approaches for microvasculature, as total tissue digest is generally avoided and detachment of

14

ECs from the intimal basement membrane is achieved by either simple explant outgrowth into tissue culture dishes [33] or by collagenase treatment with vessel perfusion [34], collagenase-covered filter paper [35], or microcarriers [36]. Alternative methods with fibrin glue have been reported [37]. Mostly, there is no need for further EC purification, since contamination with cells of the media or adventitia is limited. Protocols for endothelial isolation from macrovessels are summarized in ◘ Table 14.1.

In contrast, mechanical disruption and enzymatic digest of total tissue is required to retrieve microvessel endothelium [38–40], which relies on the abovementioned purification techniques to then separate ECs from other cell types. ◘ Table 14.2 provides an overview on published techniques for endothelial isolation according to organ, species, and vessel type, while ◘ Table 14.3 focuses on protocols established for EC acquisition from diseased tissue. Specific protocols for the isolation of lymphatic ECs have been developed [21, 41–43].

To provide examples for the most commonly applied EC isolation methods, two detailed protocols will be given for macro- and microvessels.

◘ **Table 14.1** Endothelial isolation techniques according to macrovessel type and species

| Vessel type | Species | EC source | References |
| --- | --- | --- | --- |
| Aorta | Bovine | Intimal | [35] |
| | Human | Thoracic aorta, intimal and adventitial | [45] |
| | Murine | Intimal | [46–51] |
| | Pig | Intimal | [52] |
| | Rat | Intimal | [53–57] |
| Arteries | Canine | Coronary arteries | [58] |
| | Deer | Carotid arteries | [59] |
| | Equine | Coronary arteries | [60] |
| | Human | Internal mammary artery | [61] |
| | Murine | Mesenteric artery, cerebral arteries | [49] |
| | Rabbit | Coronary arteries | [62] |
| | Rat | Coronary arteries | [63] |
| Veins | Bovine | Vena cava | [64] |
| | Canine | Jugular vein | [65] |
| | Equine | Jugular vein | [60] |
| Venules | Rat | Cremaster muscle | [66] |
| | Sheep | Postcapillary venules of mesenteric lymph nodes | [67] |

◻ **Table 14.2**    Endothelial isolation techniques according to organ and species

| Organ | Species | EC type | References |
|---|---|---|---|
| Adipose tissue | Canine | Microvascular | [68] |
| | Chicken | Microvascular | [69] |
| | Human | Microvascular | [23, 70–73] |
| | Mouse | Microvascular | [74, 75] |
| | Rat | Microvascular | [76–78] |
| Bladder | Rat | Total | [79] |
| Bone | Human | Total | [80] |
| Bone marrow | Human | Microvascular | [81] |
| | Human | Total | [82–84] |
| | Rat | Total | [85] |
| Brain | Bovine | Arterial | [86, 87] |
| | Bovine | Microvascular | [88, 89] |
| | Human | Microvascular | [88, 90–97] |
| | Monkey | Microvascular | [98] |
| | Murine | Microvascular | [96, 99–105] |
| | Pig | Microvascular | [99, 106] |
| | Rat | Arterial | [107] |
| | Rat | Microvascular | [89, 108–112] |
| | Rat | Pituitary | [113] |
| Corpus luteum | Monkey | Microvascular | [114] |
| | Porcine | Microvascular | [115] |
| | Rabbit | Microvascular | [116] |
| Decidua | Human | Microvascular | [117–120] |
| Esophagus | Human | Microvascular | [121] |
| Eye | Bovine | Choroidal | [122–124] |
| | Bovine | Retinal | [125–127] |
| | Human | Choroidal | [128] |
| | Human | Corneal | [129–133] |
| | Human | Limbal | [134] |

14

☐ **Table 14.2**   (continued)

| Organ | Species | EC type | References |
|---|---|---|---|
| | Human | Retinal | [135] |
| | Human | Schlemm's canal | [136] |
| | Murine | Retinal | [137] |
| | Rat | Retinal | [138] |
| Fetal tissue | Human | Retina microvascular | [135] |
| | Human | Skin microvascular | [139] |
| | Mouse | Heart valve | [140] |
| | Mouse | Heart ventricle | [141] |
| | Pig | Various organs | [142] |
| | Rat | Ductus arteriosus | [143] |
| Gingiva | Human | Microvascular | [144] |
| Heart | Fish (salmon, cod) | Microvascular | [145] |
| | Guinea pig | Macrovascular, coronary | [146] |
| | Guinea pig | Microvascular | [147] |
| | Human | Macro- and microvascular | [148] |
| | Human | Microvascular | [149, 150] |
| | Mouse | Microvascular | [151–155] |
| | Mouse | Valvular | [140] |
| | Pig | Valvular | [156] |
| | Rabbit | Macro- and microvascular | [157, 158] |
| | Rat | Macro- and microvascular | [159] |
| | Rat | Microvascular | [149, 160, 161] |
| | Rabbit | Microvascular | [162] |
| Intestine | Human | Microvascular | [27, 163, 164] |
| Kidney | Canine | Adrenal | [165] |
| | Human | Adrenal | [166] |
| | Human | Glomerular | [167, 168] |
| | Human | Glomerular and peritubular | [169] |
| | Mouse | Glomerular | [170, 171] |
| | Mouse | Peritubular | [172] |

(continued)

■ Table 14.2   (continued)

| Organ | Species | EC type | References |
|---|---|---|---|
| Liver | Guinea pig | Sinusoidal | [173] |
| | Human | Sinusoidal | [174] |
| | Monkey | Total | [15, 175] |
| | Mouse | Sinusoidal | [176–181] |
| | Pig | Sinusoidal | [174] |
| | Rat | Sinusoidal or total | [182–186] |
| Lung | Bovine | Arterial | [187] |
| | Equine | Arterial | [188] |
| | Guinea pig | Total | [189] |
| | Human | Arterial | [190] |
| | Human | Lymphatic | [191] |
| | Human | Microvascular | [192–196] |
| | Mouse | Microvascular | [28, 29, 151–153, 197–199] |
| | Rabbit | Microvascular | [200] |
| | Rat | Arterial | [201] |
| | Rat | Microvascular | [202, 203] |
| Lymph nodes | Ferret | Lymphatic | [204] |
| | Mouse | High endothelial venules | [205] |
| | Rat | Lymphatic | [206] |
| Myometrium | Human | Microvascular | [207] |
| Nervous system | Bovine | Cauda equina | [208, 209] |
| | Murine | Spinal cord | [210] |
| | Rat | Peripheral nerve, optic nerve | [211, 212] |
| Omentum | Equine | Microvascular | [213] |
| | Human | Microvascular | [214, 215] |
| Placenta | Human | Microvascular | [216–218] |
| | Mouse | Microvascular | [219] |
| Prostate | Rat | Microvascular | [220] |
| Skeletal muscle | Human | Microvascular | [221] |

14

◻ **Table 14.2** (continued)

| Organ | Species | EC type | References |
|---|---|---|---|
| Skin | Human | Blood microvascular and lymphatic | [31] |
| | Human | Lymphatic | [21] |
| | Human | Microvascular | [30, 139, 222–226] |
| | Mouse | Microvascular | [227] |
| | Rat | Lymphatic | [20] |
| Stomach | Human | Microvascular | [228] |
| | Rat | Microvascular | [229] |
| Thoracic duct | Rat | Lymphatic | [230] |
| Thyroid | Human | Microvascular | [231] |
| Tonsils | Human | Lymphatic | [232] |
| Umbilical cord | Human | Artery | [233] |
| | Human | Vein | [34, 233–242] |

◻ **Table 14.3** Endothelial isolation techniques according to disease and species

| Disease | Species | EC type | References |
|---|---|---|---|
| Coronary artery disease (percutaneous coronary intervention) | Human | Arterial | [243] |
| Diabetes | Mouse | Aortic and muscle endothelium | [244] |
| Emphysema | Human | Lung microvascular | [245] |
| Hypertrophic scars | Human | Microvascular | [246] |
| Rheumatoid arthritis | Human | Synovial microvascular | [247, 248] |
| Tumors | Canine | Hepatocellular carcinoma | [249] |
| | Canine | Seminoma | [249] |
| | Human | Breast cancer | [250] |
| | Human | Colorectal carcinoma | [27, 164] |
| | Human | Glioma microvascular | [90, 251] |
| | Human | Glossal lymphangioma (lymphatic) | [252] |

(continued)

**◻ Table 14.3** (continued)

| Disease | Species | EC type | References |
|---|---|---|---|
| | Human | Intramuscular hemangioma | [253] |
| | Human | Juvenile nasopharyngeal angiofi-broma | [254] |
| | Human | Various tumors | [255, 256] |
| | Mouse | Breast cancer | [257] |
| | Mouse | Colorectal carcinoma | [258] |
| | Mouse | Fibrosarcoma | [259] |
| | Mouse | Lung cancer | [26] |
| | Rat | Fibrosarcoma | [260] |
| Vessel malformations | Human | Cerebral arteriovenous | [261, 262] |
| | Human | Lymphatic | [21] |
| | Human | Oral venous | [263] |

### 14.1.1    Protocol for the Isolation of Human Umbilical Vein Endothelial Cells (HUVECs)

(Christoph Kaun and Philipp Hohensinner, Department of Medicine II, Division of Cardiology, Medical University of Vienna)

Protocol 1
- The umbilical cord is retrieved and may be stored in Hanks' balanced salt solution (HBSS) for 1–2 days at 4 °C prior to processing.
- Use sterilized isolation equipment which has been autoclaved or rinsed in a bath of 70% ethanol: forceps, scissors, hemostats, metal syringe adaptors, and cable ties.
- Prepare and sterile-filter a solution of collagenase type IV (Sigma C-5138) at 2 mg/ml in HBSS containing penicillin-streptomycin (100 U/ml) and 0.01 M Hepes at pH 7.2.
- Locate the umbilical vessels (the larger vein and two smaller arteries), and then insert the adaptor (Knopf luer lock, straight, 3 mm diameter, 80 mm length, 11G) into the lumen of the vein and fix it with a cable tie.
- Connect a sterile syringe filled with 20 ml HBSS to the adaptor and flush the vessel. Collect the waste in a petri dish on an absorbing sheet.

- Connect another syringe filled with collagenase solution to the adaptor and fill vessel slowly with collagenase solution, until it leaks from the other end. The required volume will depend on the vessel size (prepare about 10–20 ml). Close the other end of the vessel with the hemostat, and then continue gently to further fill the vessel with collagenase solution without bursting it.
- Incubate 45–60 min at room temperature.
- Open the hemostat after placing the cord end over a centrifuge tube to collect the perfusate. Flush the vessel with 20 ml HBSS as described above and add it to the perfusate.
- Collect endothelial cells by centrifugation at 500 × g for 5 min at room temperature. Remove the supernatant and resuspend the pellet in HUVEC growth medium such as EGM2 (Lonza LONCC-3162). Then transfer into an appropriate tissue culture vessel (T75), previously coated with 1% gelatin solution. Incubate at 37 °C, 5% $CO_2$, and change medium after 24–48 hours (◘ Fig. 14.1).

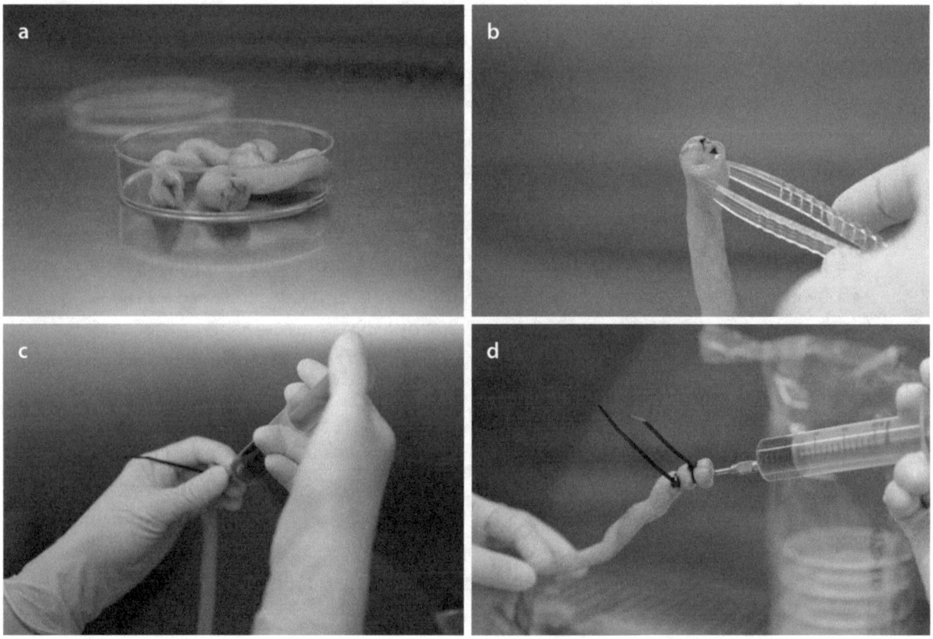


◘ **Fig. 14.1** Isolation of human umbilical vein endothelial cells (according to [44]). **a** The umbilical cord is retrieved from HBSS storage solution. **b** The umbilical vessels are located: the larger vein and the two smaller arteries. **c** The syringe adaptor is inserted into the lumen of the vein and fixed with a cable tie before flushing the vein with HBSS. **d** The umbilical vein is filled with collagenase solution, closed with a hemostat, and incubated for 45–60 min at room temperature before harvesting the endothelial cells

### 14.1.2 Protocol for the Isolation of Microvascular and Lymphatic Endothelial Cells from Human Foreskin

(Christine Brostjan, Department of Surgery, Medical University of Vienna)

#### 14.1.2.1 Cell Isolation from Skin Samples

Protocol 2
- Collect the human foreskin sample in endothelial growth medium such as EGM2-MV (Lonza LONCC-3202) and do not store for more than 24 hours at 4 °C prior to processing.
- Sterilize 2 sharp forceps with ethanol and prewarm dispase solution (Corning 354235) at 37 °C.
- Immerse skin sample in medium in a petri dish, cut to flatten the skin flap, and remove the fat tissue: use the forceps to pull the fat tissue to the side and cut it off with a sterile scalpel. The pink fat tissue is sufficiently removed when the underlying whitish tissue layer becomes apparent. Do not take longer than 10 min.
- Use a scalpel to cut the skin into small pieces of about 5–10 mm$^2$ which are transferred to a fresh petri dish, covered by 7 ml pre-warmed dispase solution and placed in a cell incubator for about 30 min (maximal 40 min).
- When the epidermis starts to detach, i.e., forms a transparent top layer which is loose at the sides, move the skin pieces to a fresh petri dish with 5 ml endothelial medium: using the two forceps, fix each piece and pull off the epidermis layer, and then transfer the remaining skin sample to yet another fresh petri dish with 5 ml endothelial medium.
- Tissue cells are then "squeezed out" by fixing the skin sample with forceps (yellow dermis layer on top) and scraping each piece five times with a regular cell scraper.
- The skin sample is then removed, and only "squeezed out" cells are collected by centrifugation for 10 min at 300 × g and room temperature. The pellet is resuspended in endothelial medium, transferred to a 30 mm culture dish, and supplemented with fibronectin at 1 µg/ml.
- Change the medium after about 2 hours, when adherent cells have settled, and thereafter every 2–3 days until islets of cobblestone-like endothelial morphology are apparent. The culture will be a mix of cell types, partly ECs but also fibroblasts and remaining epidermal cells which tend to overgrow endothelial islets when the well is reaching confluence. Hence, perform EC purification by immunomagnetic cell isolation at a stage of about 70–80% confluence (which may take 1–2 weeks of cell growth).

**14**

#### 14.1.2.2 EC Purification by Immunomagnetic Beads

Protocol 3
- Preload magnetic beads with the desired antibody: CELLection Pan Mouse IgG Kit (Thermo Fisher 11531D) for antibodies of the murine IgG isotype.
- Transfer 100 µl of bead suspension to a 1.5 ml reaction tube, place into suitable magnet (e.g., DynaMag-2), and let beads attach to the magnet for 1 min before removing the liquid.

- Wash beads with 1 ml 0.1% BSA in PBS−/− (phosphate-buffered saline without $Ca^{2+}$ and $Mg^{2+}$) by removing the tube from the magnet, dispersing the beads in buffer, and placing the tube back into the magnet for 1 min before removing the liquid.
- Resuspend the beads in 100 μl of BSA-PBS supplemented with 2 μg of the antibody of interest, e.g., antihuman CD31 for endothelial cells (Bio-Rad MCA1738) and antihuman podoplanin for lymphatic endothelial cells (Abcam ab10288).
- Transfer the reaction tube to a horizontal rotating device and let rotate for 30 min at room temperature (avoiding liquid transfer to the cap).
- Remove antibody solution and wash beads 3× with BSA-PBS as outlined above. Finally, resuspend beads in 100 μl BSA-PBS and add 0.02% sodium azide for prolonged storage at 4 °C (in upright position to avoid drying out). The concentration should range at about $4 \times 10^8$ beads/ml.
- To isolate ECs from the mixed cell culture, harvest the cells either by Accutase or by a very short incubation with 1 ml trypsin-EDTA solution to avoid loss of the cell-surface markers and immediately stop the enzymatic reaction by the addition of 5 ml 10% FCS (fetal calf serum) in RPMI1640 medium.
- Remove an aliquot of cells and determine the cell count while centrifuging the suspension for 5 min at 200 × g and 4 °C. Resuspend the cells at $1 \times 10^7$ per ml in RPMI-1%FCS (minimum volume: 100 μl). Keep cells and buffers on ice during the isolation procedure.
- Use 10 μl of antibody-coupled beads per $1 \times 10^6$ cells: Start with the selection of lymphatic ECs by anti-podoplanin beads. Add the required bead volume to a fresh 1.5 ml reaction tube, place in magnet, remove liquid, and wash with 900 μl RPMI-1%FCS (return to magnet, incubate for 1 min, and then remove the liquid).
- Resuspend the antibody-loaded beads with the cell suspension, and then rotate horizontally for 15 min at 4 °C and finally place in magnet for 1 min. While lymphatic ECs are binding to the beads, vascular ECs and other contaminating cell types will remain in suspension. Thus, harvest the solution and subject it to a second immunobead purification step with anti-CD31 coupled beads for the purification of microvascular endothelial cells.
- Lymphatic ECs bound to anti-podoplanin beads and vascular ECs bound to anti-CD31 beads will then be harvested separately: beads are resuspended in 900 μl RPMI-1%FCS, transferred to a fresh reaction tube, and placed in the magnet for 1 min. After removal of the liquid, beads are washed another 2–3× with 900 μl RPMI-1%FCS and finally resuspended at $1 \times 10^7$/ml (minimum volume, 100 μl) in RPMI-1%FCS pre-warmed to 37 °C.
- DNase solution (100 U in 2 μl) is added per originally applied 10 μl of bead suspension to cut the DNA linker between beads and antibodies and thereby "liberate" the isolated ECs. Beads are rotated horizontally for 15 min at 37 °C, thereafter vigorously pipetted and finally placed in the magnet for 1 min.
- The isolated, purified ECs are now collected from the supernatant/solution and mixed with 500 μl of endothelial growth medium. The beads are rinsed twice with 500 μl of endothelial growth medium and returned to the magnet, and additionally harvested ECs are combined and transferred to a 30 mm culture dish and then supplemented with fibronectin at 1 μg/ml.

- Cells are cultured and, if required, subjected to an additional purification procedure with either anti-CD31 or anti-podoplanin beads, when cultures reach 70–80% confluence. The isolated ECs are further subjected to propagation with endothelial growth medium.

---

**Take-Home Message**

- Isolation of primary endothelial cells requires methods specific for the particular type of vessel and tissue.
- While EC outgrowth or vessel perfusion is mostly applied to macrovessels and yields pure endothelial cultures without further purification steps, the isolation of ECs from microvessels requires tissue digestion and subsequent EC purification based on endothelial markers.

## 14.2  Transfection of Endothelial Cells

Functional characterization and mechanistic studies commonly require the selective over-expression or shutdown of a single gene of interest which is generally accomplished by transfection or transduction techniques. Comparably, gene and drug transfer methods are required for in vitro studies of vascular cells but are also of interest for the in vivo application, since therapeutic interventions targeted at the endothelium are of major clinical interest. Hence, approaches of endothelial gene transfer have been reviewed numerous times over the past decades, including several more recent summaries [264–268].

### 14.2.1  In Vitro Methods

For gene transfer into isolated primary ECs in vitro, the established transfer methods had to be optimized to be efficient and largely inert (i.e., without altering endothelial properties). However, when transfecting EC monocultures in vitro, there is no need for a cell-type-specific approach. Non-viral methods generally mediate transient expression of the transgene unless selected for spontaneous chromosomal recombination, whereas viral vectors offer both the possibility for transient or stable gene transfer. Hence, when applying non-viral methods, transgene expression is usually investigated within the first day(s) after transfection, since cellular plasmid content declines with every day and cell division.

#### 14.2.1.1  Chemical Methods of In Vitro Gene Transfer

Primary endothelial cells are more refractory to chemical transfection methods than tumor cells and other fast-growing cell lines. Thus, standard techniques based on DEAE-dextran or calcium phosphate precipitation of DNA yield low transfection efficiency when compared to liposomes [269]. Lipofection of endothelial cells, i.e., gene transfer via complexes of plasmid DNA and cationic liposomes, was applied very early on [270]. Comparison of various commercially available, lipid-based reagents documented a maximal transfection efficiency around 30% of primary human endothelial cells [271], also with more advanced formulations [272]. In a comprehensive comparison of nine commercially available chemical transfection reagents in 2010, Lipofectamine LTX (Invitrogen, Carlsbad, CA, USA) was proposed to yield best results in terms of transfection efficiency and toxicity in human umbilical vein endothelial cells [273]. Furthermore, optimized conditions for the liposome-mediated transfection of silencing RNAs were developed [274].

### 14.2.1.2   Physical Methods of In Vitro Gene Transfer

Several groups have found electroporation to be superior to other non-viral transfection methods in mediating in vitro gene transfer to endothelial cells [275, 276]. While it comes at a considerable cost of about 50% cell death [277], the surviving cells recover within 24 hours and exhibit transfection efficiencies in the range of 70–90% which are achieved with conventional electroporation devices, simple buffer compositions, and electric pulse definitions [278].

### 14.2.1.3   Viral Methods of In Vitro Gene Transfer

Viral gene transfer is consistently more efficient than non-viral approaches but holds the risk of cell activation due to virus exposure [279]. Thus, viral vectors have gradually been optimized to contain a minimum of viral components while retaining transfer efficacy. The choice of viral backbone determines the fate of transgene expression. While the adenoviral genome remains extrachromosomal and hence mediates transient gene expression over several days to weeks, retro- or lentiviral vectors as well as adeno-associated virus (AAV) are applied to achieve long-term expression due to chromosomal integration.

Adenoviral protocols for vascular gene transfer were established and are widely adopted [280, 281], since transfer efficiency is higher than for retroviral transduction which requires proliferating cells. Depending on the receptor frequency, particular adenovirus serotypes such as 35 and 49 (as opposed to the most commonly used serotype 5) have been suggested for increased infection efficiency of primary ECs [282, 283]. Furthermore, the second-generation, helper-dependent adenoviral vectors were found to be superior to first-generation vectors in terms of retaining EC physiology and avoiding endothelial activation [284]. Adenoviral gene transfer was also tested and optimized for specialized EC populations such as corneal endothelium [285] and lymphatic endothelium [286].

C-type retroviral vectors originating from, e.g., murine leukemia virus require proliferating cells for transduction. Early retroviral transfer protocols for primary ECs [287, 288] were optimized to increase transduction efficiency by either complexation with polycation DEAE-dextran [289] or by pseudotyping with envelope glycoproteins from vesicular stomatitis virus [290] and gibbon ape leukemia virus [291] which use other receptors for cell entry. Furthermore, the use of lentiviral rather than C-type oncoretroviral vectors further improved stable gene transfer into vascular cells, as these vectors are largely independent of cell division and cell entry can also be enhanced by pseudotyping with vesicular stomatitis virus G glycoprotein [292] or hantavirus glycoprotein [292]. In a direct comparison, third-generation lentiviral vectors (pseudotyped with vesicular stomatitis virus glycoprotein) proved superior to alternate adenovirus or AAV vectors for vascular gene delivery [293]. Retro- and lentiviral gene transfer was also optimized for specialized EC populations such as liver sinusoidal endothelium [276, 294, 295] and corneal endothelial cells [296]. Importantly, depending on the particular vector and protocol, retroviral transduction may result in reduced endothelial proliferation [297] and adhesion [298, 299], or ECs may remain quiescent and exhibit unaltered cell growth [300, 301].

With respect to the application of gene transfer vectors based on adeno-associated virus, the choice of AAV serotype was found to determine transcytosis versus transduction of endothelial cells: while AAV9 preferentially shows transendothelial trafficking (and was hence proposed for the crossing of the blood-brain barrier), the AAV2 efficiently transduces microvascular endothelial cells and mediates persistent transgene expression [302, 303].

### 14.2.1.4   In Vitro-Engineered ECs for In Vivo Application

The therapeutic use of in vitro-transfected endothelium has already been proposed in the 1990s in the context of vascular graft seeding [304], and the idea was further advanced by the development of stents seeded with genetically modified endothelial cells to treat or control vascular disease [291, 305, 306]. Areas of application for ex vivo endothelial gene transfer were documented by arterial transduction of rat renal allografts with an AAV vector for IL-10 gene expression, which resulted in reduced graft inflammation and neointima formation [307]. DNA-liposome complexes were applied to achieve endothelial nitric oxide synthase (eNOS) delivery to transplanted arteries and reduce transplant arteriopathy [308]. Comparably, ex vivo gene transfer of eNOS was applied to retain patency of coronary artery bypass grafts [309] and lentiviral delivery of the anti-apoptotic bcl-xL gene to corneal endothelium prolonged survival of cornea allografts [310, 311].

Alternative to mature ECs, genetically engineered endothelial progenitor cells (EPCs) [312] have been proposed for clinical application, since they are known to incorporate at sites of vessel injury and angiogenesis [313]. Thus, EPCs were successfully applied to block disease progression in tumor models [314]. For example, EPCs transduced with the chemokine CCL19 achieved significant tumor reduction in a murine ovarian cancer model which was accompanied by increased tumor-infiltrating CD8+ lymphocytes [315]. Thymidine kinase-modified EPCs were successfully employed in a glioma mouse model with ganciclovir treatment [316]. With respect to cardiovascular disease, EPCs have been employed as gene transfer vehicle to improve blood flow in ischemia models [317]. Furthermore, EPCs engineered to overexpress protein kinase B and heme-oxygenase-1 improved outcome in a mouse model of myocardial infarction [318]. Based on the increased interest in EPCs as a vehicle for vascular delivery of gene therapy [319], different viral [320] and non-viral transfection techniques were compared for gene transfer into endothelial progenitor cells and mainly revealed adenoviral, AAV, or lentiviral vectors to be of superior efficacy [321–323]. However, results also indicated that transduced EPCs may exhibit an altered phenotype which may result in unintended therapy-related side effects [324].

Similar to EPCs, blood outgrowth endothelial cells (derived from peripheral blood mononuclear cells) were shown to home preferentially to angiogenic sites and to be amenable to genetic engineering and were thus proposed to serve as tumor-specific gene therapy tool [325–327]. When administered systemically, they were found to proliferate preferentially in tumor tissue relative to other organs and reduce tumor volume in mouse models by virtue of transgene expression [328].

### 14.2.1.5   Protocol for In Vitro EC Transfection by Electroporation

(Christine Brostjan, modified version of [275])

A representative, detailed protocol for non-viral in vitro gene transfer into primary endothelial cells by electroporation is provided as follows.

Protocol 4
- Harvest and count endothelial cells 1 day prior to transfection: seed at 30,000 cells per $cm^2$ in culture flasks to trigger cell proliferation; calculate $2 \times 10^6$ cells per transfection.
- The following day, EC cultures should be 70–80% confluent. Harvest cells by trypsinization, stop the reaction with cold (4 °C) RPMI1640 medium containing 10% FCS, and centrifuge cells for 5 min at 300 × g and 4 °C.
- Resuspend ECs in RPMI-10%FCS, place on ice and determine the cell count, and then dilute to $5 \times 10^6$ cells/ml in cold RPMI-10%FCS.
- In a 1.5 ml reaction tube, mix 20 µg plasmid DNA with 400 µl ($2 * 10^6$) cells and transfer to a 4 mm electroporation cuvette (Bio-Rad).
- Electroporate immediately at 200 V, 1200 µF which should result in a pulse length of 40–45 ms (Gene Pulser Xcell™, Bio-Rad).
- Retrieve the cells from the cuvette and distribute on two 30 mm wells (adjusting the volume to 2 ml/well with endothelial growth medium and fibronectin at 1 µg/ml).
- Change the medium after 4–5 hours and let ECs recover over night before testing for transgene expression or functional changes. The expected transfection efficiency for, e.g., the reporter gene vector pEGFP-C3 ranges at 70–90% (with about 30–50% loss due to cell death).

## 14.2.2 In Vivo Methods

Selective gene transfer into endothelium in vivo is either based on local vector application [329] or requires a targeting approach which favors transfection/infection and transgene expression in endothelial cells over other cell types. Two areas of clinical use have primarily been pursued: apart from therapeutic applications in vascular disease [330], the selective targeting of tumor endothelium (versus established vasculature) is a prime area of interest [331].

### 14.2.2.1 Local Intravascular Delivery for In Vivo Gene Transfer

The local delivery of gene transfer tools to arteries and veins relies mainly on surgical techniques, local perfusion, or catheter-based strategies [305, 332]. For example, the surgical isolation of rabbit carotid arteries and an arteriotomy to deliver a vector into the arterial lumen resulted in efficient and durable transgene expression [333]. Consequently, the local delivery of an adenoviral vector expressing apolipoprotein A-I in atherosclerotic rabbit carotid arteries triggered regression of atherosclerosis [334, 335].

Blood vessels within the eye have been another prime target for localized endothelial transduction. Injection of a lentiviral vector into the anterior chamber of the eye was successfully applied to transduce corneal endothelial cells [336]. Intravitreal adenoviral gene transfer of 15-lipoxygenase-1 was successfully applied to treat angioproliferative ocular disease in a rabbit model [337].

The application of gene expression vectors via angioplasty balloon or stents to either trigger blood vessel formation or reduce ischemic injury and prevent vascular remodeling has been at the focus of arterial gene transfer but should be considered transduction of smooth muscle rather than endothelial cells, since ECs are destroyed by the injury [338–341]. Adenovirus- and AAV-coated stents were applied to iliac arteries in a rabbit model of balloon injury and resulted in transgene detection in the vessel wall for up to 4 weeks [342]. Adenoviral transfer of eNOS was superior to liposomal formulations in "gene-eluting" stents to improve re-endothelialization and control restenosis of injured iliac arteries [343]. Immobilizing AAV vectors for iNOS gene transfer via antibody-protein G bridges to arterial stents, further improved vascular transduction and inhibition of restenosis [344]. Also, pseudotyped lentiviral vectors were successfully applied to introduce therapeutic genes into balloon-injured carotid arteries [292].

A special mode of local application was attempted by labeling of gene transfer vectors such as lentivirus or adenovirus with magnetic nanoparticles or by creating magnetic DNA microspheres which can then be positioned with the use of a magnetic field to confer localized EC transduction under systemic flow conditions [345–349].

### 14.2.2.2  Chemical Methods of Targeted In Vivo Gene or Drug Transfer

Modified liposomes have been at the forefront of achieving selective gene or drug transfer into endothelium upon systemic vector delivery [350]. Liposomes with high cationic lipid content were found to preferentially accumulate in tumor endothelium compared to normal vessels and were therefore employed to transport chemotherapeutics to cancer sites [351]. They were shown to give a moderately but significantly increased therapeutic efficacy compared to the free drug [352]. However, depending on the cationic lipid content and lipid type of cationic liposomes, their affinity for ECs varied, and they were found to also accumulate in lung, dermal, and coronary ECs [353].

Thus, liposomes were modified by incorporating peptide ligands specific for endothelial receptors to increase targeted delivery. For example, polyethylene glycol (PEG)-treated liposomes conjugated with RGD-peptides to target $\alpha v\beta 3$-integrins on tumor endothelium were loaded with a vascular disrupting agent and proved significantly more potent in reducing murine melanoma growth than the free drug [354]. Comparably, liposomes modified with the APRPG peptide (and PEG conjugate) showed high affinity for tumor vessels and significantly reduced colorectal carcinoma growth in mouse models, when these liposomes were filled with an antitumor nucleoside [355–357]. Numerous studies with liposomal formulations targeted to endothelial cells (mostly via incorporated peptide ligands – but also via aptamers [358, 359]) and loaded with toxic compounds, plasmid DNA, or siRNA [360] reportedly resulted in endothelial death, intratumoral microthrombosis, reduced tumor circulation, and cancer growth with limited side effects [361–368]. More recently developed peptide-conjugated, toxin-loaded liposomes were aimed at penetrating through the tumor blood vessels to reach the tumor tissue for improved antitumor efficacy [369].

To treat atherosclerosis, negatively charged liposomes were fused with VCAM-1 antibodies and were found to selectively bind to activated endothelium. When loaded with antiproliferative and anti-inflammatory prostaglandins, these modified liposomes could prevent vascular injury and reverse atherosclerotic disease [370]. Similarly, liposomes were decorated with E-selectin antibodies [371, 372] or ligands [373], ICAM-1 antibodies [374, 375], or EC-targeting peptides for gene transfer to activated endothelium in cardiovascular disease [376]. Liposomes carrying fibrinolytic agents and engineered to employ ligand-receptor interactions in coagulation-associated cell phenotypes have been tested as therapeutics in settings of thrombosis [377].

Regarding organ-specific vascular delivery, anti-endothelial cell protein C receptor antibodies were successfully applied to target corticosteroid-loaded liposomes to human retinal endothelial cells for potential anti-inflammatory treatment of ocular disease [378]. In contrast, when liposomes were modified with a peptide motif of ApoB-100 [379] or with a hyaluronic acid conjugate [380, 381], the endothelium of the liver was preferentially transfected in mouse models.

But also mere complexes of EC-targeted oligopeptides with naked DNA (without liposomal formulation) were tested for selective gene transfer into endothelium [382]. A ligand peptide to VEGFR combined with an amphiphilic peptide resulted in efficient and endothelial-selective delivery of plasmid DNA [383]. Also, the combination of the EC-targeting REDV peptide with the cell-penetrating TAT peptide (including a nuclear localization signal) was found to promote uptake, nuclear accumulation, and expression of the encapsulated DNA [384].

### 14.2.2.3   Physical Methods of Targeted In Vivo Gene Transfer

The in vivo application of electric pulses to trigger transgene uptake is possible with specifically designed electric probes and has been tested for corneal endothelium [385] and retinal endothelium [386]. However, it should be noted that EC treatment with electric pulses damages the cytoskeleton which results in a transient loss of barrier function, i.e., an increase in permeability without substantial cell death [387, 388].

A distinct combination of chemical and physical methods has been developed to be able to systemically supply DNA expression vectors yet trigger local uptake: ultrasound exposure in the presence of microbubble echocontrast agent was found to mediate uptake of naked or liposome-complexed DNA by the endothelium [389]. Also, ultrasound was applied to locally destroy DNA-loaded albumin microbubbles (created by in vitro sonication). This technique proved an efficient method for local vascular gene transfer into perfused coronary arteries [390] and was subsequently also shown for DNA-complexed cationic microbubbles in retinal endothelium [391]. The approach has further been applied in a dog model of coronary bypass using an expression plasmid for plasminogen activator in albumin nanoparticles crosslinked to microbubbles. Local myocardial transfection was achieved by the use of ultrasound and successfully prevented graft thrombosis and restenosis [392, 393]. Microbubbles can further be targeted to endothelium when conjugated to an antibody against endothelial markers such as CD9 or CD54 [394] and were shown to confer local transfection in ultrasound exposed post-ischemic hindlimb muscle [395].

### 14.2.2.4   Viral Methods of Targeted In Vivo Gene Transfer

The target cell repertoire of a viral vector matches the host range of the original virus and is determined by the receptor expression on target cells. However, the host range can be altered by engineering viral capsid or envelope proteins to interact with distinct surface molecules on the cell type of interest. Transduction efficacy is then dependent on cell entry mediated by the novel virus-receptor interaction. Regarding viral targeting of ECs, engineered variants of adenoviral, retro-/lentiviral, and AAV vectors have been primarily tested, but further approaches included measles virus and herpes simplex virus [396].

Thus, targeting of adenovirus to endothelium was attempted by capsid modifications [397] which were primarily directed at the fiber coat protein, including positively charged lysines to mediate interaction with heparan sulfate receptors [398] or RGD sequences that allow binding to integrin receptors [398] on ECs. Moreover, the CGKRK peptide (homing to tumor vasculature) was conjugated to PEG-treated adenoviral vectors. This modification resulted in a more than 100-fold enrichment in tumor endothelium versus liver compared to the unmodified adenovirus and in superior efficacy when transferring a therapeutic gene [399].

Successful targeting of adenoviral vectors to inflamed endothelium was achieved by blocking the endogenous adenovirus receptor interaction with polyethylene glycol and concomitantly attaching an E-selectin antibody to the viral capsid [400]. Furthermore, adenoviral vectors were targeted to pulmonary arterial endothelial cells via bispecific ("bridging") antibodies against angiotensin-converting enzyme. Pulmonary hypertension was markedly decreased by tryptophan hydroxylase-1 gene transfer in a rat model [401].

Comparably, the envelope protein of retrovirus/lentivirus was engineered to selectively target an endothelial receptor for cell-specific gene transfer. The applied methods included the incorporation of the NGR sequence motif for interaction with tumor vasculature [402]. Regarding lentiviral vectors, single-chain antibodies were incorporated to recognize the endothelial cell-surface antigen CD105. The engineered constructs were found to transduce ECs as efficiently as VSV-G pseudotyped vectors but discriminated between endothelial cells and other cell types [403]; these CD105-targeted lentiviral vectors were found to specifically transduce human endothelial cells transplanted into mice [404]. When a third-generation lentiviral vector was pseudotyped with truncated Nipah virus fusion and attachment protein, endothelial transduction specificity was achieved due to the natural vascular host range of Nipah virus [405].

While regular AAV vectors have limited efficacy in infecting human endothelial cells, they could be improved by the incorporation of EC-specific peptides [406]. A random peptide library was displayed on AAV9 to select for viral vector variants with increased specificity and efficacy in EC transduction [407]. A similar peptide library approach was taken to select an AAV capsid variant with specificity for brain microvascular ECs which might serve in the treatment of neurovascular disease [408]. An AAV-based capsid mutant vector with strong endothelial cell tropism was systemically administered and showed body-wide transduction of vasculature, with a particular accumulation in lung vessels suggesting therapeutic application in pulmonary complications [409]. Also, targeting of tumor endothelium was achieved with an engineered AAV/phage vector carrying an RGD peptide which specifically delivered therapeutic transgenes to tumors [410, 411] and achieved endothelial and bystander tumor cell death in animal models.

In addition to altering viral tropism for selective vascular transduction, efforts were made to restrict transgene expression to ECs by incorporating a cell-type-specific promoter [412]. Thus, the endothelial-specific Tie2 promoter and enhancer [413, 414], the VEGFR2 promoter [415, 416], the oxidized LDL receptor promoter [414], or the prepro-endothelin-1 promoter [417] has been applied. When the preproendothelin-1 promoter was combined with a hypoxia-inducible enhancer element in a systemically delivered adenoviral vector to achieve acid sphingomyelinase overexpression in proliferating angiogenic endothelium in mice, the tumor was rendered sensitive to radiation therapy [418]. Furthermore, the VE-cadherin promoter was included in a lentiviral vector that delivered shRNA against VEGFR-2: the knockdown of VEGFR-2 in retinal endothelial cells by a subretinal injection of this lentiviral vector blocked intravitreal neovascularization in a rat retinopathy model [419].

## 14.2.2.5  Clinical Application of Vascular Gene Transfer

The extensive efforts put forth in developing efficient gene and drug transfer methods into vasculature have resulted in numerous clinical trials and several agents reaching phase III [420]. Cardiovascular gene therapy has primarily focused on coronary artery disease, heart failure, and peripheral artery disease [266]. To this point, the majority of studies have been based on the local delivery of DNA or viral vectors. While several but not all trials reached a positive endpoint, the novel targeting approaches reviewed in this book chapter may indicate the path to further advance clinical application of vascular gene transfer.

---

Take-Home Message

 ▬ Efficient in vitro gene transfer into isolated endothelial cells is achieved with optimized viral and non-viral methods (in particular electroporation) and constitutes a valuable tool for mechanistic cell studies.
 ▬ For in vivo manipulation of endothelial cells, more intricate approaches are required to selectively target ECs: modified viral and non-viral vectors recognizing endothelial surface molecules have been applied in animal models to transfer therapeutic genes or compounds and have yielded promising results which are currently transferred to clinical trials on cardio-vascular disease and cancer.

---

# References

1. Pomerat CM, Slick WC. Isolation and growth of endothelial cells in tissue culture. Nature. 1963;198:859–61.
2. Gore I, Tsutsumi H. Isolation of living endothelial cells by gelatin-film stripping of vascular walls. Stain Technol. 1969;44(3):139–42.
3. Berthiaume F. Isolation and culture of human endothelial cells. Methods Mol Med. 1999;18:253–9.
4. Mahabeleshwar GH, Somanath PR, Byzova TV. Methods for isolation of endothelial and smooth muscle cells and in vitro proliferation assays. Methods Mol Med. 2006;129:197–208.
5. Mason JC, Lidington EA, Yarwood H. Isolation and analysis of large and small vessel endothelial cells. Methods Mol Med. 2007;135:305–21.
6. van Beijnum JR, Rousch M, Castermans K, et al. Isolation of endothelial cells from fresh tissues. Nat Protoc. 2008;3(6):1085–91.
7. Hewett PW. Vascular endothelial cells from human micro- and macrovessels: isolation, characterisation and culture. Methods Mol Biol (Clifton, NJ). 2009;467:95–111.
8. Hewett PW. Isolation and culture of human endothelial cells from micro- and macro-vessels. Methods Mol Biol (Clifton, NJ). 2016;1430:61–76.
9. Yamaguchi T, Ichise T, Iwata O, et al. Development of a new method for isolation and long-term culture of organ-specific blood vascular and lymphatic endothelial cells of the mouse. FEBS J. 2008;275(9):1988–98.
10. Oosterhoff LA, Kruitwagen HS, Spee B, et al. Isolation and culture of primary endothelial cells from canine arteries and veins. J Vis Exp: JoVE. 2016;(117).
11. Chen SF, Fei X, Li SH. A new simple method for isolation of microvascular endothelial cells avoiding both chemical and mechanical injuries. Microvasc Res. 1995;50(1):119–28.
12. Zhou L, Sohet F, Daneman R. Purification of endothelial cells from rodent brain by immunopanning. Cold Spring Harb Protoc. 2014;2014(1):65–77.
13. Froehlich R, Buchanan R, Wagner R. Isolation and separation of fenestrated and continuous capillary endothelium by using magnetic beads. Microvasc Res. 1988;35(2):242–5.

14. Jackson CJ, Garbett PK, Nissen B, et al. Binding of human endothelium to Ulex europaeus I-coated Dynabeads: application to the isolation of microvascular endothelium. J Cell Sci. 1990;96(Pt 2): 257–62.

15. Gomez DE, Hartzler JL, Corbitt RH, et al. Immunomagnetic separation as a final purification step of liver endothelial cells. In Vitro Cell Dev Biol Anim. 1993;29(6):451–5.

16. Hewett PW, Murray JC. Immunomagnetic purification of human microvessel endothelial cells using Dynabeads coated with monoclonal antibodies to PECAM-1. Eur J Cell Biol. 1993;62(2):451–4.

17. Conrad-Lapostolle V, Bordenave L, Baquey C. Optimization of use of UEA-1 magnetic beads for endothelial cell isolation. Cell Biol Toxicol. 1996;12(4–6):189–97.

18. Springhorn JP. Isolation of human capillary endothelial cells using paramagnetic beads conjugated to anti-PECAM antibodies. Cold Spring Harb Protoc. 2011;2011(5):pdb prot4479.

19. Sahagun G, Moore SA, Fabry Z, et al. Purification of murine endothelial cell cultures by flow cytometry using fluorescein-labeled griffonia simplicifolia agglutinin. Am J Pathol. 1989;134(6):1227–32.

20. Thiele W, Rothley M, Schmaus A, et al. Flow cytometry-based isolation of dermal lymphatic endothelial cells from newborn rats. Lymphology. 2014;47(4):177–86.

21. Lokmic Z, Ng ES, Burton M, et al. Isolation of human lymphatic endothelial cells by multi-parameter fluorescence-activated cell sorting. J Vis Exp: JoVE. 2015;(99):e52691.

22. Voyta JC, Via DP, Butterfield CE, et al. Identification and isolation of endothelial cells based on their increased uptake of acetylated-low density lipoprotein. J Cell Biol. 1984;99(6):2034–40.

23. Springhorn JP, Madri JA, Squinto SP. Human capillary endothelial cells from abdominal wall adipose tissue: isolation using an anti-pecam antibody. In Vitro Cell Dev Biol Anim. 1995;31(6):473–81.

24. Hewett PW, Murray JC. Isolation of microvascular endothelial cells using magnetic beads coated with anti-PECAM-1 antibodies. In Vitro Cell Dev Biol Anim. 1996;32(8):462.

25. Demeule M, Labelle M, Regina A, et al. Isolation of endothelial cells from brain, lung, and kidney: expression of the multidrug resistance P-glycoprotein isoforms. Biochem Biophys Res Commun. 2001;281(3):827–34.

26. Mao Q, Huang X, He J, et al. A novel method for endothelial cell isolation. Oncol Rep. 2016;35(3): 1652–6.

27. Naschberger E, Regensburger D, Tenkerian C, et al. Isolation of human endothelial cells from normal colon and colorectal carcinoma – an improved protocol. J Vis Exp: JoVE. 2018;(134).

28. Fehrenbach ML, Cao G, Williams JT, et al. Isolation of murine lung endothelial cells. Am J Physiol Lung Cell Mol Physiol. 2009;296(6):L1096–103.

29. Gerritsen ME, Shen CP, McHugh MC, et al. Activation-dependent isolation and culture of murine pulmonary microvascular endothelium. Microcirculation (New York, NY: 1994). 1995;2(2):151–63.

30. Richard L, Velasco P, Detmar M. A simple immunomagnetic protocol for the selective isolation and long-term culture of human dermal microvascular endothelial cells. Exp Cell Res. 1998;240(1):1–6.

31. Kriehuber E, Breiteneder-Geleff S, Groeger M, et al. Isolation and characterization of dermal lymphatic and blood endothelial cells reveal stable and functionally specialized cell lineages. J Exp Med. 2001;194(6):797–808.

32. Jaffe EA. Culture and identification of large vessel endothelial cells. In: Biology of endothelial cells. Boston: Springer US; 1984. p. 1–13.

33. Montezano AC, Neves KB, Lopes RA, et al. Isolation and culture of endothelial cells from large vessels. Methods Mol Biol (Clifton, NJ). 2017;1527:345–8.

34. Jaffe EA, Nachman RL, Becker CG, et al. Culture of human endothelial cells derived from umbilical veins. Identification by morphologic and immunologic criteria. J Clin Invest. 1973;52(11):2745–56.

35. Ataollahi F, Pingguan-Murphy B, Moradi A, et al. New method for the isolation of endothelial cells from large vessels. Cytotherapy. 2014;16(8):1145–52.

36. Jin GZ, Park JH, Lee EJ, et al. Utilizing PCL microcarriers for high-purity isolation of primary endothelial cells for tissue engineering. Tissue Eng Part C Methods. 2014;20(9):761–8.

37. van Leeuwen EB, Molema G, de Jong KP, et al. One-step method for endothelial cell isolation from large human blood vessels using fibrin glue. Lab Invest. 2000;80(6):987–9.

38. Hewett PW, Murray JC. Human microvessel endothelial cells: isolation, culture and characterization. In Vitro Cell Dev Biol Anim. 1993;29A(11):823–30.

39. Scott PA, Bicknell R. The isolation and culture of microvascular endothelium. J Cell Sci. 1993;105(Pt 2):269–73.

40. Richard L, Velasco P, Detmar M. Isolation and culture of microvascular endothelial cells. Methods Mol Med. 1999;18:261–9.

**14**

41. Leak LV, Jones M. Lymphatic endothelium isolation, characterization and long-term culture. Anat Rec. 1993;236(4):641–52.
42. Garrafa E, Trainini L, Benetti A, et al. Isolation, purification, and heterogeneity of human lymphatic endothelial cells from different tissues. Lymphology. 2005;38(4):159–66.
43. Lokmic Z. Isolation, identification, and culture of human lymphatic endothelial cells. Methods Mol Biol (Clifton, NJ). 2016;1430:77–90.
44. Hohensinner PJ. The heart as a source for factors involved in homing and differentiation of progenitor cells (PhD Thesis): Medical University of Vienna; 2007.
45. Leclercq A, Veillat V, Loriot S, et al. A methodology for concomitant isolation of intimal and adventitial endothelial cells from the human thoracic aorta. PLoS One. 2015;10(11):e0143144.
46. Huang H, McIntosh J, Hoyt DG. An efficient, nonenzymatic method for isolation and culture of murine aortic endothelial cells and their response to inflammatory stimuli. In Vitro Cell Dev Biol Anim. 2003;39(1–2):43–50.
47. Lincoln DW 2nd, Larsen AM, Phillips PG, et al. Isolation of murine aortic endothelial cells in culture and the effects of sex steroids on their growth. In Vitro Cell Dev Biol Anim. 2003;39(3–4):140–5.
48. Magid R, Martinson D, Hwang J, et al. Optimization of isolation and functional characterization of primary murine aortic endothelial cells. Endothelium. 2003;10(2):103–9.
49. Choi S, Kim JA, Kim KC, et al. Isolation and in vitro culture of vascular endothelial cells from mice. Korean J Physiol Pharmacol. 2015;19(1):35–42.
50. Molina-Sanchez P, Andres V. Isolation of mouse primary aortic endothelial cells by selection with specific antibodies. Methods Mol Biol (Clifton, NJ). 2015;1339:111–7.
51. Wang JM, Chen AF, Zhang K. Isolation and primary culture of mouse aortic endothelial cells. J Vis Exp: JoVE. 2016;(118).
52. Carrillo A, Chamorro S, Rodriguez-Gago M, et al. Isolation and characterization of immortalized porcine aortic endothelial cell lines. Vet Immunol Immunopathol. 2002;89(1–2):91–8.
53. Cole OF, Fan TP, Lewis GP. Isolation, characterisation, growth and culture of endothelial cells from the rat aorta. Cell Biol Int Rep. 1986;10(6):399–405.
54. McGuire PG, Orkin RW. Isolation of rat aortic endothelial cells by primary explant techniques and their phenotypic modulation by defined substrata. Lab Invest. 1987;57(1):94–105.
55. Nicosia RF, Villaschi S, Smith M. Isolation and characterization of vasoformative endothelial cells from the rat aorta. In Vitro Cell Dev Biol Anim. 1994;30A(6):394–9.
56. Chao CF, Chao ZC, Lu MH, et al. Isolation and cultivation of aortic endothelial cells from spontaneously hypertensive rat with a modified tissue explant technique. Proc Natl Sci Counc Repub China B. 1995;19(4):208–15.
57. Oite T, Suzuki Y, Morioka T, et al. Efficient isolation of rat aortic endothelial cells by elimination of contaminating cells with a monoclonal antibody. Microvasc Res. 1995;50(1):113–8.
58. Dame MK, Yu X, Garrido R, et al. A stepwise method for the isolation of endothelial cells and smooth muscle cells from individual canine coronary arteries. In Vitro Cell Dev Biol Anim. 2003;39(10):402–6.
59. Howerth EW, Stallknecht DE. Isolation and culture of large vessel endothelium from white-tailed deer (Odocoileus virginianus). J Vet Diagn Invest. 1995;7(1):137–42.
60. Dietze K, Slosarek I, Fuhrmann-Selter T, et al. Isolation of equine endothelial cells and life cell angiogenesis assay. Clin Hemorheol Microcirc. 2014;58(1):127–46.
61. Moss SC, Bates M, Parrino PE, et al. Isolation of endothelial cells and vascular smooth muscle cells from internal mammary artery tissue. Ochsner J. 2007;7(3):133–6.
62. Cirillo P, Golino P, Ragni M, et al. A simple method for the isolation, cultivation, and characterization of endothelial cells from rabbit coronary circulation. Thromb Res. 1999;96(4):329–33.
63. Nistri S, Mazzetti L, Failli P, et al. High-yield method for isolation and culture of endothelial cells from rat coronary blood vessels suitable for analysis of intracellular calcium and nitric oxide biosynthetic pathways. Biol Proced Online. 2002;4:32–7.
64. Wechezak AR, Mansfield PB. Isolation and growth characteristics of cell lines from bovine venous endothelium. In Vitro. 1973;9(1):39–45.
65. Ford JW, Burkel WE, Kahn RH. Isolation of adult canine venous endothelium for tissue culture. In Vitro. 1981;17(1):44–50.
66. Moyer CF, Dennis PA, Majno G, et al. Venular endothelium in vitro: isolation and characterization. In Vitro Cell Dev Biol. 1988;24(4):359–68.
67. Abernethy NJ, Hay JB. Isolation, culture, and characterization of endothelial cells derived from the post-capillary venules of sheep mesenteric lymph nodes, Peyer's patches, and associated small bowel. Int Immunol. 1989;1(4):378–87.

68. Kader KN, Moore LR, Saul JM, et al. Isolation and purification of canine adipose microvascular endothelial cells. Microvasc Res. 2001;61(2):220–6.
69. Twal WO, Leach RM. Isolation and characterization of microvascular endothelial cells from chicken fat pads. In Vitro Cell Dev Biol Anim. 1996;32(7):403–8.
70. Kern PA, Knedler A, Eckel RH. Isolation and culture of microvascular endothelium from human adipose tissue. J Clin Invest. 1983;71(6):1822–9.
71. Hewett PW, Murray JC, Price EA, et al. Isolation and characterization of microvessel endothelial cells from human mammary adipose tissue. In Vitro Cell Dev Biol Anim. 1993;29A(4):325–31.
72. Hewett PW. Microvessel endothelial cells from human adipose tissues : isolation, identification, and culture. Methods Mol Med. 2001;46:213–6.
73. Springhorn JP. Isolation of human capillary endothelial cells from abdominal adipose tissue. Cold Spring Harb Protoc. 2011;2011(5):pdb prot4537.
74. Launder TM, Gegen NW, Knedler A, et al. The isolation and characterization of enriched microvascular endothelial cells from mouse adipose tissue. The induction of class II molecules of the major histocompatibility complex (MHC) by interferon-gamma (IFN-gamma). J Immunol Methods. 1987;102(1):45–52.
75. Kajimoto K, Hossen MN, Hida K, et al. Isolation and culture of microvascular endothelial cells from murine inguinal and epididymal adipose tissues. J Immunol Methods. 2010;357(1–2):43–50.
76. Wagner RC, Matthews MA. The isolation and culture of capillary endothelium from epididymal fat. Microvasc Res. 1975;10(3):286–97.
77. Bjorntorp P, Hansson GK, Jonasson L, et al. Isolation and characterization of endothelial cells from the epididymal fat pad of the rat. J Lipid Res. 1983;24(2):105–12.
78. Frye CA, Patrick CW Jr. Isolation and culture of rat microvascular endothelial cells. In Vitro Cell Dev Biol Anim. 2002;38(4):208–12.
79. Pauli BU, Anderson SN, Memoli VA, et al. The isolation and characterization in vitro of normal epithelial cells, endothelial cells and fibroblasts from rat urinary bladder. Tissue Cell. 1980;12(3):419–36.
80. Kerachian MA, Cournoyer D, Harvey EJ, et al. Isolation and characterization of human bone-derived endothelial cells. Endothelium. 2007;14(2):115–21.
81. Rafii S, Shapiro F, Rimarachin J, et al. Isolation and characterization of human bone marrow microvascular endothelial cells: hematopoietic progenitor cell adhesion. Blood. 1994;84(1):10–9.
82. Masek LC, Sweetenham JW. Isolation and culture of endothelial cells from human bone marrow. Br J Haematol. 1994;88(4):855–65.
83. Schweitzer CM, van der Schoot CE, Drager AM, et al. Isolation and culture of human bone marrow endothelial cells. Exp Hematol. 1995;23(1):41–8.
84. Almeida-Porada G, Ascensao JL. Isolation, characterization, and biologic features of bone marrow endothelial cells. J Lab Clin Med. 1996;128(4):399–407.
85. Irie S, Tavassoli M. Purification and characterization of rat bone marrow endothelial cells. Exp Hematol. 1986;14(10):912–8.
86. Machi T, Kassell NF, Scheld WM. Isolation and characterization of endothelial cells from bovine cerebral arteries. In Vitro Cell Dev Biol. 1990;26(3 Pt 1):291–300.
87. Reisner A, Olson JJ, Yang J, et al. Isolation and culture of bovine intracranial arterial endothelial cells. Neurosurgery. 1995;36(4):806–12; discussion 13.
88. Siakotos AN. The isolation of endothelial cells from normal human and bovine brain. Methods Enzymol. 1974;32:717–22.
89. Phillips P, Kumar P, Kumar S, et al. Isolation and characterization of endothelial cells from rat and cow brain white matter. J Anat. 1979;129(Pt 2):261–72.
90. Costello P, Del Maestro R. Human cerebral endothelium: isolation and characterization of cells derived from microvessels of non-neoplastic and malignant glial tissue. J Neuro-Oncol. 1990;8(3):231–43.
91. Biegel D, Spencer DD, Pachter JS. Isolation and culture of human brain microvessel endothelial cells for the study of blood-brain barrier properties in vitro. Brain Res. 1995;692(1–2):183–9.
92. Lamszus K, Schmidt NO, Ergun S, et al. Isolation and culture of human neuromicrovascular endothelial cells for the study of angiogenesis in vitro. J Neurosci Res. 1999;55(3):370–81.
93. Unger RE, Oltrogge JB, von Briesen H, et al. Isolation and molecular characterization of brain microvascular endothelial cells from human brain tumors. In Vitro Cell Dev Biol Anim. 2002;38(5):273–81.
94. Dorovini-Zis K, Prameya R, Huynh H. Isolation and characterization of human brain endothelial cells. Methods Mol Med. 2003;89:325–36.
95. Cayrol R, Haqqani AS, Ifergan I, et al. Isolation of human brain endothelial cells and characterization of lipid raft-associated proteins by mass spectroscopy. Methods Mol Biol (Clifton, NJ). 2011;686:275–95.

14

96. Navone SE, Marfia G, Invernici G, et al. Isolation and expansion of human and mouse brain microvascular endothelial cells. Nat Protoc. 2013;8(9):1680–93.
97. Rosas-Hernandez H, Cuevas E, Lantz SM, et al. Isolation and culture of brain microvascular endothelial cells for in vitro blood-brain barrier studies. Methods Mol Biol (Clifton, NJ). 2018;1727:315–31.
98. Gay F, Robert C, Pouvelle B, et al. Isolation and characterization of brain microvascular endothelial cells from Saimiri monkeys. An in vitro model for sequestration of Plasmodium falciparum-infected erythrocytes. J Immunol Methods. 1995;184(1):15–28.
99. Tontsch U, Bauer HC. Isolation, characterization, and long-term cultivation of porcine and murine cerebral capillary endothelial cells. Microvasc Res. 1989;37(2):148–61.
100. Wu Z, Hofman FM, Zlokovic BV. A simple method for isolation and characterization of mouse brain microvascular endothelial cells. J Neurosci Methods. 2003;130(1):53–63.
101. Ruck T, Bittner S, Epping L, et al. Isolation of primary murine brain microvascular endothelial cells. J Vis Exp: JoVE. 2014;(93):e52204.
102. Welser-Alves JV, Boroujerdi A, Milner R. Isolation and culture of primary mouse brain endothelial cells. Methods Mol Biol (Clifton, NJ). 2014;1135:345–56.
103. Wylot B, Konarzewska K, Bugajski L, et al. Isolation of vascular endothelial cells from intact and injured murine brain cortex-technical issues and pitfalls in FACS analysis of the nervous tissue. Cytometry A. 2015;87(10):908–20.
104. Assmann JC, Muller K, Wenzel J, et al. Isolation and cultivation of primary brain endothelial cells from adult mice. Bio Protoc. 2017;7(10):pii: e2294.
105. Crouch EE, Doetsch F. FACS isolation of endothelial cells and pericytes from mouse brain microregions. Nat Protoc. 2018;13(4):738–51.
106. Patabendige A, Abbott NJ. Primary porcine brain microvessel endothelial cell isolation and culture. Curr Protoc Neurosci. 2014;69:3.27.1–17.
107. Diglio CA, Liu W, Grammas P, et al. Isolation and characterization of cerebral resistance vessel endothelium in culture. Tissue Cell. 1993;25(6):833–46.
108. Williams SK, Gillis JF, Matthews MA, et al. Isolation and characterization of brain endothelial cells: morphology and enzyme activity. J Neurochem. 1980;35(2):374–81.
109. Diglio CA, Grammas P, Giacomelli F, et al. Primary culture of rat cerebral microvascular endothelial cells. Isolation, growth, and characterization. Lab Invest. 1982;46(6):554–63.
110. Ichikawa N, Naora K, Hirano H, et al. Isolation and primary culture of rat cerebral microvascular endothelial cells for studying drug transport in vitro. J Pharmacol Toxicol Methods. 1996;36(1):45–52.
111. Perriere N, Demeuse P, Garcia E, et al. Puromycin-based purification of rat brain capillary endothelial cell cultures. Effect on the expression of blood-brain barrier-specific properties. J Neurochem. 2005;93(2):279–89.
112. Luo J, Yin X, Sanchez A, et al. Purification of endothelial cells from rat brain. Methods in Mol Biol (Clifton, NJ). 2014;1135:357–64.
113. Chaturvedi K, Sarkar DK. Isolation and characterization of rat pituitary endothelial cells. Neuroendocrinology. 2006;83(5–6):387–93.
114. Christenson LK, Stouffer RL. Isolation and culture of microvascular endothelial cells from the primate corpus luteum. Biol Reprod. 1996;55(6):1397–404.
115. Basini G, Falasconi I, Bussolati S, et al. Isolation of endothelial cells and pericytes from swine corpus luteum. Domest Anim Endocrinol. 2014;48:100–9.
116. Bagavandoss P, Wilks JW. Isolation and characterization of microvascular endothelial cells from developing corpus luteum. Biol Reprod. 1991;44(6):1132–9.
117. Drake BL, Loke YW. Isolation of endothelial cells from human first trimester decidua using immunomagnetic beads. Hum Reprod (Oxford, England). 1991;6(8):1156–9.
118. Gallery ED, Rowe J, Schrieber L, et al. Isolation and purification of microvascular endothelium from human decidual tissue in the late phase of pregnancy. Am J Obstet Gynecol. 1991;165(1):191–6.
119. Lindenbaum ES, Langer N, Beach D. Isolation and culture of human decidual capillary endothelial cells in serum-free medium supplemented with human uterine angiogenic factor. Acta Anat. 1991;140(3):273–9.
120. Grimwood J, Bicknell R, Rees MC. The isolation, characterization and culture of human decidual endothelium. Hum Reprod (Oxford, England). 1995;10(8):2142–8.
121. Rafiee P, Ogawa H, Heidemann J, et al. Isolation and characterization of human esophageal microvascular endothelial cells: mechanisms of inflammatory activation. Am J Physiol Gastrointest Liver Physiol. 2003;285(6):G1277–92.

122. Morse LS, Sidikaro Y. Isolation and characterization of bovine choroidal microvessel endothelium and pericytes in culture. Curr Eye Res. 1990;9(7):631–42.
123. Liu X, Li W. Isolation, culture and characterization of bovine choriocapillary endothelial cells. Exp Eye Res. 1993;57(1):37–44.
124. Hoffmann S, Spee C, Murata T, et al. Rapid isolation of choriocapillary endothelial cells by Lycopersicon esculentum-coated Dynabeads. Graefes Arch Clin Exp Ophthalmol (Albrecht von Graefes Archiv fur klinische und experimentelle Ophthalmologie). 1998;236(10):779–84.
125. Schor AM, Schor SL. The isolation and culture of endothelial cells and pericytes from the bovine retinal microvasculature: a comparative study with large vessel vascular cells. Microvasc Res. 1986;32(1):21–38.
126. Antonetti DA, Wolpert EB. Isolation and characterization of retinal endothelial cells. Methods Mol Med. 2003;89:365–74.
127. Banumathi E, Haribalaganesh R, Babu SS, et al. High-yielding enzymatic method for isolation and culture of microvascular endothelial cells from bovine retinal blood vessels. Microvasc Res. 2009;77(3):377–81.
128. Browning AC, Gray T, Amoaku WM. Isolation, culture, and characterisation of human macular inner choroidal microvascular endothelial cells. Br J Ophthalmol. 2005;89(10):1343–7.
129. Engelmann K, Bohnke M, Friedl P. Isolation and long-term cultivation of human corneal endothelial cells. Invest Ophthalmol Vis Sci. 1988;29(11):1656–62.
130. Pistsov MY, Sadovnikova E, Danilov SM. Human corneal endothelial cells: isolation, characterization and long-term cultivation. Exp Eye Res. 1988;47(3):403–14.
131. Li W, Sabater AL, Chen YT, et al. A novel method of isolation, preservation, and expansion of human corneal endothelial cells. Invest Ophthalmol Vis Sci. 2007;48(2):614–20.
132. Mimura T, Yamagami S, Yokoo S, et al. Selective isolation of young cells from human corneal endothelium by the sphere-forming assay. Tissue Eng Part C Methods. 2010;16(4):803–12.
133. Choi JS, Kim EY, Kim MJ, et al. Factors affecting successful isolation of human corneal endothelial cells for clinical use. Cell Transplant. 2014;23(7):845–54.
134. Gillies PJ, Bray LJ, Richardson NA, et al. Isolation of microvascular endothelial cells from cadaveric corneal limbus. Exp Eye Res. 2015;131:20–8.
135. Xiaozhuang Z, Xianqiong L, Jingbo J, et al. Isolation and characterization of fetus human retinal microvascular endothelial cells. Ophthalmic Res. 2010;44(2):125–30.
136. Stamer WD, Roberts BC, Howell DN, et al. Isolation, culture, and characterization of endothelial cells from Schlemm's canal. Invest Ophthalmol Vis Sci. 1998;39(10):1804–12.
137. Su X, Sorenson CM, Sheibani N. Isolation and characterization of murine retinal endothelial cells. Mol Vis. 2003;9:171–8.
138. Matsubara TA, Murata TA, Wu GS, et al. Isolation and culture of rat retinal microvessel endothelial cells using magnetic beads coated with antibodies to PECAM-1. Curr Eye Res. 2000;20(1):1–7.
139. Cha MS, Rah DK, Lee KH. Isolation and pure culture of microvascular endothelial cells from the fetal skin. Yonsei Med J. 1996;37(3):186–93.
140. Miller LJ, Lincoln J. Isolation of murine valve endothelial cells. J Vis Exp: JoVE. 2014;(90).
141. Dyer LA, Patterson C. Isolation of embryonic ventricular endothelial cells. J Vis Exp: JoVE. 2013;(77).
142. Plendl J, Neumuller C, Vollmar A, et al. Isolation and characterization of endothelial cells from different organs of fetal pigs. Anat Embryol. 1996;194(5):445–56.
143. Weber SC, Gratopp A, Akanbi S, et al. Isolation and culture of fibroblasts, vascular smooth muscle, and endothelial cells from the fetal rat ductus arteriosus. Pediatr Res. 2011;70(3):236–41.
144. DeCarlo AA, Cohen JA, Aguado A, et al. Isolation and characterization of human gingival microvascular endothelial cells. J Periodontal Res. 2008;43(2):246–54.
145. Koren CW, Sveinbjornsson B, Smedsrod B. Isolation and culture of endocardial endothelial cells from Atlantic salmon (Salmo salar) and Atlantic cod (Gadus morhua). Cell Tissue Res. 1997;290(1):89–99.
146. Nees S, Gerbes AL, Gerlach E, et al. Isolation, identification, and continuous culture of coronary endothelial cells from guinea pig hearts. Eur J Cell Biol. 1981;24(2):287–97.
147. Oxhorn BC, Hirzel DJ, Buxton IL. Isolation and characterization of large numbers of endothelial cells for studies of cell signaling. Microvasc Res. 2002;64(2):302–15.
148. Grafe M, Auch-Schwelk W, Graf K, et al. Isolation and characterization of macrovascular and microvascular endothelial cells from human hearts. Am J Phys. 1994;267(6 Pt 2):H2138–48.
149. Nishida M, Carley WW, Gerritsen ME, et al. Isolation and characterization of human and rat cardiac microvascular endothelial cells. Am J Phys. 1993;264(2 Pt 2):H639–52.

**14**

150. McDouall RM, Yacoub M, Rose ML. Isolation, culture, and characterisation of MHC class II-positive microvascular endothelial cells from the human heart. Microvasc Res. 1996;51(2):137–52.
151. Marelli-Berg FM, Peek E, Lidington EA, et al. Isolation of endothelial cells from murine tissue. J Immunol Methods. 2000;244(1–2):205–15.
152. Lim YC, Luscinskas FW. Isolation and culture of murine heart and lung endothelial cells for in vitro model systems. Methods Mol Biol (Clifton, NJ). 2006;341:141–54.
153. Jin Y, Liu Y, Antonyak M, et al. Isolation and characterization of vascular endothelial cells from murine heart and lung. Methods Mol Biol (Clifton, NJ). 2012;843:147–54.
154. Pratumvinit B, Reesukumal K, Janebodin K, et al. Isolation, characterization, and transplantation of cardiac endothelial cells. Biomed Res Int. 2013;2013:359412.
155. Luo S, Truong AH, Makino A. Isolation of mouse coronary endothelial cells. J Vis Exp: JoVE. 2016;(113).
156. Gould RA, Butcher JT. Isolation of valvular endothelial cells. J Vis Exp: JoVE. 2010;(46).
157. Manduteanu I, Radu A, Simionescu M. Isolation and cultivation of rabbit endocardial endothelial cells. Preliminary data. Morphol Embryol. 1988;34(3):165–9.
158. Widmann MD, Letsou GV, Phan S, et al. Isolation and characterization of rabbit cardiac endothelial cells: response to cyclic strain and growth factors in vitro. J Surg Res. 1992;53(4):331–4.
159. Diglio CA, Grammas P, Giacomelli F, et al. Rat heart-derived endothelial and smooth muscle cell cultures: isolation, cloning and characterization. Tissue Cell. 1988;20(4):477–92.
160. He Q, Spiro MJ. Isolation of rat heart endothelial cells and pericytes: evaluation of their role in the formation of extracellular matrix components. J Mol Cell Cardiol. 1995;27(5):1173–83.
161. Gunduz D, Hamm CW, Aslam M. Simultaneous isolation of high quality cardiomyocytes, endothelial cells, and fibroblasts from an adult rat heart. J Vis Exp: JoVE. 2017;(123).
162. Simionescu M, Simionescu N. Isolation and characterization of endothelial cells from the heart microvasculature. Microvasc Res. 1978;16(3):426–52.
163. Haraldsen G, Rugtveit J, Kvale D, et al. Isolation and long term culture of human intestinal microvascular endothelial cells. Gut. 1995;37(2):225–34.
164. Naschberger E, Schellerer VS, Rau TT, et al. Isolation of endothelial cells from human tumors. Methods Mol Biol (Clifton, NJ). 2011;731:209–18.
165. Trochta OA, Jacobs RM, Jefferson BJ. Isolation and short term culture of canine adrenal microvascular endothelial cells. Can J Vet Res (Revue canadienne de recherche veterinaire). 1999;63(1):1–4.
166. Fawcett J, Harris AL, Bicknell R. Isolation and properties in culture of human adrenal capillary endothelial cells. Biochem Biophys Res Commun. 1991;174(2):903–8.
167. Striker GE, Soderland C, Bowen-Pope DF, et al. Isolation, characterization, and propagation in vitro of human glomerular endothelial cells. J Exp Med. 1984;160(1):323–8.
168. Muczynski KA, Ekle DM, Coder DM, et al. Normal human kidney HLA-DR-expressing renal microvascular endothelial cells: characterization, isolation, and regulation of MHC class II expression. J Am Soc Nephrol: JASN. 2003;14(5):1336–48.
169. McGinn S, Poronnik P, Gallery ED, et al. A method for the isolation of glomerular and tubulointerstitial endothelial cells and a comparison of characteristics with the human umbilical vein endothelial cell model. Nephrology (Carlton, VIC). 2004;9(4):229–37.
170. Akis N, Madaio MP. Isolation, culture, and characterization of endothelial cells from mouse glomeruli. Kidney Int. 2004;65(6):2223–7.
171. Rops AL, van der Vlag J, Jacobs CW, et al. Isolation and characterization of conditionally immortalized mouse glomerular endothelial cell lines. Kidney Int. 2004;66(6):2193–201.
172. Zhao Y, Zhao H, Zhang Y, et al. Isolation and epithelial co-culture of mouse renal peritubular endothelial cells. BMC Cell Biol. 2014;15:40.
173. Shaw RG, Johnson AR, Schulz WW, et al. Sinusoidal endothelial cells from normal guinea pig liver: isolation, culture and characterization. Hepatology (Baltimore, MD). 1984;4(4):591–602.
174. Gerlach JC, Zeilinger K, Spatkowski G, et al. Large-scale isolation of sinusoidal endothelial cells from pig and human liver. J Surg Res. 2001;100(1):39–45.
175. Gomez DE, Thorgeirsson UP. Lectins as tools for the purification of liver endothelial cells. Methods Mol Med. 1998;9:319–28.
176. Vidal-Vanaclocha F, Rocha M, Asumendi A, et al. Isolation and enrichment of two sublobular compartment-specific endothelial cell subpopulations from liver sinusoids. Hepatology (Baltimore, MD). 1993;18(2):328–39.
177. Cheluvappa R. Standardized isolation and culture of murine liver sinusoidal endothelial cells. Curr Protoc Cell Biol. 2014;65:2.9.1–8.

178. Meyer J, Gonelle-Gispert C, Morel P, et al. Methods for isolation and purification of murine liver sinusoidal endothelial cells: a systematic review. PLoS One. 2016;11(3):e0151945.
179. Meyer J, Lacotte S, Morel P, et al. An optimized method for mouse liver sinusoidal endothelial cell isolation. Exp Cell Res. 2016;349(2):291–301.
180. Liu J, Huang X, Werner M, et al. Advanced method for isolation of mouse hepatocytes, liver sinusoidal endothelial cells, and kupffer cells. Methods Mol Biol (Clifton, NJ). 2017;1540:249–58.
181. Cabral F, Miller CM, Kudrna KM, et al. Purification of hepatocytes and sinusoidal endothelial cells from mouse liver perfusion. J Vis Exp: JoVE. 2018;(132).
182. Knook DL, Blansjaar N, Sleyster EC. Isolation and characterization of Kupffer and endothelial cells from the rat liver. Exp Cell Res. 1977;109(2):317–29.
183. Praaning-Van Dalen DP, Knook DL. Quantitative determination of in vivo endocytosis by rat liver Kupffer and endothelial cells facilitated by an improved cell isolation method. FEBS Lett. 1982;141(2):229–32.
184. Braet F, De Zanger R, Sasaoki T, et al. Assessment of a method of isolation, purification, and cultivation of rat liver sinusoidal endothelial cells. Lab Invest. 1994;70(6):944–52.
185. Tokairin T, Nishikawa Y, Doi Y, et al. A highly specific isolation of rat sinusoidal endothelial cells by the immunomagnetic bead method using SE-1 monoclonal antibody. J Hepatol. 2002;36(6):725–33.
186. Xie G, Wang L, Wang X, et al. Isolation of periportal, midlobular, and centrilobular rat liver sinusoidal endothelial cells enables study of zonated drug toxicity. Am J Physiol Gastrointest Liver Physiol. 2010;299(5):G1204–10.
187. Ryan US, Clements E, Habliston D, et al. Isolation and culture of pulmonary artery endothelial cells. Tissue Cell. 1978;10(3):535–54.
188. MacEachern KE, Smith GL, Nolan AM. Methods for the isolation, culture and characterisation of equine pulmonary artery endothelial cells. Res Vet Sci. 1997;62(2):147–52.
189. Habliston DL, Whitaker C, Hart MA, et al. Isolation and culture of endothelial cells from the lungs of small animals. Am Rev Respir Dis. 1979;119(6):853–68.
190. Visner GA, Staples ED, Chesrown SE, et al. Isolation and maintenance of human pulmonary artery endothelial cells in culture isolated from transplant donors. Am J Phys. 1994;267(4 Pt 1):L406–13.
191. Lorusso B, Falco A, Madeddu D, et al. Isolation and characterization of human lung lymphatic endothelial cells. Biomed Res Int. 2015;2015:747864.
192. Marinescu D, Tapu V, Eskenasy A. Isolation, culture, and some morphohistochemical features of endothelial cells of the pulmonary microvasculature. Morphol Embryol. 1983;29(1):47–52.
193. Carley WW, Niedbala MJ, Gerritsen ME. Isolation, cultivation, and partial characterization of microvascular endothelium derived from human lung. Am J Respir Cell Mol Biol. 1992;7(6):620–30.
194. Hewett PW, Murray JC. Human lung microvessel endothelial cells: isolation, culture, and characterization. Microvasc Res. 1993;46(1):89–102.
195. Lou JN, Mili N, Decrind C, et al. An improved method for isolation of microvascular endothelial cells from normal and inflamed human lung. In Vitro Cell Dev Biol Anim. 1998;34(7):529–36.
196. Gaskill C, Majka SM. A high-yield isolation and enrichment strategy for human lung microvascular endothelial cells. Pulm Circ. 2017;7(1):108–16.
197. Dong QG, Bernasconi S, Lostaglio S, et al. A general strategy for isolation of endothelial cells from murine tissues. Characterization of two endothelial cell lines from the murine lung and subcutaneous sponge implants. Arterioscler Thromb Vasc Biol. 1997;17(8):1599–604.
198. Sobczak M, Dargatz J, Chrzanowska-Wodnicka M. Isolation and culture of pulmonary endothelial cells from neonatal mice. J Vis Exp: JoVE. 2010;(46).
199. Cao G, Abraham V, DeLisser HM. Isolation of endothelial cells from mouse lung. Curr Protoc Toxicol. 2014;61:24.2.1–9.
200. Carley WW, Tanoue L, Merker M, et al. Isolation of rabbit pulmonary microvascular endothelial cells and characterization of their angiotensin converting enzyme activity. Pulm Pharmacol. 1990;3(1):35–40.
201. Peng G, Wen X, Shi Y, et al. Development of a new method for the isolation and culture of pulmonary arterial endothelial cells from rat pulmonary arteries. J Vasc Res. 2013;50(6):468–77.
202. Kim NS, Kim SJ. Isolation and cultivation of microvascular endothelial cells from rat lungs: effects of gelatin substratum and serum. Yonsei Med J. 1991;32(4):303–14.
203. Magee JC, Stone AE, Oldham KT, et al. Isolation, culture, and characterization of rat lung microvascular endothelial cells. Am J Phys. 1994;267(4 Pt 1):L433–41.

14

204. Berendam SJ, Fallert Junecko BA, Murphey-Corb MA, et al. Isolation, characterization, and functional analysis of ferret lymphatic endothelial cells. Vet Immunol Immunopathol. 2015;163(3–4):134–45.
205. Cook-Mills JM, Gallagher JS, Feldbush TL. Isolation and characterization of high endothelial cell lines derived from mouse lymph nodes. In Vitro Cell Dev Biol Anim. 1996;32(3):167–77.
206. Ager A. Isolation and culture of high endothelial cells from rat lymph nodes. J Cell Sci. 1987;87(Pt 1):133–44.
207. Gargett CE, Bucak K, Rogers PA. Isolation, characterization and long-term culture of human myometrial microvascular endothelial cells. Hum Reprod (Oxford, England). 2000;15(2):293–301.
208. Kanda T, Iwasaki T, Yamawaki M, et al. Isolation and culture of bovine endothelial cells of endoneurial origin. J Neurosci Res. 1997;49(6):769–77.
209. Sano Y, Kanda T. Isolation and properties of endothelial cells forming the blood-nerve barrier. Methods Mol Biol (Clifton, NJ). 2011;686:417–25.
210. Ge S, Pachter JS. Isolation and culture of microvascular endothelial cells from murine spinal cord. J Neuroimmunol. 2006;177(1–2):209–14.
211. Argall KG, Armati PJ, Pollard JD. A method for the isolation and culture of rat peripheral nerve vascular endothelial cells. Mol Cell Neurosci. 1994;5(5):413–7.
212. Zhou L, Sohet F, Daneman R. Purification and culture of central nervous system endothelial cells. Cold Spring Harb Protoc. 2014;2014(1):44–6.
213. Bochsler PN, Slauson DO, Chandler SK, et al. Isolation and characterization of equine microvascular endothelial cells in vitro. Am J Vet Res. 1989;50(10):1800–5.
214. Chung-Welch N, Patton WF, Shepro D, et al. Two-stage isolation procedure for obtaining homogenous populations of microvascular endothelial and mesothelial cells from human omentum. Microvasc Res. 1997;54(2):121–34.
215. Winiarski BK, Acheson N, Gutowski NJ, et al. An improved and reliable method for isolation of microvascular endothelial cells from human omentum. Microcirculation (New York, NY: 1994). 2011;18(8):635–45.
216. Leach L, Bhasin Y, Clark P, et al. Isolation of endothelial cells from human term placental villi using immunomagnetic beads. Placenta. 1994;15(4):355–64.
217. Schutz M, Friedl P. Isolation and cultivation of endothelial cells derived from human placenta. Eur J Cell Biol. 1996;71(4):395–401.
218. Ugele B, Lange F. Isolation of endothelial cells from human placental microvessels: effect of different proteolytic enzymes on releasing endothelial cells from villous tissue. In Vitro Cell Dev Biol Anim. 2001;37(7):408–13.
219. Chi L, Delgado-Olguin P. Isolation and culture of mouse placental endothelial cells. Methods Mol Biol (Clifton, NJ). 2018;1752:101–9.
220. Lehr JE, Yamazaki K, Onoda J, et al. Immunomagnetic isolation of endothelial cells from normal rat prostate tissue. In Vivo (Athens, Greece). 1994;8(6):983–8.
221. Ieronimakis N, Balasundaram G, Reyes M. Direct isolation, culture and transplant of mouse skeletal muscle derived endothelial cells with angiogenic potential. PLoS One. 2008;3(3):e0001753.
222. Davison PM, Bensch K, Karasek MA. Isolation and growth of endothelial cells from the microvessels of the newborn human foreskin in cell culture. J Invest Dermatol. 1980;75(4):316–21.
223. Davison PM, Bensch K, Karasek MA. Isolation and long-term serial cultivation of endothelial cells from the microvessels of the adult human dermis. In Vitro. 1983;19(12):937–45.
224. Kraling BM, Jimenez SA, Sorger T, et al. Isolation and characterization of microvascular endothelial cells from the adult human dermis and from skin biopsies of patients with systemic sclerosis. Lab Invest. 1994;71(5):745–54.
225. Normand J, Karasek MA. A method for the isolation and serial propagation of keratinocytes, endothelial cells, and fibroblasts from a single punch biopsy of human skin. In Vitro Cell Dev Biol Anim. 1995;31(6):447–55.
226. Gupta K, Ramakrishnan S, Browne PV, et al. A novel technique for culture of human dermal microvascular endothelial cells under either serum-free or serum-supplemented conditions: isolation by panning and stimulation with vascular endothelial growth factor. Exp Cell Res. 1997;230(2):244–51.
227. Cha ST, Talavera D, Demir E, et al. A method of isolation and culture of microvascular endothelial cells from mouse skin. Microvasc Res. 2005;70(3):198–204.
228. Hull MA, Hewett PW, Brough JL, et al. Isolation and culture of human gastric endothelial cells. Gastroenterology. 1996;111(5):1230–40.

229. Jones MK, Wang H, Tomikawa M, et al. Isolation and characterization of rat gastric microvascular endothelial cells as a model for studying gastric angiogenesis in vitro. J Physiol Pharmacol. 2000;51(4 Pt 2):813–20.
230. Djoneidi M, Brodt P. Isolation and characterization of rat lymphatic endothelial cells. Microcirc Endothel Lymphat. 1991;7(4–6):161–82.
231. Patel VA, Logan A, Watkinson JC, et al. Isolation and characterization of human thyroid endothelial cells. Am J Phys Endocrinol Metab. 2003;284(1):E168–76.
232. Garrafa E, Alessandri G, Benetti A, et al. Isolation and characterization of lymphatic microvascular endothelial cells from human tonsils. J Cell Physiol. 2006;207(1):107–13.
233. Ulrich-Merzenich G, Metzner C, Bhonde RR, et al. Simultaneous isolation of endothelial and smooth muscle cells from human umbilical artery or vein and their growth response to low-density lipoproteins. In Vitro Cell Dev Biol Anim. 2002;38(5):265–72.
234. Thilo DG, Muller-Kusel S, Heinrich D, et al. Isolation of human venous endothelial cells by different proteases. Artery. 1980;8(3):259–66.
235. Morgan DM. Isolation and culture of human umbilical vein endothelial cells. Methods Mol Med. 1996;2:101–9.
236. Larrivee B, Karsan A. Isolation and culture of primary endothelial cells. Methods Mol Biol (Clifton, NJ). 2005;290:315–29.
237. Baudin B, Bruneel A, Bosselut N, et al. A protocol for isolation and culture of human umbilical vein endothelial cells. Nat Protoc. 2007;2(3):481–5.
238. Cheung AL. Isolation and culture of human umbilical vein endothelial cells (HUVEC). Curr Protoc Microbiol. 2007;Appendix 4:Appendix 4B.
239. Crampton SP, Davis J, Hughes CC. Isolation of human umbilical vein endothelial cells (HUVEC). J Vis Exp: JoVE. 2007;(3):183.
240. Kadam SS, Tiwari S, Bhonde RR. Simultaneous isolation of vascular endothelial cells and mesenchymal stem cells from the human umbilical cord. In Vitro Cell Dev Biol Anim. 2009;45(1–2):23–7.
241. Lattuada D, Roda B, Pignatari C, et al. A tag-less method for direct isolation of human umbilical vein endothelial cells by gravitational field-flow fractionation. Anal Bioanal Chem. 2013;405(2–3):977–84.
242. Lei J, Peng S, Samuel SB, et al. A simple and biosafe method for isolation of human umbilical vein endothelial cells. Anal Biochem. 2016;508:15–8.
243. Yu SY, Song YM, Li AM, et al. Isolation and characterization of human coronary artery-derived endothelial cells in vivo from patients undergoing percutaneous coronary interventions. J Vasc Res. 2009;46(5):487–94.
244. Darrow AL, Maresh JG, Shohet RV. Mouse models and techniques for the isolation of the diabetic endothelium. ISRN Endocrinol. 2013;2013:165397.
245. Mackay LS, Dodd S, Dougall IG, et al. Isolation and characterisation of human pulmonary microvascular endothelial cells from patients with severe emphysema. Respir Res. 2013;14:23.
246. Wang XQ, Liu YK, Mao ZG, et al. Isolation, culture and characterization of endothelial cells from human hypertrophic scar. Endothelium. 2008;15(3):113–9.
247. Jackson CJ, Garbett PK, Marks RM, et al. Isolation and propagation of endothelial cells derived from rheumatoid synovial microvasculature. Ann Rheum Dis. 1989;48(9):733–6.
248. Abbot SE, Kaul A, Stevens CR, et al. Isolation and culture of synovial microvascular endothelial cells. Characterization and assessment of adhesion molecule expression. Arthritis Rheum. 1992;35(4):401–6.
249. Izumi Y, Hoshino Y, Hosoya K, et al. Isolation and characterization of canine tumor endothelial cells. J Vet Med Sci. 2015;77(3):359–63.
250. Grange C, Bussolati B, Bruno S, et al. Isolation and characterization of human breast tumor-derived endothelial cells. Oncol Rep. 2006;15(2):381–6.
251. Miebach S, Grau S, Hummel V, et al. Isolation and culture of microvascular endothelial cells from gliomas of different WHO grades. J Neuro-Oncol. 2006;76(1):39–48.
252. You L, Wu M, Chen Y, et al. Isolation and characterization of lymphatic endothelial cells from human glossal lymphangioma. Oncol Rep. 2010;23(1):105–11.
253. Cho JH, Han I, Lee MR, et al. Isolation and characterization of endothelial cells from intramuscular hemangioma. J Orthop Sci. 2013;18(1):137–44.
254. Wang J, Liu Z, Hu L, et al. The isolation and characterization of endothelial cells from juvenile nasopharyngeal angiofibroma. Acta Biochim Biophys Sin. 2016;48(9):856–8.

14

255. Alessandri G, Chirivi RG, Castellani P, et al. Isolation and characterization of human tumor-derived capillary endothelial cells: role of oncofetal fibronectin. Lab Invest. 1998;78(1):127–8.
256. Hannum RS, Ojeifo JO, Zwiebel JA, et al. Isolation of tumor-derived endothelial cells. Microvasc Res. 2001;61(3):287–90.
257. Xiao L, McCann JV, Dudley AC. Isolation and culture expansion of tumor-specific endothelial cells. J Vis Exp: JoVE. 2015;(105):e53072.
258. Okaji Y, Tsuno NH, Kitayama J, et al. A novel method for isolation of endothelial cells and macrophages from murine tumors based on Ac-LDL uptake and CD16 expression. J Immunol Methods. 2004;295(1–2):183–93.
259. Modzelewski RA, Davies P, Watkins SC, et al. Isolation and identification of fresh tumor-derived endothelial cells from a murine RIF-1 fibrosarcoma. Cancer Res. 1994;54(2):336–9.
260. Utoguchi N, Dantakean A, Makimoto H, et al. Isolation and properties of tumor-derived endothelial cells from rat KMT-17 fibrosarcoma. Jpn J Cancer Res: Gann. 1995;86(2):193–201.
261. Zhang HF, Liang GB, Zhao MG, et al. An efficient and non-enzymatic method for isolation and culture of endothelial cells from the nidus of human cerebral arteriovenous malformations. Neurosci Lett. 2013;548:21–6.
262. Hao Q, Chen XL, Ma L, et al. Procedure for the isolation of endothelial cells from human cerebral arteriovenous malformation (cAVM) tissues. Front Cell Neurosci. 2018;12:30.
263. Jia J, Zhao Y, Zhang W, et al. Isolation, culture and identification of human venous malformation endothelial cells. (Zhonghua kou qiang yi xue za zhi = Zhonghua kouqiang yixue zazhi) Chin J Stomatol. 2002;37(4):284–6.
264. Karvinen H, Yla-Herttuala S. New aspects in vascular gene therapy. Curr Opin Pharmacol. 2010;10(2):208–11.
265. Sedighiani F, Nikol S. Gene therapy in vascular disease. Surg: J Royal Coll Surg Edinb Irel. 2011;9(6):326–35.
266. Halonen PJ, Nurro J, Kuivanen A, et al. Current gene therapy trials for vascular diseases. Expert Opin Biol Ther. 2014;14(3):327–36.
267. Laakkonen JP, Yla-Herttuala S. Recent advancements in cardiovascular gene therapy and vascular biology. Hum Gene Ther. 2015;26(8):518–24.
268. Cheng HS, Fish JE. Neovascularization driven by MicroRNA delivery to the endothelium. Arterioscler Thromb Vasc Biol. 2015;35(11):2263–5.
269. Tanner FC, Carr DP, Nabel GJ, et al. Transfection of human endothelial cells. Cardiovasc Res. 1997;35(3):522–8.
270. Brigham KL, Meyrick B, Christman B, et al. Expression of a prokaryotic gene in cultured lung endothelial cells after lipofection with a plasmid vector. Am J Respir Cell Mol Biol. 1989;1(2):95–100.
271. Young AT, Lakey JR, Murray AG, et al. Gene therapy: a lipofection approach for gene transfer into primary endothelial cells. Cell Transplant. 2002;11(6):573–82.
272. Brito L, Little S, Langer R, et al. Poly(beta-amino ester) and cationic phospholipid-based lipopolyplexes for gene delivery and transfection in human aortic endothelial and smooth muscle cells. Biomacromolecules. 2008;9(4):1179–87.
273. Hunt MA, Currie MJ, Robinson BA, et al. Optimizing transfection of primary human umbilical vein endothelial cells using commercially available chemical transfection reagents. J Biomol Tech: JBT. 2010;21(2):66–72.
274. Nolte A, Raabe C, Walker T, et al. Optimized basic conditions are essential for successful siRNA transfection into primary endothelial cells. Oligonucleotides. 2009;19(2):141–50.
275. Hernandez JL, Coll T, Ciudad CJ. A highly efficient electroporation method for the transfection of endothelial cells. Angiogenesis. 2004;7(3):235–41.
276. Paez-Cortez J, Montano R, Iacomini J, et al. Liver sinusoidal endothelial cells as possible vehicles for gene therapy: a comparison between plasmid-based and lentiviral gene transfer techniques. Endothelium. 2008;15(4):165–73.
277. Meulenberg CJ, Todorovic V, Cemazar M. Differential cellular effects of electroporation and electrochemotherapy in monolayers of human microvascular endothelial cells. PLoS One. 2012;7(12):e52713.
278. Yockell-Lelievre J, Riendeau V, Gagnon SN, et al. Efficient transfection of endothelial cells by a double-pulse electroporation method. DNA Cell Biol. 2009;28(11):561–6.
279. Tan PH, Xue SA, Manunta M, et al. Effect of vectors on human endothelial cell signal transduction: implications for cardiovascular gene therapy. Arterioscler Thromb Vasc Biol. 2006;26(3):462–7.

280. Nicklin SA, Baker AH. Simple methods for preparing recombinant adenoviruses for high-efficiency transduction of vascular cells. Methods Mol Med. 1999;30:271–83.
281. Takahashi T, Takahashi K, Daniel TO. High-efficiency and low-toxicity adenovirus-assisted endothelial transfection. Methods Mol Med. 1999;30:307–14.
282. Shinozaki K, Suominen E, Carrick F, et al. Efficient infection of tumor endothelial cells by a capsid-modified adenovirus. Gene Ther. 2006;13(1):52–9.
283. Dakin RS, Parker AL, Delles C, et al. Efficient transduction of primary vascular cells by the rare adenovirus serotype 49 vector. Hum Gene Ther. 2015;26(5):312–9.
284. Flynn R, Buckler JM, Tang C, et al. Helper-dependent adenoviral vectors are superior in vitro to first-generation vectors for endothelial cell-targeted gene therapy. Mol Ther. 2010;18(12):2121–9.
285. Bertelmann E. Genetic manipulation of corneal endothelial cells: transfection and viral transduction. Methods Mol Biol (Clifton, NJ). 2009;467:229–39.
286. Gashev AA, Davis MJ, Gasheva OY, et al. Methods for lymphatic vessel culture and gene transfection. Microcirculation (New York, NY: 1994). 2009;16(7):615–28.
287. Dichek DA. Retroviral vector-mediated gene transfer into endothelial cells. Mol Biol Med. 1991;8(2):257–66.
288. Dichek DA, Nussbaum O, Degen SJ, et al. Enhancement of the fibrinolytic activity of sheep endothelial cells by retroviral vector-mediated gene transfer. Blood. 1991;77(3):533–41.
289. Kahn ML, Lee SW, Dichek DA. Optimization of retroviral vector-mediated gene transfer into endothelial cells in vitro. Circ Res. 1992;71(6):1508–17.
290. Yu H, Eton D, Wang Y, et al. High efficiency in vitro gene transfer into vascular tissues using a pseudo-typed retroviral vector without pseudotransduction. Gene Ther. 1999;6(11):1876–83.
291. Koren B, Weisz A, Fischer L, et al. Efficient transduction and seeding of human endothelial cells onto metallic stents using bicistronic pseudo-typed retroviral vectors encoding vascular endothelial growth factor. Cardiovas Revasc Med. 2006;7(3):173–8.
292. Qian Z, Haessler M, Lemos JA, et al. Targeting vascular injury using Hantavirus-pseudotyped lentiviral vectors. Mol Ther. 2006;13(4):694–704.
293. Dishart KL, Denby L, George SJ, et al. Third-generation lentivirus vectors efficiently transduce and phenotypically modify vascular cells: implications for gene therapy. J Mol Cell Cardiol. 2003;35(7):739–48.
294. Totsugawa T, Kobayashi N, Maruyama M, et al. Lentiviral vector: a useful tool for transduction of human liver endothelial cells. ASAIO J (American Society for Artificial Internal Organs: 1992). 2003;49(6):635–40.
295. Paez J, Montano R, Benatuil L, et al. High efficiency and long-term foreign gene expression in cultured liver sinusoidal endothelial cells by retroviral transduction. Endothelium. 2006;13(4):279–85.
296. Valtink M, Stanke N, Knels L, et al. Pseudotyping and culture conditions affect efficiency and cytotoxicity of retroviral gene transfer to human corneal endothelial cells. Invest Ophthalmol Vis Sci. 2011;52(9):6807–13.
297. Baer RP, Whitehill TE, Sarkar R, et al. Retroviral-mediated transduction of endothelial cells with the lac Z gene impairs cellular proliferation in vitro and graft endothelialization in vivo. J Vasc Surg. 1996;24(5):892–9.
298. Sackman JE, Cezeaux JL, Reddick TT, et al. Evaluation of the effect of retroviral gene transduction on vascular endothelial cell adhesion. Tissue Eng. 1996;2(3):223–34.
299. Sackman JE, Wymore AM, Reddick TT, et al. Retroviral mediated gene transduction alters integrin expression on vascular endothelial cells. J Surg Res. 1997;69(1):45–50.
300. Jankowski RJ, Severyn DA, Vorp DA, et al. Effect of retroviral transduction on human endothelial cell phenotype and adhesion to Dacron vascular grafts. J Vasc Surg. 1997;26(4):676–84.
301. Inaba M, Toninelli E, Vanmeter G, et al. Retroviral gene transfer: effects on endothelial cell phenotype. J Surg Res. 1998;78(1):31–6.
302. Merkel SF, Andrews AM, Lutton EM, et al. Trafficking of adeno-associated virus vectors across a model of the blood-brain barrier; a comparative study of transcytosis and transduction using primary human brain endothelial cells. J Neurochem. 2017;140(2):216–30.
303. Weber-Adrian D, Heinen S, Silburt J, et al. The human brain endothelial barrier: transcytosis of AAV9, transduction by AAV2: an editorial highlight for 'trafficking of adeno-associated virus vectors across a model of the blood-brain barrier; a comparative study of transcytosis and transduction using primary human brain endothelial cells'. J Neurochem. 2017;140(2):192–4.
304. Callow AD. The vascular endothelial cell as a vehicle for gene therapy. J Vasc Surg. 1990;11(6):793–8.

14

305. Janssens SP. Applied gene therapy in preclinical models of vascular injury. Curr Atheroscler Rep. 2003;5(3):186–90.
306. Robertson KE, McDonald RA, Oldroyd KG, et al. Prevention of coronary in-stent restenosis and vein graft failure: does vascular gene therapy have a role? Pharmacol Ther. 2012;136(1):23–34.
307. Xie J, Li X, Meng D, et al. Transduction of interleukin-10 through renal artery attenuates vascular neointimal proliferation and infiltration of immune cells in rat renal allograft. Immunol Lett. 2016;176:105–13.
308. Iwata A, Sai S, Moore M, et al. Gene therapy of transplant arteriopathy by liposome-mediated transfection of endothelial nitric oxide synthase. J Heart Lung Transplant. 2000;19(11):1017–28.
309. Zhu Y, Wang HS, Li XM, et al. Establishment of a rabbit model of coronary artery bypass graft and endothelial nitric oxide synthase gene transfection. Genet Mol Res: GMR. 2015;14(1):1479–86.
310. Barcia RN, Dana MR, Kazlauskas A. Corneal graft rejection is accompanied by apoptosis of the endothelium and is prevented by gene therapy with bcl-xL. Am J Transplant. 2007;7(9):2082–9.
311. Kampik D, Ali RR, Larkin DF. Experimental gene transfer to the corneal endothelium. Exp Eye Res. 2012;95(1):54–9.
312. Brunt KR, Hall SR, Ward CA, et al. Endothelial progenitor cell and mesenchymal stem cell isolation, characterization, viral transduction. Methods Mol Med. 2007;139:197–210.
313. Asahara T. Cell therapy and gene therapy using endothelial progenitor cells for vascular regeneration. Handb Exp Pharmacol. 2007;(180):181–94.
314. Debatin KM, Wei J, Beltinger C. Endothelial progenitor cells for cancer gene therapy. Gene Ther. 2008;15(10):780–6.
315. Hamanishi J, Mandai M, Matsumura N, et al. Activated local immunity by CC chemokine ligand 19-transduced embryonic endothelial progenitor cells suppresses metastasis of murine ovarian cancer. Stem Cells (Dayton, Ohio). 2010;28(1):164–73.
316. Zhang JX, Kang CS, Shi L, et al. Use of thymidine kinase gene-modified endothelial progenitor cells as a vector targeting angiogenesis in glioma gene therapy. Oncology. 2010;78(2):94–102.
317. Choi JH, Hur J, Yoon CH, et al. Augmentation of therapeutic angiogenesis using genetically modified human endothelial progenitor cells with altered glycogen synthase kinase-3beta activity. J Biol Chem. 2004;279(47):49430–8.
318. Brunt KR, Wu J, Chen Z, et al. Ex vivo Akt/HO-1 gene therapy to human endothelial progenitor cells enhances myocardial infarction recovery. Cell Transplant. 2012;21(7):1443–61.
319. Dudek AZ. Endothelial lineage cell as a vehicle for systemic delivery of cancer gene therapy. Transl Res. 2010;156(3):136–46.
320. Stockschlaeder M, Shardakova O, Weber K, et al. Highly efficient lentiviral transduction of phenotypically and genotypically characterized endothelial progenitor cells from adult peripheral blood. Blood Coagul Fibrinolysis. 2010;21(5):464–73.
321. Kealy B, Liew A, McMahon JM, et al. Comparison of viral and nonviral vectors for gene transfer to human endothelial progenitor cells. Tissue Eng Part C Methods. 2009;15(2):223–31.
322. Dickens S, Van den Berge S, Hendrickx B, et al. Nonviral transfection strategies for keratinocytes, fibroblasts, and endothelial progenitor cells for ex vivo gene transfer to skin wounds. Tissue Eng Part C Methods. 2010;16(6):1601–8.
323. Wang Z, Fu Q, Cao J, et al. Impact of transduction towards the proliferation and migration as well as the transduction efficiency of human umbilical cord-derived late endothelial progenitor cells with nine recombinant adeno-associated virus serotypes. Biotechnol Lett. 2016;38(7):1073–9.
324. Werling NJ, Thorpe R, Zhao Y. A systematic approach to the establishment and characterization of endothelial progenitor cells for gene therapy. Hum Gene Ther Methods. 2013;24(3):171–84.
325. Dudek AZ, Bodempudi V, Welsh BW, et al. Systemic inhibition of tumour angiogenesis by endothelial cell-based gene therapy. Br J Cancer. 2007;97(4):513–22.
326. Wei J, Jarmy G, Genuneit J, et al. Human blood late outgrowth endothelial cells for gene therapy of cancer: determinants of efficacy. Gene Ther. 2007;14(4):344–56.
327. Milbauer LC, Enenstein JA, Roney M, et al. Blood outgrowth endothelial cell migration and trapping in vivo: a window into gene therapy. Transl Res. 2009;153(4):179–89.
328. Bodempudi V, Ohlfest JR, Terai K, et al. Blood outgrowth endothelial cell-based systemic delivery of antiangiogenic gene therapy for solid tumors. Cancer Gene Ther. 2010;17(12):855–63.
329. Nicklin SA, Baker AH. Efficient vascular endothelial gene transfer following intravenous adenovirus delivery. Mol Ther. 2008;16(12):1904–5.

330. Ghosh R, Walsh SR, Tang TY, et al. Gene therapy as a novel therapeutic option in the treatment of peripheral vascular disease: systematic review and meta-analysis. Int J Clin Pract. 2008;62(9): 1383–90.
331. Bazan-Peregrino M, Seymour LW, Harris AL. Gene therapy targeting to tumor endothelium. Cancer Gene Ther. 2007;14(2):117–27.
332. Willard JE, Landau C, Glamann DB, et al. Genetic modification of the vessel wall. Comparison of surgical and catheter-based techniques for delivery of recombinant adenovirus. Circulation. 1994;89(5): 2190–7.
333. Wacker BK, Bi L, Dichek DA. In vivo gene transfer to the rabbit common carotid artery endothelium. J Vis Exp: JoVE. 2018;(135).
334. Wacker BK, Dronadula N, Zhang J, et al. Local vascular gene therapy with apolipoprotein A-I to promote regression of atherosclerosis. Arterioscler Thromb Vasc Biol. 2017;37(2):316–27.
335. Wacker BK, Dronadula N, Bi L, et al. Apo A-I (Apolipoprotein A-I) vascular gene therapy provides durable protection against atherosclerosis in hyperlipidemic rabbits. Arterioscler Thromb Vasc Biol. 2018;38(1):206–17.
336. Bainbridge JW, Stephens C, Parsley K, et al. In vivo gene transfer to the mouse eye using an HIV-based lentiviral vector; efficient long-term transduction of corneal endothelium and retinal pigment epithelium. Gene Ther. 2001;8(21):1665–8.
337. Viita H, Kinnunen K, Eriksson E, et al. Intravitreal adenoviral 15-lipoxygenase-1 gene transfer prevents vascular endothelial growth factor A-induced neovascularization in rabbit eyes. Hum Gene Ther. 2009;20(12):1679–86.
338. Asahara T, Chen D, Tsurumi Y, et al. Accelerated restitution of endothelial integrity and endothelium-dependent function after phVEGF165 gene transfer. Circulation. 1996;94(12):3291–302.
339. Takeshita S, Weir L, Chen D, et al. Therapeutic angiogenesis following arterial gene transfer of vascular endothelial growth factor in a rabbit model of hindlimb ischemia. Biochem Biophys Res Commun. 1996;227(2):628–35.
340. Morishige K, Shimokawa H, Yamawaki T, et al. Local adenovirus-mediated transfer of C-type natriuretic peptide suppresses vascular remodeling in porcine coronary arteries in vivo. J Am Coll Cardiol. 2000;35(4):1040–7.
341. Fishbein I, Forbes SP, Adamo RF, et al. Vascular gene transfer from metallic stent surfaces using adenoviral vectors tethered through hydrolysable cross-linkers. J Vis Exp: JoVE. 2014;(90):e51653.
342. Sharif F, Hynes SO, McMahon J, et al. Gene-eluting stents: comparison of adenoviral and adeno-associated viral gene delivery to the blood vessel wall in vivo. Hum Gene Ther. 2006;17(7): 741–50.
343. Sharif F, Hynes SO, McCullagh KJ, et al. Gene-eluting stents: non-viral, liposome-based gene delivery of eNOS to the blood vessel wall in vivo results in enhanced endothelialization but does not reduce restenosis in a hypercholesterolemic model. Gene Ther. 2012;19(3):321–8.
344. Fishbein I, Guerrero DT, Alferiev IS, et al. Stent-based delivery of adeno-associated viral vectors with sustained vascular transduction and iNOS-mediated inhibition of in-stent restenosis. Gene Ther. 2017;24(11):717–26.
345. Chorny M, Fishbein I, Alferiev I, et al. Magnetically responsive biodegradable nanoparticles enhance adenoviral gene transfer in cultured smooth muscle and endothelial cells. Mol Pharm. 2009;6(5): 1380–7.
346. Wenzel D, Rieck S, Vosen S, et al. Identification of magnetic nanoparticles for combined positioning and lentiviral transduction of endothelial cells. Pharm Res. 2012;29(5):1242–54.
347. Voronina N, Lemcke H, Wiekhorst F, et al. Non-viral magnetic engineering of endothelial cells with microRNA and plasmid-DNA-An optimized targeting approach. Nanomedicine. 2016;12(8):2353–64.
348. Vosen S, Rieck S, Heidsieck A, et al. Improvement of vascular function by magnetic nanoparticle-assisted circumferential gene transfer into the native endothelium. J Control Release. 2016;241: 164–73.
349. Zhang T, Qu G. Magnetic nanosphere-guided site-specific delivery of vascular endothelial growth factor gene attenuates restenosis in rabbit balloon-injured artery. J Vasc Surg. 2016;63(1):226–33 e1.
350. Theoharis S, Manunta M, Tan PH. Gene delivery to vascular endothelium using chemical vectors: implications for cardiovascular gene therapy. Expert Opin Biol Ther. 2007;7(5):627–43.
351. Kalra AV, Campbell RB. Development of 5-FU and doxorubicin-loaded cationic liposomes against human pancreatic cancer: implications for tumor vascular targeting. Pharm Res. 2006;23(12): 2809–17.

14

352. Wu J, Lee A, Lu Y, et al. Vascular targeting of doxorubicin using cationic liposomes. Int J Pharm. 2007;337(1–2):329–35.
353. Dabbas S, Kaushik RR, Dandamudi S, et al. Importance of the liposomal cationic lipid content and type in tumor vascular targeting: physicochemical characterization and in vitro studies using human primary and transformed endothelial cells. Endothelium. 2008;15(4):189–201.
354. Fens MH, Hill KJ, Issa J, et al. Liposomal encapsulation enhances the antitumour efficacy of the vascular disrupting agent ZD6126 in murine B16.F10 melanoma. Br J Cancer. 2008;99(8):1256–64.
355. Asai T, Miyazawa S, Maeda N, et al. Antineovascular therapy with angiogenic vessel-targeted polyethyleneglycol-shielded liposomal DPP-CNDAC. Cancer Sci. 2008;99(5):1029–33.
356. Shimizu K, Sawazaki Y, Tanaka T, et al. Chronopharmacologic cancer treatment with an angiogenic vessel-targeted liposomal drug. Biol Pharm Bull. 2008;31(1):95–8.
357. Asai T, Oku N. Angiogenic vessel-targeting DDS by liposomalized oligopeptides. Methods Mol Biol (Clifton, NJ). 2010;605:335–47.
358. Ara MN, Matsuda T, Hyodo M, et al. An aptamer ligand based liposomal nanocarrier system that targets tumor endothelial cells. Biomaterials. 2014;35(25):7110–20.
359. Ara MN, Matsuda T, Hyodo M, et al. Construction of an aptamer modified liposomal system targeted to tumor endothelial cells. Biol Pharm Bull. 2014;37(11):1742–9.
360. Vader P, Crielaard BJ, van Dommelen SM, et al. Targeted delivery of small interfering RNA to angiogenic endothelial cells with liposome-polycation-DNA particles. J Control Release. 2012;160(2): 211–6.
361. Strieth S, Nussbaum CF, Eichhorn ME, et al. Tumor-selective vessel occlusions by platelets after vascular targeting chemotherapy using paclitaxel encapsulated in cationic liposomes. Int J Cancer. 2008;122(2):452–60.
362. Cressman S, Dobson I, Lee JB, et al. Synthesis of a labeled RGD-lipid, its incorporation into liposomal nanoparticles, and their trafficking in cultured endothelial cells. Bioconjug Chem. 2009;20(7): 1404–11.
363. Hirai M, Minematsu H, Hiramatsu Y, et al. Novel and simple loading procedure of cisplatin into liposomes and targeting tumor endothelial cells. Int J Pharm. 2010;391(1–2):274–83.
364. Sochanik A, Mitrus I, Smolarczyk R, et al. Experimental anticancer therapy with vascular-disruptive peptide and liposome-entrapped chemotherapeutic agent. Arch Immunol Ther Exp. 2010;58(3): 235–45.
365. Kluza E, Jacobs I, Hectors SJ, et al. Dual-targeting of alphavbeta3 and galectin-1 improves the specificity of paramagnetic/fluorescent liposomes to tumor endothelium in vivo. J Control Release. 2012;158(2):207–14.
366. Park K. Comparative study on liposome targeting to tumor endothelium. J Control Release. 2012;158(2):181.
367. Kawahara H, Naito H, Takara K, et al. Tumor endothelial cell-specific drug delivery system using apelin-conjugated liposomes. PLoS One. 2013;8(6):e65499.
368. Zuccari G, Milelli A, Pastorino F, et al. Tumor vascular targeted liposomal-bortezomib minimizes side effects and increases therapeutic activity in human neuroblastoma. J Control Release. 2015;211:44–52.
369. Zhou JE, Yu J, Gao L, et al. iNGR-modified liposomes for tumor vascular targeting and tumor tissue penetrating delivery in the treatment of glioblastoma. Mol Pharm. 2017;14(5):1811–20.
370. Homem de Bittencourt PI Jr, Lagranha DJ, Maslinkiewicz A, et al. LipoCardium: endothelium-directed cyclopentenone prostaglandin-based liposome formulation that completely reverses atherosclerotic lesions. Atherosclerosis. 2007;193(2):245–58.
371. Asgeirsdottir SA, Talman EG, de Graaf IA, et al. Targeted transfection increases siRNA uptake and gene silencing of primary endothelial cells in vitro–a quantitative study. J Control Release. 2010;141(2):241–51.
372. Gunawan RC, Almeda D, Auguste DT. Complementary targeting of liposomes to IL-1alpha and TNF-alpha activated endothelial cells via the transient expression of VCAM1 and E-selectin. Biomaterials. 2011;32(36):9848–53.
373. Chantarasrivong C, Ueki A, Ohyama R, et al. Synthesis and functional characterization of novel Sialyl LewisX mimic-decorated liposomes for E-selectin-mediated targeting to inflamed endothelial cells. Mol Pharm. 2017;14(5):1528–37.
374. Hua S, Chang HI, Davies NM, et al. Targeting of ICAM-1-directed immunoliposomes specifically to activated endothelial cells with low cellular uptake: use of an optimized procedure for the coupling of low concentrations of antibody to liposomes. J Liposome Res. 2011;21(2):95–105.

375. Paulis LE, Jacobs I, van den Akker NM, et al. Targeting of ICAM-1 on vascular endothelium under static and shear stress conditions using a liposomal Gd-based MRI contrast agent. J Nanobiotechnol. 2012;10:25.
376. Irvine SA, Meng QH, Afzal F, et al. Receptor-targeted nanocomplexes optimized for gene transfer to primary vascular cells and explant cultures of rabbit aorta. Mol Ther. 2008;16(3):508–15.
377. Holt B, Gupta AS. Streptokinase loading in liposomes for vascular targeted nanomedicine applications: encapsulation efficiency and effects of processing. J Biomater Appl. 2012;26(5):509–27.
378. Arta A, Eriksen AZ, Melander F, et al. Endothelial protein C-targeting liposomes show enhanced uptake and improved therapeutic efficacy in human retinal endothelial cells. Invest Ophthalmol Vis Sci. 2018;59(5):2119–32.
379. Akhter A, Hayashi Y, Sakurai Y, et al. A liposomal delivery system that targets liver endothelial cells based on a new peptide motif present in the ApoB-100 sequence. Int J Pharm. 2013;456(1):195–201.
380. Toriyabe N, Hayashi Y, Hyodo M, et al. Synthesis and evaluation of stearylated hyaluronic acid for the active delivery of liposomes to liver endothelial cells. Biol Pharm Bull. 2011;34(7):1084–9.
381. Yamada Y, Hashida M, Hayashi Y, et al. An approach to transgene expression in liver endothelial cells using a liposome-based gene vector coated with hyaluronic acid. J Pharm Sci. 2013;102(9):3119–27.
382. Seow WY, Yang YY, George AJ. Oligopeptide-mediated gene transfer into mouse corneal endothelial cells: expression, design optimization, uptake mechanism and nuclear localization. Nucleic Acids Res. 2009;37(18):6276–89.
383. Ryu DW, Kim HA, Kim S, et al. VEGF receptor binding peptide-linked amphiphilic peptide with arginines and valines for endothelial cell-specific gene delivery. J Drug Target. 2012;20(7):574–81.
384. Yang J, Li Q, Yang X, et al. Multitargeting gene delivery systems for enhancing the transfection of endothelial cells. Macromol Rapid Commun. 2016;37(23):1926–31.
385. Oshima Y, Sakamoto T, Yamanaka I, et al. Targeted gene transfer to corneal endothelium in vivo by electric pulse. Gene Ther. 1998;5(10):1347–54.
386. Matragoon S, Al-Gayyar MM, Mysona BA, et al. Electroporation-mediated gene delivery of cleavage-resistant pro-nerve growth factor causes retinal neuro- and vascular degeneration. Mol Vis. 2012;18:2993–3003.
387. Kanthou C, Kranjc S, Sersa G, et al. The endothelial cytoskeleton as a target of electroporation-based therapies. Mol Cancer Ther. 2006;5(12):3145–52.
388. Markelc B, Bellard E, Sersa G, et al. Increased permeability of blood vessels after reversible electroporation is facilitated by alterations in endothelial cell-to-cell junctions. J Control Release. 2018;276:30–41.
389. Lawrie A, Brisken AF, Francis SE, et al. Microbubble-enhanced ultrasound for vascular gene delivery. Gene Ther. 2000;7(23):2023–7.
390. Teupe C, Richter S, Fisslthaler B, et al. Vascular gene transfer of phosphomimetic endothelial nitric oxide synthase (S1177D) using ultrasound-enhanced destruction of plasmid-loaded microbubbles improves vasoreactivity. Circulation. 2002;105(9):1104–9.
391. Xu Y, Xie Z, Zhou Y, et al. Experimental endostatin-GFP gene transfection into human retinal vascular endothelial cells using ultrasound-targeted cationic microbubble destruction. Mol Vis. 2015;21: 930–8.
392. Ji J, Yang JA, He X, et al. Cardiac-targeting transfection of tissue-type plasminogen activator gene to prevent the graft thrombosis and vascular anastomotic restenosis after coronary bypass. Thromb Res. 2014;134(2):440–8.
393. Serra R, de Franciscis S. Gene therapy to prevent thrombosis and anastomotic restenosis after vascular bypass procedures. Thromb Res. 2014;134(2):215–6.
394. Barreiro O, Aguilar RJ, Tejera E, et al. Specific targeting of human inflamed endothelium and in situ vascular tissue transfection by the use of ultrasound contrast agents. J Am Coll Cardiol Img. 2009;2(8):997–1005.
395. Xie A, Belcik T, Qi Y, et al. Ultrasound-mediated vascular gene transfection by cavitation of endothelial-targeted cationic microbubbles. J Am Coll Cardiol Img. 2012;5(12):1253–62.
396. Liu Y, Deisseroth A. Tumor vascular targeting therapy with viral vectors. Blood. 2006;107(8):3027–33.
397. White KM, Alba R, Parker AL, et al. Assessment of a novel, capsid-modified adenovirus with an improved vascular gene transfer profile. J Cardiothorac Surg. 2013;8:183.
398. Kibbe MR, Murdock A, Wickham T, et al. Optimizing cardiovascular gene therapy: increased vascular gene transfer with modified adenoviral vectors. Arch Surg (Chicago Ill: 1960). 2000;135(2):191–7.

14

399. Yao X, Yoshioka Y, Morishige T, et al. Tumor vascular targeted delivery of polymer-conjugated adeno-virus vector for cancer gene therapy. Mol Ther. 2011;19(9):1619–25.
400. Ogawara K, Rots MG, Kok RJ, et al. A novel strategy to modify adenovirus tropism and enhance trans-gene delivery to activated vascular endothelial cells in vitro and in vivo. Hum Gene Ther. 2004;15(5):433–43.
401. Morecroft I, White K, Caruso P, et al. Gene therapy by targeted adenovirus-mediated knockdown of pulmonary endothelial Tph1 attenuates hypoxia-induced pulmonary hypertension. Mol Ther. 2012;20(8):1516–28.
402. Liu L, Anderson WF, Beart RW, et al. Incorporation of tumor vasculature targeting motifs into moloney murine leukemia virus env escort proteins enhances retrovirus binding and transduction of human endothelial cells. J Virol. 2000;74(11):5320–8.
403. Anliker B, Abel T, Kneissl S, et al. Specific gene transfer to neurons, endothelial cells and hematopoi-etic progenitors with lentiviral vectors. Nat Methods. 2010;7(11):929–35.
404. Abel T, El Filali E, Waern J, et al. Specific gene delivery to liver sinusoidal and artery endothelial cells. Blood. 2013;122(12):2030–8.
405. Witting SR, Vallanda P, Gamble AL. Characterization of a third generation lentiviral vector pseudo-typed with Nipah virus envelope proteins for endothelial cell transduction. Gene Ther. 2013;20(10):997–1005.
406. Merchan JA, Dean J, Azpurua F, et al. Utility of vascular endothelial specific peptides for enhance-ment of adeno-associated virus-mediated gene transfer. Int J Biomed Sci: IJBS. 2008;4(3):217–20.
407. Varadi K, Michelfelder S, Korff T, et al. Novel random peptide libraries displayed on AAV serotype 9 for selection of endothelial cell-directed gene transfer vectors. Gene Ther. 2012;19(8):800–9.
408. Korbelin J, Dogbevia G, Michelfelder S, et al. A brain microvasculature endothelial cell-specific viral vector with the potential to treat neurovascular and neurological diseases. EMBO Mol Med. 2016;8(6):609–25.
409. Lipinski DM, Reid CA, Boye SL, et al. Systemic vascular transduction by capsid mutant adeno-associated virus after intravenous injection. Hum Gene Ther. 2015;26(11):767–76.
410. Trepel M, Stoneham CA, Eleftherohorinou H, et al. A heterotypic bystander effect for tumor cell kill-ing after adeno-associated virus/phage-mediated, vascular-targeted suicide gene transfer. Mol Cancer Ther. 2009;8(8):2383–91.
411. Hajitou A. Targeted systemic gene therapy and molecular imaging of cancer contribution of the vascular-targeted AAVP vector. Adv Genet. 2010;69:65–82.
412. Dong Z, Nor JE. Transcriptional targeting of tumor endothelial cells for gene therapy. Adv Drug Deliv Rev. 2009;61(7–8):542–53.
413. Pariente N, Mao SH, Morizono K, et al. Efficient targeted transduction of primary human endothelial cells with dual-targeted lentiviral vectors. J Gene Med. 2008;10(3):242–8.
414. White SJ, Papadakis ED, Rogers CA, et al. In vitro and in vivo analysis of expression cassettes designed for vascular gene transfer. Gene Ther. 2008;15(5):340–6.
415. Song W, Dong Z, Jin T, et al. Cancer gene therapy with iCaspase-9 transcriptionally targeted to tumor endothelial cells. Cancer Gene Ther. 2008;15(10):667–75.
416. Wang Y, Xu HX, Lu MD, et al. Expression of thymidine kinase mediated by a novel non-viral delivery system under the control of vascular endothelial growth factor receptor 2 promoter selectively kills human umbilical vein endothelial cells. World J Gastroenterol. 2008;14(2):224–30.
417. Tal R, Shaish A, Rofe K, et al. Endothelial-targeted gene transfer of hypoxia-inducible factor-1alpha augments ischemic neovascularization following systemic administration. Mol Ther. 2008;16(12):1927–36.
418. Stancevic B, Varda-Bloom N, Cheng J, et al. Adenoviral transduction of human acid sphingomyelin-ase into neo-angiogenic endothelium radiosensitizes tumor cure. PLoS One. 2013;8(8):e69025.
419. Simmons AB, Bretz CA, Wang H, et al. Gene therapy knockdown of VEGFR2 in retinal endothelial cells to treat retinopathy. Angiogenesis. 2018;21:751–64.
420. Yla-Herttuala S, Baker AH. Cardiovascular gene therapy: past, present, and future. Mol Ther. 2017;25(5):1095–106.

# In Vitro Assays Used to Analyse Vascular Cell Functions

*Adrian Türkcan, David Bernhard, and Barbara Messner*

© Springer Nature Switzerland AG 2019
M. Geiger (ed.), *Fundamentals of Vascular Biology*, Learning Materials in Biosciences,
https://doi.org/10.1007/978-3-030-12270-6_15

**What Will You Learn in This Chapter?**

The usage, treatment, and analyses of isolated primary vascular cells or immortalized vascular cell lines in basic science studies are highly favourable as these techniques provide a system for fast and direct functional evaluation of cellular processes. These processes can include physiological signalling pathways, pathological disease-associated changes, as well as toxicity/pharmacological tests. Based on the architecture of the vascular wall, this chapter addresses in vitro assays using endothelial cells, smooth muscle cells, and fibroblasts. According to their physiological environment, these cells require different cell culture conditions mostly regulated by various cell culture media. For different functional test also the mimicking of the in vivo situation, e.g. culture under flow conditions or co-culture of different cell types. The in vitro assays discussed in this chapter are sorted by physiological function and corresponding cell types. Herein, we also explain how these assays can be used to determine cell biological changes associated with relevant vascular pathologies. The majority of the described assays are based on the 2D culture of cells. As 3D cultures are much better suited to mimic the in vivo situation in many cases, this chapter also includes a description of recent developments in 3D culture assay and techniques. As an extension of 3D culture towards in vivo, we also describe the usage of in vitro cultured vascular tissue as model systems to study vascular function at the end of this chapter.

## 15.1　Vascular 2D Cell Culture-Based Assays

The physiology and pathogenesis of vascular wall structure and function involve a large number of cell types, factors, and processes. In order to allow for a brief and precise summary of vascular wall relevant cell-based assay, this chapter focusses mainly on the three main cell types of the vascular wall, *endothelial cells* (ECs), *smooth muscle cells* (SMCs), and *fibroblasts* (FIBs), which build up the three main layers of the vascular wall: intima, media, and adventitia. As the three cell types dominate the character of the three vessel wall layers, the location of the cells determines their major functions, and their function determines their location. Changes in cell type location usually indicate pathogenic processes, e.g. the infiltration of SMCs into the intima. When applying in vitro assays, it needs to be kept in mind that ECs, SMCs, and FIBs, depending on the vessel type and calibre, differ in structure and phenotypes and therefore may respond in differing ways to the same stimulus [1–6]. Apart from assaying cell type-specific functions, this chapter also explains in vitro assays, which can be used to determine pathological changes as a consequence of vascular disease (summary of the assays with references in ❑ Table 15.1). Even more, all these assays can be used to search for and analyse endogenous and environmental factors, as well as drug effects.

### 15.1.1　Endothelial Cells

The endothelial monolayer forms the innermost layer of the vascular wall, which is in direct contact with the blood stream. The endothelial monolayer exerts an array of different essential functions: *barrier function*, signalling and inhibition of inflammation (adhesion of immune cells, trans-endothelial migration), inhibition of thrombus formation,

**◻ Table 15.1** Summary of techniques to study vascular functions according to the cell type using 2D cultured cells

| Culture type | Cell type | Function | Functional assay | References |
|---|---|---|---|---|
| 2D culture | Endothelial cells | Barrier function | Indirect or direct measurement of endothelial barrier permeability | [8–20, 183–185] |
| | | Vascular haemostasis | Leucocyte/platelet adhesion assay | [24–34] |
| | | Vascular activation | In vitro assays to induce endothelial activation | [35, 37] |
| | | Vascular inflammation | In vitro assays to induce vascular inflammation | [38–40] |
| | | Wound healing | Scratch assay, transmigration trans-well assay | [41–49] |
| | | Endothelial proliferation | Cell counting, detection of metabolic activity | [42, 50–64, 66] |
| | | Risk factor models | Cigarette smoke exposure | [74, 75] |
| | Smooth muscle cells | Contraction | Contractility assay | [78–87] |
| | | Smooth muscle proliferation | Direct cell counting, determination of metabolic activity | [88, 89] |
| | | Smooth muscle adhesion | Adhesion assays | [42, 90–93] |
| | | Smooth muscle migration | Scratch assay, trans-well migration assay | |
| | | Elasticity | Calcification assay | [97–100] |
| | | Pathophysiology of atherosclerosis | LDL uptake assay | [101, 102, 186] |
| | Fibroblasts | Fibroblast proliferation, migration | Direct cell counting, determination of metabolic activity, scratch assay, trans-well migration assay | [45, 108–110, 112–115] |
| | | Fibroblast angiogenesis | Angiogenesis assay | [116–120] |

regulation of vascular tone, and signalling for SMC *migration* and *proliferation*. For details please see [7]. As described in the chapter authored by Christine Brostjan, ECs can be isolated from different sources of vascular tissue. Based on the above-mentioned main functions of ECs, we now provide in vitro assays in order of the most important functions of this cell type.

### 15.1.1.1 Barrier Function Assays

The endothelium is a selective and semipermeable barrier that regulates the transfer of solutes and cells between the blood stream, the vascular wall, and organs [8]. Therefore analysing the ability of this monolayer to exert this function is highly relevant for detecting pathological changes (e.g. development of atherosclerosis) and in the study of effects of endogenous and environmental factors as well as drugs. The most frequently used EC barrier function assays were continuously reviewed and summarized and adopted for different applications (for details please see [9–11]). Generally, different assay systems and techniques are available to analyse barrier function, either using indirect determination, for example, by determination of VE-cadherin expression on the surface or the measurement of trans-endothelial electrical resistance [12–17] or – to be preferred – a direct measurement, by the application of reagents, particles, or cells that, by their diffusion/migration though the barrier, can trace endothelial permeability [10, 18–20]. Markers to trace permeability are divers, e.g. fluorescent-labelled BSA, dextran, or inulin. To provide an experimental example for one such commonly used method to analyse endothelial barrier function, a protocol for using horseradish peroxidase (HRP) as a tracer is provided (Barbara Messner and David Bernhard; Medical University of Vienna and Johannes Kepler University Linz [19]):

Protocol 1
**Materials:**
- Cells: ECs, either primary isolated human umbilical vein endothelial cells (HUVECs; for isolation protocol please see the chapter authored by Christine Brostjan), an endothelial cell line, or transfected ECs (for transfection please see the chapter authored by Christine Brostjan) can be used.
- Reagents: Culture medium appropriate for the used endothelial cell type, coating medium (for HUVECs we use 0.2% gelatine diluted in A.d.), trypsin-EDTA to detach the cells from the cell culture plate surface, and HRP.
- Equipment: Trans-well cell culture insert (pore size 8 μm, depends on cell size) and appropriate cell culture plates, sterile forceps, cell culture incubator, pipets, sterile workbench, plate reader for absorbance measurements. Trans-wells are composed of an upper and a lower well to either coculture two cell types or analyse cell migration from the upper part of the well into the lower part.

Procedure: On the first day, cells are seeded into trans-well culture inserts after coating of the trans-well inserts with gelatine solution for 30 min. All procedures have to be performed under sterile conditions, and please use an appropriate number of parallels. After coating of trans-well insert, previously detached cells (by trypsin-EDTA) are seeded into the trans-well insert and are transferred into an appropriate cell culture plate filled with sufficient cell culture medium to allow for cell submersion in culture medium from both sides. For quantification of barrier function, additional control trans-well inserts are needed: trans-well inserts without cells (representing 100% permeability) and trans-well inserts with cells and without tracer as blank (negative) reference. After cell seeding, let them adhere overnight. On the next day, culture medium is replaced by fresh medium. To allow for the formation of a dense barrier, the cells should be cultured for at least 72 h before the assay is started. Replenish culture

15

medium every 48 h. After treating the cells according to the study hypothesis, the tracer horseradish peroxidase (HRP) is added into the trans-well insert. After an incubation time, which needs to be adjusted to your experimental conditions, the amount of HRP that has diffused though the (leaky) endothelial layer is determined by adding a HRP substrate (e.g. ABTS). The change in absorbance determined by photometry (plate reader) correlates with the permeability of the monolayer.

### 15.1.1.2  Endothelial Dysfunction, Activation, and Inflammation

Nowadays it is well known that the endothelial monolayer is central and essential for the proper function of the entire vascular wall and is highly relevant to physiological and pathophysiological processes of the blood stream-vessel wall-organ (and reverse) axis. The relevance of ECs for these processes has long been unclear and much has only been discovered in recent years [5, 21, 22]. Maintenance of vascular homeostasis is necessary for normal function of this tissue and organ. Accordingly, a healthy endothelium is able to properly react to changes in the physiological state of the system, e.g. by its activation in infection or inflammation. An altered and inappropriate response to physiological and pathological stimuli will cause or aggravate diseases. *Endothelial dysfunction* is a well-established marker in the clinical setting and plays a major role in experimental in vitro studies and testing.

As a consequence of vascular damage, the coagulation system is activated [23] – a process that is essential for vascular healing but also a basis for vascular disease and thrombosis. The activation of the vascular damage response machinery can be mimicked using in vitro assays. To do so, ECs (e.g. HUVECs) are cultured to confluence in the culture plate of choice. Next, platelets are isolated as already described [24, 25]. Then, isolated platelets have to be stained to detect their adhesion after incubation with the ECs. Whereas platelets were stained using radioactive dyes in earlier days [26], modern protocols use fluorescent dyes which can be analysed either by a flow cytometer or plate readers. Widely used dyes are calcein acetoxymethyl ester (Calcein-AM) and rhodamine 6G and 5-CFDA AM (5-chloromethylfluorescein diacetate). After desired treatment of ECs, cells are incubated with labelled platelets for the times chosen. After washing out the non-adherent platelets, the amount of platelets which adhere to the endothelium can be determined by quantifying the fluorescence signal by flow cytometry or using a plate reader [27–30].

Aside from platelets, the adhesion of other cell, like leucocytes, can be determined in vitro using the same approach [20, 27]. Of note, adhesion of immune cells can also be analysed under flow conditions to better mimic the in vivo situation [31]. Immune cells labelled by fluorescent dyes (or unlabelled) can be monitored and counted (automated counting by software is nowadays available) by live-cell fluorescence imaging as described [32–34]. Flow chambers used in these protocols are commercially available.

A prerequisite for leucocyte adhesion and subsequent transmigration through the endothelium is *endothelial activation*. Endothelial activation can be monitored by in vitro assays. Endothelial activation includes the expression of adhesion molecules, for example, VCAM-1 (vascular cell adhesion molecule 1) and E-selectin (endothelial-leukocyte adhesion molecule) or by the secretion of chemokines such as IL-8 and MCP1 (monocyte chemoattractant protein 1) [21]. The amount of secreted chemokines is easily detectable using commercially available enzyme-linked immunosorbent assays (ELISAs) [35]. Cell-based assays to determine the expression of adhesion molecules mainly use an approach whereby cells are incubated with a specific inducer of adhesion molecule expression, like TNF-alpha or NF-κB [36]. A standard protocol for the detection of adhesion molecule expression on ECs to deduce the athero-protective potential of a substance is given below (Barbara Messner, Iris Zeller, and David Bernhard; Medical University of Vienna and Johannes Kepler University Linz [37]):

Protocol 2
**Materials:**
- Cells: ECs, either primary isolated human umbilical vein endothelial cells (HUVECs; for isolation protocol please see the chapter authored by Christine Brostjan), an endothelial cell line, or transfected ECs (for transfection please see the chapter authored by Christine Brostjan) can be used.
- Reagents: Culture medium appropriate for the used endothelial cell type, coating medium (for HUVECs we use 0.2% gelatine diluted in A.d.), trypsin-EDTA to detach cells from the cell culture plate surface, paraformaldehyde for cell fixation, primary and secondary antibodies (HRP-labelled) suitable for ELISA, substrate ABTS for detection, XTT viability assay kit, TNF-alpha (tumour necrosis factor-alpha), and BSA (bovine serum albumin).
- Equipment: 96-well plate cell culture incubator, absorbance measurement equipment, sterile forceps, pipets, and sterile workbench.

Procedure:
ECs are cultured in 96-well plates until confluence. After incubation of cells with TNF-alpha (the corresponding control with solvent), the cells are treated with the substance of interest. To exclude an influence of the substance on the viability of the cells, the viability of cells during the assay has to be determined during the whole experiment simultaneously by using an XTT-based assay. Of note, to ensure that the assay works, a positive control is needed (e.g. an agent that significantly inhibits the adhesion molecule expression of the cells). After the incubation, cells are fixed using 4% paraformaldehyde for 3 min following intensive washing steps with PBS. It is essential not to exceeding this time, since masking of adhesion molecule is the consequence. After blocking of unspecific binding sites with 1% BSA (in PBS), cells are incubated with the primary antibody against cell surface adhesion molecules. Labelling of fixed cells with a secondary antibody coupled to HRP and subsequent addition of the substrate ABTS will result in colour change, which can be detected by absorbance measurement. Taking into account the viability analyses, the effectiveness of the tested substance can be determined in relation to the control containing TNF-alpha alone.

After the attachment of immune cells to the endothelial monolayer, they transmigrate into the sub-endothelial space. This process is object of therapeutic approaches which can be monitored in vitro [38, 39]. As summarized by William A. Muller, this assay can be performed under static as well as under flow conditions [40].

### 15.1.1.3  Wound Healing, Proliferation, and Angiogenesis Assays

The ability to proliferate and migrate enables *wound healing* and the development of new vessels. Proliferation, migration, and sprouting assays can be performed in vitro to analyse endothelial function and dysfunction as well as for drug testing.

The scratch assay (wound healing assay) is used to investigate the migration ability of ECs, bearing in mind that also proliferation takes place [41]. To perform a scratch assay, ECs are grown to confluence, and thereafter a wound area is introduced into the monolayer by using a pipet tip or a cell scraper. After having determined the degree of density of the cell layer (e.g. primary cells, cell line, cells at low or higher passage, cells from diseased tissue) monitored by repeated microscopic observation, filling of the scraped area by cell migration can be determined. Quantification of migration can be performed by measuring the distance between the moved cells and by the quantification of the area repopulated by ECs (or the corresponding area free of cells) or the time needed to close the wound. These in vitro assays can be used to characterize diseased cells compared to a corresponding control or to test the pro- or anti-migratory effect of a substance (depending on the scientific question). Despite the simplicity of the assay, the user has to consider the difficulties in creating scratches of equal size within and between wells and the associated difficulties in assay evaluation as well as differences in cell confluence based on cell seeding variability [42–47]. To overcome the difficulty of creating equal-sized scratches, a silicone insert for gap creation can be used, and wound closure can be observed using life cell imaging as described by Jonkman et al. [48]. Moreover staining of cells can help with image analysis [49].

Cell proliferation is a central parameter in health and disease and therefore also of the vasculature. The assessment of cell proliferation is a very basic but extremely important tool for in vitro studies. Proliferation analyses are important in the comparison of diseased tissue compared to healthy tissue or in analysing the effects of drugs and agents. Generally, proliferation assays are versatile and easy to perform. Nevertheless, choosing the right assay for the specific question is important. The simplest of these methods is counting cells (e.g. monitoring cell culture using a haemocytometer [42]). Staining of cells with trypan blue helps to distinguish between healthy and dead cells [42].

Proliferation can further be assessed by quantifying DNA replication, e.g. by using agents that are integrated into the DNA in the course of its replication such as tritium-labelled thymidine or the non-radioactive bromodeoxyuridine incorporation assay (BrdU assay) [50–58]. Another well-established method is the labelling of cell membranes using CFSE and the subsequent quantification of the signal by flow cytometry. Indirect measurements of cell proliferation include the detection of metabolic activity, for example, the colorimetric detection using XTT [59–61] or MTT [62–64] assays and followed by quantification using a spectrophotometer. The decision of which of these indirect assays is performed has to be drawn carefully, as substances that are tested may interfere with MTT/XTT metabolization, thereby producing false-positive or false-negative results [65, 66]. This is also true for comparing proliferation of healthy versus diseased cells as metabolic activity is no direct measure for proliferation [42].

Endothelial proliferation and migration is essential for sprouting and the formation of new vessels, which plays a central role in physiological and pathophysiological processes. The proliferation and migration can be monitored as described above. More complex assays to analyse *angiogenesis* are described in the below section on *3D culture*-based assays.

### 15.1.1.4    Exposure Models for Endothelial Cells

In the in vivo situation, ECs, and cells of the human body in general, are exposed to various factors and compounds that originate from the body itself (cytokines, hormones, etc.) or that stem from the environment (nutrition, pollutants, etc.). These factors can be divided into (i) factors that are essential for cells, (ii) factors that do not play a major role and have no negative influence, and (iii) factors that harm cells. As the lack of essential factors and the presence of harmful factors may cause disease, it is relevant to study their impact on the body and accordingly on the vascular system. These factors can be defined as physical factors (e.g. blood pressure, mechanical damage, radiation) and chemical as well as biological factors (e.g. harmful compounds, toxic compounds, vitamins, cholesterol). Depending on the factor, it is essential to know via which pathway and in which form the factor reaches and enters the cell of interest. For example, toxic metal ions are often transported in a protein bound form (to metallothioneins [67–69]). As the number of such factors is enormous, in this part of the chapter, we exemplify such an approach in describing the exposure of ECs to the cardiovascular risk factor cigarette smoke.

Cigarette smoke is an extremely complex mixture of chemicals (over 4000 different chemicals) of very different compound classes, ranging from ions, via polycyclic aromatic hydrocarbons, to particulate matter [70]. Cigarette smoke chemicals reach the circulation mainly via the inhalation pathway (compound-type and size-dependent filtration by lung tissue) but also via the oral and nasal mucosa (hydrophobic compounds), via the gastrointestinal tract, and even via the skin (third-hand smoke [71–73]). Compounds are further modified by enzymes, e.g. by the CYP (cytochrome) enzyme family. Further, it is also the individual smoking behaviour that plays a significant role in the extent of exposure and the composition of chemicals that are taken up. In summary, and this applies for many other factors as well (nutrition, lifestyle, etc.), it must be clear that the analysis of such factors in vitro can only be an approximation. Nevertheless it is the duty of the researcher to generate models that mimic the exposure situation in vivo as close as possible in the in vitro situation.

In order to generate a cigarette smoke chemical exposure model for the vascular system, we generated a *cigarette smoke sampling device*, performed chemical fingerprinting of smoke extracts (which were generated with hydrophilic and hydrophobic solvents), and compared and adjusted the composition of smoke extracts to smokers blood [74]. The rationale of doing so was that the smoke chemicals in smokers' blood are exactly what ECs (in persons who smoke) are exposed to. Importantly we applied these extracts to primary ECs and conducted tissue culture analyses for comparison.

The results, e.g. an oxidation disruption of the endothelial microtubule system leading to cell contraction in the in vitro system [75], were also analysed in and compared to vascular tissue culture results and to findings in in vivo animal models. Ultimately, after having considered all relevant factors (e.g. in vivo compound concentrations) in the in vitro model, it is also the comparison of results on the molecular, cellular, tissue, animal, and human study data level that will allow for a valid verification or falsification of results and concepts, including the quality of the in vitro model chosen.

## 15.1.2    Smooth Muscle Cells

Vascular smooth muscle cells (VSMCs) predominantly reside in the medial layer of blood vessels in the physiological state of the cardiovascular system. As their name implies, VSMCs possess contractile abilities, which are crucial for circulatory function such as blood pressure homeostasis, organ perfusion, and haemostasis. For this purpose, VSMCs express a unique set of contractile proteins that can be used for phenotyping purposes and to draw conclusions about their functional state. To be able to draw substantiated conclusions about SMC functionality, the relative expression of ideally more than one SMC marker (alpha-smooth muscle actin, smooth muscle myosin heavy chain, SM22 alpha (TAGLN, transgelin), smoothelin, calponin, or desmin) should be determined. VSMCs also play an important role in local signalling events where they alter EC, FIBs, and immune cell function with a set of paracrine signalling molecules. While VSMCs are present in a quiescent *contractile* state under physiological conditions, their switch to a functionally active form during pathological changes is an important characteristic of VSCMs that can be investigated by a variety of functional assays. It is important to keep in mind that VSMCs bear significant differences depending on their vascular origin, i.e. aorta, coronary artery, and veins; their embryological origin, e.g. VSMCs in the ascending aorta; and the organ system which they are involved in, e.g. retinal versus hepatic circulation [76]. Investigating VSMC functionality has also become an important part of iPSC research where VSMCs derived from iPSCs are frequently used for in vitro studies or even the testing of clinical applications [77].

### 15.1.2.1 Contractility Assays

The classic agonist-antagonist-stimulated contractility assay for VSMCs is performed by plating cells on a cell culture dish or slide suitable for microscopy. After allowing cells to adhere, cells are rinsed and supplemented with fresh medium containing an agonist for cellular contraction, commonly carbachol at a concentration of $10^{-5}$ M. Microscopic images of individual cells in representative fields of view are taken using phase contrast at 0 min (baseline) before carbachol stimulation and after 30 min of incubation. Differences in total cellular area are quantified using image processing software such as ImageJ. The experiment is repeated using an antagonist or relaxant, commonly $10^{-4}$ M atropine. With modern microscopes and stage incubators, time-lapse or live-cell imaging can be used to investigate contractility in a more time-dependent manner [78–80]. For the improvement of cellular visibility, cells can be stained with a cytoplasmic fluorescent probe for viable cells, e.g. Calcein-AM prior to contraction [81, 82].

A second method for the quantification of VSMC contractility is the collagen gel contraction assay. VSMCs are mixed with a collagen gel matrix at a given cell number per volume, distributed in a cell culture well of appropriate size, covered with VSMC cell culture medium, and allowed to solidify at 37 °C. Following solidification, gels are gently detached in order for them to be freely floating on the medium or attached at a small portion, dependent on quantification method used [79, 83]. Analogous to the solid contraction assay, cells can now be stimulated using carbachol or other agonists. Microscopic images are taken of entire gels at baseline and at set time points following stimulation. The change in cross-sectional area of the gel is representative of the contractibility of the number of VSMCs [84]. Alternatively, fluorescent or coloured beads can be incorporated into the gel, which functions as guiding structures in a time-lapse experiment setup. Contraction can be followed at real time by measuring the displacement of theses markers relative to time [79]. Gel contractility assay can also be purchased as standardized assay kits, where only the desired cells need to be added [85].

Lastly, a difference in impedance can be used to quantify cellular contraction, as the change in cytoskeletal arrangement and cellular morphology leads to an altered flow of electricity [86, 87].

### 15.1.2.2 VSMC Proliferation, Adhesion, and Migration

As VSMCs switch from a contractile to a synthetic *phenotype* which often occurs in response to pathological stimuli, their proliferation, adhesion, and migration abilities change. Therefore, these three attributes are important functional markers for VSMC status.

Similar to analysing proliferation of EC, VSMC proliferation is frequently measured using direct cell counting methods such as the haemocytometer or trypan blue staining or standardized metabolic assays such as XTT or MTT assays [88]. Alternatively, BrdU incorporation assays are useful in the quantification of VSMC proliferation [89]. As it is the case in EC proliferation, it is crucial to determine the appropriate method for the experiment at hand to ensure reliable proliferation data.

VSMC attachment can be measured using an *adhesion* assay, where adherent cells are stained using different cellular dyes or commercially available kits [90]. Whatever dye is chosen to visualize and quantify the adherent cells, the general protocol for a cell adhesion assay is as follows:

Cell culture ware is pre-coated with the corresponding matrix protein, e.g. collagen, fibronectin, or human Matrigel© (BD or Corning) as indicated for the coating substrates. After incubation, pre-coated wells are rinsed and blocked using 1% BSA for 1 h at 37 °C to block unspecific binding sites. VSMCs are suspended in supplemented DMEM, plated onto the pre-coated wells, and allowed to adhere for 60 min. Following incubation, wells are gently washed twice with PBS$^{-/-}$ to remove non-adherent VSMCs. Adherent cells are fixed using 4% paraformaldehyde for 15 min, rinsed, and stained with violet blue, crystal violet, Giemsa solution, or toluidine blue. Photometrical analysis (at 560 nm crystal violet) of stained wells allows for quantification of adherent cells [91–93].

Migratory ability of VSMCs can be measured (as described with ECs) using a wound scratch assay. For that experimental design, it should be kept in mind that over longer periods, i.e. 24-h proliferation, might play a role and can skew scratch assay results. A second method for measuring migration is the trans-well migration assay or Boyden chamber assay [94–96] (sample images are shown in ☐ Fig. 15.1). A general protocol for the trans-well migration assay is as follows (Barbara Messner and Adrian Tuerkcan; Medical University of Vienna and Ludwig Maximilians University Munich):

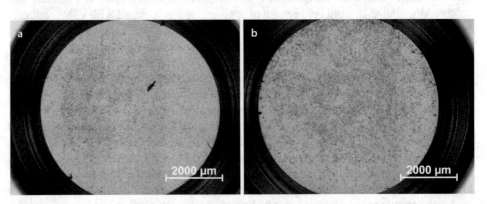

☐ **Fig. 15.1** *Transmigration assay* of toluidine blue stained VSMCs using trans-wells. **a** shows a cell line with weak transmigration ability, and in contrast the cells in **b** show an increased transmigration ability

## Protocol 3
**Materials:**
- Primary isolated VSMCs or a VSMC cell line. The trans-well migration can also be performed with transfected VSMCs to test for changes of individual proteins on migratory ability.
- Cell culture plates (6-, 12-, 24-, 48-well) and their corresponding modified Boyden chamber culture plate inserts. For VSMCs a large pore size of 8 μm can be used.
- Serum-free smooth muscle cell culture medium
- Human Matrigel© or other appropriate coating substrate
- Cold smooth muscle cell medium
- Pre-cooled pipetting tips
- If applicable, neutralizing antibodies, inhibitory molecules, etc.
- 4% paraformaldehyde
- 1% aqueous solution of toluidine blue (bicarbonate, etc.)
- Cotton swab
- Deionized $H_2O$ for rinsing

Protocol:
Matrigel© is allowed to thaw on ice at 4 °C overnight to avoid jellification. Trans-well inserts and pipetting tips are pre-cooled to 4 °C overnight. Trans-well inserts are placed into a corresponding well plate and rinsed three times with cold smooth muscle cell medium. Matrigel© is diluted in cold DMEM without supplements (1:6), and 30 μl are pipetted into every trans-well insert using the pre-cooled pipetting tips. Pre-coated trans-well inserts are allowed to incubate at 37 °C until Matrigel© has solidified (30 min). During the incubation period, VSMCs are obtained by trypsinization and resuspended in serum-free smooth muscle cell medium. Surplus liquid medium from the Matrigel© is gently removed after solidification. VSMCs are now plated at the desired number onto the Matrigel© (e.g. 5000 VSMCs per 48-well insert resuspended in 150 μl of cell culture medium). Total cell culture medium in trans-well inserts ads up to 180 μl. The bottom well is filled with cell culture medium (±chemoattractant/drug if applicable) until the bottom of the trans-well insert is submerged (300 μl per 48-well). Cells are incubated 60 min at 37 °C. After the incubation period, cell culture medium is removed from the wells, and cells are fixated for 15 min by adding 50 μl 4% paraformaldehyde into the trans-well and adding 300 μl into the bottom well. Paraformaldehyde is discarded and inserts are rinsed three times in $H_2O$. VSMCs are stained in 1% toluidine blue for 30 min by adding 50 μl to each trans-well and placing them into the bottom well filled with 300 μl of staining solution. Trans-wells are rinsed three times in fresh $H_2O$. Using the cotton swap, non-migrating cells and Matrigel© are removed from the topside of the trans-well inserts. Repeat until cotton swap does not stain blue. Removal should be performed gently to avoid deforming the trans-well membrane which will complicate image processing. Trans-well inserts are placed unto a clean microscope slide, and images are taken using a magnification of 40x or more (the use of an inverse microscope allows for higher magnification without the trans-well interfering with the objective's position). For quantification VSMCs in five or more fields of view are counted.

### 15.1.2.3  Calcification Assay

The VSMC *calcification* assay was first published by Shioi A et al. [97]. VSMCs are cultured to reach confluency when cell culture medium is replaced by calcification medium which is high glucose (4.5 g/l) DMEM with 15% FBS, 8 mmol/L $CaCl_2$, 10 mmol/L sodium pyruvate, 1 µmol/l insulin, 50 µg/ml ascorbic acid, 100 U/ml penicillin, 100 µg/ml streptomycin, 10 mmol/l β-glycerophosphate, and optional 100 nmol/L dexamethasone [98, 99]. Calcification medium is replenished every 2–3 days and should be made fresh every 2 weeks. After 7–14 days (depending on the experimental setup), VSMC calcification medium is removed, and cells are fixed using paraformaldehyde and stained using a $Ca^{2+}$-van Kossa staining. Calcium deposition is quantified using several fields of view and measuring $Ca^{2+}$ crystal deposition. Alternatively, calcium deposition can be measured by decalcifying VSMCs with 0.6 N HCL for 24 h, followed by photometric quantification of the released calcium in the supernatant with the *o*-cresolphthalein method [100]. Additionally, protein content of the decalcified VSMCs is performed using a standard protein quantification assay, e.g. BCA protein assay to normalize calcium content to total protein.

### 15.1.2.4  LDL Uptake Assay

*LDL uptake*, which is a hallmark of atherosclerosis, is mainly attributed to macrophages within the atherosclerotic plaque leading to foam cell formation. Nevertheless, VSMCs can also participate in LDL uptake leading to local cell death and inflammation. LDL uptake by VSMCs can be measured using the following protocol: initially, one needs to determine which type of LDL is appropriate for the experiment, e.g. serum LDL, oxidized LDL, acetylated LDL, etc. The LDL is linked to a fluorescent probe, e.g. DiI (1,1'-dioctadecyl-3,3,3',3'-tetramethylindocarbocyanine perchlorate) or Alexa Fluor dyes, by incubating LDL overnight at 37 °C with DiI in DMSO followed by extraction of the fluorescent LDL molecules by ultracentrifugation [101, 102]. VSMCs are cultured on microscopic chamber slides for the desired time, and fluorescently labelled LDL is added to the culture medium at a concentration of 20 µg/ml and incubated for 24 h. Following the incubation period, cells are fixated using 4% paraformaldehyde and counterstained using DAPI. LDL uptake can then be quantified as fluorescent signal per total cell count [102, 103].

### 15.1.3  Fibroblasts

The outermost layer of the vascular wall, named adventitia, is designated as the collagen-rich tissue layer connecting the innermost vascular wall layers with the residual tissue. Aside from extracellular matrix components, this layer also contains FIBs [104, 105]. Within the adventitial layer as well as in the case of injury and following wound healing, FIBs are responsible for the production of collagen [104, 106, 107].

### 15.1.3.1  Wound Healing, Proliferation, and Angiogenesis Assays

Functional assays using FIBs include proliferation and wound scratch assays as already described in the section dealing with ECs [45, 108–111]. The above-mentioned assays can equally be used for FIBs. Of note, as the role of FIBs in various *vascular functions and dysfunctions* is a comparably new idea, functional assays for this cell type are still rare. If functional in vitro tests regarding the *extracellular matrix* production capacity of FIBs are planned, culture media needs to be supplemented with ascorbic acid to reach effective collagen I biosynthesis and deposition [112, 113]. Of note, the in vitro differentiation of cultured FIBs into myo-FIBs, characterized by the expression of smooth muscle actin, must be considered especially regarding wound closure models [114, 115].

FIBs also play an essential role in angiogenesis [107, 116]. In vitro 2D models for angiogenesis include the co-culture of FIBs with ECs to monitor the development of a vascular bed. Angiogenesis is induced by the supplementation of cell culture medium with VEGF (vascular endothelial growth factor). In contrast to the 3D models of angiogenesis (explained below), this assay does not require the usage of a 3D matrix [117–120].

## 15.2  Assays Using 3D Culture of Vascular Cells

Although 2D cell culture and related assays are predominant, the usage of 3D cell culture to analyse vascular functions has become more and more important in the recent years. As the physiology and pathophysiology of vascular tissues rely on cell-to-cell contact and the extracellular matrix, biochemical interactions, as well as mechanical interactions between cells, the usage of 3D cell culture models and associated functional tests is highly interesting for studying vascular functions and dysfunctions [118, 119, 121, 122]. Aside from the application of conditioned media (cell culture supernatants containing mediators produced by the cells of interest) to mimic the exchange of soluble factors (e.g. proliferation inducing, proliferation inhibiting, and others) [123], *co-culture* of cells is also suitable to study interactions. To do so, vascular cells, in the majority of cases ECs and SMCs, can be cocultured using a so-called trans-wells, where one cell type is cultured in the upper well and the other in the lower well [124]. Therefore the influence of one cell on the other cell type can be analysed, for example, the analysis of vasoactive substances applied onto ECs and their subsequent effects on SMCs. Nevertheless, such co-culture models are not real 3D models, as the latter are characterized by the usage of extracellular matrix as a scaffold to mimic real three-dimensionality [124]. 3D cell culture assays include the three-dimensional culture of a single cell type (e.g. ECs) or more cell types (e.g. ECs together with SMCs) with the help of a 3D matrix of natural or artificial scaffolds. In the last years, different 3D models for studying vascular functions have been developed and are explained in the following section (summary of the assays with references in ◘ Table 15.2). The majority of these assays are used to analyse angiogenesis in vitro.

◘ **Table 15.2**  Summary of 3D culture systems for vascular cells

| Culture type | Model system | Test system | References |
|---|---|---|---|
| 3D culture | Tube formation | Proliferation, migration, angiogenesis | [43, 125–133] |
| | Spheroid culture | Vascularization/blood vessel maturation/drug screening | [90, 134–145] |
| | Hanging drop assay | Angiogenesis/fibrosis | [146–149] |

## 15.2.1  Capillary Tube Formation Assay

In vivo, proliferating and migrating ECs are able to form lumen-containing tubes to enable blood flow. This process can be mimicked in vitro to study physiological and pathological factors of angiogenesis as well as for *pharmacological studies* [125–129]. Confluent endothelial cell layers plated on a collagen or fibrin containing basement membrane matrix (e.g. Matrigel©) can be induced to form capillary-like tubes as shown already 30 years ago by Kubota et al. [130]. *Sprouting* will be induced by pro-angiogenetic factors such as VEGF and bFGF (basic fibroblast growth factor) [131, 132]. A detailed protocol for a tube formation assay using animal brain ECs can be found in Guo et al. [43]. Nevertheless, this assay is also controversially discussed as other non-endothelial cells can also form tubes within basement matrixes [133].

## 15.2.2  Spheroid Assay

Spheroid-based cell culture represents the growth of cells in small aggregates in which they produce their own extracellular matrix to imitate the in vivo microenvironment [134, 135]. Aside from studying physiological and pathological functions in cell aggregates, this method can also be used for drug screening [136]. This 3D cell culture is a fast-upcoming technique in different research areas, for example, a model of tumour vascularization, and a lot of changes/improvements were introduced in the last years to increase reproducibility as well as the application in high-throughput screening techniques [137]. Of note, this 3D culture method does allow for culture of all vascular cells (ECs, SMCs, FIBs) and also for co-culture of these cell types to analyse their interaction [138]. Thus it is possible to form vascular spheroids to model a blood vessel containing an endothelial monolayer as well as a multilayer smooth muscle cell sheet [139, 140].

A *spheroid assay* using ECs can be used to study angiogenesis per se [141], as well as the effect of pro- and anti-angiogenic factors as shown by Heiss et al. [142], and it can be adopted for specialized research areas [143]. Combining the two vascular cell types ECs and SMCs, physiological blood vessel maturation can be studied in a three-dimensional manner [139, 140]. The same applies for the spheroid-based co-culture of ECs with FIBs to study angiogenesis [144, 145]. Nevertheless, also this in vitro assay has some problems which are difficult to control: the diffusion of oxygen and nutrients is limited to a depth of 150–200 μm; waste material will be accumulated within the centre which results in the building of a necrotic core [138].

## 15.2.3  Hanging Drop Assay

In relation to the spheroid assay, the newly evolving technique of the *hanging drop* assay has to be mentioned. This assay can be seen as a modified form of the spheroid assay. In contrast to the spheroid assay, cell aggregates within the hanging drop have no contact to synthetic cell culture material, and gravity promotes the formation of cell spheroids. Using this technique sequential layering of cells is feasible [146]. Like spheroids, hanging drops containing one or more types of vascular cells can be used to assess angiogenesis [147, 148]. Similarly, this assay can be further applied to study cardiac fibrosis, when ECs, SMCs, cardiac myocytes, and cardiac FIBs are cocultured in hanging drops as shown by Figtree et al. [149].

## 15.3 Assays Using Whole Tissue or Organ Culture for Analysing Vascular Function

The extension of 2D and 3D culture regarding in vitro methods to study vascular functions represents the culture of whole tissue or parts of organs. In addition to the microenvironment created using 3D culture, whole *tissue culture* provides the natural *tissue architecture* and extracellular matrix components [150, 151]. Culture of whole vessel tissue can be used to study primarily angiogenesis [152] and endothelial function/dysfunction [153], but also to model the pathophysiology of vessel diseases [154]. In the following section, the most frequently used methods will be explained, although many others exist or are currently under development (summary of the assays with references in ◨ Table 15.3).

### 15.3.1 Aortic Ring Assay

Aside from 2D cell culture and 3D spheroid cultures, the usage of tissue parts to study angiogenesis becomes more and more interesting, especially regarding the replacement of in vivo animal studies. This ex vivo assay enables us to study pro- and anti-angiogenic mechanisms and factors relatively simple. This assay is based on the dissection of vessel pieces of animal or human origin from surrounding tissue and embedding it in extracellular matrix without the addition of growth factors (e.g. addition of serum) [155–159]. Outgrowth of micro-vessels is induced by injuries related to the dissection procedure, and these newly forming vessels are composed of a mixture of cells (ECs, pericytes, FIBs, and macrophages) [155, 159]. Additionally, vessel outgrowth can be induced by supplementation of growth medium with angiogenetic factors, like VEGF or bFGF and others, although these factors are also produced by the vessel explant itself [159]. A detailed protocol for the *aortic ring assay*, useful manipulation techniques, and analyses methods can be found in Baker et al. [152], and Nicosia et al. provided a useful and detailed summary of the aortic ring assay [160]. Quantification of angiogenesis by the aortic ring assay includes image acquisition and quantification of the number of out-grown vessels as well as their distance from tissue, whereby the analysis methods improved over time and histological stainings as well as electron microscopic visualization were added in the last years [161, 162]. Obviously this assay is suitable to test new substances and their effectiveness in promoting or inhibiting angiogenesis [163–168].

◨ **Table 15.3** Summary of organ culture models to study vessel function

| Culture type | Model system | Test system | References |
|---|---|---|---|
| Organ culture | Angiogenesis | Aortic ring assay | [152, 155–168] |
| | Physiology, pathophysiology, drug screening | Vessel culture (animal, human) | [153, 154, 169–182] |

### 15.3.2  Vessel Culture Models

The cultivation of whole pieces of tissue offers the possibility of analysing physiological and pathological processes in the whole tissue network, as well as the effectiveness of different substances apart from those that influence angiogenesis. Examples of processes to be studied by vessel *organ culture* methods include atherosclerotic changes, neointima formation, contraction and relaxation studies and toxicological studies [169], as well as mechanisms of cerebral ischemia [170] and inflammation [171]. Protocols for culture of vein segments have been established in the early 1990s by Pederson et al. [172] and Soyombo et al. [173]. Based on the study hypothesis, culture conditions can be adopted by the researcher, e.g. the addition of high concentrations of FCS (foetal calf serum, above 20%) induces the proliferation and migration of SMCs to model intimal hyperplasia after bypass operations [169]. Of note, importantly at the beginning of each experiment, a piece of the vessel of choice (artery, vein, human or animal origin) has to be collected to obtain baseline measurements. As mentioned above, the culture of vein segments can be used to mimic and study pathological changes like *intimal hyperplasia* occurring after bypass surgery or other pathologies [154, 174–177] and to test pharmacological active agents [178–180]. Following the culturing of vessels, experimental setups include an array of histological stainings, detection of differential expression of various cell markers (e.g. Ki67 for proliferation of SMCs in intimal hyperplasia), detection of potentially occurring cell death (e.g. by performing a TUNEL assay to detect DNA strand breaks), and visualization by electron microscopy.

Likewise, also endothelial function and dysfunction can be tested not only using isolated cells in *2D culture* but also using vessel organ culture under serum-free conditions. As a positive control, the endothelial monolayer can be removed, e.g. by Triton X-100. Readout measurement in this case is the endothelium-dependent vasodilatation using tissue baths as described [153, 181]. Aside from small vessels, like veins and arteries, organ culture of larger vessels is an upcoming technique, especially using animal tissues. Pathological changes affecting the function of the aorta are, for example, calcifications, which can be studied not only in 2D culture as mentioned above but also in organ culture models as described by Akiyoshi et al. [182].

**15**

> **Take-Home Message**
>
> This chapter describes only a part of the different in vitro methods that are used today and does not claim to be exhaustive and is therefore only intended to provide an overview. In addition, the methods described should be seen as a rough guideline, and therefore they should also invite the scientist to adapt them in the course of a study according to the circumstances. Like so much in science, the various in vitro methods are a rapidly developing field, so you have to keep up to date to be able to incorporate new developments into your project. Ultimately, research methods and data must not be seen as valid and stable as data from validated routine clinical laboratories, accordingly, always seek to confirm results by other different methods and techniques.

# References

1. Ferrer M, Encabo A, Conde MV, Marin J, Balfagon G. Heterogeneity of endothelium-dependent mechanisms in different rabbit arteries. J Vasc Res. 1995;32:339–46. https://doi.org/10.1159/000159108.
2. Ghitescu L, Robert M. Diversity in unity: the biochemical composition of the endothelial cell surface varies between the vascular beds. Microsc Res Tech. 2002;57:381–9. https://doi.org/10.1002/jemt.10091.
3. Hill CE, Phillips JK, Sandow SI. Heterogeneous control of blood flow amongst different vascular beds. Med Res Rev. 2001;21:1–60.
4. Rhodin JAG. Architecture of the vessel wall. Compr Physiol 2014, Supplement 7: handbook of physiology, the cardiovascular system, vascular smooth muscle: 1–31. First published in print 1980. Wiley Online Library. https://doi.org/10.1002/cphy.cp020201.
5. Sandoo A, van Zanten JJ, Metsios GS, Carroll D, Kitas GD. The endothelium and its role in regulating vascular tone. Open Cardiovasc Med J. 2010;4:302–12. https://doi.org/10.2174/1874192401004010302.
6. Thorin E, Shatos MA, Shreeve SM, Walters CL, Bevan JA. Human vascular endothelium heterogeneity. A comparative study of cerebral and peripheral cultured vascular endothelial cells. Stroke. 1997;28:375–81.
7. Félétou M. The endothelium: Part 1: multiple functions of the endothelial cells—focus on endothelium-derived vasoactive mediators. San Rafael: Morgan & Claypool Life Sciences; 2011. Available from: https://www.ncbi.nlm.nih.gov/books/NBK57149. https://doi.org/10.4199/C00031ED1V01Y201105ISP019.
8. Mehta D, Malik AB. Signaling mechanisms regulating endothelial permeability. Physiol Rev. 2006;86:279–367. https://doi.org/10.1152/physrev.00012.2005.
9. Yuan SYR, Rigor RR. Ch. Chapter 3. In: Regulation of endothelial barrier function. San Rafael: Morgan & Claypool Life Sciences; 2010.
10. Wang Y, Alexander JS. Analysis of endothelial barrier function in vitro. Methods Mol Biol. 2011;763:253–64. https://doi.org/10.1007/978-1-61779-191-8_17.
11. Ho YT, et al. A facile method to probe the vascular permeability of nanoparticles in nanomedicine applications. Sci Rep. 2017;7:707. https://doi.org/10.1038/s41598-017-00750-3.
12. Tschugguel W, et al. High precision measurement of electrical resistance across endothelial cell monolayers. Pflugers Arch. 1995;430:145–7.
13. Chen H-R, Yeh T-M. In vitro assays for measuring endothelial permeability by Transwells and electrical impedance systems. Bio-protocol. 2017;7:e2273. https://doi.org/10.21769/BioProtoc.2273.
14. Kazakoff PW, McGuire TR, Hoie EB, Cano M, Iversen PL. An in vitro model for endothelial permeability: assessment of monolayer integrity. In Vitro Cell Dev Biol Anim. 1995;31:846–52. https://doi.org/10.1007/BF02634568.
15. Srinivasan B, et al. TEER measurement techniques for in vitro barrier model systems. J Lab Autom. 2015;20:107–26. https://doi.org/10.1177/2211068214561025.
16. Kustermann S, et al. A real-time impedance based screening assay for drug-induced vascular leakage. Toxicol Sci. 2014;138:333–43. https://doi.org/10.1093/toxsci/kft336.
17. Kiseleva RY, et al. Vascular endothelial effects of collaborative binding to platelet/endothelial cell adhesion molecule-1 (PECAM-1). Sci Rep. 2018;8:1510. https://doi.org/10.1038/s41598-018-20027-7.
18. Martins-Green M, Petreaca M, Yao M. An assay system for in vitro detection of permeability in human "endothelium". Methods Enzymol. 2008;443:137–53. https://doi.org/10.1016/S0076-6879(08)02008-9.
19. Messner B, et al. Cadmium is a novel and independent risk factor for early atherosclerosis mechanisms and in vivo relevance. Arterioscler Thromb Vasc Biol. 2009;29:1392–8. https://doi.org/10.1161/ATVBAHA.109.190082.
20. Shin HS, et al. Bacterial lipoprotein TLR2 agonists broadly modulate endothelial function and coagulation pathways in vitro and in vivo. J Immunol. 2011;186:1119–30. https://doi.org/10.4049/jimmunol.1001647.
21. Gimbrone MA Jr, Garcia-Cardena G. Endothelial cell dysfunction and the pathobiology of atherosclerosis. Circ Res. 2016;118:620–36. https://doi.org/10.1161/CIRCRESAHA.115.306301.
22. Khazaei M, Moien-Afshari F, Laher I. Vascular endothelial function in health and diseases. Pathophysiology. 2008;15:49–67. https://doi.org/10.1016/j.pathophys.2008.02.002.

23. Rumbaut ER, Thiagarajan P. Platelet-vessel wall interactions in hemostasis and thrombosis. Colloq Ser Integr Syst Physiol Mol Funct. 2010;2(1):1–75. Morgan and Claypool Publishers. https://doi.org/10.4199/C00007ED1V01Y201002ISP004.

24. Hoffman M, Monroe DM, Roberts HR. A rapid method to isolate platelets from human blood by density gradient centrifugation. Am J Clin Pathol. 1992;98:531–3.

25. Watson SP, Authi KS, editors. Platelets: a practical approach. Oxford/New York: IRL Press at Oxford University Press; 1996. ISBN: 0199635374.

26. Curwen KD, Kim HY, Vazquez M, Handin RI, Gimbrone MA Jr. Platelet adhesion to cultured vascular endothelial cells. A quantitative monolayer adhesion assay. J Lab Clin Med. 1982;100:425–36.

27. Verheul HM, et al. Vascular endothelial growth factor-stimulated endothelial cells promote adhesion and activation of platelets. Blood. 2000;96:4216–21.

28. Kojima H, et al. CD226 mediates platelet and megakaryocytic cell adhesion to vascular endothelial cells. J Biol Chem. 2003;278:36748–53. https://doi.org/10.1074/jbc.M300702200.

29. Bombeli T, Schwartz BR, Harlan JM. Adhesion of activated platelets to endothelial cells: evidence for a GPIIbIIIa-dependent bridging mechanism and novel roles for endothelial intercellular adhesion molecule 1 (ICAM-1), alphavbeta3 integrin, and GPIbalpha. J Exp Med. 1998;187:329–39.

30. Gaugler MH, Vereycken-Holler V, Squiban C, Aigueperse J. PECAM-1 (CD31) is required for interactions of platelets with endothelial cells after irradiation. J Thromb Haemost. 2004;2:2020–6. https://doi.org/10.1111/j.1538-7836.2004.00951.x.

31. Burns MP, DePaola N. Flow-conditioned HUVECs support clustered leukocyte adhesion by coexpressing ICAM-1 and E-selectin. Am J Physiol Heart Circ Physiol. 2005;288:H194–204. https://doi.org/10.1152/ajpheart.01078.2003.

32. Kucik DF. Measurement of adhesion under flow conditions. Curr Protoc Cell Biol. 2009;Chapter 9:Unit 9.6. https://doi.org/10.1002/0471143030.cb0906s43.

33. Mulki L, Sweigard JH, Connor KM. Assessing leukocyte-endothelial interactions under flow conditions in an ex vivo autoperfused microflow chamber assay. J Vis Exp. 2014; https://doi.org/10.3791/52130.

34. Zahr A, et al. Endomucin prevents leukocyte–endothelial cell adhesion and has a critical role under resting and inflammatory conditions. Nat Commun. 2016;7:10363. https://doi.org/10.1038/ncomms10363. https://www.nature.com/articles/ncomms10363#supplementary-information.

35. Song L, et al. Crocetin inhibits lipopolysaccharide-induced inflammatory response in human umbilical vein endothelial cells. Cell Physiol Biochem. 2016;40:443–52. https://doi.org/10.1159/000452559.

36. Mayer T, et al. Cell-based assays using primary endothelial cells to study multiple steps in inflammation. Methods Enzymol. 2006;414:266–83. https://doi.org/10.1016/S0076-6879(06)14015-X.

37. Zeller I, et al. Inhibition of cell surface expression of endothelial adhesion molecules by ursolic acid prevents intimal hyperplasia of venous bypass grafts in rats. Eur J Cardiothorac Surg. 2012;42:878–84. https://doi.org/10.1093/ejcts/ezs128.

38. Ayres-Sander CE, et al. Transendothelial migration enables subsequent transmigration of neutrophils through underlying pericytes. PLoS One. 2013;8:e60025. https://doi.org/10.1371/journal.pone.0060025.

39. Chakraborty S, Ain R. Nitric-oxide synthase trafficking inducer is a pleiotropic regulator of endothelial cell function and signaling. J Biol Chem. 2017;292:6600–20. https://doi.org/10.1074/jbc.M116.742627.

40. Muller WA, Luscinskas FW. Assays of transendothelial migration in vitro. Methods Enzymol. 2008;443:155–76. https://doi.org/10.1016/S0076-6879(08)02009-0.

41. Lampugnani MG. Cell migration into a wounded area in vitro. Methods Mol Biol. 1999;96:177–82. https://doi.org/10.1385/1-59259-258-9:177.

42. Goodwin AM. In vitro assays of angiogenesis for assessment of angiogenic and anti-angiogenic agents. Microvasc Res. 2007;74:172–83. https://doi.org/10.1016/j.mvr.2007.05.006.

43. Guo S, et al. Assays to examine endothelial cell migration, tube formation, and gene expression profiles. Methods Mol Biol. 2014;1135:393–402. https://doi.org/10.1007/978-1-4939-0320-7_32.

44. Oommen S, Gupta SK, Vlahakis NE. Vascular endothelial growth factor A (VEGF-A) induces endothelial and cancer cell migration through direct binding to integrin {alpha}9{beta}1: identification of a specific {alpha}9{beta}1 binding site. J Biol Chem. 2011;286:1083–92. https://doi.org/10.1074/jbc.M110.175158.

45. Monsuur HN, et al. Methods to study differences in cell mobility during skin wound healing in vitro. J Biomech. 2016;49:1381–7. https://doi.org/10.1016/j.jbiomech.2016.01.040.

46. Yue PY, Leung EP, Mak NK, Wong RN. A simplified method for quantifying cell migration/wound healing in 96-well plates. J Biomol Screen. 2010;15:427–33. https://doi.org/10.1177/1087057110361772.

15

47. Ammann KR, et al. Collective cell migration of smooth muscle and endothelial cells: impact of injury versus non-injury stimuli. J Biol Eng. 2015;9:19. https://doi.org/10.1186/s13036-015-0015-y.

48. Jonkman JE, et al. An introduction to the wound healing assay using live-cell microscopy. Cell Adhes Migr. 2014;8:440–51. https://doi.org/10.4161/cam.36224.

49. Wang S, et al. Control of endothelial cell proliferation and migration by VEGF signaling to histone deacetylase 7. Proc Natl Acad Sci U S A. 2008;105:7738–43. https://doi.org/10.1073/pnas.0802857105.

50. Messele T, et al. Nonradioactive techniques for measurement of in vitro T-cell proliferation: alternatives to the [(3)H]thymidine incorporation assay. Clin Diagn Lab Immunol. 2000;7:687–92.

51. Ezaki T, et al. Time course of endothelial cell proliferation and microvascular remodeling in chronic inflammation. Am J Pathol. 2001;158:2043–55. https://doi.org/10.1016/S0002-9440(10)64676-7.

52. Wang S, et al. Regulation of endothelial cell proliferation and vascular assembly through distinct mTORC2 signaling pathways. Mol Cell Biol. 2015;35:1299–313. https://doi.org/10.1128/MCB.00306-14.

53. Shu Q, Li W, Li H, Sun G. Vasostatin inhibits VEGF-induced endothelial cell proliferation, tube formation and induces cell apoptosis under oxygen deprivation. Int J Mol Sci. 2014;15:6019–30. https://doi.org/10.3390/ijms15046019.

54. Abdel-Malak NA, et al. Angiopoietin-1 promotes endothelial cell proliferation and migration through AP-1-dependent autocrine production of interleukin-8. Blood. 2008;111:4145–54. https://doi.org/10.1182/blood-2007-08-110338.

55. Logie JJ, et al. Glucocorticoid-mediated inhibition of angiogenic changes in human endothelial cells is not caused by reductions in cell proliferation or migration. PLoS One. 2010;5:e14476. https://doi.org/10.1371/journal.pone.0014476.

56. Poirier O, et al. Inhibition of apelin expression by BMP signaling in endothelial cells. Am J Physiol Cell Physiol. 2012;303:C1139–45. https://doi.org/10.1152/ajpcell.00168.2012.

57. Pearson LJ, Yandle TG, Nicholls MG, Evans JJ. Intermedin (adrenomedullin-2): a potential protective role in human aortic endothelial cells. Cell Physiol Biochem. 2009;23:97–108. https://doi.org/10.1159/000204098.

58. Sakao S, et al. Initial apoptosis is followed by increased proliferation of apoptosis-resistant endothelial cells. FASEB J. 2005;19:1178–80. https://doi.org/10.1096/fj.04-3261fje.

59. Thoppil RJ, et al. TRPV4 channel activation selectively inhibits tumor endothelial cell proliferation. Sci Rep. 2015;5:14257. https://doi.org/10.1038/srep14257. https://www.nature.com/articles/srep14257#supplementary-information.

60. Hwang SH, et al. Effects of gintonin on the proliferation, migration, and tube formation of human umbilical-vein endothelial cells: involvement of lysophosphatidic-acid receptors and vascular-endothelial-growth-factor signaling. J Ginseng Res. 2016;40:325–33. https://doi.org/10.1016/j.jgr.2015.10.002.

61. Duah E, et al. Cysteinyl leukotrienes regulate endothelial cell inflammatory and proliferative signals through CysLT2 and CysLT1 receptors. Sci Rep. 2013;3:3274. https://doi.org/10.1038/srep03274. https://www.nature.com/articles/srep03274#supplementary-information.

62. Denizot F, Lang R. Rapid colorimetric assay for cell growth and survival. Modifications to the tetrazolium dye procedure giving improved sensitivity and reliability. J Immunol Methods. 1986;89:271–7.

63. Ma J, et al. Inhibition of endothelial cell proliferation and tumor angiogenesis by up-regulating NDRG2 expression in breast cancer cells. PLoS One. 2012;7:e32368. https://doi.org/10.1371/journal.pone.0032368.

64. Liu XL, Hu X, Cai WX, Lu WW, Zheng LW. Effect of granulocyte-Colony stimulating factor on endothelial cells and osteoblasts. Biomed Res Int. 2016;2016:8485721. https://doi.org/10.1155/2016/8485721.

65. Ahmad S, Ahmad A, Schneider KB, White CW. Cholesterol interferes with the MTT assay in human epithelial-like (A549) and endothelial (HLMVE and HCAE) cells. Int J Toxicol. 2006;25:17–23. https://doi.org/10.1080/10915810500488361.

66. Trevisi L, Pighin I, Bazzan S, Luciani S. Inhibition of 3-(4,5-dimethylthiazol-2-yl)-2,5-diphenyltetrazolium bromide (MTT) endocytosis by ouabain in human endothelial cells. FEBS Lett. 2006;580:2769–73. https://doi.org/10.1016/j.febslet.2006.04.040.

67. Klaassen CD, Liu J, Diwan BA. Metallothionein protection of cadmium toxicity. Toxicol Appl Pharmacol. 2009;238:215–20. https://doi.org/10.1016/j.taap.2009.03.026.

68. Thirumoorthy N, et al. A review of metallothionein isoforms and their role in pathophysiology. World J Surg Oncol. 2011;9:54. https://doi.org/10.1186/1477-7819-9-54.

69. Rahman MT, Haque N, Abu Kasim NH, De Ley M. In: Nilius B, et al., editors. Reviews of physiology, biochemistry and pharmacology, vol. 173: Springer International Publishing; 2017. p. 41–62.

70. Services., U. D. o. H. a. H. in How tobacco smoke causes disease: the biology and behavioral basis for smoking-attributable disease: a report of the surgeon general; 2010.

71. Northrup TF, et al. Thirdhand smoke: state of the science and a call for policy expansion. Public Health Rep. 2016;131:233–8. https://doi.org/10.1177/003335491613100206.

72. Diez-Izquierdo A, et al. Update on thirdhand smoke: a comprehensive systematic review. Environ Res. 2018;167:341–71. https://doi.org/10.1016/j.envres.2018.07.020.

73. Tillett T. Thirdhand smoke in review: research needs and recommendations. Environ Health Perspect. 2011;119:a399. https://doi.org/10.1289/ehp.119-a399b.

74. Bernhard D, et al. Development and evaluation of an in vitro model for the analysis of cigarette smoke effects on cultured cells and tissues. J Pharmacol Toxicol Methods. 2004;50:45–51. https://doi.org/10.1016/j.vascn.2004.01.003.

75. Bernhard D, et al. Cigarette smoke metal-catalyzed protein oxidation leads to vascular endothelial cell contraction by depolymerization of microtubules. FASEB J. 2005;19:1096–107. https://doi.org/10.1096/fj.04-3192com.

76. Michel JB, Li Z, Lacolley P. Smooth muscle cells and vascular diseases. Cardiovasc Res. 2012;95:135–7. https://doi.org/10.1093/cvr/cvs172.

77. Dash BC, Jiang Z, Suh C, Qyang Y. Induced pluripotent stem cell-derived vascular smooth muscle cells: methods and application. Biochem J. 2015;465:185–94. https://doi.org/10.1042/BJ20141078.

78. Vazao H, das Neves RP, Graos M, Ferreira L. Towards the maturation and characterization of smooth muscle cells derived from human embryonic stem cells. PLoS One. 2011;6:e17771. https://doi.org/10.1371/journal.pone.0017771.

79. Rodriguez LV, et al. Clonogenic multipotent stem cells in human adipose tissue differentiate into functional smooth muscle cells. Proc Natl Acad Sci U S A. 2006;103:12167–72. https://doi.org/10.1073/pnas.0604850103.

80. Yun SJ, et al. Akt1 isoform modulates phenotypic conversion of vascular smooth muscle cells. Biochim Biophys Acta. 2014;1842:2184–92. https://doi.org/10.1016/j.bbadis.2014.08.014.

81. Vo E, Hanjaya-Putra D, Zha Y, Kusuma S, Gerecht S. Smooth-muscle-like cells derived from human embryonic stem cells support and augment cord-like structures in vitro. Stem Cell Rev. 2010;6:237–47. https://doi.org/10.1007/s12015-010-9144-3.

82. Wanjare M, Kuo F, Gerecht S. Derivation and maturation of synthetic and contractile vascular smooth muscle cells from human pluripotent stem cells. Cardiovasc Res. 2013;97:321–30. https://doi.org/10.1093/cvr/cvs315.

83. Benoit C, Gu Y, Zhang Y, Alexander JS, Wang Y. Contractility of placental vascular smooth muscle cells in response to stimuli produced by the placenta: roles of ACE vs. non-ACE and AT1 vs. AT2 in placental vessel cells. Placenta. 2008;29:503–9. https://doi.org/10.1016/j.placenta.2008.03.002.

84. Do KH, et al. Angiotensin II-induced aortic ring constriction is mediated by phosphatidylinositol 3-kinase/L-type calcium channel signaling pathway. Exp Mol Med. 2009;41:569–76. https://doi.org/10.3858/emm.2009.41.8.062.

85. Wu T, et al. Identification of BPIFA1/SPLUNC1 as an epithelium-derived smooth muscle relaxing factor. Nat Commun. 2017;8:14118. https://doi.org/10.1038/ncomms14118.

86. Wilson JL, et al. Unraveling endothelin-1 induced hypercontractility of human pulmonary artery smooth muscle cells from patients with pulmonary arterial hypertension. PLoS One. 2018;13:e0195780. https://doi.org/10.1371/journal.pone.0195780.

87. Steinbach SK, et al. Directed differentiation of skin-derived precursors into functional vascular smooth muscle cells. Arterioscler Thromb Vasc Biol. 2011;31:2938–48. https://doi.org/10.1161/ATVBAHA.111.232975.

88. Hsieh HL, et al. Thrombin induces EGF receptor expression and cell proliferation via a PKC(delta)/c-Src-dependent pathway in vascular smooth muscle cells. Arterioscler Thromb Vasc Biol. 2009;29:1594–601. https://doi.org/10.1161/ATVBAHA.109.185801.

89. Gennaro G, Menard C, Michaud SE, Deblois D, Rivard A. Inhibition of vascular smooth muscle cell proliferation and neointimal formation in injured arteries by a novel, oral mitogen-activated protein kinase/extracellular signal-regulated kinase inhibitor. Circulation. 2004;110:3367–71. https://doi.org/10.1161/01.CIR.0000147773.86866.CD.

90. Baron JH, Moiseeva EP, de Bono DP, Abrams KR, Gershlick AH. Inhibition of vascular smooth muscle cell adhesion and migration by c7E3 Fab (abciximab): a possible mechanism for influencing restenosis. Cardiovasc Res. 2000;48:464–72.

91. Sala-Newby GB, George SJ, Bond M, Dhoot GK, Newby AC. Regulation of vascular smooth muscle cell proliferation, migration and death by heparan sulfate 6-O-endosulfatase1. FEBS Lett. 2005;579:6493–8. https://doi.org/10.1016/j.febslet.2005.10.026.

15

92. Huang S, Sun Z, Li Z, Martinez-Lemus LA, Meininger GA. Modulation of microvascular smooth muscle adhesion and mechanotransduction by integrin-linked kinase. Microcirculation. 2010;17:113–27. https://doi.org/10.1111/j.1549-8719.2009.00011.x.
93. Witzenbichler B, et al. Regulation of smooth muscle cell migration and integrin expression by the Gax transcription factor. J Clin Invest. 1999;104:1469–80. https://doi.org/10.1172/JCI7251.
94. Fegley AJ, Tanski WJ, Roztocil E, Davies MG. Sphingosine-1-phosphate stimulates smooth muscle cell migration through galpha(i)- and pi3-kinase-dependent p38(MAPK) activation. J Surg Res. 2003;113:32–41.
95. Goueffic Y, et al. Hyaluronan induces vascular smooth muscle cell migration through RHAMM-mediated PI3K-dependent Rac activation. Cardiovasc Res. 2006;72:339–48. https://doi.org/10.1016/j.cardiores.2006.07.017.
96. Poon M, et al. Rapamycin inhibits vascular smooth muscle cell migration. J Clin Invest. 1996;98: 2277–83. https://doi.org/10.1172/JCI119038.
97. Shioi A, et al. Beta-glycerophosphate accelerates calcification in cultured bovine vascular smooth muscle cells. Arterioscler Thromb Vasc Biol. 1995;15:2003–9.
98. Trion A, Schutte-Bart C, Bax WH, Jukema JW, van der Laarse A. Modulation of calcification of vascular smooth muscle cells in culture by calcium antagonists, statins, and their combination. Mol Cell Biochem. 2008;308:25–33. https://doi.org/10.1007/s11010-007-9608-1.
99. Wada T, McKee MD, Steitz S, Giachelli CM. Calcification of vascular smooth muscle cell cultures: inhibition by osteopontin. Circ Res. 1999;84:166–78.
100. Jono S, Nishizawa Y, Shioi A, Morii H. Parathyroid hormone-related peptide as a local regulator of vascular calcification. Its inhibitory action on in vitro calcification by bovine vascular smooth muscle cells. Arterioscler Thromb Vasc Biol. 1997;17:1135–42.
101. Reynolds GD, St Clair RW. A comparative microscopic and biochemical study of the uptake of fluorescent and 125I-labeled lipoproteins by skin fibroblasts, smooth muscle cells, and peritoneal macrophages in culture. Am J Pathol. 1985;121:200–11.
102. Viola M, et al. Oxidized low density lipoprotein (LDL) affects hyaluronan synthesis in human aortic smooth muscle cells. J Biol Chem. 2013;288:29595–603. https://doi.org/10.1074/jbc.M113.508341.
103. Wang H, et al. 17beta-estradiol promotes cholesterol efflux from vascular smooth muscle cells through a liver X receptor alpha-dependent pathway. Int J Mol Med. 2014;33:550–8. https://doi.org/10.3892/ijmm.2014.1619.
104. Maiellaro K, Taylor WR. The role of the adventitia in vascular inflammation. Cardiovasc Res. 2007;75:640–8. https://doi.org/10.1016/j.cardiores.2007.06.023.
105. Majesky MW, Dong XR, Hoglund V, Mahoney WM Jr, Daum G. The adventitia: a dynamic interface containing resident progenitor cells. Arterioscler Thromb Vasc Biol. 2011;31:1530–9. https://doi.org/10.1161/ATVBAHA.110.221549.
106. Rey FE, Pagano PJ. The reactive adventitia: fibroblast oxidase in vascular function. Arterioscler Thromb Vasc Biol. 2002;22:1962–71.
107. Kendall RT, Feghali-Bostwick CA. Fibroblasts in fibrosis: novel roles and mediators. Front Pharmacol. 2014;5:123. https://doi.org/10.3389/fphar.2014.00123.
108. Chai X, et al. Hypoxia induces pulmonary arterial fibroblast proliferation, migration, differentiation and vascular remodeling via the PI3K/Akt/p70S6K signaling pathway. Int J Mol Med. 2018;41: 2461–72. https://doi.org/10.3892/ijmm.2018.3462.
109. Cai XJ, et al. Adiponectin inhibits lipopolysaccharide-induced adventitial fibroblast migration and transition to myofibroblasts via AdipoR1-AMPK-iNOS pathway. Mol Endocrinol. 2010;24:218–28. https://doi.org/10.1210/me.2009-0128.
110. Liu G, Eskin SG, Mikos AG. Integrin alpha(v)beta(3) is involved in stimulated migration of vascular adventitial fibroblasts by basic fibroblast growth factor but not platelet-derived growth factor. J Cell Biochem. 2001;83:129–35.
111. Liu Y, et al. AGEs increased migration and inflammatory responses of adventitial fibroblasts via RAGE, MAPK and NF-kappaB pathways. Atherosclerosis. 2010;208:34–42. https://doi.org/10.1016/j.atherosclerosis.2009.06.007.
112. Boyera N, Galey I, Bernard BA. Effect of vitamin C and its derivatives on collagen synthesis and cross-linking by normal human fibroblasts. Int J Cosmet Sci. 1998;20:151–8. https://doi.org/10.1046/j.1467-2494.1998.171747.x.
113. Schwarz RI. Collagen I and the fibroblast: high protein expression requires a new paradigm of post-transcriptional, feedback regulation. Biochem Biophys Rep. 2015;3:38–44. https://doi.org/10.1016/j.bbrep.2015.07.007.

114. Forte A, Della Corte A, De Feo M, Cerasuolo F, Cipollaro M. Role of myofibroblasts in vascular remodelling: focus on restenosis and aneurysm. Cardiovasc Res. 2010;88:395–405. https://doi.org/10.1093/cvr/cvq224.

115. Coen M, Gabbiani G, Bochaton-Piallat ML. Myofibroblast-mediated adventitial remodeling: an underestimated player in arterial pathology. Arterioscler Thromb Vasc Biol. 2011;31:2391–6. https://doi.org/10.1161/ATVBAHA.111.231548.

116. Newman AC, Nakatsu MN, Chou W, Gershon PD, Hughes CC. The requirement for fibroblasts in angiogenesis: fibroblast-derived matrix proteins are essential for endothelial cell lumen formation. Mol Biol Cell. 2011;22:3791–800. https://doi.org/10.1091/mbc.E11-05-0393.

117. Bishop ET, et al. An in vitro model of angiogenesis: basic features. Angiogenesis. 1999;3:335–44.

118. Huh D, Hamilton GA, Ingber DE. From 3D cell culture to organs-on-chips. Trends Cell Biol. 2011;21:745–54. https://doi.org/10.1016/j.tcb.2011.09.005.

119. van Duinen V, Trietsch SJ, Joore J, Vulto P, Hankemeier T. Microfluidic 3D cell culture: from tools to tissue models. Curr Opin Biotechnol. 2015;35:118–26. https://doi.org/10.1016/j.copbio.2015.05.002.

120. Richards M, Mellor H. In vitro coculture assays of angiogenesis. In: Martin S, Hewett P, editors. Angiogenesis protocols. Methods in molecular biology, vol. 1430. New York: Humana Press; 2016. https://doi.org/10.1007/978-1-4939-3628-1_10.

121. Duval K, et al. Modeling physiological events in 2D vs. 3D cell culture. Physiology (Bethesda). 2017;32:266–77. https://doi.org/10.1152/physiol.00036.2016.

122. Edmondson R, Broglie JJ, Adcock AF, Yang L. Three-dimensional cell culture systems and their applications in drug discovery and cell-based biosensors. Assay Drug Dev Technol. 2014;12:207–18. https://doi.org/10.1089/adt.2014.573.

123. Fillinger MF, Sampson LN, Cronenwett JL, Powell RJ, Wagner RJ. Coculture of endothelial cells and smooth muscle cells in bilayer and conditioned media models. J Surg Res. 1997;67:169–78. https://doi.org/10.1006/jsre.1996.4978.

124. Sanchez-Palencia DM, Bigger-Allen A, Saint-Geniez M, Arboleda-Velasquez JF, D'Amore PA. Coculture assays for endothelial cells-mural cells interactions. Methods Mol Biol. 2016;1464:35–47. https://doi.org/10.1007/978-1-4939-3999-2_4.

125. Arnaoutova I, George J, Kleinman HK, Benton G. The endothelial cell tube formation assay on basement membrane turns 20: state of the science and the art. Angiogenesis. 2009;12:267–74. https://doi.org/10.1007/s10456-009-9146-4.

126. Troyanovsky B, Levchenko T, Mansson G, Matvijenko O, Holmgren L. Angiomotin: an angiostatin binding protein that regulates endothelial cell migration and tube formation. J Cell Biol. 2001;152:1247–54.

127. Nacev BA, Liu JO. Synergistic inhibition of endothelial cell proliferation, tube formation, and sprouting by cyclosporin A and itraconazole. PLoS One. 2011;6:e24793. https://doi.org/10.1371/journal.pone.0024793.

128. Sakurai T, et al. Stimulation of tube formation mediated through the prostaglandin EP2 receptor in rat luteal endothelial cells. J Endocrinol. 2011;209:33–43. https://doi.org/10.1530/JOE-10-0357.

129. Stratman AN, et al. Endothelial cell lumen and vascular guidance tunnel formation requires MT1-MMP-dependent proteolysis in 3-dimensional collagen matrices. Blood. 2009;114:237–47. https://doi.org/10.1182/blood-2008-12-196451.

130. Kubota Y, Kleinman HK, Martin GR, Lawley TJ. Role of laminin and basement membrane in the morphological differentiation of human endothelial cells into capillary-like structures. J Cell Biol. 1988;107:1589–98.

131. Davis GE, Black SM, Bayless KJ. Capillary morphogenesis during human endothelial cell invasion of three-dimensional collagen matrices. In Vitro Cell Dev Biol Anim. 2000;36:513–9. https://doi.org/10.1290/1071-2690(2000)036<0513:CMDHEC>2.0.CO;2.

132. Montesano R, Vassalli JD, Baird A, Guillemin R, Orci L. Basic fibroblast growth factor induces angiogenesis in vitro. Proc Natl Acad Sci U S A. 1986;83:7297–301.

133. Smith EJ, Staton CA. Tubule formation assays. In: Staton CA, Lewis C, Bicknell R, editors. Angiogenesis assays – a critical appraisal of current techniques: Wiley Online Libary; 2007. p. 65–87. https://doi.org/10.1002/9780470029350.ch4.

134. Fennema E, Rivron N, Rouwkema J, van Blitterswijk C, de Boer J. Spheroid culture as a tool for creating 3D complex tissues. Trends Biotechnol. 2013;31:108–15. https://doi.org/10.1016/j.tibtech.2012.12.003.

15

135. Lin RZ, Chang HY. Recent advances in three-dimensional multicellular spheroid culture for biomedical research. Biotechnol J. 2008;3:1172–84. https://doi.org/10.1002/biot.200700228.
136. Friedrich J, Seidel C, Ebner R, Kunz-Schughart LA. Spheroid-based drug screen: considerations and practical approach. Nat Protoc. 2009;4:309. https://doi.org/10.1038/nprot.2008.226.
137. Blacher S, et al. Cell invasion in the spheroid sprouting assay: a spatial organisation analysis adaptable to cell behaviour. PLoS One. 2014;9:e97019. https://doi.org/10.1371/journal.pone.0097019.
138. Cui X, Hartanto Y, Zhang H. Advances in multicellular spheroids formation. J R Soc Interface. 2017;14 https://doi.org/10.1098/rsif.2016.0877.
139. Fleming PA, et al. Fusion of uniluminal vascular spheroids: a model for assembly of blood vessels. Dev Dyn. 2010;239:398–406. https://doi.org/10.1002/dvdy.22161.
140. Korff T, Kimmina S, Martiny-Baron G, Augustin HG. Blood vessel maturation in a 3-dimensional spheroidal coculture model: direct contact with smooth muscle cells regulates endothelial cell quiescence and abrogates VEGF responsiveness. FASEB J. 2001;15:447–57. https://doi.org/10.1096/fj.00-0139com.
141. Welch-Reardon KM, et al. Angiogenic sprouting is regulated by endothelial cell expression of slug. J Cell Sci. 2014;127:2017–28. https://doi.org/10.1242/jcs.143420.
142. Heiss M, et al. Endothelial cell spheroids as a versatile tool to study angiogenesis in vitro. FASEB J. 2015;29:3076–84. https://doi.org/10.1096/fj.14-267633.
143. Dittrich A, et al. Key proteins involved in spheroid formation and angiogenesis in endothelial cells after long-term exposure to simulated microgravity. Cell Physiol Biochem. 2018;45:429–45. https://doi.org/10.1159/000486920.
144. Eckermann CW, Lehle K, Schmid SA, Wheatley DN, Kunz-Schughart LA. Characterization and modulation of fibroblast/endothelial cell co-cultures for the in vitro preformation of three-dimensional tubular networks. Cell Biol Int. 2011;35:1097–110. https://doi.org/10.1042/CBI20100718.
145. Kunz-Schughart LA, et al. Potential of fibroblasts to regulate the formation of three-dimensional vessel-like structures from endothelial cells in vitro. Am J Physiol Cell Physiol. 2006;290:C1385–98. https://doi.org/10.1152/ajpcell.00248.2005.
146. Zuppinger C. 3D culture for cardiac cells. Biochim Biophys Acta. 2016;1863:1873–81. https://doi.org/10.1016/j.bbamcr.2015.11.036.
147. Kelm JM, et al. VEGF profiling and angiogenesis in human microtissues. J Biotechnol. 2005;118:213–29. https://doi.org/10.1016/j.jbiotec.2005.03.016.
148. Pfisterer L, Korff T. In: Martin SG, Hewett PW, editors. Angiogenesis protocols. New York: Springer; 2016. p. 167–77.
149. Figtree GA, Bubb KJ, Tang O, Kizana E, Gentile C. Vascularized cardiac spheroids as novel 3D in vitro models to study cardiac fibrosis. Cells Tissues Organs. 2017;204:191–8. https://doi.org/10.1159/000477436.
150. Resau JH, Sakamoto K, Cottrell JR, Hudson EA, Meltzer SJ. Explant organ culture: a review. Cytotechnology. 1991;7:137–49.
151. Al-Lamki RS, Bradley JR, Pober JS. Human organ culture: updating the approach to bridge the gap from in vitro to in vivo in inflammation, Cancer, and stem cell biology. Front Med. 2017;4 https://doi.org/10.3389/fmed.2017.00148.
152. Baker M, et al. Use of the mouse aortic ring assay to study angiogenesis. Nat Protoc. 2011;7:89–104. https://doi.org/10.1038/nprot.2011.435.
153. Alm R, Edvinsson L, Malmsjo M. Organ culture: a new model for vascular endothelium dysfunction. BMC Cardiovasc Disord. 2002;2:8.
154. Mekontso-Dessap A, et al. Vascular-wall remodeling of 3 human bypass vessels: organ culture and smooth muscle cell properties. J Thorac Cardiovasc Surg. 2006;131:651–8. https://doi.org/10.1016/j.jtcvs.2005.08.048.
155. Nicosia RF, Ottinetti A. Growth of microvessels in serum-free matrix culture of rat aorta. A quantitative assay of angiogenesis in vitro. Lab Investig. 1990;63:115–22.
156. Masson VV, et al. Mouse aortic ring assay: a new approach of the molecular genetics of angiogenesis. Biol Proced Online. 2002;4:24–31. https://doi.org/10.1251/bpo30.
157. Zhu WH, Iurlaro M, MacIntyre A, Fogel E, Nicosia RF. The mouse aorta model: influence of genetic background and aging on bFGF- and VEGF-induced angiogenic sprouting. Angiogenesis. 2003;6:193–9. https://doi.org/10.1023/B:AGEN.0000021397.18713.9c.

158. Aplin AC, Fogel E, Zorzi P, Nicosia RF. The aortic ring model of angiogenesis. Methods Enzymol. 2008;443:119–36. https://doi.org/10.1016/S0076-6879(08)02007-7.
159. Nicosia RF, Zorzi P, Ligresti G, Morishita A, Aplin AC. Paracrine regulation of angiogenesis by different cell types in the aorta ring model. Int J Dev Biol. 2011;55:447–53. https://doi.org/10.1387/ijdb.103222rn.
160. Nicosia RF. The aortic ring model of angiogenesis: a quarter century of search and discovery. J Cell Mol Med. 2009;13:4113–36. https://doi.org/10.1111/j.1582-4934.2009.00891.x.
161. Blacher S, et al. Improved quantification of angiogenesis in the rat aortic ring assay. Angiogenesis. 2001;4:133–42.
162. Zhu WH, Nicosia RF. The thin prep rat aortic ring assay: a modified method for the characterization of angiogenesis in whole mounts. Angiogenesis. 2002;5:81–6.
163. Kruger EA, et al. Endostatin inhibits microvessel formation in the ex vivo rat aortic ring angiogenesis assay. Biochem Biophys Res Commun. 2000;268:183–91. https://doi.org/10.1006/bbrc.1999.2018.
164. Matsubara K, Mori M, Matsuura Y, Kato N. Pyridoxal 5′-phosphate and pyridoxal inhibit angiogenesis in serum-free rat aortic ring assay. Int J Mol Med. 2001;8:505–8.
165. Carnevale ML, Bergdahl A. Study of the anti-angiogenic effects of cardiolipin by the aortic ring assay. Can J Physiol Pharmacol. 2015;93:1015–9. https://doi.org/10.1139/cjpp-2015-0016.
166. Stati T, et al. beta-Blockers promote angiogenesis in the mouse aortic ring assay. J Cardiovasc Pharmacol. 2014;64:21–7. https://doi.org/10.1097/FJC.0000000000000085.
167. Giustarini D, Tsikas D, Rossi R. Study of the effect of thiols on the vasodilatory potency of S-nitrosothiols by using a modified aortic ring assay. Toxicol Appl Pharmacol. 2011;256:95–102. https://doi.org/10.1016/j.taap.2011.07.011.
168. Salahdeen HM, Idowu GO, Yemitan OK, Murtala BA, Alada AR. Calcium-dependent mechanisms mediate the vasorelaxant effects of Tridax procumbens (Lin) aqueous leaf extract in rat aortic ring. J Basic Clin Physiol Pharmacol. 2014;25:161–6. https://doi.org/10.1515/jbcpp-2013-0030.
169. Ozaki H, Karaki H. Organ culture as a useful method for studying the biology of blood vessels and other smooth muscle tissues. Jpn J Pharmacol. 2002;89:93–100.
170. Ahnstedt H, Stenman E, Cao L, Henriksson M, Edvinsson L. Cytokines and growth factors modify the upregulation of contractile endothelin ET(A) and ET(B) receptors in rat cerebral arteries after organ culture. Acta Physiol (Oxf). 2012;205:266–78. https://doi.org/10.1111/j.1748-1716.2011.02392.x.
171. Waldsee R, Eftekhari S, Ahnstedt H, Johnson LE, Edvinsson L. CaMKII and MEK1/2 inhibition time-dependently modify inflammatory signaling in rat cerebral arteries during organ culture. J Neuroinflammation. 2014;11:90. https://doi.org/10.1186/1742-2094-11-90.
172. Pederson DC, Bowyer DE. Endothelial injury and healing in vitro. Studies using an organ culture system. Am J Pathol. 1985;119:264–72.
173. Soyombo AA, Angelini GD, Bryan AJ, Jasani B, Newby AC. Intimal proliferation in an organ culture of human saphenous vein. Am J Pathol. 1990;137:1401–10.
174. Slomp J, et al. Nature and origin of the neointima in whole vessel wall organ culture of the human saphenous vein. Virchows Arch. 1996;428:59–67.
175. Del Rizzo DF, Moon MC, Werner JP, Zahradka P. A novel organ culture method to study intimal hyperplasia at the site of a coronary artery bypass anastomosis. Ann Thorac Surg. 2001;71:1273–9; discussion 1279-1280.
176. Xiao Y, Liu Q, Han HC. Buckling reduces eNOS production and stimulates extracellular matrix remodeling in arteries in organ culture. Ann Biomed Eng. 2016;44:2840–50. https://doi.org/10.1007/s10439-016-1571-0.
177. Lim CS, Kiriakidis S, Paleolog EM, Davies AH. Cell death pattern of a varicose vein organ culture model. Vascular. 2013;21:129–36. https://doi.org/10.1177/1708538113478413.
178. Wilson DP, Saward L, Zahradka P, Cheung PK. Angiotensin II receptor antagonists prevent neointimal proliferation in a porcine coronary artery organ culture model. Cardiovasc Res. 1999;42:761–72.
179. Reisinger U, et al. Leoligin, the major lignan from Edelweiss, inhibits intimal hyperplasia of venous bypass grafts. Cardiovasc Res. 2009;82:542–9. https://doi.org/10.1093/cvr/cvp059.
180. Zheng JP, et al. Vasomotor dysfunction in the mesenteric artery after organ culture with cyclosporin A. Basic Clin Pharmacol Toxicol. 2013;113:370–6. https://doi.org/10.1111/bcpt.12105.
181. Nilsson D, et al. Endothelin receptor-mediated vasodilatation: effects of organ culture. Eur J Pharmacol. 2008;579:233–40. https://doi.org/10.1016/j.ejphar.2007.09.031.

15

182. Akiyoshi T, et al. A novel organ culture model of aorta for vascular calcification. Atherosclerosis. 2016;244:51–8. https://doi.org/10.1016/j.atherosclerosis.2015.11.005.
183. Aragon-Sanabria V, et al. VE-cadherin disassembly and cell contractility in the endothelium are necessary for barrier disruption induced by tumor cells. Sci Rep. 2017;7:45835. https://doi.org/10.1038/srep45835.
184. Benn A, Bredow C, Casanova I, Vukicevic S, Knaus P. VE-cadherin facilitates BMP-induced endothelial cell permeability and signaling. J Cell Sci. 2016;129:206–18. https://doi.org/10.1242/jcs.179960.
185. Hordijk PL, et al. Vascular-endothelial-cadherin modulates endothelial monolayer permeability. J Cell Sci. 1999;112(Pt 12):1915–23.
186. Wang S, Liang B, Viollet B, Zou MH. Inhibition of the AMP-activated protein kinase-alpha2 accentuates agonist-induced vascular smooth muscle contraction and high blood pressure in mice. Hypertension. 2011;57:1010–7. https://doi.org/10.1161/HYPERTENSIONAHA.110.168906.

# The Porcine Coronary Artery Ring Myograph System

*Diethart Schmid and Thomas M. Hofbauer*

© Springer Nature Switzerland AG 2019
M. Geiger (ed.), *Fundamentals of Vascular Biology*, Learning Materials in Biosciences,
https://doi.org/10.1007/978-3-030-12270-6_16

**What You Will Learn in This Chapter**

In this chapter, you will learn the methodological principles of the *in vitro* porcine coronary artery ring model. Knowledge on endogenous vasoconstrictors and vasodilators relevant for coronary arteries will be extended with physiological principles regarding their myogenic tone and explanation of the necessity of experimental precontraction for studies that characterize potential vasodilators. Design and proper usage of the myograph system are explained in detail. You will then learn about buffers suitable for transporting the explanted pig heart and for perfusing isolated coronary artery rings and how to expose coronary arteries from the heart and prepare coronary artery rings, mount the rings onto the myograph, adjust passive stretch and preload, equilibrate specimens and assess their viability, and ensure proper precontraction before application of experimental substances or other treatments. At the end of this chapter, original data of two exemplary experiments using a vasodilator from the class of potassium channel openers are shown.

## 16.1 Introduction

The heart, especially its muscular part, the myocardium, is a metabolically highly active tissue. The high demand for energy-rich nutrients and oxygen is met by extraction of these substances from blood flowing through the myocardium via the coronary circulation [1]. Approximately 60–70% of the oxygen content in the arterial blood is extracted and utilized in the myocardium at single passage through the myocardial microcirculation. The main function of the coronary arteries is to dilate properly and deliver more blood to the myocardium in case of increasing heart rate and force, e.g., during vigorous physical exercise. This ability is also called adequate coronary flow reserve.

Given the lively importance of proper heart function, any dysfunction of the coronary arteries may have great impact on morbidity and mortality of patients. Such dysfunction may be the result of atherosclerotic plaques in the vessel wall, constituting a hallmark of coronary artery disease (CAD) [2]. Complications of CAD, like myocardial infarction, are leading causes of death especially in the Western world [3]. Consequently, a large number of scientific groups conduct research concerning the coronary circulation, with special emphasis on coronary arteries. The aim of these investigations is mainly to reveal pathophysiological mechanisms of diseases and to identify potential pharmacological targets.

Arteries may be studied *in vivo* using laboratory animals or *ex vivo* in isolated hearts. However, both *in vivo* and *ex vivo*, the arteries are influenced by a variety of factors, hormones, and local mediators produced by different tissues, mainly the muscle and connective tissue cells. Consequently, under these conditions, it is very difficult to pinpoint distinct properties of the vessels themselves and the reaction to a particular vasoactive substance. In order to overcome these shortfalls and eliminate confounding influences, methods to study isolated arteries have been developed.

A classical approach is to isolate, prepare, and study ring segments from arteries *in vitro* by holding these segments between two hooks inserted into the vessel lumen [4]. A thread fixed to one hook connects to a force-measuring transducer that converts the wall strain of the vessel into a proportional electrical signal. This signal can then be amplified and recorded on a chart recorder or, as it is done nowadays, on a data acquisition computer system.

In this chapter, we will describe the components and design of a modernized coronary artery ring myograph system. We will also explain how a typical experiment is performed, from obtaining the heart and performing the surgical preparations of the coronary artery specimens to data display and storage. We will also discuss how a typical experiment is planned, especially the question of selecting proper precontraction strategies.

## 16.2  Physiological Principles: In Vivo Myogenic Tone and In Vitro Precontraction

Arteries that are perfused under *in vivo* physiological conditions with arterial blood are usually exposed to rather high hydrostatic pressures in the lumen of up to 120 mmHg or even more that cause a transmural pressure. Arteries usually respond to this stimulus with a myogenic contraction, also called Bayliss effect [5]. This results in a considerable basal myogenic tone. Mechanistically, this is caused by a mechanosensitive receptor on the luminal side of the plasma cell membrane of smooth muscle cells. Recently, this receptor was identified to be identical with the angiotensin II receptor [6]. Its activation leads to an increase in cytoplasmic calcium ions and myosin light-chain activation and contraction.

Spontaneous myogenic tone is usually not observed in isolated ring vessel preparations, even though they are stretched to a representative wall tension value by application of a proper preload after mounting. This is a considerable problem, especially when studying substances that may cause vasodilation. In this case, a dilatory effect may be masked and not detectable. In order to overcome the problem, coronary artery ring segments need to be precontracted before application of a potential vasodilator. Precontraction refers to the (in most cases, submaximally) contracted state of a vessel. Aside from being necessary for the assessment of vasodilators, it has been suggested to increase a vessel's susceptibility to additional contractile stimuli.

Precontraction may be achieved using two different strategies (◻ Fig. 16.1). One would be to add a depolarizing agent like potassium chloride (KCl), causing opening of L-type

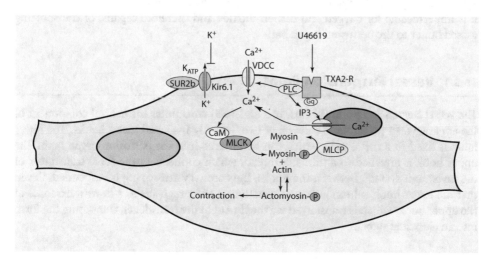

◻ **Fig. 16.1**  Mechanisms mediating precontraction of a coronary artery smooth muscle cell. Actomyosin-P phosphorylated actomyosin, CaM Calmodulin, $G_q$ G-protein stimulating PLC, IP3 inositol triphosphate, $K_{ATP}$ ATP-sensitive potassium channel, Kir6.1 potassium inwardly rectifying channel 6.1, pore-forming unit of the $K_{ATP}$ channel, MLCK myosin light-chain kinase, MLCP myosin light-chain phosphatase, Myosin-P phosphorylated myosin light chain, PLC phospholipase C, SUR2b sulfonylurea receptor 2b, a regulatory component of the $K_{ATP}$ channel, TXA2-R thromboxane A2 receptor, U46619 thromboxane A2 analogue, VDCC voltage-dependent calcium channel

voltage-dependent calcium channels and resulting in elevation of cytosolic calcium at the beginning of experiments. One study found potassium-induced precontraction to activate serotonin receptors, which lead to strongly accentuated vasoconstriction upon addition of serotonin [7].

Another strategy would be the application of vasoconstrictors that are recognized by $G_q$ protein-coupled receptors on smooth muscle cells, the activation of which would lead to inositol triphosphate binding to and calcium release from the endoplasmic reticulum, which represents an intracellular store for calcium. As a result, cytosolic calcium will increase as in the case of depolarizing agents as explained above. For example, mild precontraction using the thromboxane A2 analogue U46619 "primed" arteries to contract further in response to α2 adrenoceptor signaling [8–10].

## 16.3 Setup of Apparatus and Materials

The whole setup consists of the following basal components:
- One or several vessel suspension points with the coronary artery ring specimen
- One or more water-jacketed perfusion organ bath(s)
- Gassing flask in water bath, pumps, and tubing
- Force transducer(s)
- Electronic transducer amplifier(s)
- Analog to digital (A/D) converter
- Data acquisition and processing software with computer

◘ Figures 16.2, 16.3, and 16.4 show simplified graphics and an actual photograph of the apparatus, respectively. Tygon® polymer was generally used as a tubing material, because it is impermeable for oxygen and carbon dioxide and therefore capable of transporting gassed buffer to the perfusion organ bath.

### 16.3.1 Vessel Suspension Point

The vessel suspension point (◘ Figs. 16.2 and 16.5) constitutes the central component of the apparatus. In principle, it consists of two, roughly L-shaped metal hooks. The lower hook is fixed to a metal pole, which can be lowered into the perfusion organ bath. The upper hook is attached to a thin silk thread, which connects to the force transducer of the myograph system. Between the hooks, the coronary artery ring is suspended. Given that the lower hook is immovable, contraction of the ring leads to a downward force of the upper hook, which is transferred via the thread to the transducer, translating the force into an electrical signal.

### 16.3.2 Water-Jacketed Perfusion Organ Bath

The perfusion organ bath was manufactured out of glass by a glassblower according to our drawings (for schematic, see ◘ Fig. 16.2). It has an inner chamber connected to an inlet at the bottom and an outlet on the upper left end which is also reachable from the top opening. It has a double wall creating an outer chamber that has an inlet on the lower left end and an outlet on the upper right end. The outer chamber is fed and drained by prewarmed

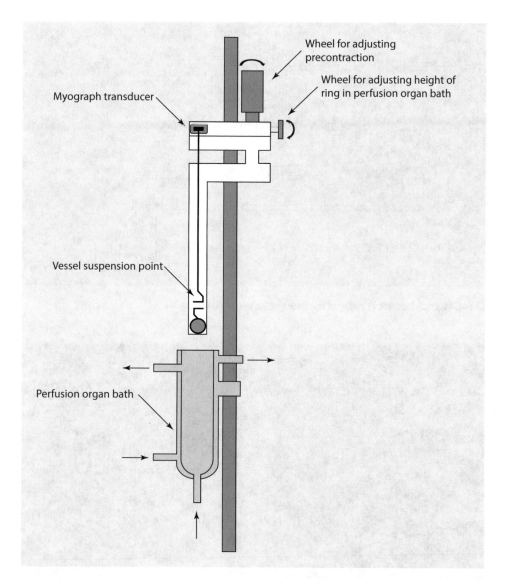

**□ Fig. 16.2** Schematic drawing of the apparatus

water flowing from and to the perfusion water bath in order to warm up the perfusion organ bath to 37 °C.

## 16.3.3 Gassing Flask in Water Bath, Pumps, and Tubing

Perfusion organ bath buffer (for recipe, see □ Table 16.2) is gassed with carbogen (95% $O_2$, 5% $CO_2$) using a fine pore sparger in a flask standing in a water bath adjusted to 37 °C. Rotary perfusor pumps transport the organ buffer with a rate of 2 ml per minute through the lower opening, thereby filling up the organ bath from the bottom, while another pump (>2.1 ml per minute) aspirates the buffer (and air) from the upper opening, thereby keeping a constant fluid level.

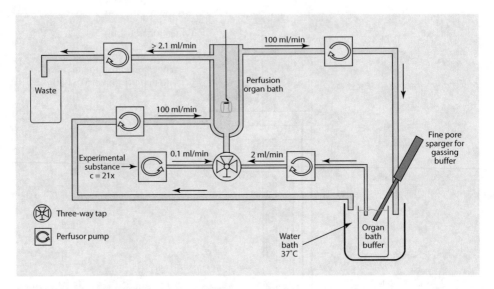

**Fig. 16.3**  Schematic drawing of the flow system and used tubing

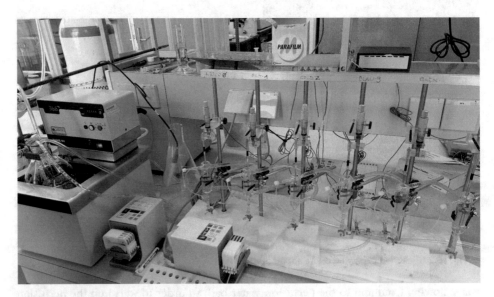

**Fig. 16.4**  Overview of the instrumental setup. The core setup consists of six flow chambers (center to right) diffused in parallel with perfusion organ buffer. Buffer is prewarmed in a water bath (left) and kept at a pH of 7.40 by gassing with carbogen. Experimental compounds are added using a perfusor pump (front)

### 16.3.4 **Force Transducer**

The force transducer translates the downward force applied to the hook into a proportional small electrical signal. We use the FORT10-100 force transducer from WPI ( Fig. 16.6a) that has a force range of up to 100 mN (approximately 10 grams). It needs a stabilized

🔲 **Fig. 16.5** Close-up of perfusion organ bath. A single perfusion organ bath is depicted in the left photograph. The enlarged section shows a close-up of one porcine coronary artery ring that is mounted onto the vessel suspension point

🔲 **Fig. 16.6** Force transducer. **a** Force transducer. **b** Schematic of strain gauge circuit

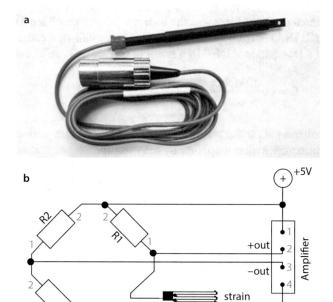

constant voltage supply of up to 10 V DC for its operation. Inside the housing, there is a Wheatstone resistance bridge. Using this bridge, a change in the electrical resistance of the resistance strain gauge is translated into a small change in output voltage. The linearity error of this transducer is less than 0.2% at full scale. The output sensitivity is 0.275 mV/mN (old model: 0.22 mV/mN) at 10 V supply. There are very small voltages that need to be amplified using an electronic transducer amplifier in order to be able to display and save the time course of the force on a computerized chart recorder. A schematic of the strain gauge circuit is given in ◻ Fig. 16.6b.

## 16.3.5 Electronic Transducer Amplifier

An electronic transducer amplifier that is ready to use is commercially available, e.g., a Bridge8 Transducer Amplifier Module (World Precision Instruments). However, this amplifier is rather expensive. As a cost-effective alternative, electronics may be soldered by the experimenter on a breadboard using the schematics given in the following section. Else, a ready-made amplifier may be obtained from the authors at cost price upon request.

The schematics of the electronic circuit are shown in ◻ Fig. 16.7. The amplifier requires two normal commercial 12 V DC power supplies (min. 1 Watt) connected externally for its operation. They are connected on pin 1 (+) and 2 (−) and pin 4 (+) and 3 (−) to the power supply connector. Integrated circuits 2 and 3 (IC2 [7805] and IC3 [7905]) are so-called positive and negative voltage regulators that produce low-noise symmetric, stabilized voltages of +5 V and −5 V, respectively. For their proper function, both need a 100 nF capacitor (C1 and C2) to avoid oscillations. These stabilized voltages are used for the supply of both the force transducer and the instrumentation amplifier. The instrumentation amplifier is IC1 INA114 from Texas Instruments. It amplifies the voltage difference between input pins (V IN+ minus V IN−) by a factor of G. Factor G can be calculated as follows:

$$G = 1 + \frac{50\,\text{kOhm}}{R_G}$$

whereby $R_G$ is the resistance of an external resistor between the pins $R_G 1$ and $R_G 2$ of the instrumentation amplifier. By selecting the proper resistor, the gain of the amplifier can be set. The external resistor may set any gain between 1 and 10,000. In our case, we expect input voltages from the force transducer of up to 100 mN * 0.275 mV/mN = 27.5 mV. In order to produce an output voltage of 2.75 V, which would be optimal for the analog to digital (A/D) converter, we need an amplification gain (G) of 100. By rearranging the formula given above, we get:

$$R_G = \frac{50\,\text{kOhm}}{G-1} = \frac{50000}{100-1} = 505 \text{ Ohm}$$

We select 500 Ohm for the resistor $R_G$. Since resistors with this value are not available, we switch two 1000 Ohm resistors R1 and R2 in parallel, obtaining the desired 500 Ohm:

$$R_G = \frac{1}{\dfrac{1}{R1} + \dfrac{1}{R2}} = \frac{1}{\dfrac{1}{1000} + \dfrac{1}{1000}} = 500 \text{ Ohm}$$

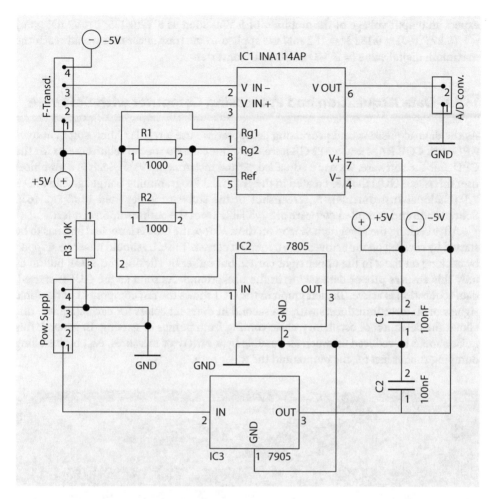

■ **Fig. 16.7** Schematic of the electronic circuit

The voltage between pin Vout of the instrumentation amplifier and ground (GND) is the properly amplified voltage that can now be connected to the A/D converter. Since the A/D converter cannot accept voltages that are zero or negative, an offset voltage of about 0.5 V at pin Vref at the instrumentation amplifier needs to be set using the R3 precision potentiometer. The software later on subtracts this offset again and converts voltages to force values.

### 16.3.6 Analog/Digital Converter

As an A/D converter of our setup, we employ a PCI 6024E A/D converter PCI plug-in card (National Instruments) for a standard personal computer. This card has 16 analog input channels, a maximal sample rate of 200 kSamples/sec and an A/D converter accuracy of 12 bits. The maximum input voltage for the analog inputs is 5 V. Since the force transducer can be overloaded by up to 80–100% (up to 180–200 mN), we can expect an output voltage of the amplifier of 5 V as soon as 5 V/(0.275*1 mV/mN*gain) = 5/(0.275*100) = 0.182 N = 182 mN are applied to the transducer that would reach the maximum digital value (= $2^{12}-1$) of the A/D converter.

### 16.3.7 Data Acquisition and Processing Computer with Software

As the data acquisition and processing computer, we use a regular Microsoft Windows 7 PC with 4 GB RAM and a 512 GB hard disk. There are no special requirements for the CPU. For the software, we use a so-called virtual instrument (VI), which is a graphical user interface (GUI) that we created in the graphical Programming Language LabVIEW 7.1 (National Instruments). A screenshot of the software is depicted in ❑ Fig. 16.8. Source code and compiled software are available from the authors upon request.

After clicking the Myograph.vi icon on the desktop, the GUI is launched and has to be started by clicking on the arrow in the upper left corner. The GUI should never be stopped by clicking on the x in the upper right corner, but rather by clicking the STOP button as only this ensures proper data saving in the background. As soon as the GUI is started, data acquisition is active. The left graph in the GUI shows the fast component of the force signals of all force transducers within a second in different colors for each channel. This allows the detection of oscillatory noise coming from pumps or gassing. In general, this noise should be reduced as much as possible by a variety of measures, e.g., by installing dumping rubber feet for the pumps and the water bath.

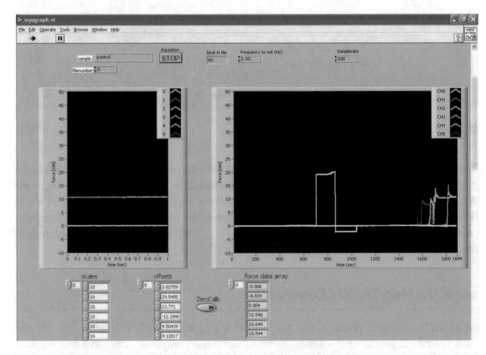

❑ **Fig. 16.8** Screenshot of the graphical user interface of the virtual instrument. The center of the right graph (slow phase) shows the calibration event (around second 650–850, ZeroCalib, and hooking of a 20 mN calibration weight). Then, coronary artery rings were mounted and prestretched to 10–15 mN (right side of slow phase; colored lines indicate separate rings)

The right graph shows the slow component of the force signals that is internally calculated by averaging the left "one-second" traces. The trigger frequency for this averaging can be adjusted using the "frequency to set" input box in the top area of the GUI. In the background, these averaged force data are stored into CSV files on the hard disk. After each 1800 time points, a new file with a new file number is automatically generated. The actual file number can be read in the file number box. The general form of the file name is YYYYMMDDFnB_sampledescr.csv, where Y, M, and D are the actual date, Fn is the file number, and sampledescr is the sample description which can be changed in the sample box of the GUI.

After powering up the setup including the amplifiers, the force transducers require 15 minutes to warm up. The threads of the force transducers with the upper hook should hang freely, without touching any structure. Under these conditions, the zero calibration can be made by clicking the "ZeroCalib" button which automatically chooses proper values for the offsets that are indicated in the offset box of the GUI. After this event, if all traces stay on the zero line for at least 10 minutes, sufficient stability is reached, and the coronary rings may be mounted.

## 16.3.8 Preparation of Organ Buffers

Prior to starting the experiment, two different buffers must be prepared.

Modified Krebs-Henseleit (MKH) buffer is based on a recipe originally used for liver tissue [11] but since then has been modified for use in various organ systems. In our setting, bicarbonate has been replaced by HEPES, since bicarbonate buffers are prone to spontaneous alkalization. The MKH buffer serves as transport medium of porcine hearts after explantation. The recipe is given in ◻ Table 16.1. Prior to use, pH must be adjusted to 7.4 at room temperature. It is recommended to always prepare the buffer freshly on the day before experiments.

For the perfusion organ bath, a Krebs-Henseleit buffer (KHB) without HEPES and insulin, but with bicarbonate, is used. The recipe is given in ◻ Table 16.2. Due to high turnaround of buffer, it is recommended to prepare higher amounts. Depending on the experimental setup, between 2–5 L is required.

◻ **Table 16.1**    Modified Krebs-Henseleit (MKH) buffer

| Compound | Molecular weight (g/mol) | Concentration | Amount per L |
|---|---|---|---|
| NaCl | 58.44 | 118 mM | 6.9 g |
| HEPES | 238.3 | 20 mM | 4.77 g |
| KCl | 74.56 | 4 mM | 0.298 g |
| $CaCl_2$ | 147.01 | 1.25 mM | 0.184 g |
| $NaH_2PO_4$ | 119.98 | 1.2 mM | 0.144 g |
| $MgSO_4$ | 120.36 | 1.2 mM | 0.144 g |
| D-Glucose | 180.16 | 4.5 mM | 0.811 g |
| Insulin aspart (100 IE/ml) | 5831.65 | 10 U/l | 100 µl |

◻ **Table 16.2**    Perfusion organ bath buffer (KHB)

| Compound | Molecular weight (g/mol) | Concentration | Amount per 2 L |
|---|---|---|---|
| NaCl | 58.44 | 118 mM | 13.8 g |
| NaHCO$_3$ | 84.01 | 20 mM | 3.360 g |
| KCl | 74.56 | 4.7 mM | 0.596 g |
| CaCl$_2$ | 147.01 | 1.25 mM | 0.368 g |
| NaH$_2$PO$_4$ | 119.98 | 1.2 mM | 0.288 g |
| MgSO$_4$ | 120.36 | 1.2 mM | 0.288 g |
| D-Glucose | 180.16 | 5.0 mM | 1.8 g |

❓ **Question Q1:** Recapitulate the basic structure of the myograph system.

## 16.4 Experimental Procedure

### 16.4.1 Obtaining and Transport of the Porcine Heart

Porcine hearts are obtained from a local slaughter house. Immediately after slaughter, the heart is explanted, taking care not to injure the coronary arteries. The organ is transferred directly to a 2 L borosilicate glass container filled with approximately 750 mL of ice-cold MKH buffer, and the container is put into a polystyrene or cool box filled with precooled thermal packs. Remaining MKH buffer is then filled into the glass container to prevent drying-out of tissue. The opening can be temporarily sealed using parafilm. Once in the laboratory, the heart is placed on a 245 × 245 mm square dish covered with dry tissues to ensure proper handling and grip during preparation.

### 16.4.2 Preparation of Coronary Artery Rings

Various studies have investigated the coronary anatomy and physiology of pigs as compared to that of humans [12–14], which is, apart from minor differences, very similar. Concerning the choice of which coronary artery should be used, no clear consensus exists; while a majority of studies have used the left descending artery, in our hands, the right coronary artery is, in general, easier to prepare and yields adequately sized ring specimens.

Using atraumatic tweezers and scalpels, epicardial fat and connective tissue are carefully removed from the coronary sulcus to expose the artery. Preparation of the vessel starting at its medial part and then following its course to the coronary ostium is recommended to facilitate efficient harvest of a section with a large enough diameter. During preparation, the tissue must be kept humid by frequently applying buffer via a pipette. An exemplary preparation of a coronary artery is given in ◻ Fig. 16.9.

**◘ Fig. 16.9** Exemplary preparation of a porcine coronary artery. The right coronary artery (arrow) was prepared for one exemplary set of experiments

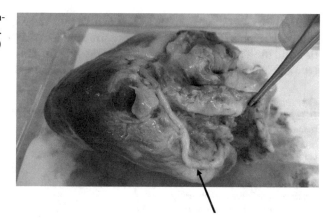

Using a scalpel, rings with a width of 4–5 mm are generated. Width should not vary significantly between rings to avoid variability of the following measurements. Sections with side branches should be discarded. Rings can be stored in a glass container filled with prewarmed MKH buffer until further use; however, it is recommended to mount rings within 10–15 minutes in order to maintain viability and functionality.

### 16.4.3 Mounting of Rings onto the Vessel Suspension Point

For measurements, rings are mounted onto the two metal hooks of the vessel suspension point (see enlarged section of ◘ Fig. 16.5). Using a hand wheel, rings are then lowered into the perfusion organ bath, so that it is completely immersed in buffer. Since buffer and experimental substances are introduced at the bottom of the perfusion organ bath, time delay between start of application and onset of effect can be modulated by the height of the vessel suspension point in the flow chamber. Proper mounting of the ring is crucial, i.e., hooks must run in parallel; otherwise, abrupt sliding of the ring into its ideal position can occur, leading to artificial jumps in tracings (see ◘ Fig. 16.10).

### 16.4.4 Equilibration and Adjusting Passive Stretch

After mounting rings onto the vessel suspension point and immersion into the perfusion organ bath, passive stretch must be applied to the rings (usually ~10–15 mN for coronary artery rings with a length of 4–5 mm). Using the respective wheel, one can adjust the height of the upper hook, thus applying strain on the ring and increasing baseline contractile force. Once prestretched, rings reach their contractile steady state after 10–20 minutes. Depending on the type and size of used vessel ring, different levels of passive stretch may be required.

The practical importance of passive stretch for experiments is highlighted in ◘ Fig. 16.11. Passive stretch was adjusted to defined levels, after which 20 mM KCl were added. Using a perfusor pump that connects to the lower inlet of the interior perfusion organ bath chamber, continuous flow of KCl was initiated. On the y-axis, active developed force is plotted, i.e., the difference between the maximum force and the level of passive stretch. ◘ Figure 16.12 depicts the time course of force of one exemplary ring, including stretching to the new preload, a short equilibration phase, and subsequent active contraction in response to KCl. This experiment demonstrates that the experimenter should determine the amount of passive stretch that results in submaximal active contraction and use a preload that is about 10–15% lower for subsequent experiments, given that type and length of coronary artery rings remain unchanged.

**◻ Fig. 16.10** Faulty tracing due to inadequately mounted artery ring. After adding potassium chloride, a potent vasoconstrictor, contractile force increased. Due to insufficient mounting of the ring, signal dropped abruptly until the ring slid into place, after which the potassium chloride-mediated effect was again observed

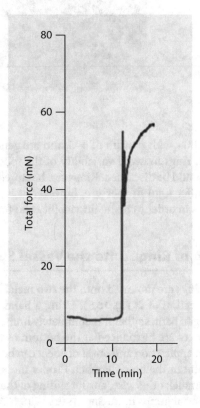

**◻ Fig. 16.11** Importance of passive stretch on ring functionality. Active developed force is calculated as the difference between maximum force and the level of passive stretch

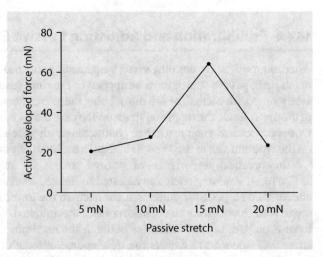

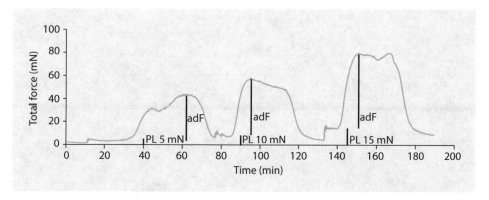

**Fig. 16.12** Time course of passive stretch experiment. Using the same coronary artery ring, contraction was induced using 20 mM KCl. With each consecutive stimulation, preload (PL) was pre-adjusted to 5 mN (first contraction), 10 mN (second contraction), and 15 mN (third contraction) prior to addition of 20 mM KCl. Active developed contractile force (adF) after application was monitored

## 16.4.5 Validation of Ring Viability

Before the start of actual experiments, viability of rings must be assessed. This can be done by applying a vasoconstrictor like U46619 (100 nM); alternatively, in case that preload optimization was not performed, 30–40 mM KCl can be used. Upon addition of vasoconstrictors, rings react with a prompt and accentuated contraction (see ☐ Fig. 16.13). After termination of stimulation using KCl, contractile force returns to baseline within 20–30 minutes. Another important validation of ring integrity is the absence of spontaneous vasospasm. Lacerated or otherwise wounded coronary artery rings exhibit rhythmic contractions or spontaneous and irreversible contracture. Any ring not fulfilling these two criteria should be discarded.

**?** **Question Q2:** Why is precontraction of vessels required to assess the vasodilative effect of a substance in isolated coronary artery segments?

**?** **Question Q3:** Which strategies are required to ensure coronary artery ring viability and integrity?

## 16.4.6 Measurement of Changes of the Contractile Force Induced by Substance of Interest

Any resulting contraction of the coronary artery rings is translated into an electrical signal which is documented in real time and can be followed via the GUI on the computer display. After equilibration and validation of viability, experimental substances can be added.

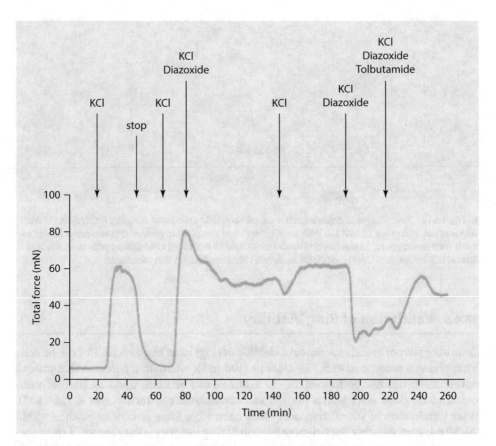

**Fig. 16.13** Tracing of one exemplary experiment using potassium channel modulators. Arrows indicate the time point of addition of respective compounds

## 16.5 Exemplary Analyses

### 16.5.1 Potassium Channel Modulation

High concentrations of potassium reduce the ion's transmembrane gradient; this causes a depolarization of the membrane potential leading to calcium influx and subsequent activation of the contractile apparatus [15]. Diazoxide is a potent ATP-sensitive potassium ($K^+_{ATP}$) channel opener [16, 17], inducing $K^+$ efflux. This in turn decreases activity of voltage-gated calcium channels, leading to reduced contractility [18]. Contrarily, tolbutamide is a $K^+_{ATP}$ channel blocker [19] which leads to vasoconstriction by increasing intracellular calcium influx, thus activating the contractile apparatus.

In ◙ Fig. 16.13 a tracing of an exemplary experiment testing potassium channel modulators is given. First, KCl (20 mM) was used as a potent vasoconstrictor. After washout, contractive force returned to baseline. Then, KCl was again added to induce vasoconstriction, proving continued viability of artery rings. After 20 min, diazoxide (200 μM) was added. Indeed, diazoxide partially antagonized potassium chloride-induced vasoconstriction. To reverse this effect, tolbutamide (200 μM) was added.

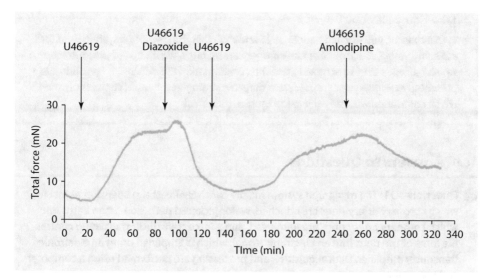

**Fig. 16.14** Tracing of one exemplary experiment, using thromboxane A2 analog U46619 as a vaso-constrictor and diazoxide and amlodipine as vasodilators. Arrows indicate the time point of addition of respective compounds

### 16.5.2 **Influence of Thromboxane**

Thromboxane A2 (TXA2) is an arachidonic acid metabolite that has a variety of physiological functions, among them regulation of smooth muscle cell tone via $G_q$ protein coupling and subsequent calcium influx leading to constriction [20–22]. Amlodipine is a calcium channel blocker [23], inhibiting voltage-gated calcium channels [24, 25], and is used therapeutically for treatment of arterial hypertension [26] and vasospastic angina [27].

◻ Figure 16.14 shows a tracing of a separate experiment investigating the precontractile effect of TXA2. TXA2 is very unstable in solution; hence, a synthetic stable analogue, U46619, was used for experimental applications. After a time delay of approximately 10 minutes, 100 nM U46619 induced vasoconstriction. Note the slower increase compared to the effect of KCl (◻ Fig. 16.13). Similar to the experiment described in 5.1, addition of diazoxide elicited vasodilation. A delayed onset of vasoconstriction could be observed after a second U46619 stimulus, which reflects a retarded washout of diazoxide, most likely, due to its rather lipophilic nature. Addition of amlodipine antagonized the vasoconstrictive effect of U46619. In conclusion, we observed that precontraction with U46619 more closely resembles physiologic conditions, given that compared to KCl, no massive depolarization ensues. Hence, U46619 more closely reflects physiological membrane potential changes and is therefore recommended to study the effect of potential vasodilators.

❓ **Question Q4:** Explain the fundamental difference between precontraction using KCl and the TXA2 analogue U46619.

---

**Take-Home Message**

In this chapter, you learned how to build and maintain the myograph system used for assessing vasoreactivity of isolated coronary artery rings. We discussed precontraction, which methods exist to acquire it, and their fundamental differences. To conclude, we presented exemplary results using two different strategies for vessel contraction and which compounds can be used to antagonize this effect.

---

## 16.6  Answers to Questions

✅ **Question #Q1:** The myograph system mainly consists of vessel suspension points to which coronary artery rings are attached, water-jacketed perfusion organ baths into which coronary artery rings are immersed, and a force transducer translating contractile force of the rings into an electronic signal, which is amplified using an electronic transducer amplifier. Data acquisition and processing are performed using a computer and myograph software.

✅ **Question #Q2:** In contrast to in vivo arteries, isolated arterial segments do not develop an autonomous muscular tone, even though the vessel wall is stretched by application of passive preload. Because of this, vasodilative effects, in the absence of active precontraction, could not be demonstrated or identified for unknown substances.

✅ **Question #Q3:** Coronary artery rings must, upon stimulation with a vasoconstrictor like KCl, respond with a contractile force that returns to baseline within 20–30 minutes. Rings must further not exhibit spontaneous vasospasms or contracture.

✅ **Question #Q4:** Precontraction using KCl results in a massive reduction of the transmembrane potassium gradient, thereby depolarizing the membrane potential. The ensuing calcium influx activates the contractile apparatus. Since hyperkalemia is a pathological state (e.g., during renal failure massive cell death or under certain pharmacological treatment regimens), other compounds inducing precontraction are preferred for functional experiments with coronary artery rings. TXA2 induces vasoconstriction via $G_q$ protein coupling and subsequent calcium influx. Also note the difference in slope of the force curve between ◻ Figs. 16.13 and 16.14.

Acknowledgments    This work was supported by the "Medizinisch-Wissenschaftlicher Fonds des Bürgermeisters der Stadt Wien" grant # 2047 issued to D. Schmid, by a grant # H-1143 from the Hochschuljubiläumsstiftung issued to Ivo Volf, and by the Austrian Science Funds (SFB-54). We thank Christian Plass, MD, for assembling the amplifier electronics on a breadboard and Mr. Alfred Mika for manufacturing the mechanical parts.

## References

1.  Pape H, Kurtz A, Silbernagl S. Physiologie. 8th ed. Stuttgart/New York: Thieme; 2018.
2.  Bonow R, Mann D, Zipes D, Libby P, editors. Braunwald's heart disease – a textbook of cardiovascular medicine. 9th ed. Philadelphia: Saunders, imprint of Elsevier; 2012.

3. Ibanez B, James S, Agewall S, Antunes MJ, Bucciarelli-Ducci C, Bueno H, Caforio ALP, Crea F, Goudeve-nos JA, Halvorsen S, Hindricks G, Kastrati A, Lenzen MJ, Prescott E, Roffi M, Valgimigli M, Varenhorst C, Vranckx P, Widimský P. 2017 ESC guidelines for the management of acute myocardial infarction in patients presenting with ST-segment elevation. Eur Heart J. 2017;39(2):119–77. https://doi.org/10.1093/eurheartj/ehx393.

4. Mulvany M, Halpern W. Mechanical properties of vascular smooth muscle cells in situ. Nature. 1976;260:643–5.

5. Bayliss W. On the local reactions of the arterial wall to changes of internal pressure. J Physiol. 1902;28(3):220–31.

6. Mederos y Schnitzler M, Storch U, Gudermann T. Mechanosensitive G $_{q/11}$ protein-coupled receptors mediate myogenic vasoconstriction. Microcirculation. 2016;23(8):621–5. https://doi.org/10.1111/micc.12293.

7. Smith JR, Kim C, Kim H, Purdy RE. Precontraction with elevated concentrations of extracellular potassium enables both 5-HT 1B and 5-HT 2A "silent" receptors in rabbit ear artery. J Pharmacol Exp Ther. 1999;289(1):354–60.

8. Bhattacharya B, Roberts RE. Enhancement of α2-adrenoceptor-mediated vasoconstriction by the thromboxane-mimetic U46619 in the porcine isolated ear artery: role of the ERK – MAP kinase signal transduction cascade. Br J Pharmacol. 2003;139(1):156–62. https://doi.org/10.1038/sj.bjp.0705208.

9. Maclean MR, Mcgrath JC. Effects of pre-contraction with endothelin-1 on alpha 2-adrenoceptor- and (endothelium-dependent) neuropeptide Y-mediated contractions in the isolated vascular bed of the rat tail. Br J Haematol. 1990;101(1):205–11.

10. Roberts RE. Pre-contraction with the thromboxane-mimetic U46619 enhances P2X receptor-mediated contractions in isolated porcine splenic artery. Purinergic Signal. 2012;8(2):287–93. https://doi.org/10.1007/s11302-011-9284-1.

11. Krebs H, Henseleit K. Untersuchungen über die Harnstoffbildung im Tierkörper. Physiol Chemie. 1932;11(18):757–9.

12. Gómez FA, Ballesteros LE. Morphologic expression of the left coronary artery in pigs. An approach in relation to human heart. Rev Bras Cir Cardiovasc. 2014;29(2):214–20. https://doi.org/10.5935/1678-9741.20140027.

13. Gómez FA, Ballesteros LE. Anatomic study of the right coronary artery in pigs: feature review in comparison with the human artery. Int J Morphol. 2013;31(4):1289–96. https://doi.org/10.4067/S0717-95022013000400023.

14. Sahni D, Kaur GD, Harjeet JI. Anatomy & distribution of coronary arteries in pig in comparison with man. Indian J Med Res. 2008;127(6):564–70.

15. de Oliveira MAB, Brandi AC, dos Santos CA, Botelho PHH, Cortez JLL, Braile DM. Modes of induced cardiac arrest: hyperkalemia and hypocalcemia – literature review. Rev Bras Cir Cardiovasc. 2014;29(3):432–6. https://doi.org/10.5935/1678-9741.20140074.

16. Atalik KE, Kiliç M, Doğan N. Role of the nitric oxide on diazoxide-induced relaxation of the calf cardiac vein and coronary artery during cooling. Fundam Clin Pharmacol. 2009;23(3):271–7. https://doi.org/10.1111/j.1472-8206.2009.00671.x.

17. Quayle JM, Nelson MT, Standen NB. ATP-sensitive and inwardly rectifying potassium channels in smooth muscle. Physiol Rev. 1997;77(4):1165–232. https://doi.org/10.1152/physrev.1997.77.4.1165.

18. Nakai T, Ichihara K. Effects of diazoxide on norepinephrine-induced vasocontraction and ischemic myocardium in rats. Biol Pharm Bull. 1994;17(10):1341–4. https://doi.org/10.1248/cpb.37.3229.

19. Bijlstra P, Russel F, Thien T, Lutterman J, Smits P. Effects of Tolbutamide on vascular ATP-sensitive potassium channels in humans. Horm Metab Res. 1996;28(9):512–6. https://doi.org/10.1055/s-2007-979843.

20. Nakahata N. Thromboxane A2: physiology/pathophysiology, cellular signal transduction and pharmacology. Pharmacol Ther. 2008;118(1):18–35. https://doi.org/10.1016/j.pharmthera.2008.01.001.

21. Yamamoto K, Ebina S, Nakanishi H, Nakahata N. Thromboxane A2 receptor-mediated signal transduction in rabbit aortic smooth muscle cells. Gen Pharmacol Vasc Syst. 1995;26(7):1489–98. doi:0306362395000259 [pii].

22. Kinsella BT, O'Mahony DJ, Fitzgerald GA. The human thromboxane A2 receptor alpha isoform (TP alpha) functionally couples to the G proteins Gq and G11 in vivo and is activated by the isoprostane 8-epi prostaglandin F2 alpha. J Pharmacol Exp Ther. 1997;281(2):957–64. http://www.ncbi.nlm.nih.gov/pubmed/9152406.

23. Godfraind T. Discovery and development of calcium channel blockers. Front Pharmacol. 2017;8:286. https://doi.org/10.3389/fphar.2017.00286.

24. Matlib MA. Relaxation of potassium chloride-induced contractions by amlodipine and its interaction with the 1,4-dihydropyridine-binding site in pig coronary artery. Am Heart J. 1989;64(17):51I–7I.
25. Matlib MA, French JF, Grupp IL, Van Gorp C, Grupp G, Schwartz A. Vasodilatory action of amlodipine on rat aorta, pig coronary artery, human coronary artery, and on isolated Langendorff rat heart preparations. J Cardiovasc Pharmacol. 1988;12(Suppl 7):S50–4.
26. Williams B, Mancia G, Spiering W, Agabiti Rosei E, Azizi M, Burnier M, Clement DL, Coca A, de Simone G, Dominiczak A, Kahan T, Mahfoud F, Redon J, Ruilope L, Zanchetti A, Kerins M, Kjeldsen SE, Kreutz R, Laurent S, Lip GYH, McManus R, Narkiewicz K, Ruschi DI. 2018 ESC/ESH guidelines for the management of arterial hypertension. Eur Heart J. 2018;39(33):3021–104. https://doi.org/10.1097/HJH.
27. Harris JR, Hale GM, Dasari TW, Schwier NC. Pharmacotherapy of vasospastic angina. J Cardiovasc Pharmacol Ther. 2016;21(5):439–51. https://doi.org/10.1177/1074248416640161.

# Proteomics in Vascular Biology

*Maria Zellner and Ellen Umlauf*

© Springer Nature Switzerland AG 2019
M. Geiger (ed.), *Fundamentals of Vascular Biology*, Learning Materials in Biosciences,
https://doi.org/10.1007/978-3-030-12270-6_17

## What You Will Learn in This Chapter

The *proteome* is the entire set of proteins of a biological sample, and *proteomics* is the large-scale qualitative (protein composition) and quantitative analysis of the proteome. A collection of dedicated biochemical methods is combined with protein databases to identify the proteins and to allocate them to biological pathways. The work process involves two major steps, the separation of the proteins (*electrophoresis* and *liquid chromatography*) and the protein identification by *mass spectrometry*. In medical research, proteomics is mainly applied to discover proteins that play a role in pathological processes; thus it is also a valuable tool in vascular biology. These disease-related proteins are generally defined as *biomarkers* and are of great interest to the diagnosis and treatment of patients. In vascular research, biological samples are, for example, plasma, serum, platelets, endothelial cells and vascular smooth muscle cells. This chapter will provide an overview of this technology and will demonstrate its application.

---

**Info Box 17.1**

**Why Do We Perform Proteomics?**

A. *To characterize the protein composition of a biological sample*
   - One focus in vascular disease research has been the characterization of the protein composition of the coronary thrombus of myocardial infarction (MI) patients [3, 45].

B. *To identify the degree of influence of a certain condition on the protein composition of a biological sample*
   - The knockout of protein C inhibitor in male mice, which are infertile as a result, affected only three proteins in the testis proteome [64].
   - The knockout of acetylcholinesterase in mice had strong influence on their brain *proteome*, changing 221 proteins. Their phenotype included delay of growth, immature external ears and persistent body tremor [31].

C. *To identify biomarkers, which show altered abundance levels in a particular condition (e.g. in disease, upon in vivo or in vitro treatment, etc.)*
   Altered *protein abundances* can be caused by changes in protein synthesis and/or protein degradation:
   - The proteome analysis of human monocytes showed a very prominent induction of interleukin 1beta (IL-1β), upon in vitro LPS treatment. Additionally, it was shown that the degradation of this normally short-lived protein was reduced during glutamine depletion, whereas the degradation rate of long-lived cytoskeletal proteins such as β-actin was unaffected [66].

   Altered protein abundances can be caused by protein modifications that are specific to a particular state or condition of the tested samples (e.g. protein phosphorylation regulates platelet reactivity):
   - Inhibition of platelets by a stable prostacyclin analogue strongly changed the *phosphorylation proteome*, affecting 360 proteins [5].

D. *To identify biological processes and pathways that are affected by certain conditions, by applying pathway analysis software to the corresponding, altered proteomes*

The plasma proteome of patients treated with torcetrapib, an inhibitor of the cholesteryl ester transfer protein, revealed changes in biological pathways related to immune, inflammatory and endocrine functions. These were previously shown to be associated with increased cardiovascular and mortality risk of patients treated with this drug [60].

**17**

## 17.1 Proteomics and Biomarkers

The term proteomics was coined only 20 years ago, in 1997 [26]. Besides identification of biomarkers predicting the risk of disease or diagnosing disease, proteomics may also determine markers that infer the efficacy of therapeutic concepts in patients.

Proteins are the essential effectors of genes and play major roles in all life processes at all organizational levels (cells, tissues and organs) and thus are also involved in disease development. Causal events may be malfunctions induced by, e.g. altered *post-translational modifications (PTM)* or abnormal expression levels. In proteomics, these changes are reflected by significantly different *proteoform* (▶ Info Box 17.3) abundances, when proteomes of unaffected and pathological samples are compared. The respective distinctive feature, e.g. the protein expression level or protein modification, may be defined as biomarker.

> **Info Box 17.2**
> The general definition of a biomarker is "a feature of a sample that can be objectively measured and evaluated as a specific indicator of normal biological processes, pathogenic processes or pharmacological responses to therapeutic intervention" [6].

Favourable sources for diagnostic biomarkers are blood (e.g. plasma, serum) or other easily available body fluids such as urine and saliva. In vascular biology, platelets, an easily accessible blood fraction, are another interesting clinical proteomics specimen, because they are the major therapeutic target in preventing atherothrombotic events [32]. For example, antiplatelet drugs are the main therapy against cardiovascular thrombus formation. However, they show patient-dependent potency, which may lead to recurrent thrombotic events in affected patients. This interindividual variability in the platelet response was observed for aspirin and clopidogrel, which also correlated with the risk of subsequent thrombotic events [10]. Several discovery studies of the platelet proteome identified biomarker candidates, which may predict the response to antiplatelet treatment, like decreased levels of glyceraldehyde 3-phosphate dehydrogenase [33] and increased levels of the glycoprotein IIIa in platelets of aspirin-resistant patients [14].

### 17.1.1 Routine Protein Biomarkers in Vascular Disease

A perfect biomarker is 100% sensitive (correctly identified positive samples) and 100% specific (correctly identified negatives); thus it shows perfect accuracy – in addition, it can be quantified in easily accessible specimens. The following examples illustrate that diagnostic biomarkers of acute diseases are in general easier to identify and more accurate than prognostic biomarkers that would provide more information on the causes and course of the disease.

A classic example for a high-quality, diagnostic biomarker is high-sensitivity (hs) troponin that is a central marker for myocardial necrosis. For that reason it identifies acute myocardial infarction (MI) and is easily quantifiable by a routine immunoassay system in blood serum [8]. Its *specificity* was shown to be in the range of 94%, and the *sensitivity* to rule out acute MI was 99% [8]. Another example is B-type natriuretic peptide (BNP). Its increase indicates heightened ventricular stretch and wall tension, which reflects primary left ventricular dysfunction [25]. However, its positive predictive value is low (44–57%), and the negative predictive value is relatively high (94–98%). Therefore, it is used to exclude chronic heart failure rather than to diagnose its presence [40]. The characterization of prognostic biomarkers is more challenging because the definition of these study cohorts is more difficult compared to the acute status of cardiovascular disease (CVD). Thus, the accuracy of biomarkers for identifying presymptomatic stages is in general lower. Currently, the plasma levels of cardiac troponins [34] and C-reactive protein (CRP) [23] are used to predict the risk of developing CVD. While hs troponin is highly accurate in diagnosing acute MI (see above), its prognostic ability is limited. A slight, continuous increase indicates subclinical myocardial damage and higher risk of subsequent heart failure [34]. Furthermore, increased levels of CRP are related to systemic and vascular inflammation, which precedes CVD [23]. However, the specificity of CRP for the prediction of CVD is low, because it is also increased during other inflammatory-linked conditions, namely, infections, smoking [7], cancer [2] and Alzheimer's disease [15, 34, 53].

The general low specificity of CRP, the shortcomings of BNP to detect the causes of chronic heart failure [46] and the need to precisely characterize the pathology and early events leading to vascular diseases and thrombotic events require the identification of additional biomarkers.

### 17.1.2 Diagnostic Biomarkers: From Proteomics to Clinical Routine

The translation of diagnostic biomarkers from the *discovery phase* to clinical practice is generally a very laborious, time-consuming and cost-intensive procedure. This development process is characterized by five central phases [43]. The first one is the discovery phase, which has screening character and is preclinical and exploratory in design. Today, it is mostly carried out with *omics* studies (proteomics, genomics or metabolomics) that identify protein, genetic or metabolic biomarkers. In proteomics, the proteomes of patient samples are compared to those of healthy age- and sex-matched controls. This discovery stage is one of the major applications for proteomics in clinical research, and the used technologies are described below. The second phase comprises two stages: the development of a clinical assay for the potential biomarker and the *verification* of the potential biomarker's diagnostic performance. The clinical assay is established because the proteomics techniques (*2D electrophoresis* and *shotgun mass spectrometry*) are not suited for clinical routine due to their technical complexity and low throughput. The new assay must show high sensitivity (detects the biomarker) and high specificity (no cross-reaction with other compounds) and must have high-throughput capabilities. In addition to high accuracy, it must show high precision, which is the reproducibility of the assay results within and between laboratories. In most cases, the method of choice will be immuno-based, e.g. enzyme-linked immunosorbent assay (ELISA), multiplex protein biochips or beads or immunoturbidimetry. In the second stage of phase two, the potential biomarker is analysed in a new study cohort using the newly developed assay. Ideally, its ability to discriminate between health and disease is verified, while taking into account potential confounders, such as sex and age. In the third phase, retrospective studies are performed, which evaluate clinical and longitudinal data. At this stage, the development focuses more and more on clinical challenges, for example, whether the biomarker assay is capable of diagnosing the disease at earlier stages. The forth phase assesses the prospective diagnostic accuracy of the assay. The fifth phase evaluates the routine use of the biomarker assay with respect to earlier or more accurate diagnosis and more targeted treatment of the disease, which should result in the reduction of mortality, morbidity and disability in the population [43].

## 17.2  Proteomics and Statistics

In general, the proteomes of two groups or more groups of samples, representing different conditions (e.g. healthy versus diseased; untreated versus treated), are compared in the first phase of discovering biomarkers. The degree of the difference between the two groups, the intra-group variability and the sample size determine the *statistical significance of the results.*

However, proteomics techniques typically analyse thousands of proteins simultaneously, which inflates the number of *false positives* (type I error) – presumed biomarkers may actually be none. Several methods that aim to *correct for this multiple testing* are too strict (e.g. *Bonferroni correction*) and also eliminate the true positives, which is equivalent to the loss of power and an increased type II error (*false negative*) [12]. Thus, in the discovery phase, in which it is important to find true positive biomarker candidates, other more tolerant adjustments are used, namely, the *Benjamini-Hochberg correction* [24] and the Storey correction [54]. Even these adjustments are considered too strict by some, and alternatives have been proposed [41, 49]. Another possibility would be to define the discovery phase as *explorative study*, which is usually evaluated without statistical corrections but includes the risk for false-positive results, which may be revealed in a following verification study with newly recruited study individuals. In any case, it is advisable to define the sample size in the discovery phase of a proteomics study as high as possible for a characterization of robust biomarkers [22].

## 17.3  Sample Preparation and Reduction of Sample Complexity in Vascular Proteomics

The vascular system is manifold in its cellular heterogeneity and its distribution throughout the whole body. A deregulation of vascular functions is associated with pathologies such as arterial and venous thrombosis, atherosclerotic cardiovascular disease and aneurism. Systematic analysis of malfunctions in vascular biology is a central driving force in medical research. To this end, the sample must be clearly defined, and a highly standardized sample preparation procedure is essential for obtaining reliable proteome maps. Protein samples can be extracted from body fluids, cells or subcellular compartments.

However, tissues or even blood samples are too heterogeneous to allow reasonable and reliable proteome analysis. Accordingly, target cell populations have to be isolated from these biological sources, using fluorescence-activated cell sorting (FACS) or laser capture microdissection (LCM). FACS is a specialized technology of flow cytometry that separates the cell type of interest from the remaining biological fluid via its specific light scattering and fluorescent characteristics. LCM isolates the cellular subpopulation of interest from tissue cells under direct microscopic visualization. For example, it was used to obtain vascular smooth muscle cells from the aortic wall of MI patients and non-MI individuals. Their proteomes were compared, and the top change in MI was an increased abundance of superoxide dismutase 1. Subsequent database-assisted pathway analysis indicated a strong induction of hypoxia signalling [63]. Additionally, the optimal protein extraction and sample preparation methods depend on the used downstream proteome technology and are, therefore, an important field in proteome research [30, 67].

Summing up, the use of clearly defined samples and standardized sample preparation procedures is essential for performing conclusive proteomic analyses. However, the large number of proteins and their dynamic range poses additional challenges for the proteomics techniques. The human genome encodes approximately 20,230 protein-coding genes, which produce about 50,000 different mRNAs that are translated into possibly 100,000 proteoforms (see ▶ Info Box 17.3) [20]. To date, more than 10,000 proteins have been identified in plasma [37]. In addition, the dynamic concentration range of the proteins in samples is huge. There are highly concentrated proteins that may disturb the analytical process and mask other lower-abundant proteins, whereas the levels of very low-abundant proteins are below the detection limit. Plasma, for instance, has a total protein concentration of approximately 75 mg/ml. Half of it is albumin, at a concentration of 35–45 mg/ml, followed by transferrin, 2–3.6 mg/ml, and fibrinogen, 2–4.5 mg/ml. In contrast, the concentrations of very low-abundant proteins such as the cytokines, TNF alpha and interleukin 6, lie in the pg/ml range [4].

Thus, the procedures used for adjusting the sample's dynamic range of protein concentrations determine which proteins can be analysed. For instance, the high-abundant protein in a plasma sample can be removed by immunoprecipitation to allow a more comprehensive analysis of lower-abundant proteins [4]. Furthermore, the complexity of a cellular extract can be reduced by enriching subcellular compartments such as nuclei, cytosol, mitochondria or mixed microsomes, using differential centrifugation. The result is a detailed subcellular proteome analysis [11]. In addition, proteins or peptides with certain PTMs, e.g. phosphorylations [47] or *glycosylations* [42], can be specifically enriched by antibodies, lectin or $TiO_2$ [58]. These pre-analytical preparation steps result in an in-depth proteomic characterization of the *phosphoproteome* or *glycoproteome*.

After sample purification and preparation are completed, the next step involves either 2-DE or LC (see next paragraph) that fractionates the sample in order that MS, the last technique in the proteomics process, can finally identify the individual proteins.

## 17.4 Overview of Proteomics Technologies for the Characterization of Protein Biomarkers

In everyday laboratory practice, a protein of interest (= potential biomarker) is analysed by targeted immunologically based techniques such as immunoblotting (Western blot), ELISA or flow cytometry. However, these biochemical methods focus on a very small selection of proteins in a biological sample. Proteomics-based technologies facilitate the qualitative and quantitative profiling of hundreds to thousands of proteins and their proteoforms (▶ Info Box 17.3).

Two fundamental, technical steps are necessary to study the proteome qualitatively and quantitatively: first, the extracted proteins in the sample are separated to further reduce the complexity. This is done either by electrophoretic gel-based methods such as *2D gel electrophoresis* followed by a proteolytic digest or by liquid chromatography of the enzymatically digested protein sample (◘ Fig. 17.1). Normally, enzymatic digestion of proteins is done using trypsin, because of its high specificity and ease of handling. Trypsin is a serine protease, which cleaves at the carboxyl side of arginine and lysine. However, sometimes the cleavage by trypsin can interfere with the amino acid sequence of interest, for instance, a region potentially carrying a PTM. In this case, alternative proteases such as chymotrypsin, LysC, LysN, AspN, GluC or ArgC [18] can be used. The proteolytically cleaved peptides are separated by *liquid chromatography* and are then to be analysed by *mass spectrometry* (MS). In the second step, MS analyses the masses of these peptides with which the original proteoforms can be deduced (◘ Fig. 17.1). The two most important MS methods that measure peptide masses are *electrospray ionization (ESI)* [61] and *matrix-assisted laser desorption/ionization (MALDI)* [27]. This analytical procedure for protein identification of unknown proteins is called *peptide mass fingerprinting* (PMF). The peptide masses obtained by MS are compared to calculated peptide masses (software MASCOT) derived by in silico digestions of protein sequence entries in primary sequence databases (e.g. SwissProt, NCBIprot). Peptide mass fingerprinting is possible if the genome sequence of the sample donor species is known. However, PMF generates only a first survey spectrum of each particular protein. To further increase the reliability of the protein identification, an MS2 (alternative terms: MS/MS or *tandem MS* spectrum) is generated. The mass spectrometer isolates individual peptides, fragments them in a collision cell into pieces and records the fragments in a second mass spectrum scan. This causes rated break points, which are compared to in silico-derived rated break points, and improves the accuracy of protein identifications derived from that particular peptide [50]. For this reason, PMF and the latter peptide fragment masses are usually used in combination to reliably identify proteins in complex proteomic samples [20].

## 17.4.1 Proteome Analysis by Two-Dimensional Gel Electrophoresis

The development of two-dimensional gel electrophoresis (2-DE) was the birth of proteomics [44]. The first reports on a protein separation using this technique were published by O'Farrell [39] and Klose [28] in 1975. This biochemical method separates protein mixtures in two dimensions; first, in the horizontal direction, according to their *isoelectric*

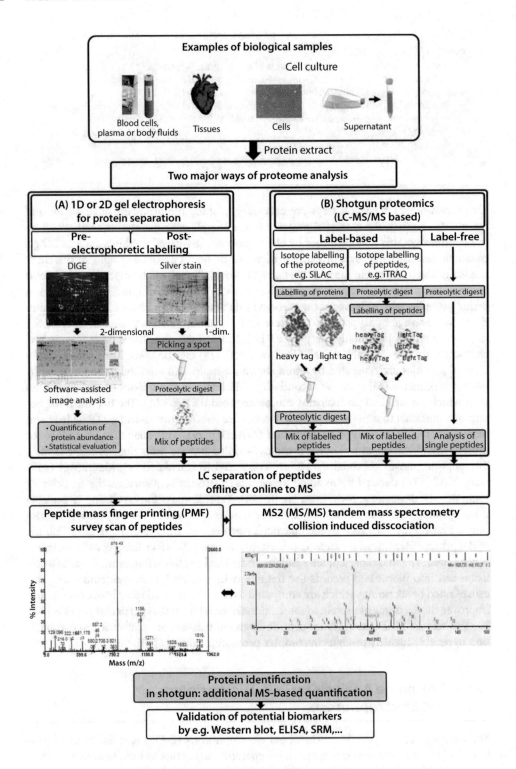

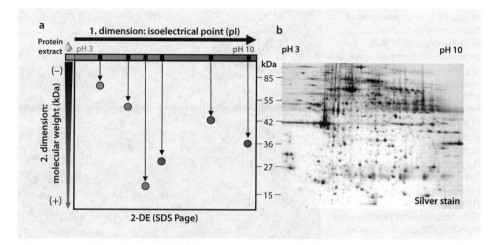

**◼ Fig. 17.2**    Basic principle of two-dimensional gel electrophoresis. **A** Schematic representation of two-dimensional gel electrophoresis. The two-dimensional separation of proteins starts with isoelectric focusing of the protein extract. The proteins in a complete focused IPG strip (pH 3–10) are uncharged because they have reached their isoelectric point (black rectangle). Afterwards the focused proteins in the IPG strip are loaded with SDS to negatively charge the proteins. Then this SDS-treated focused IPG strip is placed on the top of an SDS PAGE gel, and the proteins migrate out of the IPG strip into the SDS-PAGE gel. Each protein moves with a different velocity through the acryl amid gel matrix depending on its size. Finally, the two-dimensionally separated proteins have the shape of a spot (coloured circles). **B** Example of a silver-stained two-dimensional gel of a human platelet protein extract. Fifty µg of platelet protein extract was separated in the pH range 3–10 according to their isoelectric point (pI) and subsequently according to their molecular weight (120–15 kDa). The final 2-DE gel was stained with mass-compatible silver stain [51]

*point*, and then, in the vertical, by their *molecular weight* (MW). The result is a polyacrylamide gel containing separated proteoforms shaped to spots (◼ Fig. 17.2) [59]. Since it is quite rare that two proteoforms have the same pI and MW, most spots correspond to one specific proteoform, while bands in the common 1D electrophoresis usually contain several proteoforms or proteins.

Despite the excellent resolution of 2-DE, the 800 to 1000 proteoform spots on the gel correspond to only about 500 different proteins. This agrees well with the most abundantly

**◼ Fig. 17.1**    Workflow of the two main proteomics analysis systems. Proteins are extracted from a biological sample (e.g. blood cells, body fluids, such as plasma, tissues, cell lines, cell culture or supernatants). **a** In gel-based proteome analysis, the extracted proteins are separated by 1D or 2D gel electrophoresis. Protein extracts can be labelled before the electrophoretic separation (pre-electrophoretic labelling, DIGE) or after electrophoresis (post-electrophoretic labelling, silver stain) by staining of 1D or 2D gels. Protein spots of interest are selected using software-assisted image analysis. These protein spots are picked from a preparative 2-DE gel (usually silver-stained), and peptides are extracted by proteolytic digest and identified by MS (these peptides are usually separated by LC before injection to MS). **b** For shotgun proteomics the proteins can be labelled before or after the proteolytic digest (label-based) or the proteins remain unlabelled (label-free). Then, the extracted proteins are proteolytically digested in solution. These complex peptide mixtures are also separated by LC, which usually last longer than those for 2D spot digests. During the MS analysis, the masses and fragment masses of the peptides are acquired. These mass data are used in combination with databases to identify the original proteins. In shotgun proteomics, the intensity of the peptide signals is used for quantification

**◘ Fig. 17.3** Two examples of two-dimensional differential gel electrophoresis (2D DIGE): **a** Comparative proteome analysis of platelet protein extract from mouse (12 µg protein labelled with Cy5 – red) and human (12 µg protein labelled with Cy3 – green). **b** Comparative proteome analysis of a lupus anticoagulant (LA) patient (12 µg protein labelled with Cy3 – green) and an age- and sex-matched healthy control (12 µg protein labelled with Cy5 – red). Labelled proteins from both 2D DIGE gels (A+B) were focused according to their isoelectric point (pI) in a pH range 4–7 and subsequently according to their molecular weight (120–15 kDa). Yellow protein spots represent similar abundances of the two proteomes. Marked protein spots were picked from a preparative (150 µg platelet protein) silver-stained 2-DE gel, tryptically digested and identified by mass spectrometry

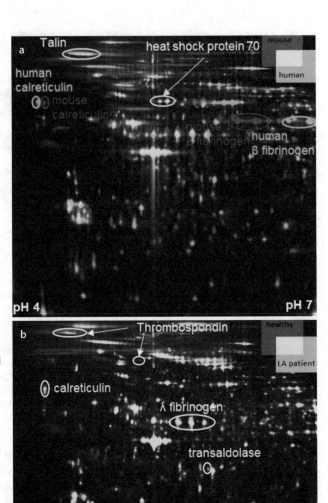

expressed genes, which account for approximately 10% of the translated genes [44], and shows its detection bias against lower-abundant proteins. The high number of proteoform spots, as compared to the effectively identified proteins, is the result of PTMs and splicing that increase the variation of the MWs and/or pIs of the proteoforms. For instance, thrombospondin is separated into several proteoform spots with different pIs and MWs in a 2-DE map of platelets (◘ Fig. 17.3b).

Analyses of 2-DE gels require the detection of the proteoform spots, which depends on either pre-electrophoretic or post-electrophoretic protein staining. However, not all staining procedures are compatible with MS. *Post-electrophoretic staining* is done after two-dimensional separation of the proteins using colloidal Coomassie, silver nitrate or fluorescence dyes (SYPRO™ Ruby, SYPRO™ Orange, Deep Purple™).

*Pre-electrophoretic staining* is carried out before the first dimension of the 2-DE, using mostly cyanine-based fluorescent dyes that are attached to the protein via covalent binding. The dye-to-protein ratio is usually kept low (at 3% or below, "minimal labelling") to ensure that only a single lysine residue in each protein is tagged and the pI and MW of the protein are little affected. Dyes of similar structure but different spectral properties are used to label up to four different samples that are run on the same 2-DE gel [13, 36]. Typically, one of these samples is an internal standard, which allows precise within-gel and between-gel comparisons of samples – a major technical challenge in proteomics without these measures. Two normalization procedures reduce the technical variability to approximately 7% [62], which is an excellent value in the proteomics field. This proteomics 2-DE technology is called *2D difference gel electrophoresis (2D DIGE)* [56] and was developed by Unlü et al. in 1997 [56] which was a further breakthrough in 2-DE and quantitative proteomics.

■ Figure 17.3a shows an example from our research laboratory that compares a human platelet proteome with that of mice (performed by Maria Zellner and Alice Assinger). Mouse platelets are labelled in red and the human platelets are labelled in green. Overlapping spots from the two species turn yellow if their proportion is close to one. Human and mouse protein-coding genes are about 85% identical [38]. However, the 2D DIGE gel in ■ Fig. 17.3a demonstrates that the difference between the two species is larger on the proteome level. For example, β-fibrinogen and calreticulin differ strongly in their pIs, which is indicated by the severe shifts of the respective spots in the horizontal direction. On the other hand, talin and the heat shock protein 70 show similar pIs, MWs and abundances. However, in most clinical proteomics studies, the proteomes differ mostly in the abundance of the protein spots but not in the protein spot pattern (■ Fig. 17.3b). The example shows a 2D DIGE gel that compares a platelet sample of a lupus anticoagulant patient with thrombosis history, labelled in green, to that of an age- and sex-matched healthy control, labelled in red. Green spots indicate elevated proteoform levels in the patient and represent potential biomarkers for an increased thrombosis risk [21].

Another variant of pre-electrophoresis labelling is in vivo radiolabelling (metabolic labelling), using $^{35}S$, $^{14}C$, $^{3}H$ or, in the case of phosphoproteins, $^{32}P$ or $^{33}P$. After electrophoresis, an autoradiography is prepared by drying the gel and exposing it to X-ray film or phosphor screens. With this method, only newly synthesized proteins are detected as 2D resolved proteoform spots [66].

### 17.4.1.1  Imaging and Evaluation

Qualitative and quantitative analyses of 2-DE gels require precise detection and digitization of the protein spot signals at the respective wavelengths. There are three major types of equipment to do this: flatbed scanners for, e.g. silver and Coomassie stains, CCD camera systems for detection of chemiluminescence and laser scanners for fluorescence signals.

Evaluation of digitized 2-DE gels is made using dedicated software (DeCyder, GE Healthcare; Delta2D, Decodon; Melanie 9, Melanie™) and comprises three main steps: spot detection, the matching of these spots and spot quantification. During the spot detection process, the position and boundary of each individual protein spot have to be defined reproducibly. This process ensures exact quantification and reproducible matching of each spot across all gels and samples. However, across-gel matching is often challenged by protein pattern distortion between experiments due to small variations in sample preparation or electrophoresis conditions. To reduce these effects and for normalization purposes, an *internal standard* is applied on every 2-DE gel run. This standard should be a pool of all samples included in the study and is always labelled with the same dye, usually Cy2. If it is impossible to do this, the minimal requirements for an internal standard are that it must be from the same biological source as the samples and it must be available in sufficient quantity to run it in parallel with every sample in the study. Under optimal conditions, the internal standard's spot pattern indicates whether biological variation of the sample or technical variation caused the sample spot pattern of the 2-DE gel.

Quantitative analysis of the samples is done by running two normalization steps. First, based on the assumption that most of the proteoform spots remain unchanged, a model frequency histogram curve is used to calculate the mean spot intensity of each single sample on a 2D map. Then the normalized abundance of each protein spot in a sample is calculated as a ratio between this spot and the corresponding mean spot intensity. In a second normalization step, the abundance of each protein spot in a sample is calculated as a ratio between the sample spot and the corresponding internal standard spot [1]. Finally, these quantitative and standardized abundance values of the protein spots are statistically evaluated. Significant, condition-related protein spots are picked from a stained (e.g. MS-compatible silver stain), preparative 2-DE gel. Afterwards, the respective proteoform is enzymatically digested within the piece of gel (usually trypsin), and the peptides are extracted into solution. This peptide mixture is subjected to LC before injection into MS for protein identification.

In the following example, a 2D DIGE study was applied to identify plasma biomarkers for aortic stenosis (AS). The plasma proteome of six AS patients was compared to that of six healthy controls. The statistical analysis of about 800 protein spots revealed reduced levels of apolipoprotein AI (APOA1), apolipoprotein E (APOE), paraoxonase 1, (PON1), complement C3 (CO3) and alpha-2-HS glycoprotein (FETUA) in the patients. These results were confirmed by the MS-based validation method selected reaction monitoring (SRM) in an independent study set of six AS patients and six matched, healthy controls [19].

At the time of writing this book chapter, only a limited number of clinical vascular disease-related studies had been published that use human plasma 2-DE proteomics. Additionally, the sample sizes in these studies were small, which limits the statistical power and validity of these data. However, the small sample sets may be explained by the high manual workload, which is much lower in semiautomated gel-free-based MS (shotgun) proteomics techniques.

## 17.4.2 Proteome Analysis by Shotgun Proteomics

*Shotgun proteomics* typically analyses complex protein mixtures by sequential, enzymatic digestion of the biological sample, fractionation of the resulting peptides by LC (instead of 2D gel electrophoresis) and MS analysis.

This gel-free *mass spectrometry* technique has emerged as a core tool in proteome analysis. Rapid technological advances of mass spectrometers have made shotgun proteomics the method of choice in proteome analysis, replacing gel-based methods. The immediate identification of proteins in an injected biological sample is a big advantage over 2-DE gel-based proteome analysis.

Two key challenges remain, the huge dynamic range of protein concentrations in plasma and cellular lysates and the need to keep proteins and peptides in solution during the analytical process.

The enzymatic digestion of the protein sample in gel-free proteome analysis differs from that in gel-based methods. In shotgun proteomics, an in-solution digest of the whole biological sample is usually done, whereas gel-based MS identification employs an in-gel digest of picked 1D bands or 2D spots. This in-gel digestion step results in a considerable loss of sample material compared to in-solution digestion.

In LC, the peptides are separated according to their interaction with two phases, a gradient of aqueous to organic solvents and a hydrophobic C18 stationary phase. The ratio determines the retention times of the peptides in the column. The workflow for protein identification in shotgun proteomics is very similar to that of gel-based proteomics. However, the complexity of the shotgun MS proteomics analysis is higher. Peptides eluted from LC are ionized by electrospray and injected into MS for mass analysis. Typically, mass spectrometers record three pieces of information about each peptide: its mass (used in PMF), its ion intensity and a list of its fragments (called tandem MS spectrum, MS/MS or MS$^2$). The peptide mass and fragment masses are used for protein identification, whereas the ion intensities serve for quantification of the particular protein. Today, an average of 8000 proteins can be identified with one optimized LC run in about 4 hours, using high-quality tandem mass spectrometry equipment [35]. Shotgun proteomics has a broader dynamic range for the detection of proteins than 2-DE but provides less information about the intact protein. In addition to the qualitative analysis of the protein composition of a biological sample, shotgun proteomics is used for the quantification of injected peptides. With this approach, disease-related abundance changes of clinical biomarkers can be measured.

Quantitative measurements of peptide abundances are frequently performed in combination with stable isotope labelling. When proteins are metabolically labelled, it is called stable isotope labelling with isotope-labelled amino acids in cell culture *SILAC*; when peptides are the target, it is termed *iTRAQ* isobaric tags for relative and absolute quantitation. In SILAC, the cell or a whole organism is labelled by metabolic incorporation of amino acids tagged with stable (non-radioactive) heavy isotopes, such as $^{13}$C- or $^{12}$C-arginine. Usually, three biological conditions can be compared, using different isotopic forms of arginine or lysine. In iTRAQ-based quantification, these chemical tags are attached to all peptides in a protein digest via the free amines at the peptide N-terminus and in lysine residues [57]. Usually, either peptide sample, SILAC or iTRAQ labelled, is then pooled and analysed in a shotgun run. The MS can assign each peptide to the respective sample due to the slight difference in mass. The pooling of samples improves the comparability between different samples, a concept which is also applied in 2-DE. As in 2-DE, reproducibility between runs must be high, and the variation of quantitative characteristics has to be kept low. Therefore, an internal standard is used to improve the reliability of the qualitative and quantitative analysis system.

However, the simplest method of quantifying peptides is to compare the same peptides between different sample runs, using spectral counting or MS precursor intensities. This "label-free" quantification has gained interest in the last years, because the improved robustness, resolution of LC-MS systems and quantification accuracy have reduced between-run variation [55].

A procedure related to LC-MS is *selected reaction monitoring (SRM)*, which is a targeted, more exact, quantification method of selected proteins. It achieves its high quantification accuracy by combining external calibrator proteins or peptides with internal standard peptides derived from endogenous proteins with low biological variation [17]. The technical settings of this method provide a lower resolution for complex protein mixtures but a higher quantification capacity for profiling a particular protein [16, 29]. SRM is an alternative to immunological methods and is therefore increasingly used for validating those proteins that had been identified in the discovery phase (second phase in the translation of a clinical biomarker).

In the following study, 135 MI cases and 135 matched controls were analysed in the discovery phase, using iTRAQ shotgun proteomics. Its aim was to identify plasma biomarkers for the new onset of cardiovascular disease. Plasma concentrations of 861 proteins were determined and revealed several MI-associated candidates, namely, glycoprotein 5, CD5 antigen-like, myoglobin, protein kinase C inhibitor protein 1, C-reactive protein, cyclophilin A and contactin-1. In the following validation phase including 336 atherosclerotic cardiovascular disease cases and control pairs, 59 of the top MI-related plasma biomarkers were validated by SRM. C-reactive protein and CD5 antigen-like were finally confirmed as biomarkers for atherosclerotic cardiovascular disease risk, and additional risk biomarkers were described, namely, alpha-1-acid glycoprotein 1, paraoxonase 1 and tetranectin [65].

In general, the selection of a particular proteomics technology is independent of the biomedical or biological research question. The decision for one of the proteomics technologies depends to a great part on the equipment and know-how of the respective research team. Most of the proteomics technologies would provide synergist proteome insights into the sample of interest, but financial costs and time expenditure are too large to apply different proteomics technologies in one proteomics study. However, the expertise of the authors of this chapter is mainly in 2-DE, which is why it focuses on this proteomics method.

---

**Take-Home Message**

- Biological samples such as saliva, plasma or cellular extracts contain several thousand proteins.
- "Protein isoform" refers to all protein molecules that originate from the same gene but are different because of genetic mutations (e.g. SNPs) or changes on the transcriptional level, e.g. splicing.
- "Protein species" refer to all protein molecules that differ in their chemical modifications, e.g. phosphorylation and glycosylation. "Proteoform" includes both protein isoforms and protein species.
- The number of proteoforms in a protein sample that originate from the respective gene is higher than that of the related, spliced RNA transcripts because it includes post-translational protein modifications.
- Proteomics is a selection of different technologies that analyse the protein composition of a biological protein sample as comprehensively as possible.
- Currently, it is not possible to detect every protein of a complex biological sample within one single, analytical proteomic run, because the protein concentrations show a high dynamic range (e.g. blood plasma or cellular extracts). However, the range of analysed proteins can be controlled, at least partially, by varying the sample preparation procedures.
- Usually, proteomic profiles of cellular extracts or body fluids are generated. Alternatively, subproteomes of subcellular compartments or of proteins with specific post-translational modifications (e.g. phosphorylation) may be analysed and focus on specific, cellular compartments, mechanisms and pathways.

17

- Proteomics is an important technology in medical research for the characterization of biomarkers.
- A biomarker is a feature of a sample that can be objectively measured and evaluated as a specific and sensitive indicator of normal biological processes, pathogenic processes or pharmacological responses to therapeutic intervention.
- In proteomics, two or more groups with well-defined characteristics are compared. Proteoforms that are sensitive for a certain condition and specific for the noncondition (e.g. health vs. disease, treatment vs. nontreatment) are potential biomarkers.
- The two main processes in a proteome analysis are the fractionation of a sample into proteins (2-DE) or peptides (LC) and the identification of the initial sample proteins.
- Gel-based methods separate proteins, while liquid chromatography fractionates proteolytically cleaved peptides.
- Proteins in proteome samples are usually identified by determining the masses of proteolytically derived peptides with mass spectrometry.
- The two most important MS methods for measuring peptide masses are electrospray ionization (ESI) and matrix-assisted laser desorption/ionization (MALDI).
- Peptide masses (PMF) and their peptide fragment masses (MS/MS) are usually used in combination to reliably identify and quantify proteins in complex protein samples.

## References

1. Alban A, David SO, Bjorkesten L, Andersson C, Sloge E, Lewis S, Currie I. A novel experimental design for comparative two-dimensional gel analysis: two-dimensional difference gel electrophoresis incorporating a pooled internal standard. Proteomics. 2003;3(1):36–44. https://doi.org/10.1002/pmic.200390006.
2. Allin KH, Nordestgaard BG. Elevated C-reactive protein in the diagnosis, prognosis, and cause of cancer. Crit Rev Clin Lab Sci. 2011;48(4):155–70. https://doi.org/10.3109/10408363.2011.599831.
3. Alonso-Orgaz S, Moreno-Luna R, Lopez JA, Gil-Dones F, Padial LR, Moreu J, de la Cuesta F, Barderas MG. Proteomic characterization of human coronary thrombus in patients with ST-segment elevation acute myocardial infarction. J Proteome. 2014;109:368–81. https://doi.org/10.1016/j.jprot.2014.07.016.
4. Baker ES, Liu T, Petyuk VA, Burnum-Johnson KE, Ibrahim YM, Anderson GA, Smith RD. Mass spectrometry for translational proteomics: progress and clinical implications. Genome Med. 2012;4(8):63. https://doi.org/10.1186/gm364.
5. Beck F, Geiger J, Gambaryan S, Veit J, Vaudel M, Nollau P, Kohlbacher O, Martens L, Walter U, Sickmann A, Zahedi RP. Time-resolved characterization of cAMP/PKA-dependent signaling reveals that platelet inhibition is a concerted process involving multiple signaling pathways. Blood. 2014;123(5):e1–e10. https://doi.org/10.1182/blood-2013-07-512384.
6. Biomarkers Definitions Working G. Biomarkers and surrogate endpoints: preferred definitions and conceptual framework. Clin Pharmacol Ther. 2001;69(3):89–95. https://doi.org/10.1067/mcp.2001.113989.
7. Bittoni MA, Focht BC, Clinton SK, Buckworth J, Harris RE. Prospective evaluation of C-reactive protein, smoking and lung cancer death in the Third National Health and Nutrition Examination Survey. Int J Oncol. 2015;47(4):1537–44. https://doi.org/10.3892/ijo.2015.3141.
8. Boeddinghaus J, Twerenbold R, Nestelberger T, Badertscher P, Wildi K, Puelacher C, du Fay de Lavallaz J, Keser E, Rubini Gimenez M, Wussler D, Kozhuharov N, Rentsch K, Miro O, Martin-Sanchez FJ, Morawiec B, Stefanelli S, Geigy N, Keller DI, Reichlin T, Mueller C, Investigators A. Clinical validation of

a novel high-sensitivity cardiac troponin I assay for early diagnosis of acute myocardial infarction. Clin Chem. 2018;64(9):1347–60. https://doi.org/10.1373/clinchem.2018.286906.

9. Breitbart RE, Andreadis A, Nadal-Ginard B. Alternative splicing: a ubiquitous mechanism for the generation of multiple protein isoforms from single genes. Annu Rev Biochem. 1987;56:467–95. https://doi.org/10.1146/annurev.bi.56.070187.002343.

10. Cattaneo M. Resistance to antiplatelet drugs: molecular mechanisms and laboratory detection. J Thromb Haemost. 2007;5(Suppl 1):230–7. https://doi.org/10.1111/j.1538-7836.2007.02498.x.

11. Cox B, Emili A. Tissue subcellular fractionation and protein extraction for use in mass-spectrometry-based proteomics. Nat Protoc. 2006;1(4):1872–8. https://doi.org/10.1038/nprot.2006.273.

12. Dunn OJ. Multiple comparisons among means. J Am Stat Assoc. 1961;56:52–64.

13. DyeAGNOSTICS 2D Protein Labeling Kits. https://www.dyeagnostics.com/site/products/refraction-2d/. Accessed 31 Aug 2018.

14. Floyd CN, Goodman T, Becker S, Chen N, Mustafa A, Schofield E, Campbell J, Ward M, Sharma P, Ferro A. Increased platelet expression of glycoprotein IIIa following aspirin treatment in aspirin-resistant but not aspirin-sensitive subjects. Br J Clin Pharmacol. 2014;78(2):320–8. https://doi.org/10.1111/bcp.12335.

15. Fu S, Ping P, Zhu Q, Ye P, Luo L. Brain natriuretic peptide and its biochemical, analytical, and clinical issues in heart failure: a narrative review. Front Physiol. 2018;9:692. https://doi.org/10.3389/fphys.2018.00692.

16. Gallien S, Bourmaud A, Kim SY, Domon B. Technical considerations for large-scale parallel reaction monitoring analysis. J Proteome. 2014;100:147–59. https://doi.org/10.1016/j.jprot.2013.10.029.

17. Gianazza E, Tremoli E, Banfi C. The selected reaction monitoring/multiple reaction monitoring-based mass spectrometry approach for the accurate quantitation of proteins: clinical applications in the cardiovascular diseases. Expert Rev Proteomics. 2014;11(6):771–88. https://doi.org/10.1586/14789450.2014.947966.

18. Giansanti P, Tsiatsiani L, Low TY, Heck AJ. Six alternative proteases for mass spectrometry-based proteomics beyond trypsin. Nat Protoc. 2016;11(5):993–1006. https://doi.org/10.1038/nprot.2016.057.

19. Gil-Dones F, Darde VM, Alonso-Orgaz S, Lopez-Almodovar LF, Mourino-Alvarez L, Padial LR, Vivanco F, Barderas MG. Inside human aortic stenosis: a proteomic analysis of plasma. J Proteome. 2012;75(5):1639–53. https://doi.org/10.1016/j.jprot.2011.11.036.

20. Gstaiger M, Aebersold R. Applying mass spectrometry-based proteomics to genetics, genomics and network biology. Nat Rev Genet. 2009;10(9):617–27. https://doi.org/10.1038/nrg2633.

21. Hell L, Lurger K, Gebhart S, Koder S, Ay C, Pabinger I, Maria Z. Differences in the platelet proteome between lupus anticoagulant positive individuals with or without thrombotic manifestations and healthy controls. 2017. http://www.professionalabstracts.com/isth2017/iplanner/#/presentation/856.

22. Hernandez B, Parnell A, Pennington SR. Why have so few proteomic biomarkers "survived" validation? (sample size and independent validation considerations). Proteomics. 2014;14(13-14):1587–92. https://doi.org/10.1002/pmic.201300377.

23. Hingorani AD, Sofat R, Morris RW, Whincup P, Lowe GD, Mindell J, Sattar N, Casas JP, Shah T. Is it important to measure or reduce C-reactive protein in people at risk of cardiovascular disease? Eur Heart J. 2012;33(18):2258–64. https://doi.org/10.1093/eurheartj/ehs168.

24. Hochberg Y, Benjamini Y. More powerful procedures for multiple significance testing. Stat Med. 1990;9(7):811–8.

25. Iwanaga Y, Nishi I, Furuichi S, Noguchi T, Sase K, Kihara Y, Goto Y, Nonogi H. B-type natriuretic peptide strongly reflects diastolic wall stress in patients with chronic heart failure: comparison between systolic and diastolic heart failure. J Am Coll Cardiol. 2006;47(4):742–8. https://doi.org/10.1016/j.jacc.2005.11.030.

26. James P. Protein identification in the post-genome era: the rapid rise of proteomics. Q Rev Biophys. 1997;30(4):279–331.

27. Karas M, Hillenkamp F. Laser desorption ionization of proteins with molecular masses exceeding 10,000 daltons. Anal Chem. 1988;60(20):2299–301.

28. Klose J. Protein mapping by combined isoelectric focusing and electrophoresis of mouse tissues. A novel approach to testing for induced point mutations in mammals. Humangenetik. 1975;26(3):231–43.

29. Lange V, Picotti P, Domon B, Aebersold R. Selected reaction monitoring for quantitative proteomics: a tutorial. Mol Syst Biol. 2008;4:222. https://doi.org/10.1038/msb.2008.61.

17

30. Leon IR, Schwammle V, Jensen ON, Sprenger RR. Quantitative assessment of in-solution digestion efficiency identifies optimal protocols for unbiased protein analysis. Mol Cell Proteomics. 2013;12(10):2992–3005. https://doi.org/10.1074/mcp.M112.025585.

31. Lin HQ, Wang Y, Chan KL, Ip TM, Wan CC. Differential regulation of lipid metabolism genes in the brain of acetylcholinesterase knockout mice. J Mol Neurosci. 2014;53(3):397–408. https://doi.org/10.1007/s12031-014-0267-x.

32. Marcone S, Dervin F, Fitzgerald DJ. Proteomic signatures of antiplatelet drugs: new approaches to exploring drug effects. J Thromb Haemost. 2015;13(Suppl 1):S323–31. https://doi.org/10.1111/jth.12943.

33. Mateos-Caceres PJ, Macaya C, Azcona L, Modrego J, Mahillo E, Bernardo E, Fernandez-Ortiz A, Lopez-Farre AJ. Different expression of proteins in platelets from aspirin-resistant and aspirin-sensitive patients. Thromb Haemost. 2010;103(1):160–70. https://doi.org/10.1160/TH09-05-0290.

34. McEvoy JW, Chen Y, Ndumele CE, Solomon SD, Nambi V, Ballantyne CM, Blumenthal RS, Coresh J, Selvin E. Six-year change in high-sensitivity cardiac troponin T and risk of subsequent coronary heart disease, heart failure, and death. JAMA Cardiol. 2016;1(5):519–28. https://doi.org/10.1001/jamacardio.2016.0765.

35. Meissner F, Mann M. Quantitative shotgun proteomics: considerations for a high-quality workflow in immunology. Nat Immunol. 2014;15(2):112–7. https://doi.org/10.1038/ni.2781.

36. Miller I, Crawford J, Gianazza E. Protein stains for proteomic applications: which, when, why? Proteomics. 2006;6(20):5385–408. https://doi.org/10.1002/pmic.200600323.

37. Nanjappa V, Thomas JK, Marimuthu A, Muthusamy B, Radhakrishnan A, Sharma R, Ahmad Khan A, Balakrishnan L, Sahasrabuddhe NA, Kumar S, Jhaveri BN, Sheth KV, Kumar Khatana R, Shaw PG, Srikanth SM, Mathur PP, Shankar S, Nagaraja D, Christopher R, Mathivanan S, Raju R, Sirdeshmukh R, Chatterjee A, Simpson RJ, Harsha HC, Pandey A, Prasad TS. Plasma proteome database as a resource for proteomics research: 2014 update. Nucleic Acids Res. 2014;42(Database issue):D959–65. https://doi.org/10.1093/nar/gkt1251.

38. National Human Genome Research Institute. Why mouse matters. 2010. https://www.genome.gov/10001345/importance-of-mouse-genome/. Accessed 11 Sept 2018.

39. O'Farrell PH. High resolution two-dimensional electrophoresis of proteins. J Biol Chem. 1975;250(10):4007–21.

40. Pan Y, Li D, Ma J, Shan L, Wei M. NT-proBNP test with improved accuracy for the diagnosis of chronic heart failure. Medicine (Baltimore). 2017;96(51):e9181. https://doi.org/10.1097/MD.0000000000009181.

41. Pascovici D, Handler DC, Wu JX, Haynes PA. Multiple testing corrections in quantitative proteomics: a useful but blunt tool. Proteomics. 2016;16(18):2448–53. https://doi.org/10.1002/pmic.201600044.

42. Patrie SM, Roth MJ, Kohler JJ. Introduction to glycosylation and mass spectrometry. Methods Mol Biol. 2013;951:1–17. https://doi.org/10.1007/978-1-62703-146-2_1.

43. Pepe MS, Etzioni R, Feng Z, Potter JD, Thompson ML, Thornquist M, Winget M, Yasui Y. Phases of biomarker development for early detection of cancer. J Natl Cancer Inst. 2001;93(14):1054–61.

44. Rabilloud T, Lelong C. Two-dimensional gel electrophoresis in proteomics: a tutorial. J Proteome. 2011;74(10):1829–41. https://doi.org/10.1016/j.jprot.2011.05.040.

45. Ramaiola I, Padro T, Pena E, Juan-Babot O, Cubedo J, Martin-Yuste V, Sabate M, Badimon L. Changes in thrombus composition and profilin-1 release in acute myocardial infarction. Eur Heart J. 2015;36(16):965–75. https://doi.org/10.1093/eurheartj/ehu356.

46. Roberts E, Ludman AJ, Dworzynski K, Al-Mohammad A, Cowie MR, McMurray JJ, Mant J. Failure NGDGfAHThe diagnostic accuracy of the natriuretic peptides in heart failure: systematic review and diagnostic meta-analysis in the acute care setting. BMJ. 2015;350:h910. https://doi.org/10.1136/bmj.h910.

47. Rocchetti MT, Papale M, Gesualdo L. Two-dimensional gel electrophoresis approach for CTL phosphoproteome analysis. Methods Mol Biol. 2014;1186:243–51. https://doi.org/10.1007/978-1-4939-1158-5_13.

48. Schluter H, Apweiler R, Holzhutter HG, Jungblut PR. Finding one's way in proteomics: a protein species nomenclature. Chem Cent J. 2009;3:11. https://doi.org/10.1186/1752-153X-3-11.

49. Serang O, Käll L. Solution to statistical challenges in proteomics is more statistics, not less. J Proteome Res. 2015;14(10):4099–103. https://doi.org/10.1021/acs.jproteome.5b00568.

50. Shadforth I, Crowther D, Bessant C. Protein and peptide identification algorithms using MS for use in high-throughput, automated pipelines. Proteomics. 2005;5(16):4082–95. https://doi.org/10.1002/pmic.200402091.

51. Shevchenko A, Wilm M, Vorm O, Mann M. Mass spectrometric sequencing of proteins silver-stained polyacrylamide gels. Anal Chem. 1996;68(5):850–8.
52. Smith LM, Kelleher NL, Consortium for Top Down P. Proteoform: a single term describing protein complexity. Nat Methods. 2013;10(3):186–7. https://doi.org/10.1038/nmeth.2369.
53. Song IU, Chung SW, Kim YD, Maeng LS. Relationship between the hs-CRP as non-specific biomarker and Alzheimer's disease according to aging process. Int J Med Sci. 2015;12(8):613–7. https://doi.org/10.7150/ijms.12742.
54. Storey JD. A direct approach to false discovery rates. J R Stat Soc Ser B. 2002;64(Part 3):479–98.
55. Trudgian DC, Ridlova G, Fischer R, Mackeen MM, Ternette N, Acuto O, Kessler BM, Thomas B. Comparative evaluation of label-free SINQ normalized spectral index quantitation in the central proteomics facilities pipeline. Proteomics. 2011;11(14):2790–7. https://doi.org/10.1002/pmic.201000800.
56. Unlu M, Morgan ME, Minden JS. Difference gel electrophoresis: a single gel method for detecting changes in protein extracts. Electrophoresis. 1997;18(11):2071–7. https://doi.org/10.1002/elps.1150181133.
57. Unwin RD. Quantification of proteins by iTRAQ. Methods Mol Biol. 2010;658:205–15. https://doi.org/10.1007/978-1-60761-780-8_12.
58. Wang MC, Lee YH, Liao PC. Optimization of titanium dioxide and immunoaffinity-based enrichment procedures for tyrosine phosphopeptide using matrix-assisted laser desorption/ionization time-of-flight mass spectrometry. Anal Bioanal Chem. 2015;407(5):1343–56. https://doi.org/10.1007/s00216-014-8352-0.
59. Westermeier TN, Höpker HR. Proteomics in practice: a guide to successful experimental design. 2nd ed: Wiley Online Library; 2008. https://onlinelibrary.wiley.com/doi/book/10.1002/9783527622290.
60. Williams SA, Murthy AC, DeLisle RK, Hyde C, Malarstig A, Ostroff R, Weiss SJ, Segal MR, Ganz P. Improving assessment of drug safety through proteomics: early detection and mechanistic characterization of the unforeseen harmful effects of Torcetrapib. Circulation. 2018;137(10):999–1010. https://doi.org/10.1161/CIRCULATIONAHA.117.028213.
61. Wilm M, Shevchenko A, Houthaeve T, Breit S, Schweigerer L, Fotsis T, Mann M. Femtomole sequencing of proteins from polyacrylamide gels by nano-electrospray mass spectrometry. Nature. 1996;379(6564):466–9. https://doi.org/10.1038/379466a0.
62. Winkler W, Zellner M, Diestinger M, Babeluk R, Marchetti M, Goll A, Zehetmayer S, Bauer P, Rappold E, Miller I, Roth E, Allmaier G, Oehler R. Biological variation of the platelet proteome in the elderly population and its implication for biomarker research. Mol Cell Proteomics. 2008;7(1):193–203. https://doi.org/10.1074/mcp.M700137-MCP200.
63. Wongsurawat T, Woo CC, Giannakakis A, Lin XY, Cheow ESH, Lee CN, Richards M, Sze SK, Nookaew I, Kuznetsov VA, Sorokin V. Distinctive molecular signature and activated signaling pathways in aortic smooth muscle cells of patients with myocardial infarction. Atherosclerosis. 2018;271:237–44. https://doi.org/10.1016/j.atherosclerosis.2018.01.024.
64. Yang H, Wahlmuller FC, Uhrin P, Baumgartner R, Mitulovic G, Sarg B, Geiger M, Zellner M. Proteome analysis of testis from infertile protein C inhibitor-deficient mice reveals novel changes in serpin processing and prostaglandin metabolism. Electrophoresis. 2015;36(21–22):2837–40. https://doi.org/10.1002/elps.201500218.
65. Yin X, Subramanian S, Hwang SJ, O'Donnell CJ, Fox CS, Courchesne P, Muntendam P, Gordon N, Adourian A, Juhasz P, Larson MG, Levy D. Protein biomarkers of new-onset cardiovascular disease: prospective study from the systems approach to biomarker research in cardiovascular disease initiative. Arterioscler Thromb Vasc Biol. 2014;34(4):939–45. https://doi.org/10.1161/atvbaha.113.302918.
66. Zellner M, Gerner C, Munk Eliasen M, Wurm S, Pollheimer J, Spittler A, Brostjan C, Roth E, Oehler R. Glutamine starvation of monocytes inhibits the ubiquitin-proteasome proteolytic pathway. Biochim Biophys Acta. 2003;1638(2):138–48.
67. Zellner M, Winkler W, Hayden H, Diestinger M, Eliasen M, Gesslbauer B, Miller I, Chang M, Kungl A, Roth E, Oehler R. Quantitative validation of different protein precipitation methods in proteome analysis of blood platelets. Electrophoresis. 2005;26(12):2481–9. https://doi.org/10.1002/elps.200410262.

17

Printed in the United States
By Bookmasters